W9-ASD-642

Chemical Defenses of Arthropods

Chemical Defenses of Arthropods

MURRAY S. BLUM

Department of Entomology
The University of Georgia
Athens, Georgia

1981

ACADEMIC PRESS
A Subsidiary of Harcourt Brace Jovanovich, Publishers
New York London Toronto Sydney San Francisco

ACADEMIC PRESS, INC.
111 Fifth Avenue, New York, New York 10003

United Kingdom Edition published by
ACADEMIC PRESS, INC. (LONDON) LTD.
24/28 Oval Road, London NW1 7DX

Library of Congress Cataloging in Publication Data

Blum, Murray Sheldon, Date.
Chemical defenses of arthropods.

Bibliography: p.
Includes index.
1. Insects--Physiology. 2. Arthropoda--Physiology.
3. Animal defenses. 4. Exocrine glands--Secretions.
5. Insect venom. I. Title.
QL495.B57 595.2'0457 81-7925
ISBN 0-12-108380-2 AACR2

PRINTED IN THE UNITED STATES OF AMERICA

81 82 83 84 9 8 7 6 5 4 3 2 1

To Ann

Contents

Preface

The motto of many arthropods may well be "I'll survive through natural products chemistry." The dawning has been slow, but it is now evident to both biologists and chemists that these invertebrates have a remarkable biosynthetic virtuosity when it comes to producing defensive compounds for utilization against their omnipresent enemies. In particular, research conducted in the last three decades has demonstrated that a wide variety of arthropods synthesize an incredible diversity of natural products in their exocrine glands. Many of these compounds are unique natural products that are often limited in their known arthropod distribution to a few species. In short, arthropods have already proved to be full of chemical surprises in spite of the fact that the defensive products of relatively few of these animals have been subjected to analytical scrutiny. It seems likely that the best is yet to come. While vigorous collaborative undertakings between chemists and biologists have made this insect chemical revolution possible, they have also exposed our lack of comprehension of the *modi operandi* of these defensive secretions. Furthermore, although we now know a considerable amount about the natural products chemistry of species in several arthropod taxa, for the most part we are at a loss to explain why particular species generate such idiosyncratic products in their defensive glands. We hope that this flagrant lacuna in our knowledge of arthropod chemical ecology will be filled by biologists who will pursue these invertebrates in the field, where their chemical defenses are utilized.

This volume was written in order to analyze the significant progress that has characterized the fairly recent and numerous developments in the study of arthropod chemical defenses. In addition, an attempt has been made to indicate major topics that can be fruitfully investigated in probing a multitude of questions that now characterize this rapidly expanding area of chemical ecology. In analyzing the subject of arthropod chemical defenses, I became convinced that these invertebrates are wondrous animals, a view that I hope I am at least moderately successful in imparting to both biologists and chemists.

I would like to express my sincere appreciation to my colleagues J. M. Brand, J. G. MacConnell, S. S. Duffey, T. H. Jones, the late H. H. Ross, and C. G. Jones, for reading parts of the manuscript and offering constructive advice. H. M. Fales, W. R. Tschinkel, J. W. Wheeler, and J. E. Weatherston kindly provided unpublished data for which I am grateful. I am particularly thankful to the editors of Academic Press who offered patient encouragement and editorial aid throughout the course of the preparation of this book. And finally, my thanks go to my wife, Ann, who tolerated my incessant grumblings as an author while she acted as the outstanding editor that she is. She shares my love for the diminutive and remarkable invertebrates that are described, and she always shows it.

Murray S. Blum

Chapter 1

The Many Faces of Defensive Secretions

Things are seldom what they seem. The defensive exudates of arthropods are no exception to this generalization. These secretions sometimes originate from unlikely sources and their known modes of action are often as multifarious as their compositions. Furthermore, the possessors of these highly adaptive discharges must often tolerate the presence of toxic compounds in these exudates which often cover a large area of the producer's body. Even at this early juncture in our comprehension of the *raison d'être* of arthropod defensive secretions, it is obvious that these animals have evolved an incredible variety of mechanisms for optimizing the effectiveness of their chemical defensive systems. It appears that many arthropods have admirably exploited their potential as natural product chemists as a means of deterring the ubiquitous and omnipresent predators with which they share their world.

I. VARIED SOURCES OF DEFENSIVE EXUDATES

Although the deterrent secretions of arthropods generally originate in exocrine glands which are readily classified as defensive organs (e.g., the poison glands of hymenopterans), other glandular and nonglandular discharges, which are normally identified with other functions, have also frequently been adapted to serve the role of highly effective defensive exudates. Furthermore, in some cases the products of exocrine glands are in admixture with small quantities of products which are obviously of nonexocrine origin. Enteric discharges are frequently produced by some insects and in addition to ingested food, these regurgitations may be fortified with blood (Cuénot, 1896b). Indeed, blood in itself is often discharged reflexively as a defensive vehicle, and in some cases it contains highly toxic compounds which have been synthesized *de novo* by the arthropod. Therefore, it seems imprudent at this time to generalize about the composition of these defensive secretions except to note that compounds of both intrinsic and extrinsic origin have been utilized by arthropods for chemical defense.

A. Salivary Secretions

The labial (salivary) glands of many arthropods have been converted into important defensive organs. In some cases the compounds synthesized in these glands possess no obvious digestive roles and it appears that these salivary products have been evolved to function primarily as allomonal or pheromonal agents.

1. *Salivary Venoms ("Spitting")*

Many groups of predatory arthropods (e.g., Reduviidae, Asilidae) utilize salivary venoms to immobilize their prey as one of the special functions of external digestion. These salivary secretions, which appear to be rich in digestive enzymes, can obviously function defensively against both invertebrate and vertebrate aggressors. In at least one case, the salivary venom can be forcibly spat for a considerable distance and this protein-rich secretion, aimed with great accuracy, may function admirably to deter vertebrate predators if it strikes sensitive tissues on the head.

The assassin bug *Platymeris rhadamanthus* reacts to disturbances with a stereotyped series of behavioral reactions which usually result in the forcible discharge of copious quantities of salivary venom (Edwards, 1962). Spitting does not require physical contact since sudden movements in the vicinity of the bug resulting in changes in incident light may trigger expectoration. An adult of *P. rhadamanthus* can eject its saliva up to a distance of 30 cm and a stimulated bug may spit once or twice or if highly excited, 15 successive times at a rate of 3–5 spits/second. A single series of salivary ejaculations may contain up to 2 mg of saliva containing from 9 to 20% solids. Rostral deflection enables the bug to achieve a "firing arc" of about 65° as the penultimate segment of the rostrum is deflected over and to one side of the body. Deflection of the terminal segment of the rostrum while each jet of saliva is ejaculated insures great accuracy in aiming the discharges toward the source of the disturbance (Edwards, 1962).

The saliva of *P. rhadamanthus* is enriched with a trypsin-like protease, hyaluronidase, and phospholipase (Edwards, 1961). Contact of this enzyme-rich solution with the eye or nose membranes of vertebrates results in intense local pain, vasodilation, and edema; these physiological reactions serve admirably to deter vertebrate predators such as reptiles, birds, and monkeys. The aposematic coloration of *P. rhadamanthus,* two blood red patches on the wings contrasting to jet black coloration elsewhere, should make this reduviid conspicuous to even the most hyperopic vertebrate predators. However, this spitting reduviid has a vertebrate parallel in *Naja nigricollis,* the spitting cobra, which can similarly eject a salivary venom at potential predators with great accuracy. The parallel evolution in insects and reptiles of the same defensive mechanism of spitting clearly emphasizes the probability that the same novel and effective systems to deter predators may be evolved in disparate taxa.

2. *Entspannungsschwimmen*

The aquatic bug *Velia capraii* has adapted its presumably proteinaceous saliva to effect a remarkable escape reaction. In response to a variety of stimuli, this hemipteran will discharge its saliva onto the water, a reaction that effectively lowers the surface tension of the water behind the bug. This enables *V. capraii* to rapidly propel itself across the water surface and thus puts considerable distance between itself and the source of the disturbance (Linsenmair and Jander, 1963). This escape reaction occurs after saliva is discharged posteriorly from the rostrum and the bug may be propelled 10–25 cm by the contracting water surface on which it is riding. *Entspannungsschwimmen* has been independently evolved by the staphylinids *Stenus bipunctatus* and *S. comma* but in this case the surface tension of the water is lowered by pygidial gland products rather than a salivary gland secretion. The exudate of the beetles arises from two pairs of pygidial glands and is primarily composed of three monoterpenes—isopipertenol, 1,8-cineole, and 6-methyl-5-hepten-2-one (Schildknecht, 1970) and a piperidine (Schildknecht *et al.*, 1975b, 1976). The monoterpenes are surface active and have been demonstrated to propel a surrogate "beetle" through the water in the same manner as *S. bipunctatus.* However, the main spreading agent is stenusine, *N*-ethyl-3-(2-methylbutyl)piperidine, a product from the larger pair of pygidial glands.

3. *Entangling Saliva*

Many syrphid larvae feed on aphids, a predatory habit that frequently produces confrontations with aggressive ants. The formicids guard their aphid wards assiduously, receiving in exchange droplets of honeydew which the latter provide upon appropriate tactile stimulation. Syrphid larvae, however, have evolved an effective defense against the ants that attempt to interfere with their daily aphid repast. When attacked by an ant, a syrphid larva (*Syrphus* sp.) arches its body in order to position its mouthparts on the body of the assailant. The larva then discharges a drop of viscous fluid onto the ant and the latter immediately releases its mandibular grip on the syrphid and attempts to remove the exudate which now entangles it (Eisner, 1972). The oral discharge originates in the salivary glands and almost certainly represents a proteinaceous "glue."

Although syrphid larvae have probably adapted the products of a digestive organ to function as a physical deterrent to predators, it remains to be seen whether this salivary secretion may be diluted enterically in order to provide digestive enzymes. It is possible that in the Arthropoda viscous salivary secretions may be utilized with some frequency as entangling agents *per se.* The salivary glands of many insect species are composed of paired lobes which often contain viscous proteinaceous constituents which, if secreted externally, could easily entangle small predators. Edwards (1962) reported that the salivary venom of the reduviid *Platymeris rhadamanthus* consisted of a viscid protein mixture which was diluted by the watery accessory gland secretion of the salivary apparatus during the discharge process. The external secretion of salivary proteins *in the absence of an aqueous diluent* may have provided arthropods with a readily available glue to be utilized against small predators.

4. *Salivary Gland Natural Products*

When restrained, adults of the earwig *Labidura riparia* discharge a pungent secretion from the mouth which is strongly repellent to ants (M. S. Blum, unpublished data, 1978). We have determined that this defensive material originates in the capacious salivary glands extending through the thorax and into the abdomen. These glands are usually turgid with a clear yellow secretion that contains several volatile compounds. The salivary secretion of *L. riparia* does not appear to be composed of a typical proteinaceous mixture and it is not unlikely that one of the primary functions of the gland is to serve as the source of a repellent exudate. Other examples of the salivary glands synthesizing volatile exocrine products indicate that these organs may not be an uncommon source of defensive natural products in the Arthropoda.

Many species of termites appear to utilize proteins and *p*-benzoquinones derived from the salivary glands as part of their defensive exudate. These protein–quinone mixtures react to form rubbery products which will be discussed in the next section.

Male bumblebees scent mark many territorial sites with cephalic secretions which are strongly odoriferous. These secretions consist of acyclic mono-, sesqui-, and diterpene alcohols and acetates, as well as straight chain alcohols, esters, aldehydes, and hydrocarbons (Calam, 1969; Kullenberg *et al.*, 1970; Svensson and Bergström, 1977). Some of these compounds are identical to well-known arthropod defensive compounds and it would be surprising if these aposematically colored insects did not secrete these cephalic products when molested. Recently, the source of these natural products has been established as the cephalic lobes of the salivary glands (Kullenberg *et al.*, 1973). These results further document the salivary glands of insects as a potentially rich source of natural products and emphasize the importance of establishing with certainty the morphological sources of defensive compounds identified in the cephalic secretions of arthropods.

B. Physical Deterrents of Exocrine Origin

The exudates from arthropod exocrine glands often become very viscous after being discharged and these secretions may immediately deter the would-be attacker especially if the discharge fouls its mouthparts. These arthropod "glues" appear to be derived from a variety of compounds which, in some cases, may react with the primary resinous constituent(s) in the exudate. In some cases adhesive secretions are normally present on parts of an arthropod's body and these viscid products appear to retain their properties for a considerable length of time. In addition to the viscous components, these exudates may also be enriched with low molecular-weight compounds that may increase the deterrency of the secretion or for that matter may possess other functions as well. Obviously, viscid exocrine secretions constitute a potpourri that, however, share at least one property that enables them to be grouped together, they are all of proved effectiveness as defensive secretions. Resinous secretions which are derived primarily from nonexocrine sources (e.g., blood) will be discussed separately.

1. *Spiders*

The scytodid *Scytodes thoracica* captures its prey by ejecting a viscid secretion and it has been suggested that this glue can also function as a defensive secretion (Monterosso, 1928). This suggestion has been confirmed by McAlister (1960) who examined the defensive behavior of the spider *S. intricata* in the presence of the scorpion *Centruroides vittatus.* After the scorpion had grasped the spider with a palp, the latter ejected a copious amount of secretion from its chelicerae toward the source of tactile disturbance. The scorpion's chelicerae, palps, and cephalathorax were bathed by the viscid secretion and the palp which had originally restrained the spider was glued to the substrate. Usually the sprayed scorpion displayed an immediate but awkward escape reaction, primarily because its legs were often glued together. Attempts by the irritated scorpions to sting the offending glue resulted in their telsons becoming entangled in the viscous secretion. The scytodids were none the worse for their brief encounters with the aggressive scorpions.

2. *Centipedes*

Viscous threads are discharged from the posterior legs of lithobiid centipedes; these sticky exudates effectively entangle predatory spiders and ants (Verhoeff, 1925). The viscid and proteinaceous secretions originating from ventral segmental glands of geophilid centipedes effectively deters small predators such as ants (Jones *et al.*, 1976a). In addition to entangling these insects, these secretions are highly repellent because of the presence of cyanogenic compounds.

3. *Millipedes*

Viscous secretions are discharged from the ozopores of millipedes in two unrelated families but in both cases volatile exocrine products play a major role in augmenting the deterrent efficacy of the exudates. When disturbed, the glomerid millipede *Glomeris marginata* coils itself into a ball of cuticular-plated armor in much the same manner as an armadillo. If prodded, the coiled diplopod discharges from middorsal glandular pores a viscous proteinaceous secretion which can both entangle and repel small arthropods. This exudate is fortified with two quinazolinone alkaloids (Y. C. Meinwald *et al.*, 1966; Schildknecht *et al.*, 1966c, 1967c) which are both toxic and distasteful. In combination with the proteinaceous glue, this secretion presents a formidable obstacle to predators.

The sticky whitish secretion of the polyzoniid millipede *Polyzonium rosalbum* is similarly enriched with alkaloids which render this exudate highly effective against ants and other invertebrate predators. The major compound in the secretion, 6,6-dimethyl-2-azaspirol[4,4]non-1-ene (Smolanoff *et al.*, 1975b), is very volatile and provides a highly effective repellent to approaching ants. On the other hand, the minor constituent, a spirocyclic nitro compound, nitropolyzonamine (Meinwald *et al.*, 1975), may contribute to the viscosity of the secretion.

4. *Cockroaches*

The posterior abdominal tergites and cerci of many species of cockroaches are covered with a viscous secretion which retains its adhesive properties for a considerable period of time (Nayler, 1964; Brossut and Roth, 1977). Mucous-like secretions have been observed on larvae of species of *Blatta, Blattella, Periplaneta, Supella, Parcoblatta, Neostylopyga, Eurycotis, Ectobius, Loboptera* (Roth and Stahl, 1956), and *Pseudoderopeltis* (Nayler, 1964). In *Blatta orientalis* and *Pseudoderopeltis bicolor* the secretions are proteinaceous (Roth and Stahl, 1956; Nayler, 1964). The secretion of *B. orientalis,* which consists of 90% protein combined with 10% of a polysaccharide (Roth and Stahl, 1956), readily deters small predatory arthropods (Plattner *et al.*, 1972).

Both larval and adult cockroaches spend considerable time in cleaning the tergal exudate, an indication that the presence of an unfouled adhesive coating is highly adaptive for these insects. Nayler (1964) suggested that because of its location, the secretion would be initially encountered by a predator from which the cockroach was fleeing. Although vertebrate predators readily ate these insects, small invertebrates were temporarily repelled if their mouthparts were fouled with the viscid exudate. Carabid beetles (*Thermophilum* and *Haplothachelus* spp.), centipedes (*Arthrorhabdus* and *Cormocephalus* spp.), and ants, all of which occur in the same habitat as *Pseudoderopeltis,* were usually repelled by the cockroaches after seizing the latter, and the blattids had ample time to escape while their would-be predators cleaned their mouthparts.

5. *Termites*

Rubberlike or resinous defensive exudates are discharged by soldiers of termite species in both highly evolved and primitive genera. Moore (1968) established that soldiers of *Mastotermes darwiniensis,* the only extant species in the primitive family Mastotermitidae, secrete a colorless and mobile fluid from their buccal cavity when disturbed. This secretion is soon converted to a dark rubberlike material which presumably functions to entangle foes. The deterrency of the secretion is undoubtedly augmented by the presence of *p*-benzoquinone accompanied by a trace of toluquinone (Moore, 1968). The rubberlike component formed *in situ* after discharge from the soldiers may represent a sclerotized product resulting from the reaction of the quinones with salivary proteins which are simultaneously discharged from the buccal cavity.

Recently, sticky secretions fortified with quinones have been detected in the cephalic exudates of several species of termites in the highly specialized family Termitidae. Maschwitz and Tho (1974) reported that viscous constituents accompanied benzoquinone in the exudate of *Hypotermes obscuriceps* whereas toluquinone was present in the secretions of *Microtermes globicola* and two *Odontotermes* species. The composition of the rubbery defensive product in the secretion of *Odontotermes badius* has been studied by Wood *et al.* (1975). The dark aqueous exudate of this species, which becomes sticky on exposure to air, contains *p*-benzoquinone and protein. Each soldier ejects about 5 μg of benzoquinone and the total secretion (1.6 mg) constitutes nearly 10% of its body weight. If the secretion was held in moist air it remained unchanged, indicating that a simple reaction between the quinone and protein does not occur. It seems most likely that the hardening of the secretion may result from protein denaturation subsequent to drying. The salivary protein was unusual in containing about 12% cysteine, an amino acid that has been found in high concentrations only in proteins such as keratin (Wood *et al.*, 1975).

Macrotermes natalensis and *Globitermes sulphureus* are also reported to produce exudates which darken and become sticky after exposure to air (Noirot, 1969). However, Maschwitz (1975) reports that soldiers of *G. sulphureus* actually rupture when attacked, thus liberating a yellow and sticky fluid which entangles adversaries.

The secretion of the rhinotermetid *Coptotermes lacteus* consists of a suspension of aliphatic hydrocarbons in an aqueous mucopolysaccharide. The secretion becomes sticky after being emitted, but can be easily reconstituted by adding water. Saturated normal hydrocarbons in the range C_{22}–C_{27} dominated the lipid phase and the mucopolysaccharide was composed mainly of glucosamine and possibly glucose units (Moore, 1968).

Termites in several other genera are also reported to eject secretions which become viscous on exposure to air without darkening. For example, the exudates from soldiers of *Pseudacanthotermes spiniger* and *Protermes prorepens* are liberated as white mobile fluids which rapidly become viscous (Noirot, 1969). Whether these secretions are based on mucopolysaccharides or proteins has not been determined.

Moore (1964) reported that the frontal gland secretions from soldiers of three species of *Nasutitermes* are rapidly converted from a mobile fluid to a resinous exudate after exposure to air. These secretions, which are rich in monoterpene hydrocarbons, are dominated by α- and β-pinene. More recent analyses have demonstrated that the exudates ejected by *Nasutitermes* soldiers also contain limonene, terpinolene, and α-phellandrene (Moore, 1968; Vrkoč *et al.*, 1978; Prestwich, 1977). The resinous component in the secretion consists of a mixture of related terpenoids containing one or more acetoxy groups. The mandibulate soldiers of *Amitermes vitiosus* secrete copious amounts of a sticky secretion which is also based on terpenoids. The exudate of this species is dominated by α-phellandrene and, in addition, contains α-pinene, β-pinene, limonene, terpinolene, and myrcene. The monoterpene hydrocarbons may also function as alarm pheromones (Moore, 1968), but there is little doubt that these compounds considerably augment the defensive "punch" of the sticky constituent in the secretion. Ernst (1959) had observed that the secretion of *Nasutitermes* soldiers entangled and sometimes killed ants, and he concluded that the exudates functioned purely as mechanical deterrents. However, it is now seems likely that the rapid knockdown effect on ants exerted by these nasute secretions is chiefly attributable to the fortifying monoterpenes.

The frontal gland secretions of some *Nasutitermes* and *Trinervitermes* species contain diterpenes that contribute to the viscousness of these exudates. These compounds, the trinervitenes, constitute novel diterpenoid constituents whose known distribution is limited to some species in the subfamily Termitinae (Prestwich *et al.*, 1976a,b; Prestwich, 1978; Vrkoč *et al.*, 1978). Although the diterpenes are somewhat toxic to insects (Hrdý *et al.*, 1977), the precise defensive role that these compounds play remains to be determined. It has been suggested that the mixture of hydrophobic monoterpenes and hydrophilic trinervitenes constitute an ideal defensive glue because of the surface wetting properties of the former and the high viscosity of the latter (Prestwich, 1979).

Nutting *et al.* (1974) studied the behavior of soldiers of the nasute termite *Tenuirostritermes tenuirostris* when confronted with ants. The termite soldiers rapidly closed on the ants and immobilized the formicids with rapid discharges from their cephalic projections. The sticky secretions were delivered while the soldiers jerked back and forth or oscillated. Their behavior resulted in throwing bits or loops into the sticky discharge, the ants frequently becoming engulfed by these secretions. Simultaneous attacks by several termites were invariably fatal to the ants, and it seems probable that the monoterpenes in the exudate, α-pinene, myrcene, and limonene, were responsible for the rapid immobilization of the ants. Although each soldier is capable of discharging an aliquot of only about 0.002 μl of secretion, the combined action of the multiple discharges from several soldiers provides these termites with a highly effective defense against their omnipresent formicid adversaries.

6. *Aphids*

When tactually stimulated, many species of aphids discharge droplets of fluid from tubular organs located on the fifth and sixth abdominal tergites. Usually, within seconds, these sticky exudates harden to form a waxy plaque. Several functions have been ascribed to the secretions from these organs, the cornicles, but it now seems fairly certain that these exudates are utilized as defensive secretions as first proposed by Busgen (1891). However, whereas the allomonal function of the cornicular exudate was suggested many years ago, it has only been recently determined that this secretion sometimes possesses an important pheromonal role which considerably augments its defensive value.

Dixon (1958) observed that larvae and adults of the aphid *Microlophium evansi* would frequently discharge droplets from their cornicles when attacked by larvae of the coccinellid *Adalia decempunctata*. If the mouthparts of the coccinellid are coated with the discharge, it rapidly hardens to form a caste. Under these circumstances, the aphid may escape its temporarily immobilized predator. Dixon (1958) demonstrated that aphids generally escaped from beetle larvae that were experimentally waxed with the cornicular secretion. On the other hand, aphids were generally eaten by the beetle larvae if the secretion did not entangle their mouthparts, a development that did not occur if the aphid was attacked from the front. Although the cornicular secretion may not be highly effective against specialized predators (e.g., coccinellids, nabids), its deterrent value against generalized predators may be considerably greater. Similarly, the secretion may not be especially repellent to specialized parasitoids (Goff and Nault, 1974), although the shriveled bodies of these hymenopterans have occasionally been found fixed to the bodies of aphids by plaques of cornicular wax (Edwards, 1966).

The exudate from the cornicles consists of lipid droplets suspended in an aqueous carrier (Edwards, 1966; Strong, 1967). The droplets, which range from 30–275 μm in diameter, generally harden in less than 30 seconds at humidities less than 50% (Strong, 1967). Edwards (1966) has suggested that the rapid crystallization of the lipids after a droplet is exposed to air may result from either (a) evaporation of the solvent which keeps the lipids in a liquid state or, (b) contact with a seeding nucleus causing crystallization of lipids which are normally in a stable liquid-crystalline state. The latter explanation, which involves supercooling, is clearly favored by Edwards, but is not supported by the results of Strong (1967) who demonstrated that the conversion of the lipids from a liquid to a crystalline solid was dependent on the evaporation of water from the droplets. Rapid evaporation would undoubtedly cause some supercooling to occur, but crystallization would primarily reflect supersaturation resulting from water loss. Indeed, Strong (1967) suggests that a monomolecular lipid layer may actually coat the aqueous carrier and thus prolong hardening time.

The cornicular secretions of the aphids *Myzus persicae* and *Acyrthosiphon pisum* were analyzed by Strong (1967) and shown to consist almost exclusively of triglycerides. The exudate of *A. pisum* contains three triglycerides whereas that of *M. persicae* contains two triglycerides which are dominated by myristic acid. Significantly, the secretion of *A. pisum* contains a small amount of hydrocarbon which is probably responsible for the important pheromonal role which was subsequently demonstrated to be possessed by the cornicular exudates of many aphid species.

Nault *et al.* (1973) observed that when nabid predators attacked individuals of the aphids *Aphis coreopidis*, *Acyrthosiphon pisum*, *A. solani*, and *Myzus persicae*, the secretion of the cornicular fluid resulted in proximate aphids either moving from the scene of the encounter or even dropping off the leaves. Bowers *et al.* (1972) identified this alarm pheromone as *trans*-β-farnesene and Edwards *et al.* (1973) characterized the same compound in the secretion of *Myzus persicae*. The secretion of this sesquiterpene hydrocarbon results in the rapid dispersion of clustered aphids (Nault *et al.*, 1973; Montgomery and Nault, 1977a,b; Wientjens *et al.*, 1973) and effectively reduces the probability that a voracious predator will have easy access to an aphid bonanza. Thus, even if the cornicular exudate is not an especially effective physical deterrent to aphid predators, it constitutes a highly adaptive defensive secretion because of its role as a rapid releaser of alarm and dispersal activity, especially in nonmyrmecophilous aphids (Nault *et al.*, 1976). The recent identification of germacrene A as the alarm pheromone of the aphids *Therecaphis riehmi* and *T. maculata* (Bowers *et al.*, 1977; Nishino *et al.*, 1977) demonstrates that sesquiterpene hydrocarbons are utilized by a wide variety of aphids as signaling agents.

7. *Ants*

Viscous defensive secretions characterize the exudates derived from a wide variety of different glands of ants. The anal glands, structures which are well developed in species in the subfamily Dolichoderinae, often produce a cyclopentanoid monoterpene dialdehyde, iridodial, which rapidly polymerizes on exposure to air (Cavill and Ford, 1960). In general, the anal gland secretions of many dolichoderine species constitute viscous exudates which effectively entangle small predators such as other species of ants. The anal gland secretion of a neotropical species of *Hypoclinea* is discharged as a copious froth that is immediately converted to a viscous exudate which can temporarily immobilize *Camponotus* spp. on contact (M. S. Blum, unpublished data, 1974). It has also been suggested that the polymerization of iridodial serves to fix the volatile carbonyl compounds that function as chemical releasers of alarm activity (Pavan, 1959).

Kugler (1979) has demonstrated that the pygidial gland of the myrmicine *Pheidole biconstricta* is the source of a viscous secretion that entangles small predators. In addition to this defensive function, the exudate also releases alarm behavior. It has been suggested that pygidial glands, which are present in many genera in the Myrmicinae (Kugler, 1978), are homologous to the anal glands of the Dolichoderinae. Thus, it appears that glandular cells opening onto the pygidium are relatively widespread in the Formicidae, frequently serving as the source of viscid and pheromonal exudates with important defensive roles.

The mandibular glands of two *Camponotus* spp., *C. saundersi* and *C.* sp. nr. *saundersi*, have been adapted to serve as novel defensive organs. These capacious glands, which contain a viscous yellow fluid, extend through the entire body of minor workers of these two formicine species. When mechanically disturbed, the workers contract their gasters until they rupture at an intersegmental fold. The liberated mandibular gland contents, which are extremely sticky, completely immobilize attacking ants (Maschwitz and Maschwitz, 1974). The phenomenon has been named autothysis as an apt description for this formicid self-sacrifice.

Maschwitz (1974) described a glandular exudate from a myrmicine species which is utilized as a potent viscous deterrent for small arthropod predators. The metapleural gland reservoirs of *Crematogaster inflata* are hypertrophied and when these ants are attacked, they discharge large quantities of a sticky secretion that rapidly entangles their adversaries. This glandular exudate also contains a volatile chemical releaser of alarm behavior. The utilization of the metapleural glands as defensive organs against metazoan predators appears to be exceptional development among ants especially since in other species these glands are known to be a source of organic acids (Schildknecht and Koob, 1971) with no known deterrent function against animals.

C. Secretory Froths

In general, the defensive exudates of arthropods seep from the glandular orifice(s) or are discharged as well-aimed sprays (Eisner *et al.*, 1959). In some cases, an arthropod may deliver its exocrine products by utilizing either method of discharge. The millipede *Orthocricus arboreus* responds to gentle tactile stimulation by allowing its quinone-rich exudate to ooze from the ozopores nearest to the site of irritation; the secretion can be forcibly ejected for up to 10 cm if the diplopod is strongly stimulated (Woodring and Blum, 1965). However, in addition to oozing and spraying, some insects discharge their glandular products as a copious froth which may be accompanied by an explosive sound. These secretory froths are often fortified with toxic and distasteful compounds and it is perhaps not surprising that their producers often constitute highly aposematic species. Almost without exception these froth-producing species are both warningly colored and very sluggish, a combination that almost ensures that they will remain as conspicuous objects even after they are disturbed.

1. Grasshoppers

Adults of the pyrgomorphid grasshopper (*Poekilocerus bufonius*) constitute one of the most distinctive species of insects that feeds on milkweeds. This species is dark bluish gray with yellow spots and possesses orange hind wings. When disturbed, adults discharge a viscous secretion from a bilobed gland which opens between the first two abdominal tergites (Fishelson, 1960). The exudate flows down a shallow groove on either side of the body until it passes over the second abdominal spiracles where it is mixed with air. The resultant froth is milky white and appears rainbow tinted in sunshine, contrasting vividly with the dark background color of the grasshopper (von Euw *et al.*, 1967). Thus, the froth's color augments the aposematism of *P. bufonius*. The foam can form a bubble as large as 1.5 cm (Fishelson, 1960) and therefore can cover an extensive area of the insect. In addition, the secretion possesses a pungent odor which may be easily detected for several meters by potential predators (von Euw *et al.*, 1967).

Fishelson (1960) has demonstrated that the secretion of *P. bufonius* completely protects this insect from a wide variety of both invertebrate and vertebrate predators. When injected into white mice, the secretion produced rapid paralysis followed by respiratory failure. The abdominal exudate contains about 1% histamine (dihydrochloride), a concentration comparable to that found in hymenopterous venoms (von Euw *et al.*, 1967). However, the main deterrent and toxic compounds present in the secretion are two cardenolides which are sequestered from the milkweed food plant. The secretion of each adult contains about 0–2 mg each of both calactin and calotropin, two highly emetic cardenolides. Grasshoppers reared on nonasclepiadaceous plants contained very low levels of these compounds, demonstrating the importance of the food plant to the composition of the secretion. Thus, in a sense, *P. bufonius* has appropriated the plant's chemical defenses and utilized them as a highly effective deterrent system of its own.

Another grasshopper, the acridid *Romalea microptera*, also secretes a froth when disturbed, but both the origin and chemistry of the secretion differ considerably from that produced by *P. bufonius*. The secretion is produced in glandular tissue located in the tracheae leading to the paired mesothoracic spiracles and is forcibly discharged in admixture with exhaled air (Eisner, 1972). The emission of large bubbles of brown froth is accompanied by a loud hissing sound and the strongly odoriferous secretion has been demonstrated to be an excellent repellent for ants (Eisner *et al.*, 1971a). The simultaneous production of the froth and an audible sound greatly enhances the aposematism of *R. microptera*—yellow and black with red wings which are suddenly revealed as flash coloration when the dark elytra are elevated by a disturbed grasshopper.

The compounds present in the secretion of *R. microptera* constitute a diversity of low molecular weight natural products which, in all probability, are primarily derived from the ingested food plants. The major constituent in the exudate is an allenic sesquiterpene, romallenone (Meinwald *et al.*, 1968a), a compound that probably represents a degradation product of a carotenoid (Isoe *et al.*, 1971). Romallenone does not possess any demonstrable repellent properties (Eisner *et al.*, 1971a). On the other hand, the main volatile compound in the secretion, phenol, along with *p*-cresol, guaiacol, *p*-benzoquinone, verbenone, isophorone, and 2,6,6-trimethylcyclohex-2-ene-1,4-dione, are characteristic plant natural products that may function as effective repellents for predators. Indeed, the phenolics, along with benzoquinone, are well-known components of arthropod defensive secretions. Thus, *R. microptera*, like its milkweed-eating counterpart *P. bufonius*, appears to sequester plant natural products and subsequently utilizes them against its own aggressors.

In addition to these plant-derived phenolics and terpenes, adults of *R. microptera* sequester a phenol which apparently is not produced by its host plant. Eisner *et al.* (1971a) identified 2,5-dichlorophenol in the froth of grasshoppers that had been collected in an area previously treated with the herbicide 2,4-dichlorophenoxyacetic acid (2,4-D). Since this dichlorophenol could not be detected in the froth of insects collected in an untreated area, it has been suggested that this compound is derived from 2,4-D or degraded 2,4-D ingested with the plants. The ability of both fungi and higher plants to rearrange chlorine from a 2,4- to a 2,5-substitution pattern is well established, and it is assumed that *Romalea* adults possess a similar metabolic capability (Eisner *et al.*, 1971a). However, it is also possible that 2,4-D was degraded and rearranged either by fungi or the host plant *before* it was ingested by the acridids. Furthermore, if *R. microptera* is viewed as an efficient sequestrator of low molecular-weight phenolics, then the presence of the dichlorophenol in the defensive froth is not necessarily surprising. Whatever the sequestrative *modus operandi* of this insect, it may be eminently capable of rapidly channeling a diversity of ingested oxygenated compounds to the tracheal gland.

Jones and Blum (1979) demonstrated that the secretions of individuals of *R. microptera* exhibited considerable qualitative and quantitative variation. It was further noted that each grasshopper fed on a wide range of host plants, raising the possibility that no two individuals have the same diet during their developmental periods. Since the defensive secretions are primarily composed of compounds sequestered from the host plants, it can be anticipated that both qualitative and quantitative variation in the secretions will reflect the chemical diversity of the plants fed upon.

Frothing, sometimes accompanied by audible sounds, is characteristic of grasshoppers in several other genera. In all cases, these orthopterans are both brilliantly aposematic and relatively sluggish and are capable of exhibiting flash coloration when disturbed by exposing their brightly colored wings. Numerous investigations have amply demonstrated that their frothy exudates render them unpalatable to virtually all vertebrate predators (Carpenter, 1938). Species in the genera *Phymateus*, *Dictyophorus* and *Aulacris* discharge froth from thoracic glands which is both strongly odoriferous and very distasteful to vertebrates (Marshall, 1902;

Carpenter, 1921, 1938). Although the chemistry of these orthopterous secretions has not been investigated, it will not be surprising if sequestered plant natural products fortify these froths.

2. ***Moths***

Frothing is also characteristic of many species of Arctiidae; these insects are typically both aposematic and sluggish. The generation of froth is frequently accompanied by a hissing sound, thus enhancing the aposematism. Furthermore, these lepidopterous froths often possess pungent odors and often are discharged from brightly colored areas on or proximate to the prothorax (Carpenter, 1938). At this juncture there is no evidence that these froths contain plant natural products sequestered by the herbivorous larval stages of these moths. Thus, although both aposematic moths and grasshoppers have independently evolved chemical defense systems with many common characteristics, it is possible that the defensive compounds utilized by the former are primarily or exclusively synthesized *de novo* whereas those of the latter may mainly represent compounds sequestered from plants.

Frothing has been detected in arctiid species in the genera *Rhodogastria* (Carpenter, 1938), *Composia* (Dyar, 1891), *Pericopsis* (Dyar, 1915), *Dyschema* (Seitz, 1925), *Arctia* (Rothschild and Haskell, 1966), *Belemniastis* (Blest *et al.*, 1963), *Utethesia* (Eisner, 1970), and *Apantesis* (Rothschild, 1972). Froths are also produced by *Psuedohypsa* and *Amphicallia* spp., taxa in the family Hypsidae, as well as species in the zygaenid genera *Erasmia* (Carpenter, 1938) and *Zygaena* (Lane, 1959). There is often a latent period between the time of stimulation and discharge of froth, probably indicating that it is adaptively favorable to conserve the secretion unless the moth is seriously molested.

Bisset *et al.* (1960) have demonstrated that the prothoracic (cervical) gland secretion is enriched with a choline ester which appears to be identical to β,β-dimethylacrylylcholine. The choline content of the prothoracic secretion was about 500 μg/ml, demonstrating that the defensive glands were a rich source of this pharmacologically active ester. The same ester is reported to be present in the secretion of *Utethesia bella* (Rothschild and Morley, in Rothschild and Haskell, 1966).

The froth from an unidentified African arctiid contained a high concentration of proteins (M. S. Blum and A. M. Young, unpublished data, 1980). If macromolecular constituents are characteristic of these secretions, it will be worthwhile to investigate their gustatory and pharmacological properties vis-à-vis predatory vertebrates.

The froths discharged by at least some species of arctiids also contain blood (Rothschild and Haskell, 1966; Eisner, 1970). The mechanism by which blood can be liberated as part of a glandular secretion is unknown, and it has not been demonstrated that blood augments the repellent efficacy of the froth. Indeed, it is possible that proteins detected in these arctiid secretions are derived from the blood. Obviously, at this juncture, both the chemistry and origin of the components in these lepidopterous froths must be considered as *terra incognita*.

3. *Ants*

Maschwitz (1964) demonstrated that disturbed workers of *Crematogaster* spp. produce a froth at the tip of the gaster. Species in other genera of myrmicine ants are known to elaborate an abdominal froth, and it now seems evident that this is an adaptation correlated with the evolution of a unique sting apparatus. Buren (1958) pointed out that the tip of the sting of *Crematogaster* spp. is spatulate and thus unable to function as a hypodermic organ. He also noted that the venom accumulated on the spatulate portion of the sting and suggested that this enlargement provided an increased surface area on which the poison gland products can accumulate. In essence, the spatulate sting of *Crematogaster* spp. is a structure that has been adapted to function as a topical applicator ("paint brush") for the venom.

We have observed that workers of *C. lineolata* can effectively repel workers of the aggressive fire ant *Solenopsis invicta* by smearing the latter with venomous froth that is applied with the spatulate portion of the sting. In addition, the venom of another species, *C. peringueyi*, possesses considerable toxicity when topically applied to the integument of termites (M. S. Blum and R. M. Crewe, unpublished data, 1974). This topical toxicity is in marked contrast to many myrmicine venoms which have been similarly examined. For the most part, myrmicine venoms consist of protein-rich solutions with no demonstrable toxicity when applied to the lipophilic cuticles of arthropods. Therefore, the venoms of *Crematogaster* spp. appear to be enriched with nonpolar toxicants that can rapidly penetrate the cuticle of arthropod predators. In contrast to the defensive froths generated by grasshoppers and moths, those produced by *Crematogaster* workers are not secreted in admixture with tracheal air. These ant froths are derived from the venom apparatus and there is no evidence to indicate that air is added to the venom as it is discharged from the sting shaft (M. S. Blum and H. R. Hermann, unpublished data, 1978).

D. Blood as a Defensive Agent

Many arthropods respond to molestation by discharging either blood or exocrine secretions in admixture with blood. In some cases, the blood is fortified with natural products synthesized *de novo* by the animal which render the discharge very unpalatable to predators whereas in other instances toxic compounds ingested from plants are present as blood-borne constituents. Whether the blood acts as a simple vehicle for dissolving or emulsifying these plant and animal natural products remains to be determined. However, it will not prove surprising if the complex of enzymes, lipids, etc., which constitute normal components of the blood, play an important role in maximizing the deterrent potential of the exudate.

Reflex Bleeding

Most of the principles pertaining to reflex bleeding in adult insects were adumbrated by Cuénot (1896b) who studied this phenomenon in a wide variety of coleopterous species as well as a few orthopterans. Cuénot clearly established that the discharge of blood constituted a defensive reaction which was highly effective in deterring aggressive carabid beetles as well as frogs, toads, and lizards. It is almost certain that the blood discharged by all species of beetles examined by Cuénot was fortified with compounds which rendered it distasteful, emetic, or vesicatory to selected predators. In contrast to the blood discharged by these autohemorrhagic species, those of a wide range of insects appear to function purely as physical agents of deterrency in the absence of any especially distasteful blood-borne natural products.

a. Blood as a Purely Physical Deterrent. Larvae of the chrysomelids *Diabrotica undecimpunctata* and *D. balteata* autohemorrhage at two sites on their bodies in response to tacticle stimulation. Reflex bleeding occurs at the intersegmental membrane between the head and prothorax as well as that between the last two abdominal segments (Wallace and Blum, 1971). Autohemorrhage, which usually only occurs at the site nearest the point of stimulation, can result in losses of blood up to 13% of the wet weight of a larva without resulting in any apparent harm to the insect. The blood clots almost instantaneously and effectively repels ants, often binding them together. On the other hand the larvae are readily eaten by mice and it appears that the blood does not render these insects unpalatable to mammals (Wallace and Blum, 1971).

Adults of the firefly *Photinus pyralis* exhibit reflex bleeding from specialized weak spots in the cuticle along the elytra and antennal sockets (Blum and Sannasi, 1974). Unless a very strong stimulus is applied, bleeding is restricted to sites proximate to the region of tactile stimulation. This autohemorrhagic system effectively blankets the dorsum of the beetle with a blood bath. The ventral area of the lampyrid is enveloped by the concave elytra, thus ensuring that predators will encounter the battery of primed bleeding sites through which is discharged viscous and rapid-clotting blood. Autohemorrhage protects these beetles from small predators such as ants by entangling them in the coagulating hemolymph. On the other hand, the lampyrids are readily eaten by some species of frogs. However, the presence of emetic steroidal pyrones (lucibufagins) in the blood of *Photinus* spp. (Meinwald *et al.*, 1979; Goetz *et al.*, 1979) demonstrates that the hemolymph of these insects is enriched with compounds that are effective deterrents against some vertebrate predators.

Disturbed adults of two species of stoneflies, *Pteronarcys proteus* and *Peltoperla maria,* autohemorrhage at the intersegmental membranes of the coxal and tibiofemoral joints of the legs (Benfield, 1974). The odorless and bland blood clotted much more rapidly when mechanically "seeded" than when it was allowed to accumulate at the bleeding site. Contact with ants resulted in the blood clotting within seconds and in the process these formicids were entangled. The blood did not prevent deer mice from ingesting the stoneflies with apparent gusto. Freshly emerged adults of *P. proteus* can forcibly eject blood up to 25 cm and these discharges are accompanied by a loud "popping" sound that may temporarily distract surprised vertebrate predators, enabling the stoneflies to escape (Benfield, 1974).

The rapidly clotting discharges of these beetles and stoneflies appear to have been evolved as effective deterrents against small invertebrate predators such as ants. However, the bloody exudates of many other autohemorrhagic species are enriched with potent repellents that generally function admirably against even the most aggressive vertebrate predators.

b. Blood Enriched with Arthropod Natural Products. Over eighty years ago it was clearly established that the blood discharged reflexively by beetles in several unrelated families as well as some species of grasshoppers was fortified with compounds which rendered it very distasteful and/or odoriferous (Cuénot, 1894, 1896a). Cuénot (1896b) further demonstrated that predators rapidly learned to discriminate against these autohemorrhagic species as food items and concluded that these blood-borne natural products were the *raison d'être* of bloods' effectiveness as gustastory repellents.

Cantharadin, the terpenoid anhydride (Gadamer, 1914) synthesized exclusively by species of beetles in the family Meloidae, was considered by Cuénot (1890) to be responsible for the repellent properties of meloid blood against a wide variety of predators. Meloids bleed reflexively from the femorotibial joints, and will generally only discharge blood from the leg which is tactually stimulated (Carrel and Eisner, 1974). Cantharidin is sometimes referred to as Spanish fly because of its derivation from adults of the large meloid *Lytta vesicatoria* that is commonly found in Spain; it is a potent vesicant with a putative function as an aphrodisiac. It is also found in the eggs (Selander and Mathieu, 1969) and larvae (Meyer *et al.*, 1968) of meloids but is not synthesized by adult female beetles although it is present in their blood (Schlatter *et al.*, 1968). Recent investigations demonstrate that the repellent properties of meloid blood are indeed attributable to the presence of cantharidin, as concluded by Cuénot (1890, 1894, 1896b).

Carrel and Eisner (1974) reported that cantharadin markedly reduced the acceptibility of sucrose solutions for ants. In addition, when cantharidin emulsions were applied to the mandibles of predatory beetles (*Calosoma prominens*), these insects exhibited immediate cleansing behavior by ploughing their mouthparts into the substrate. The astringency of this compound in combination with its ability to produce both gastrointestinal and renal lesions in vertebrates undoubtedly contribute to its observed efficacy as a deterrent for reptiles and mammals. On the other hand, Cuénot (1896b) reported that amphibians readily ate meloids and certain groups of predatory insects are known to take these beetles as normal prey (Selander, 1960). However, it would be extraordinary if selected predators had not evolved mechanisms for tolerating even the most apparently distasteful and toxic defensive compounds.

Adults of ladybird beetles (Coccinellidae) discharge blood from the femorotibial joints when tactually stimulated (Cuénot, 1894, 1896b; Hollande, 1911). The blood of these beetles is astringent and often very odoriferous, and constitutes a highly effective defense against both insect and vertebrate predators. The basis for the repellency of coccinellid blood is clearly correlated with the presence of a family of alkaloids which is synthesized *de novo* by these insects (Tursch *et al.*, 1975).

Pasteels *et al.* (1973) and Pasteels (1977) have only detected alkaloids in the blood of aposematic beetles, thus demonstrating an apparent correlation between warning coloration and the presence of these compounds. The alkaloids produce a spectrum of repellency responses against both insect and vertebrate predators. Ants (*Myrmica rubra*) would generally reject freshly killed beetles as food items and were repelled by aqueous solutions of the alkaloids at concentrations ranging from 10^{-3} to 10^{-4} *M*. In general, these compounds were more effective than *p*-benzoquinone as repellents for the ants. European quail were more variable in their responses to beetles although it was noted that coccinellids lacking alkaloids were always eaten by these birds (Pasteels *et al.*, 1973).

Autohemorrhage of adults of the coccinellid *Epilachna varivestis* also effectively deters small predators such as ants as a result of the rapid clotting of the blood (Happ and Eisner, 1961), the formicids being rapidly entangled as the blood coagulates. It is also possible that natural products in the blood contribute to the deterrent properties of the discharge which induces immediate cleansing behavior in the formicids. *Epilachna* is a genus in the subfamily Epilachninae, a taxon in which alkaloids have been detected in several species of one of two genera examined (Pasteels *et al.*, 1973).

Blood is also reflexively discharged by some species of larval coccinellids and it is likely that reflex bleeding will be found to be widespread among immature ladybirds. Kendall (1971) demonstrated that larvae of *Exochomus quadripustulatus* and *Adalia bipunctata* autohemorrhage in response to traumatic stimuli; the discharged blood rapidly coagulates. Although not strictly reflex bleeding, blood is readily discharged from the fragile spines which invest the dorsum of larvae of *Epilachna varivestis* (Happ and Eisner, 1961); ants are effectively repelled or entangled by the discharge. It has not been established whether larval blood contains alkaloids such as are found in adults, but it would not prove surprising if this was the case.

A surprising variety of additional insect groups also display autohemorrhage. Cuénot (1894, 1896b) has determined that adults of species in the chrysomelid genera *Timarcha* and *Galeruca* discharge droplets of blood from their mouths when tactually stimulated. The bitter blood is very effective in repelling both insect and vertebrate predators. Unlike the blood of coccinellid species, that of chrysomelids is not reported to be odorous (Cuénot, 1894). On the other hand, Darlington (1938) reported that the blood liberated by lycid beetles (*Thonalmus* spp.) was very mild to the taste. Nevertheless, these aposematic insects were readily rejected by anolid lizards.

Stimulated grasshoppers in the allied genera *Eugaster* and *Ephippiger* discharge blood which is obviously fortified with very repellent compounds. The brachypterous species *Eugaster guyoni* can autohemorrhage in response to a visual stimulus and the blood can be accurately ejected for distances between 40 to 50 cm (Vosseler, 1893). The blood is liberated from joints on the legs situated between the coxa and trochanter. The blood is very irritating especially if it impinges on the conjunctiva, and can produce lesions on the skin. On the other hand, *Ephippiger brunneri* autohemorrhages at the base of each elytron and the astringent blood constitutes an effective defense against lizards (Cuénot, 1894, 1896b). The defensive exudate of larvae, but not adults, of the pyrgomorphid grasshopper *Poekilocerus bufonius* contains blood in addition to the cardenolides derived from its host plant (von Euw *et al.*, 1967).

As mentioned previously (Section I,D,1a), the blood reflexively discharged by some species of lampyrids (*Photinus* spp.) contains steroids that have been demonstrated to repel some predatory species (Meinwald *et al.*, 1979; Goetz *et al.*, 1979). The chemistry of these compounds, the first insect-derived bufadienolides, is discussed in Chapter 18, Section XI.

c. Blood Enriched with Plant Natural Products. Certain species of lygaeids have evolved an extraordinary defensive system predicated on the storage of selected plant compounds in specialized reservoirs (spaces) containing a fluid with proteins similar to those present in blood (Duffey and Scudder, 1974). Although this space fluid differs from blood in several characteristics (Graham and Staddon, 1974), for convenience it will be treated here as a blood filtrate. Adults of *Oncopeltus fasciatus*, *O. sandarachatus*, and *Lygaeus kalmii* possess dorsolateral thoracic (meso- and metathorax) and abdominal (sterna 2–7) spaces which contain a fluid which is very rich in cardenolides sequestered from the milkweeds upon which these insects feed (Duffey and Scudder, 1974). This dorsolateral space complex is also present in species in nine other lygaeid genera (Scudder and Duffey, 1972).

The cardenolides are about 100 times more concentrated in the dorsolateral fluid than they are in the blood (Duffey and Scudder, 1974). The fluid in these lateral spaces is thus a far more efficient sequestrator of these plant compounds than the blood that is freely circulating in the hemocele. Furthermore, both larvae and adults of *O. fasciatus* preferentially sequester polar cardenolides, some of which may be metabolized *in vivo* from less polar compounds. Therefore, the cardenolide content of the fluid in the dorsolateral spaces does not faithfully mirror the cardenolide composition of the host plant (Duffey and Scudder, 1974). These lygaeids have utilized a bloodlike fluid, isolated in the dorsolateral spaces, for the preferential uptake and concentration of cardenolides as a major evolutionary development in their chemical defensive system.

Graham and Staddon (1974) have obtained pharmacological evidence for the existence of a histamine-like substance in the dorsolateral fluid secreted by *O. fasciatus*.

Finally, it is not unlikely that plant natural products may be present in the blood which is discharged reflexively by chrysomelids in the genera *Timarcha* and *Galeruca*. Cuénot (1896b) has demonstrated that reflex bleeding occurs through the mouths of these beetles, and it would not be surprising if ingested plant constituents were regurgitated simultaneously. The addition of repellent or distasteful plant-derived compounds to the expelled blood would increase considerably the deterrent effectiveness of the latter. This possibility could be easily tested.

d. Blood as Part of a Glandular Secretion. In several insect species blood normally constitutes part of the exudate which is secreted from defensive glands. Neither the exact role that blood plays as a component of these secretions or the means by which blood is discharged as part of an exocrine exudate are known, but it may be assumed that these special cases of autohemorrhage are at least somewhat adaptive.

Blood appears to be a normal constituent of the froths emitted from the cervical glands of various species of Arctiidae. A pharmacologically active choline ester (β,β-dimethylacrylylcholine) fortifies the blood-rich secretions of at least two species, *Arctia caja* and *Utethesia bella* (Rothschild and Haskell, 1966; Rothschild *et al.*, 1970b), and similar compounds may be characteristic products of these exocrine glands. Similarly, the histamine-rich exudate from the abdominal glands of larvae of the grasshopper *Poekilocerus bufonius* is enriched with blood (von Euw *et al.*, 1967) and it is not unlikely that the defensive secretions of other orthopterans are also fortified with blood-borne constituents.

Blood may have a widespread occurrence as a concomitant of the exocrine products of diverse taxonomic groups of arthropods. Its apparent scattered distribution in these defensive secretions may simply reflect the fact that its presence has not been sought.

E. Plant Natural Products in Exocrine Secretions

A surprising variety of arthropods incorporate ingested plant natural products into both exocrine and nonexocrine defensive secretions. Indeed, in some instances a species may enrich more than one type of defensive exudate with compounds derived from plants (Duffey, 1970; Duffey and Scudder, 1974). However, it cannot be assumed that arthropods are nonselective in their uptake of plant natural products, since it has been demonstrated that some species preferentially sequester certain ingested compounds and indeed may metabolize them before they are transferred to the reservoirs of the defensive glands (Scudder and Duffey, 1972; Duffey and Scudder, 1974).

The metathoracic scent gland secretion and larval midorsal gland fluid of *Oncopeltus fasciatus* are rich sources α,β-unsaturated aldehydes that are typical hemipterous defensive compounds (Games and Staddon, 1973a,b). In addition, the exudates of these glands contain cardenolides derived from the asclepiadaceous host plants (Duffey and Scudder, 1974). Although plant natural products are seldom looked for in the defensive secretions from arthropod exocrine glands, there are no strong grounds for believing that some of these compounds may not be sequestered in the glandular fluids.

F. Plant Natural Products in Enteric Exudates

The propensity of arthropods to discharge substances from their digestive tracts in response to traumatic stimuli is well documented. Many insects readily regurgitate or defecate when disturbed, and it has been emphasized that these enteric discharges can constitute highly effective repellents because of the plant natural products that they contain (Eisner, 1970). The regurgitates of grasshoppers are very irritating when applied to the conjunctiva of vertebrates (Curasson, 1934) and have been reported to be very toxic when injected (Freeman, 1968). However, the significance of any results obtained by injecting these regurgitates into the body cavity of vertebrates is dubious at best, if for no other reason than the artificiality of the anticipated anaphylactic reaction that would follow the intravenous introduction of foreign proteins. Suffice it to say, the regurgitates of grasshoppers are not naturally injected into predators in the first place.

Eisner (1970) has demonstrated that the regurgitates of two grasshoppers, *Romalea microptera* and *Brachystola magna*, effectively repel aggressive ant workers. Furthermore, the application of these enteric discharges to acceptable insect parts rendered them highly repellent to the ants. It has also been observed that regurgitates from grasshoppers (*R. microptera*) that had fed on their natural food plants were very repellent to ants whereas these derived from grasshoppers which had eaten unnatural plants were not (Eisner, 1970). Furthermore, there is evidence that insects may enterically inactivate the repellent compounds present in a food plant which is not normally fed upon. Therefore, it has been suggested that it may be adaptive for insects to select food plants whose repellent natural products are not detoxified in the digestive tract (Eisner,

1970). However, it is also possible that arthropods may have evolved the capacity to enterically alter the ingested plant compounds to more toxic products which could then serve admirably as potent repellents when liberated by regurgitation or defacation.

Many plants produce potentially toxic compounds which are present in inactive forms such as glycosides, esters, etc. Upon ingestion and trituration, plant-derived enzymes may rapidly convert these compounds to more toxic forms. Furthermore, if these plant natural products are enterically metabolized by the host's enzymes, additional new compounds could be generated. On the other hand, if these toxins are then discharged in a regurgitate, they can constitute potent deterrents that have been generated with little apparent metabolic cost to the herbivore. In a sense, this hypertoxication of plant compounds is analogous to the utilization of prodrugs in the pharmaceutical industry. Many drugs are administered in inactive forms that are converted *in vivo* to compounds possessing greater physiological activity. For arthropods such metabolic transformations would be highly adaptive since they would enable these animals to appropriate the "latent" chemical defenses of plants and utilize them to generate, *on demand,* chemical repellents.

The whole question of the toxicity and repellency of ingested plant natural products vis-à-vis the regurgitates of herbivores is clouded by our lack of information on both enteric transformations and the toxicity and repellency of specific plant natural products. Conceivably, plant compounds may be detoxified, converted to compounds of the same relative toxicity (isotoxication), or transformed to compounds of greater toxicity (hypertoxication). Indeed, these metabolic alterations need not be mutually exclusive, since all these possibilities could be realized with ingested compounds present in a single plant species. Furthermore, many compounds are probably metabolically unaltered during their passage through the digestive tract and thus represent some of the unaltered chemical defenses of the host plant. In some cases, insects not only appropriate the plant's defense chemicals, but store them unaltered and utilize them as defensive secretions in much the same way as are the *de novo* synthesized exudates of other insects.

Larvae of the sawfly *Neodiprion sertifer* sequester the constituents present in the resin of the pines which they feed upon (Eisner *et al.*, 1974b). The compounds are stored in two large pouches at the level of the esophagus, and are muscled so that their contents can be forcibly regurgitated. The resinous discharge contains α-pinene, β-pinene, the monomethyl ester of pinafolic acid, and a variety of resin acids including abietic, neoabietic, levopimaric, pimaric, palustric, and dehydroabietic. The compounds comprise a resinous exudate that is a highly effective chemical defense against ants. Similarly, eucalypt oils are stored in a single diverticular pouch in the forget of sawfly larvae (Pergidae) and, as in the case of *N. sertifer*, are utilized for defense (Morrow *et al.*, 1976). If the acquisitions of these defensive systems of sawfly larvae does not involve a very large energy expenditure, they would constitute highly adaptive systems for these species.

The role of defecation for chemical defense in arthropods has not really been studied in any detail but it will probably be found to be significant. For example, Schildknecht and Weis (1962b) have identified ammonia as the main constituent in the anal droplets of adult silphid beetles and it is probable that this compound is produced in the hind gut of the carrion-feeding beetles either *de novo* or by microorganisms. Both larvae and adults of the lygaeid *Oncopeltus fasciatus* defecate when stimulated tactually, and the feces are rich in cardenolides acquired from the milkweed host (Duffey, 1977). The frequency with which arthropods defecate when tactually stimulated is indicative of the potential defensive value of anal discharges.

Defecation may prove to be especially adaptive as a defense during a particularly vulnerable period in the adult arthropod's life. Freshly emerged adults of the arctiid *Estigmene acraea* discharge their meconium when traumatically stimulated and this material is very repellent to ants (M. S. Blum, unpublished data, 1975). The meconium of *E. acraea*, which consists of end products of nitrogenous metabolism and other degradation products accumulated during the pupal instar, contains volatile compounds which may be partly responsible for the repellency of this discharge. It is not unlikely that endopterygote insects have utilized the potentially toxic waste products accumulated during the pupal stage as a chemical defense that can be utilized when they are most susceptible to predation as adults. This possible utilization of defecation for chemical defense should be easily amenable to experimental evaluation.

II. CONSERVATION OF SECRETION

The defensive secretions of arthropods constitute highly adaptive deterrents for predators and a variety of devices have been evolved in order to conserve these exocrine discharges. Many of these invertebrates only discharge their secretions after repeated molestation, relying initially on other defensive mechanisms for frustrating the aggressive actions of their tormentors. A variety of millipede species roll into a compact ball or spiral when molested and this behavior, in combination with an extremely hard deflective cuticle, often will serve to discourage sustained attacks by small predators. Continued molestation of the coiled diplopod usually results in the discharge of the defensive glands (Y. Meinwald *et al.*, 1966). Similarly, the staphylinid beetle *Drusilla canaliculata* only discharges its extraordinary tergal gland products as a "last resort" (Brand *et al.*, 1973a).

Recently, Blum *et al.* (1977b) observed that secretory conservation may be much more characteristic of some instars than others. Early-instar larvae of the papilionid *Parides arcas* readily evert their osmeteria and secrete defensive compounds from the gland lying at the base of this forked organ. On the other hand, larvae in the penultimate and ultimate instar only discharge their osmeterial products in response to sustained traumatic stimuli. The secretory behavior of these different larval instars appears to be highly adaptive. Small larvae appear to be much more susceptible to attack by small predators (e.g., ants) and a readily available and effective defensive secretion would constitute a *sine qua non* for repelling these attackers. On the other hand, larger larvae rely on other defensive devices such as copious enteric discharges and the osmeterial secretion appears to be brought into play when all else fails. Many arthropods that possess a series of defensive glands will, in effect, conserve their secretions by only discharging from the glands which are proximate to the point of tactile stimulation. For example, millipedes discharge only the paired glands adjacent to the region of its body subjected to disturbance (Eisner, 1970). Similarly, larvae of the chrysomelid *Chrysomela scripta*, which are armed with both paired thoracic and abdominal glands, discharge their salicylaldehyde-rich secretions only from glands nearest the point of stimulation (Wallace and Blum, 1969). Happ and Eisner (1961) have demonstrated that reflex bleeding in the coccinellid *Epilachna varivestis* only occurs from those legs that are grasped by predators, thus ensuring that the bloody defensive exudate will be frugally discharged and only at the site of direct molestation.

In at least a few cases, the expressed secretion may be reclaimed by inverted invagination of the storage organ, presumably to be utilized again on demand. Garb (1915) observed that the exudate from the everted paired tubercles of the larva of *Melasoma lapponica* can be withdrawn back into the tubercle by muscular contraction. Similarly, larvae in other chrysomelid genera (e.g., *Chrysomela* spp.) also suck back their glandular products, thus effectively conserving the salicylaldehyde-rich exudates (Wallace and Blum, 1969; Eisner, 1970).

Blood discharged reflexively may be reclaimed by certain autohemorrhagic species. Chrysomelid species in the genera *Timarcha* and *Galeruca* generally reflex bleed from the mouth, often while they are feigning death while lying on their backs (Cuénot, 1896b). If undisturbed, the droplet of blood is withdrawn into the mouth, possibly to be resorbed into the hemocele.

In general, arthropods only discharge a fraction of the fluid present in the reservoirs of their defensive glands and thus retain a considerable reserve of exocrine products for subsequent utilization. Garb (1915) noted that larvae of the beetle *Melasoma lapponica* only secreted a small portion of the products present in a defensive gland in response to a traumatic stimulus.

Sometimes the defensive products of exocrine glands can be conserved by being mixed with a carrier. Eisner *et al.* (1971b, 1977) reported that the secretions of the opilionids (Cosmetidae) *Vonones sayi*, *Poecilaemella eutypta*, *P. quadripunctata*, and *Cynorta astora* are mixed with regurgitated water from the mouth before being applied to the body of a predator by the harvestmen's forelegs. Although these opilionids produce limited amounts of *p*-benzoquinones or cresols in their small glands, the utilization of enteric fluid as a carrier enables these arthropods to conserve their defensive secretions so effectively that in the case of *V. sayi* enough quinone is available to charge 30 regurgitations capable of repelling an estimated 1500 individual ant attacks.

III. REGENERATION OF SECRETORY CONSTITUENTS

The general conservativeness of arthropods vis-à-vis their defensive secretions may reflect the inability of these animals to rapidly regenerate these important deterrent exudates. The results obtained from scattered investigations indicate that this may indeed be the case.

Dethier (1939) observed that the froth produced by the arctiid *Apantesis arge* was only partially regenerated 3–12 hr after the initial secretion had occurred. Even after 12 hr very little secretion was discharged after the moths were tactually stimulated.

H. E. Eisner *et al.* (1967) examined the cyanogenetic potential of two species of millipedes and demonstrated that both resynthesized their defensive products very slowly. The polydesmoids *Apheloria corrugata* and *Pseudopolydesmus serratus* have mean yields of 114 μg and 41 μg/animal of hydrogen cyanide respectively, when first stimulated. Eight days after some individuals of *A. corrugata* had been stimulated to depletion, only 15 μg/animal of hydrogen cyanide were produced. Thus, the millipedes are able to produce and discharge less than 10% of their initial capacity about one week after their cyanogenetic glands had been depleted. Even after two months, only 20–100 μg of HCN were obtained from the millipedes, and the yield was about the same 2–3 months later. Individuals of *Pseudopolydesmus* appeared to also recharge their glands slowly (H. E. Eisner *et al.*, 1967). S. E. Duffey (personal communication, 1974) observed that regeneration of the quinone-rich secretion of the millipede *Rhinocricus holomelanus* was also very slow.

The defensive exudate of larvae of the pyrogomorphid grasshopper *Poekilocerus bufonius* is completely regenerated in 8–14 days (von Euw *et al.*, 1967). This secretion is fortified with cardenolides that are ingested during feeding on asclepiadaceous plants and presumably the reloading of the abdominal glands partially reflects the sequestration of these plant natural products during the period of secretory regeneration.

Larvae of the chrysomelid *Paropsis atomaria* appear to be exceptional in being able to recharge their abdominal defensive glands in 24 hr (Moore, 1967). The cyanogenic secretion, which also contains benzaldehyde and glucose, is probably derived from a glycoside. About 0–2 mg of secretion can be generated in a 24-hr period.

IV. TWO-PHASE SECRETIONS

Many arthropod secretions consist of two phases that may either be characterized by a discharge with distinct super- and subnatant layers or, in some cases, by suspensions that are resolved only after standing. For the most part, the chemistry of each of the two phases present in these arthropod exudates has not been determined, and published analyses represent average values for the constituents in the secretion as a whole. On the other hand, compounds dissolved in aqueous layers may be lost when the exudates are dried over hydroscopic salts, thus resulting in incomplete analyses of components present in the discharges. Beyond these analytical considerations, it seems obvious that only through an appreciation of the chemistry of each of the phases will it be possible to comprehend the possible functions possessed by each of these phases.

Blum *et al.* (1960) observed that the primary constituent present in the secretion of the pentatomid *Oebalus pugnax* was *n*-tridecane. This hydrocarbon consisted of about 60% of the secretion and formed a clear supernatant which markedly contrasted with the enal-rich yellow subnatant layer. Similarly, Gilby and Waterhouse (1965) demonstrated that *n*-tridecane dominated the upper phase (ca. 70%) of the secretion of the pentatomid *Nezara viridula.* For the most part, only trace amounts of the 18 polar compounds that dominated the upper phase were present in the subnatant layer. Longer-chain aldehydes [e.g., (*E*)-2-decenal] were considerably more soluble in the nonpolar alkane supernatant than their shorter-chain homologs [e.g., (*E*)-2-octenal]. (*E*)-2-Octenal partitioned between the denser polar layer and supernatant hydrocarbon layer in the ratio 10:1 (Gilby and Waterhouse, 1965). It may be significant that the secretions of adult bugs in the family Coreidae, which consist of a single phase, do not contain any hydrocarbon constituents (Waterhouse and Gilby, 1964).

It has been recently demonstrated that the aqueous phase of the two-phase secretion of the coreid, *Leptoglossus phyllopus,* contains enzymes that produce the more reactive constituents found in the organic phase (Aldrich *et al.*, 1978). Highly reactive aldehydes (e.g., hexanal) are generated in the impermeable cuticular reservoir of the metathoracic defensive gland from precursors that are hydrolyzed and subsequently oxidized. Thus, the two-phase secretion of this coreid reflects the presence of an aqueous phase that contains the enzymatic catalysts which produce the reactive aldehydes that dominate the organic phase of the defensive exudate. In all probability the *raison d'être* of the two-phase secretion in the median reservoir of the lygaeid *Oncopeltus fasciatus* (Everton and Staddon, 1979) is comparable to that of *L. phyllopus.* As is the case for the coreid, the final secretion of *O. fasciatus* is rich in aldehydes that are presumably derived from esters produced in one of the accessory glands.

The abdominal gland secretions of tenebrionid beetles are generally biphasic with a quinone-rich layer generally constituting the supernatant. Hurst *et al.* (1964) identified glucose in the aqueous subnatant phase of the secretion of *Eleodes longicollis* and Meinwald and Eisner (1964) noted that octanoic acid was also present in this layer. Tschinkel (1975b) analyzed the defensive exudates of a wide range of tenebrionid species and observed that in the presence of alkenes, the quinonoid-rich phase was supernatant, whereas in the absence of these hydrocarbons, the quinones constituted a subnatant phase. Tschinkel believes that for tenebrionids that headstand before discharging their secretions, the presence of alkenes has great selective value since the irritating quinones are supernatant and are probably ejected in greater proportion than the hydrocarbons.

Cockroaches in the genus *Polyzosteria* eject a secretion that is generally biphasic with the upper phase being thoroughly dominated by (*E*)-2-hexenal (Wallbank and Waterhouse, 1970). The lower phase, on the other hand, is aqueous and contains 6.5% gluconic acid in equilibrium with its α- and δ-lactones, as well as glucose. The aqueous secretory phase of cockroach species in the genus *Eurycotis* is also enriched with gluconic acid in equilibrium with its lactones (Dateo and Roth, 1967a,b).

In contrast to the subnatant phases present in *Polyzosteria* and *Eurycotis* defensive secretions, those of cockroaches in the genus *Platyzosteria* lack gluconic acid and its lactones (Waterhouse and Wallbank, 1967). The aqueous lower phase of these secretions contains trace amounts of compounds that may be phospholipids as well as a small amount of reducing sugars. The upper phase is thoroughly dominated by 2-methylenebutanal (ca. 94%) and in addition, contains minor amounts of related compounds; the aqueous phase contains only traces of these volatile aldehydes. Initially discharged secretion contains a predominance of the upper phase whereas the lower phase predominates in exudates collected toward the end of a discharge. After frequent milkings, the proportion of aqueous lower phase increases considerably (Waterhouse and Wallbank, 1967).

Larvae of the notodontid *Heterocampa manteo* discharge a secretion that contains about 27% formic acid but in addition, minor amounts of two ketonic constituents are present as a suspension (Eisner *et al.*, 1972). 2-Undecanone and 2-tridecanone constitute about 1.4% of the secretion and at this concentration these compounds exceed the limits of saturation in the aqueous formic acid.

The venom of the fire ant *Solenopsis invicta* (=*saevissima*) consists of a main alkaloidal phase in which fine droplets are suspended (Blum *et al.*, 1959). The major phase of the venom, which constitutes about 95% of the discharge, consists of a series of 2,6-dialkylpiperidines that appear to be responsible for the toxic properties of the venom (MacConnell *et al.*, 1971). On the other hand, the suspended droplets represent an aqueous phase which contains proteinaceous constituents that may be identified with the allergenic properties of this secretion (Brand *et al.*, 1972; Baer *et al.*, 1979). In the case of hypersensitive individuals, the minor nonalkaloidal components of the venom may ultimately constitute the most physiologically deleterious elements in this secretion.

Two-phase secretions may result when the immiscible products of two exocrine glands are simultaneously evacuated through a common orifice. This appears to be the case for the venoms of formicine ants which are composed of a mixture of the constituents derived from both the Dufour's and poison glands. The poison gland secretions contain highly concentrated solutions of aqueous formic acid (Osman and Brander, 1961) whereas the Dufour's gland exudates are composed of hydrocarbons, ketones, aldehydes, esters, and alcohols (Bergström and Löfqvist, 1968, 1970). The long-chain constituents synthesized in the Dufour's gland have a limited solubility in the highly polar formic acid solution and as a consequence, suspensions of the Dufour's constituents in the venom are the rule. In a sense, these formicine secretions are comparable to those generated by notodontid larvae (Eisner *et al.*, 1972; Weatherston *et al.*, 1979), since both contain aqueous solutions of formic acid in which are suspended less polar constituents. Indeed, 2-tridecanone is common to both notodontid and some formicine exudates (Bergström and Löfqvist, 1972).

It seems obvious that although one of the phases in a biphasic exocrine discharge may constitute a minor proportion of the total secretion, its presence may render the secretion highly adaptive as a predator deterrent. Thus, arthropod defensive exudates must be analyzed both chemically and physically in order to comprehend their deterrent *raison d'être*.

V. MINOR SECRETORY COMPONENTS

The defensive secretions of arthropods almost invariably contain minor or trace constituents whose structures are usually not elucidated. However, although these minor components may often represent only a few percent or less of the total exudate, their role in promoting the deterrent efficacy of the discharge may be considerable. In addition, the identification of minor glandular constituents may illuminate biosynthetic pathways for the main exocrine products. In short, to ignore the chemistry of minor compounds in a defensive secretion may result in failure to comprehend the adaptiveness of the discharge vis-à-vis predators at a variety of phyletic levels. Indeed, a minor compound may possess considerable value as an antibiotic against microorganisms or may appreciably augment the deterrent potency of major constituent as a vertebrate repellent. Such possibilities can only be considered within an analytical framework that encompasses both major and minor products generated in a defensive gland.

Eisner *et al.* (1961) provided a persuasive caveat about the necessity of identifying minor constituents in defensive secretions when they studied the deterrent function of the constituents in the defensive secretion of the whip scorpion *Mastigoproctus giganteus*. The defensive exudate of this arachnid contains a concentrated solution of acetic acid (84%), but in addition is enriched with 5% octanoic acid. The latter compound admirably promotes the effectiveness of the C_2 acid as a topical irritant by acting as a spreading agent and lipoid solvent. In essence, octanoic acid disrupts the lipophilic cuticle of predatory arthropods and enables the major polar constituent to penetrate into the interior of the body and thus gain access to sensitive tissues. Thus, the minor acidic constituent guarantees the success of the major acidic product as a tissue irritant. The hydrocarbons synthesized in the Dufour's gland of formicine ants very likely play the same role for the formic acid-rich exudates with which they are secreted in admixture.

The Dufour's gland secretions of formicine ants constitute some of the most complex mixtures of compounds which have been detected as products of arthropod exocrine glands. Significantly, most of the volatile compounds, often numbering more than 40, constitute a potpourri of high-boiling trace constituents that accompany the alkanes dominating these exudates. Bergström and Löfqvist (1971) have emphasized that these trace natural products may represent key elements in the alarm–defense–recognition system of the ants by acting as specific labeling agents for alien species, especially in the area of the nest. Labeled with highly specific compounds such as all-(*E*)-farnesyl acetate (Bergström and Löfqvist, 1972) or all-(*E*)-geranylgeraniol (Bergström and Löfqvist, 1973), intruders become chemical beacons to be attacked by the aggressive and alarmed ants. The selective advantage to formicine species of producing such a variety of trace constituents in their Dufour's gland must be considerable.

The defensive exudates of many notodontid larvae consist of aqueous formic acid solutions that are discharged from a gland opening ventrally on the neck. The formic acid rich secretion of *Heterocampa manteo* contains a 1.4% solution of 2-undecanone and 2-tridecanone (Eisner *et al.*, 1972) whereas that of *Schizura concinna* contains 2-tridecanone, decyl acetate, and dodecyl acetate (Weatherston *et al.*, 1979). The ketones and acetates considerably enhance the repellency of formic acid against arthropod predators by virtue of their own irritating properties and possibly because they may promote the penetration of the acid through the lipophilic arthropod cuticle. Notwithstanding their presence as minor glandular products, these minor concomitants of formic acid represent important elements in the chemical defensive systems of the larvae.

The defensive exudates of four species of cockroaches in the genus *Platyzosteria* are dominated by 2-methylenebutanal (ethyl acrolein), a compound that constitutes over 90% of the observed volatiles in all secretions (Waterhouse and Wallbank, 1967). However, it is very probable that minor constituents play a key role in promoting the effectiveness of the discharge as a deterrent against arthropod predators. Both 2-methylene-1-butanol (ca. 1.5%) and 2-methylene-1-butanol dimer (ca. 3%) probably contribute valuable spreading properties to the mixture and thus serve to reduce the rate of evaporation of ethyl acrolein from the cuticle of a predator. Therefore, these minor constituents may provide a vital defensive function by insuring that the main exocrine product will be stabilized on the predator's cuticle long enough for it to penetrate and exert its toxic action.

When tactually stimulated, polydesmoid millipedes discharge an exudate that often contains more than 95% of a single organic compound, benzaldehyde. However, this aromatic aldehyde is often accompanied by 1–2% of phenolics such as phenol and/or guaiacol (Blum *et al.*, 1971; Duffield *et al.*, 1974; Duffey *et al.*, 1977). The phenolics appear to be present in the outer reaction chamber and are thus liberated with the initial discharges. Indeed, the phenolic odors of these secretions are evident before the benzaldehydic note becomes dominant. The presence of these phenolics in the reaction chamber may be highly adaptive, since they can constitute an "early warning system" to predators and thus result in an overall conservation of the secretion. Furthermore, these minor compounds may be highly effective in inhibiting microbial growth in the reaction chamber which opens directly to the exterior (Duffey *et al.*, 1977). Therefore, minor phenolic concomitants of benzaldehyde may actually represent key defensive compounds since they may function as repellents to predatory animals while at the same time ensuring that the reaction chamber in which benzaldehyde is generated (along with HCN) does not become the bacterial brew for which it seems ideally suited.

It seems no exaggeration to state that to ignore the chemistry of the minor defensive compounds will guarantee that the precise selective value of these important secretions will continue to remain virtual *terra incognita*.

VI. AUTODETOXICATION OF DEFENSIVE COMPOUNDS

Arthropods have evolved the ability to cope with a variety of toxic compounds which may be present in their blood (e.g., cantharidin) or more or less insulated in impermeable cuticular reservoirs. However, the producers of these compounds are frequently exposed to large doses of their defensive products which may penetrate their own cuticle or spiracles after they are secreted. Obviously, although mechanisms have been evolved to reduce the autointoxicative effects of these natural products, very few investigations have been undertaken in order to illuminate the *raison d'être* of this tolerance.

Hall *et al.* (1971) studied the resistance of polydesmoid millipedes to HCN, one of their primary defensive products. These millipedes are considerably more resistant to HCN vapors or injected cyanide than noncyanide-producing arthropods but are no more tolerant to anaerobiosis than cockroaches. On the other hand, the cytochrome oxidase system of the millipede *Euryurus leachii*, as measured by succinate oxidation, was considerably more resistant than those of the cockroaches *Blaberus discoidalis* and *Blattella germanica* to both sodium cyanide and sodium azide. Hall *et al.* (1971) believe that the natural cyanide tolerance of the millipede is due to the presence of a naturally resistant terminal oxidase as opposed to an excess of cytochrome oxidase.

Adult zygaenid moths, which release HCN from crushed tissues at all body stages, are also very resistant to cyanide but the basis for this tolerance is unknown (Jones *et al.*, 1962).

Duffey *et al.* (1974) demonstrated that the millipede *Harpaphe heydeniana* detoxifies some of its cyanide by first converting it to β-cyanoalanine which is then transformed to asparagine. This detoxification mechanism may be normally operative in terms of small quantities of HCN that are slowly produced in the storage chamber and gain access to the body. However, the millipede *Oxidus gracilis*, in addition to producing β-cyanoalanine and asparagine, contains a high titer of rhodanese which transforms the HCN into the harmless thiocyanate ion (Duffey and Blum, 1977). It is possible that rhodanese constitutes the primary means of detoxifying the high levels of HCN which the millipede encounters after discharging its cyanogenic secretion, since thiocyanate production considerably exceeds that of the other two detoxication products.

Oxidus gracilis, a paradoxsomatid millipede, produces phenol in its reaction chamber from tyrosine (Duffey and Blum, 1977). Injected phenol can be rapidly converted to tyrosine by tyrosine phenol lyase, the same enzyme that produces phenol in this species. In addition, phenol is detoxified by being converted to phenyl glycoside and possibly a small amount of arbutin. Duffey and Blum (1977) have also demonstrated that *O. gracilis* produces guaiacol from tyrosine and vice versa. Thus, tyrosine constitutes a key detoxifying compound for the phenolics produced by polydesmoid millipedes.

The sequestration of toxic compounds ingested from plants may involve preliminary detoxication before the compounds are actually sequestered in specific sites. Although this topic has received virtually no attention, recent experiments with the milkweed bug *Oncopeltus fasciatus* suggest that cardenolides which are quite toxic to this insect may be metabolized to less toxic compounds prior to sequestration (Duffey *et al.*, 1978). Whereas the polar cardenolide ouabain is facilely sequestered in the dorsolateral space fluid of the adult, digitoxin is first converted (hydroxylated?) to more polar compounds before being transferred and stored. Thus, this lygaeid has the potential of metabolizing nonpolar toxins (cardenolides) to more polar compounds and then secondarily utilizing them as part of its own chemical defenses. Whether detoxification–sequestration is widespread among insects feeding on plants fortified with toxic natural products remains to be seen. However, it is not unlikely that the multitude of insect herbivores associated with toxic plants have exploited the plants' natural products by metabolizing these compounds to products which they can then easily sequester as part of their own defensive systems.

PART I

The Chemistry of Defensive Secretions

Part I

Introduction

The diversity of compounds that have been identified as exocrine products of arthropods has emphasized the biosynthetic versatility that these invertebrates possess. Nevertheless, since the chemistry of the defensive allomones of only a relatively small proportion of arthropods has been examined, these glandular constituents have been well characterized for species in no more than a few arthropod taxa. In these, the products of exocrine glands almost invariably consist of mixtures of compounds and, in many cases, only the primary constituents have been identified in characterized secretions. The roles of minor constituents in augmenting the value of defensive exudates have been well established, and also underscore the importance of thoroughly analyzing a secretion as a prerequisite to comprehending its defensive *modus operandi.*

In general, most defensive compounds are identical to well-known organic compounds with relatively simple structures. Obviously, chemical complexity is not a necessary property of an effective antagonistic allomone, and it has been thoroughly demonstrated that very simple organic substances will serve admirably to repel the molestations of potential predators. However, recent investigations on the chemical defenses of species in arthropod taxa which had not been previously studied show that these animals synthesize a remarkable variety of compounds, and the natural products potential of these invertebrates probably has not been fully illuminated. Even in groups that appear to produce characteristic defensive secretions, flagrant chemical exceptions may prevail, and at this juncture exocrine glands of arthropods should be regarded as a potential storehouse of chemical surprises.

In this volume, the defensive compounds of arthropods have been grouped in classes according to their main functionalities and are sequentially listed based on their carbon numbers. Although in some cases this system of arrangement results in lumping together a veritable potpourri of compounds, it seems justified as a reasonable alternative to presenting an inordinately large number of categories of compounds. An empirical formula index at the end of this volume is available for convenient reference to all compounds cited in the tables and in Chapter 18.

Chapter 2

Hydrocarbons

Arthropods, and insects in particular, may constitute the hydrocarbon chemists *par excellence* in the animal kingdom. More than 110 hydrocarbons are known to be produced in the exocrine glands of these invertebrates, and for the most part these compounds consist of relatively simple alkanes and alkenes in the range C_7–C_{28} (see Table 2.2 and Chapter 18). The virtuosity of arthropods as producers of hydrocarbon allomones is not necessarily surprising, since they already possess the ability to synthesize a large variety of cuticular hydrocarbons in ectodermal tissues. Inasmuch as exocrine glands constitute ectodermal invaginations, the potential of these glands for hydrocarbon biosynthesis should be well developed. Further evidence of the proficiency of arthropods as hydrocarbon synthesizers is provided by the fact that far more of these compounds are produced than are allomones in any other chemical class.

The defensive exudates of opilionids and insects in eight orders contain hydrocarbons (Table 2.1), and the species in three-fourths of these taxa have been demonstrated to produce a variety of these compounds. Whereas a total of only three hydrocarbons have been identified in homopterous and neuropterous secretions, a minimum of 10 of these compounds (Lepidoptera) has been detected in the glandular products of species in any of the other orders. Species of opilionids, true bugs, and beetles each produce about 12% of the characterized hydrocarbons, but the distribution of these allomones within these three orders is very uneven. Alkanes and alkenes are only known as opilionid defensive products because of their detection in the secretions of one phalangiid species in a single population. In the Hemiptera, 12 hydrocarbons are distributed between species in five families, but half of these compounds are limited

Table 2.1

Distribution of Hydrocarbons in the Defensive Secretions of the Arthropoda[a]

Class	Order	Family
Arachnida	Opiliones (Phalangida)	Phalangiidae
Insecta	Dictyoptera	
	Isoptera	Termitidae
		Rhinotermitidae
	Homoptera	Aphididae
	Hemiptera	Cydnidae
		Pentatomidae
		Plataspididae
		Pyrrhocoridae
		Rhopalidae
	Neuroptera	Chrysopidae
	Coleoptera	Carabidae
		Cerambycidae
		Staphylinidae
		Tenebrionidae
	Lepidoptera	Papilionidae
	Hymenoptera	Andrenidae
		Colletidae
		Halictidae
		Apidae
		Formicidae

[a] Also see Chapter 18.

to the exudates of species in one family. The same is true of coleopterous secretions. Nearly 70% of the hydrocarbons identified in beetle exudates are produced by tenebrionids (Table 2.2).

Termites, bees, and ants are the major hydrocarbon producers in the Arthropoda. Thirty-five percent of these compounds have been characterized from isopterous secretions, particularly of species in the family Termitidae. Although bees generate about 20 hydrocarbons, they are a poor hymenopterous second to the ants which synthesize about two-thirds (74) of the alkanes and alkenes detected as defensive compounds. The Dufour's gland of ants appears to be the ultimate hydrocarbon factory in the Arthropoda, producing more than 80% of these ant-derived compounds. Most of the hydrocarbons produced by bees are also synthesized by ants, and an extraordinary variety of branched and unbranched compounds in the range C_9–C_{28} have been characterized as formicid natural products (Table 2.2). Since hydrocarbons have been detected as exocrine products in all species of ants analyzed, it appears that these insects will continue to be an excellent source of diverse alkanes and alkenes.

AN OVERVIEW

The abundance of hydrocarbons produced by insects is rather misleading since many of these compounds have a very circumscribed distribution in their glandular exudates. A few compounds such as *n*-tridecane and *n*-pentadecane occur in the secretions of species in many families (Table 2.2), but these allomones are exceptions. Several hydrocarbons with rather widespread distributions (e.g., α-pinene) are only characteristic products of species in one order (e.g., Dictyoptera), and these compounds are rather unusual products for the species in the other orders. Many of the alkanes and alkenes are unique natural products that have only been detected as defensive compounds of a few or several species in one family. For example, about 30% of the identified hydrocarbons are limited to the exocrine products of ants, and nearly 20% of these compounds are restricted to termitid exudates.

The lack of aromatic hydrocarbons in arthropod defensive secretions is conspicuous, only toluene and *p*-cymene having been infrequently detected as allomones of termites and beetles (Table 2.2 and Chapter 18). Although the majority of these compounds are saturated, a large variety of alkenes are produced which includes mono-, di-, tri-, and tetraunsaturated alkenes. About 25% of the identified hydrocarbons are terminally unsaturated, although a variety of Δ^2, Δ^4, Δ^7, Δ^8, and Δ^9, alkenes have also been detected. Some of the ant-derived alkenes are methyl-branched, although most of the branched compounds are saturated. Nearly 25% of the characterized hydrocarbons are either mono- or dimethyl-branched compounds, and the majority of these hydrocarbons are 3- and 5-methylalkanes in the range C_{10}–C_{20}. All of these hydrocarbons are natural products of ants (Table 2.2).

More than a quarter of these defensive compounds are terpenes with about 75% of these isoprenoids being produced by termitid soldiers. A real potpourri of mono-, sesqui-, and diterpenes are synthesized by aphids, papilionid larvae, termite soldiers, bees, and ants, and many of these compounds have very restricted arthropod distributions (Table 2.2 and Chapter 18). Indeed, these terpenoid allomones are among the most idiosyncratic compounds synthesized by arthropods.

Table 2.2 Hydrocarbons in arthropod defensive secretions

Name and Formula	Occurrence	Glandular Source	Comments
Toluene 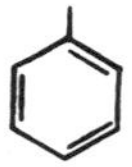 C_7H_8 MW 92	Coleoptera Cerambycidae: _Stenocentrus ostricilla_ and _Syllitus grammicus_ (Moore and Brown, 1971b).	Mandibular glands.	It constitutes about 75% of the defensive secretions of these two species. An unusual pit-and-tongue organ at the base of each mandible dispenses the secretion which is stored in reservoirs extending nearly to the base of the abdomen.
1-Nonene  C_9H_{18} MW 126	Coleoptera Tenebrionidae: _Eleodes longicollis_ (Hurst _et al_., 1964), _Eleodes_ spp., _Psorodes_ spp., _Schelodontes_ spp., _Cratidus_ spp., _Neobaphion_ spp., _Notibius_ spp., _Gonopus_ spp., _Melanopterus_ spp., _Alobates_ spp., _Toxicum_ spp., _Zadenos_ spp., _Blapstinus_ spp., and _Parastizopus_ spp. (Tschinkel, 1975a,b).	Paired abdominal sternal glands.	It is usually a relatively minor component which often accompanies the C_{11} and C_{13} alkenes.

n-Nonane  C_9H_{20} MW 128	Hymenoptera Formicidae: Formica sanguinea, F. fusca, F. rufibarbis (Bergström and Löfqvist, 1968), F. nigricans, F. rufa, F. polyctena (Bergström and Löfqvist, 1973), and Camponotus ligniperda (Bergström and Löfqvist, 1971).	Dufour's gland of workers.	It is a minor component in the secretions of all species.
α-Pinene  $C_{10}H_{16}$ MW 136	Dictyoptera Termitidae: Nasutitermes exitiosus, N. graveolus, N. longipennis, N. magnus, N. triodiae, N. walkeri, Tumulitermes pastinator, Amitermes laurensis, A. vitiosus (Moore, 1964, 1968), and Tenuirostritermes tenuirostris (Nutting et al., 1974).	Termitidae: Frontal gland of soldiers.	It is the major constituent in the secretions of the nasute soldiers.
	Hymenoptera Formicidae: Myrmicaria natalensis (Brand et al., 1974), Pristomyrmex pungens, Pheidole nodus, and Lasius spathepus (Hayashi et al., 1973b).	Formicidae: Poison gland of M. natalensis workers; unknown for other species.	

(continued)

Table 2.2 (Continued)

Compound	Taxon and species	Gland source	Comments
β- Pinene $C_{10}H_{16}$ MW 136	Dictyoptera Termitidae: Nasutitermes exitiosus, N. graveolus, N. longipennis, N. magnus, N. triodiae, N. walkeri, Tumulitermes pastinator, Amitermes laurensis, and A. vitiosus (Moore, 1964, 1968). Hymenoptera Formicidae: Pristomyrmex pungens, Pheidole nodus, Crematogaster laboriosa, Lasius spathepus, (Hayashi et al., 1973b), and Myrmicaria natalensis (Brand et al., 1974).	Termitidae: Frontal gland of soldiers. Formicidae: Poison gland of M. natalensis workers; unknown for other species.	It is generally the second most abundant compound in the secretions of Nasutitermes species and a trace constituent in the exudates of Amitermes species. It is a major constituent in most of the myrmicine species.
Limonene (Dipentene) $C_{10}H_{16}$ MW 136	Opiliones Phalangiidae: Phalangium opilio (Blum et al., 1973e). Dictyoptera Termitidae: Nasutitermes exitiosus, N. graveolus, N. longipennis, N. magnus, N. triodiae, N. walkeri, Tumulitermes laurensis, A. vitiosus, Drepanotermes rubriceps (Moore, 1968), and Tenuirostritermes tenuirostris (Nutting et al., 1974).	Phalangiidae: Paired glands located on the flanks of the prosoma between the first and second pair of coxae. Termitidae: Frontal gland of soldiers. Formicidae: Poison gland of M. natalensis workers; unknown for other species.	It is produced only by females of P. opilio. The d-isomer of this terpene comprises about 80% of the secretion of M. natalensis. This is an unusual poison gland secretion since myrmicine venoms are generally identified with proteins. It is present in moderate concentrations in the Nasutitermes and Amitermes secretions whereas the secretion of D. rubriceps consists of

	Hymenoptera Formicidae: _Myrmicaria natalensis_ (Quilico _et al_., 1962), _Pristomyrmex pungens_, and _Crematogaster laboriosa_ (Hayashi _et al_., 1973b).		more than 90% of this terpene.
α - Terpinene $C_{10}H_{16}$ MW 136	Hymenoptera Formicidae: _Myrmicaria natalensis_ (Brand _et al_., 1974).	Poison gland of workers.	It is a minor constituent.
Camphene $C_{10}H_{16}$ MW 136	Hymenoptera Formicidae: _Pristomyrmex pungens_, _Crematogaster laboriosa_ (Hayashi _et al_., 1973b), and _Myrmicaria natalensis_ (Brand _et al_., 1974).	Poison gland of _M. natalensis_ workers; unknown for other species.	It is a minor component in all species.
Sabinene $C_{10}H_{16}$ MW 136	Hymenoptera Formicidae: _Myrmicaria natalensis_ (Brand _et al_., 1974).	Poison gland of workers.	It is a trace constituent.

(continued)

Table 2.2 (Continued)

Terpinolene  $C_{10}H_{16}$ MW 136	Dictyoptera Termitidae: *Nasutitermes longipennis*, *N. magnus*, *N. walkeri*, *Amitermes herbertensis*, *A. laurensis*, *A. vitiosus*, and *Drepanotermes rubriceps* (Moore, 1968). Hymenoptera Formicidae: *Myrmicaria natalensis* (Brand *et al.*, 1974).	Termitidae: Frontal gland of soldiers. Formicidae: Poison gland of workers.	The secretion of *A. herbertensis* consists of almost pure terpinolene whereas in the other species this compound is a trace or minor constituent.
α-Phellandrene  $C_{10}H_{16}$ MW 136	Dictyoptera Termitidae: *Nasutitermes graveolus*, *N. longipennis*, *N. magnus*, *Amitermes herbertensis*, *A. laurensis*, *A. vitiosus*, and *Drepanotermes rubriceps* (Moore, 1968). Hymenoptera Formicidae: *Myrmicaria natalensis* (Brand *et al.*, 1974).	Termitidae: Frontal gland of soldiers. Formicidae: Poison gland of workers.	It is a trace or minor component of the secretions of all species except *A. laurensis* and *A. vitiosus*.
Myrcene  $C_{10}H_{16}$ MW 136	Dictyoptera Termitidae: *Amitermes vitiosus* (Moore, 1968) and *Tenuirostritermes tenuirostris* (Nutting *et al.*, 1974). Hymenoptera Formicidae: *Myrmicaria natalensis* (Brand *et al.*, 1974).	Termitidae: Frontal gland of soldiers. Formicidae: Poison gland of workers.	It is an atypical defensive component of termite soldiers produced by only 2 of 12 termitid species and only 1 of 3 species of *Amitermes*. It is a minor component in the secretion of *T. tenuirostris*.

Decene $C_{10}H_{20}$ MW 140	Hymenoptera Formicidae: *Anoplolepis custodiens* (Schreuder and Brand, 1972).	Dufour's gland of workers.	The location of the double bond has not been determined for this trace constituent.
n-Decane $C_{10}H_{22}$ MW 142	Coleoptera Carabidae: *Idiochroma dorsalis* (Schildknecht *et al*., 1968c), *Poecilus cupreus*, *Pterostichus niger*, *P. macer*, *P. vulgaris*, *P. metallicus*, *Amara similata*, and *A. familiaris* (Schildknecht *et al*., 1968a). Hymenoptera Formicidae: *Formica sanguinea*, *F. fusca*, *F. rufibarbis* (Bergström and Löfqvist, 1970), *Camponotus ligniperda* (Bergström and Löfqvist, 1971), and *C. herculeanus* (Bergström and Löfqvist, 1972).	Carabidae: Pygidial glands. Formicidae: Dufour's gland of workers and females.	It is a minor constituent in the secretion of both the carabids and ants. It is apparently a rather characteristic component of formicine species.
3-Methylnonane $C_{10}H_{22}$ MW 142	Hymenoptera Formicidae: *Formica nigricans* (Bergström and Löfqvist, 1973) and *Anoplolepis custodiens* (Schreuder and Brand, 1972).	Dufour's gland of workers.	It appears to be an atypical trace constituent which has not been detected in the secretions of other species of *Formica*.

(*continued*)

Table 2.2 (Continued)

Compound	Source	Gland	Remarks
1-Undecene  $C_{11}H_{22}$ MW 154	Cpiliones Phalangiidae: *Phalangium opilio* (Blum *et al.*, 1973e). Coleoptera Tenebrionidae: *Eleodes longicollis* (Hurst *et al.*, 1964), *Eleodes* spp., *Alphitobius* spp., *Psorodes* spp., *Meracantha* spp., *Schelodontes* spp., *Gonopus* spp., *Anomalipus* spp., *Melanopterus* spp., *Alobates* spp., *Toxicum* spp., *Coelocnemis* spp., *Nyctobates* spp., *Cibdelis* spp., *Merinus* spp., *Zadenos* spp., *Cratidus* spp., *Neobaphion* spp., *Blapstinus* spp., *Gonocephalum* spp., *Parastizopus* spp., and *Platydema* spp. (Tschinkel, 1975a,b). Staphylinidae: *Bledius mandibularis*, *B. spectabilis*, (Wheeler *et al.*, 1972a), and *Drusilla canaliculata* (Brand *et al.*, 1973b).	Phalangiidae: Paired glands located on the flanks of the prosoma between the first and second pair of coxa. Tenebrionidae: Paired abdominal sternal glands. Staphylinidae: Pygidial glands (*Bledius* spp.) and tergal gland (*D. canaliculata*).	It is a major constituent in the exudates of *Bledius* spp. and tenebrionids in the genera *Gonopus*, *Cibdelis*, and *Cratidus*. It is an important component in the secretions of tenebrionids in the tribes Opatrini and Helopini.

2-Undecene  $C_{11}H_{22}$ MW 154	Hymenoptera Formicidae: *Formica nigricans*, *F. rufa*, *F. polyctena* (Bergström and Löfqvist, 1973), *Camponotus ligniperda* (Bergström and Löfqvist, 1971), *C. herculeanus* (Bergström and Löfqvist, 1972), and *Anoplolepis custodiens* (Schreuder and Brand, 1972).	Dufour's gland of workers.	It is a minor or trace component in the secretions of all species. Ozonolysis of the alkenes of the *Formica* species indicates that the double bond is located between positions 2 and 3. The location of the double bond in the alkenes produced in the Dufour's glands of *Camponotus* and *Anoplolepis* has not been determined.
n-Undecane  $C_{11}H_{24}$ MW 156	Opiliones Phalangiidae: *Phalangium opilio* (Blum *et al*., 1973e).	Phalangiidae: Paired glands located on the flanks of the prosoma between the first and second pair of coxae.	It is a major and characteristic component in the secretions of formicine ants which in some cases also functions as an alarm pheromone.
	Hemiptera Pentatomidae: *Nezara viridula* (Gilby and Waterhouse, 1964).	Pentatomidae: Metasternal scent gland of adults.	Although this hydrocarbon appears to be characteristic of *Pterostichus* species, it does not generally have a widespread distribution in the defensive exudates of carabids.
	Coleoptera Carabidae: *Idiochroma dorsalis* (Schildknecht *et al*., 1968c), *Poecilus cupreus*, *Pterostichus niger*, *P. macer*, *P. vulgaris*, *P. melas*, *P. metallicus*, *Amara similata*, and *A. familiaris* (Schildknecht *et al*., 1968d).	Carabidae: Pygidial glands Staphylinidae: Adult tergal gland situated between the sixth and seventh abdominal tergites.	It is a minor constituent in the secretion of *N. viridula* and apparently an atypical compound in the chemical

(*continued*)

Table 2.2 (Continued)

	Staphylinidae: Drusilla canaliculata (Brand et al., 1973b).	Formicidae: Dufour's gland of workers and females.	defense arsenals of hemipterans. It is the major hydrocarbon produced in the tergal gland of D. canaliculata.
	Hymenoptera Formicidae: Lasius umbratus (Quilico et al., 1957a), L. fuliginosus (Bernardi et al., 1967), L. alienus (Regnier and Wilson, 1969; Bergström and Löfqvist, 1970), L. niger, L. flavus, L. carniolicus (Bergström and Löfqvist,1970), L. sitkaensis, L. neoniger, L. nearcticus, L. speculiventris (Wilson and Regnier, 1971), Acanthomyops claviger (Regnier and Wilson, 1968), A. latipes, A. subglaber (Wilson and Regnier, 1971), Formica rufa (Schall, 1892), F. sanguinea, F. fusca, F. rufibarbis (Bergström and Löfqvist, 1968), F. neogagates, F. subsericea, F. schaufussi, F. exsectoides, F. rubicunda, F. pergandei, F. subintegra (Wilson and Regnier, 1971), F. nigricans, F. rufa, F. polyctena (Bergström and Löfqvist, 1973), F. japonica (Hayashi et al., 1973a), Polyergus rufescens (Bergström and Löfqvist, 1968), Camponotus americanus, C. pennsylvanicus (Ayre and Blum,		

	1971), *C. herculeanus* (Ayre and Blum, 1971; Bergström and Löfqvist, 1972), *C. ligniperda* (Bergström and Löfqvist, 1971), *C. noveboracensis* (Wilson and Regnier, 1971), *C. intrepidus* (Brophy *et al.*, 1973), *C. japonicus*, *C. obscuripes* (Hayashi *et al.*, 1973a), *Anoplolepis custodiens* (Schreuder and Brand, 1972), and *Pheidole lamellidens* (Hayashi *et al.*, 1973a).		
Dodecadiene $C_{12}H_{22}$ MW 166	Hymenoptera Formicidae: *Formica nigricans* and *F. polyctena* (Bergström and Löfqvist, 1973).	Dufour's gland of workers.	The locations of the double bonds have not been determined.
1-Dodecene  $C_{12}H_{24}$ MW 168	Hymenoptera Formicidae: *F. nigricans*, *F. rufa*, *F. polyctena* (Bergström and Löfqvist, 1973), *Camponotus ligniperda* (Bergström and Löfqvist, 1971), *C. herculeanus* (Bergström and Löfqvist, 1972), *C. intrepidus* (Brophy *et al.*, 1973), and *Anoplolepis custodiens* (Schreuder and Brand, 1972).	Dufour's gland of workers.	It is a trace component which appears to be absent from the Dufour's secretions of other formicine species. The location of the double bond has only been ascertained for the alkene of *C. intrepidus*. Since the major alkenes produced by the *Formica* species possess a double bond in position 3 and 4, it

(continued)

Table 2.2 (Continued)

			is possible that these species produce 3-dodecene instead.
n-Dodecane $C_{12}H_{26}$ MW 170	Hemiptera Pentatomidae: *Nezara viridula* (Gilby and Waterhouse, 1964), *Musgraveia sulciventris* (Gilby and Waterhouse 1967), and *Apodiphus amygdali* (Everton *et al*., 1974).	Pentatomidae: Metasternal scent gland of adults and dorsal abdominal glands of larvae.	It is a minor component in the exudates of all species. Although it is produced in the abdominal glands of larvae of *D. intermedius*, it is not detectable in the metasternal scent gland secretion of adults.
	Cydnidae: *Macrocystus* sp. (Baggini *et al*., 1966).	Cydnidae: Metasternal scent gland of adults.	
	Pyrrhocoridae: *Dysdercus intermedius* (Calam and Youdeowei, 1968).	Pyrrhocoridae: Dorsal abdominal glands of larvae.	
	Hymenoptera Formicidae: *Oecophylla longinoda* (Bradshaw *et al*., 1973), *Formica fusca*, *F. rufibarbis*, (Bergström and Löfqvist, 1968), *F. nigricans*, *F. rufa*, *F. polyctena* (Bergström and Löfqvist, 1973), *F. japonica*, *Pheidole lamellidens* (Hayashi *et al*., 1973a), *Lasius alienus* (Bergström and Löfqvist, 1970), *Anoplolepis custodiens* (Schreuder and Brand, 1972), *Camponotus ligniperda* (Bergström and Löfqvist, 1971), *C. herculeanus* (Bergström and Löfqvist, 1972), *C. intrepidus* (Brophy *et al*., 1973),	Formicidae: Dufour's gland of workers.	

	Pogonomyrmex rugosus, and *P. barbatus* (Regnier *et al.*, 1973).		
3-Methylundecane $C_{12}H_{26}$ MW 170	Hymenoptera Formicidae: *Lasius carniolicus* (Bergström and Löfqvist, 1970), *Formica fusca* (Bergström and Löfqvist, 1968), *F. nigricans*, *F. rufa*, *F. polyctena* (Bergström and Löfqvist, 1973), *Camponotus ligniperda* (Bergström and Löfqvist, 1971), *C. herculeanus* (Bergström and Löfqvist, 1972), *C. intrepidus* (Brophy *et al.*, 1973), *Pogonomyrmex rugosus*, and *P. barbatus* (Regnier *et al.*, 1973).	Dufour's gland of workers.	It is a minor or trace component in the secretions of all four formicine species.
5-Methylundecane $C_{12}H_{26}$ MW 170	Hymenoptera Formicidae: *Formica nigricans*, *F. rufa*, *F. polyctena* (Bergström and Löfqvist, 1973), *Camponotus intrepidus* (Brophy *et al.*, 1973), and *Pogonomyrmex rugosus* (Regnier *et al.*, 1973).	Dufour's gland of workers.	It is a trace constituent.

(*continued*)

Table 2.2 (Continued)

6-Methylundecane  $C_{12}H_{26}$ MW 170	Hymenoptera Formicidae: _Pogonomyrmex rugosus_ (Regnier _et al_., 1973).	Dufour's gland of workers.	It is a minor constituent.
4,7-Tridecadiene  $C_{13}H_{24}$ MW 180	Coleoptera Staphylinidae: _Drusilla canaliculata_ (Brand _et al_., 1973b).	Staphylinidae: Adult tergal gland situated between the sixth and seventh abdominal tergites.	The positions of the double bonds have not been established for the _Camponotus_ hydrocarbon.
	Hymenoptera Formicidae: _Camponotus ligniperda_ (Bergström and Löfqvist, 1971).	Formicidae: Dufour's gland of workers.	
1-Tridecene  $C_{13}H_{26}$ MW 182	Opiliones Phalangiidae: _Phalangium opilio_ (Blum _et al_., 1973e).	Phalangiidae: Paired glands opening on the flanks of the prosoma between the first and second pair of coxae.	The most common alkene in the defensive secretions of tenebrionines. Generally a minor or trace constituent in the exudates of insects. The location of the double bond has been established only for the alkene in the tenebrionine discharges. It is a major constituent in the tenebrionid tribes Litoborini and Diapirini.
	Neuroptera Chrysopidae: _Chrysopa oculata_ (Blum _et al_., 1973d).	Chrysopidae: Paired glands opening on the frontal margin of the prothorax.	
	Coleoptera Tenebrionidae: _Gnathocerus_ spp., _Eurynotus_ spp., _Coelocnemis_ spp., _Centronopus_ spp., _Zadenos_ spp., _Scarus_ spp., _Blaps_ spp., _Amphidora_ spp., _Cratidus_ spp., _Eleodes_ spp., _Embaphion_ spp., _Neobaphion_	Tenebrionidae: Paired abdominal sternal glands.	

Compound	Source	Gland	Comments
	spp., *Gonocephalum* spp., *Parastizopus* spp., *Platydema* spp., *Diaperis* spp., *Metaclisa* spp., and *Neomida* spp., (Tschinkel, 1975b) *Eleodes longicollis* (Hurst *et al*., 1964), and *Eleodes* spp. (Tschinkel, 1975a,b).		
4-Tridecene $C_{13}H_{26}$ MW 182	Coleoptera Staphylinidae: *Drusilla canaliculata* (Brand *et al*., 1973b), and *Lomechusa strumosa* (Blum *et al*., 1971).	Staphylinidae: Adult tergal gland situated between the sixth and seventh abdominal tergites.	It is the second most abundant hydrocarbon in the secretions.
	Hymenoptera Formicidae: *Formica fusca* (Bergström and Löfqvist, 1968), *F. nigricans*, *F. rufa*, *F. polyctena* (Bergström and Löfqvist, 1973), *Lasius flavus* (Bergström and Löfqvist, 1970), *Camponotus ligniperda* (Bergström and Löfqvist, 1971), *C. herculeanus* (Bergström and Löfqvist, 1972), *C. americanus*, *C. pennsylvanicus* (Ayre and Blum, 1971), *C. intrepidus* (Brophy *et al*., 1973), *Anoplolepis custodiens* (Schreuder and Brand, 1972), and *Oecophylla longinoda* (Bradshaw *et al*., 1973).	Formicidae: Dufour's gland of workers.	The position of the double bond has only been determined for the compound in the secretion of *C. ligniperda* and three of the *Formica* species in which it is one of the major alkenes present.

(continued)

Table 2.2 (Continued)

Compound	Species	Gland source	Remarks
n-Tridecane  $C_{13}H_{28}$ MW 184	Opiliones Phalangiidae: *Phalangium opilio* (Blum *et al*., 1973e).	Phalangiidae: Paired glands located on the flanks of the prosoma between the first and second pair of coxae.	The most widespread hydrocarbon in the defensive secretions of insects. It frequently comprises more than 50% of the two-phase secretion of pentatomids and it is the only volatile compound detected in extracts of the plataspidid *C. cephalicus*. It is only produced by the third abdominal gland of *D. intermedius* larvae (Youdeowei and Calam, 1969). It is usually the second most abundant hydrocarbon produced in the Dufour's gland of formicine ants. This compound appears to be a rather characteristic component of the secretions of formicine species although it has been detected as a trace constituent in the Dufour's gland secretions of ant species in two other subfamilies.
	Hemiptera Pentatomidae: *Oebalus pugnax* (Blum *et al*., 1960), *Euschistus servus* (Blum and Traynham, 1962), *Biprorulus bibax*, *Musgraveia sulciventris* (Park and Sutherland, 1962), *Nezara viridula* (Gilby and Waterhouse, 1964), *Tessaratoma aethiops* (Baggini *et al*., 1966), *Apodiphus amygdali* (Everton *et al*., 1974).	Pentatomidae: Metasternal scent gland of adults of all species and dorsal abdominal glands of larval *T. aethiops* and *A. amygdali*.	
	Cydnidae: *Macrocystus* sp. (Baggini *et al*., 1966).	Cydnidae: Metasternal scent gland of adults.	
	Plataspididae: *Ceratocoris cephalicus* (Baggini *et al*., 1966).	Plataspididae: Metasternal scent gland of adults.	
	Pyrrhocoridae: *Dysdercus intermedius* (Calam and Youdeowei, 1968).	Pyrrhocoridae: Dorsal abdominal glands of larvae.	
	Coleoptera Carabidae: *Craspedophorus* sp. (Moore and Wallbank, 1968), *Poecilus cupreus*, *Pterostichus niger*, *P. macer*, *P. vulgaris*, *P. melas*, *P. metallicus*, *Amara similata*, and *A. familiaris* (Schildknecht *et al*., 1968d).	Carabidae: Pygidial glands. Staphylinidae: Dorsal abdominal tergal gland or pygidial glands (*H. semirufus* and *P. politus*).	

Staphylinidae: *Lomechusa strumosa* (Blum *et al*., 1971), *Drusilla canaliculata* (Brand *et al*., 1973b), *Hesperus semirufus*, and *Philonthus politus* (Bellas *et al*., 1974).	Formicidae: Dufour's gland of workers.
Hymenoptera Formicidae: *Formica sanguinea*, *F. fusca*, *F. rufibarbis* (Bergström and Löfqvist, 1968), *F. subintegra*, *F. subsericea* (Regnier and Wilson, 1971), *Lasius fuliginosus* (Bernardi *et al*., 1967), *Lasius alienus* (Regnier and Wilson, 1969; Bergström and Löfqvist, 1970), *F. nigricans*, *F. rufa*, *F. polyctena* (Bergström and Löfqvist, 1973), *F. japonica* (Hayashi *et al*., 1973a), *Lasius niger*, *L. flavus*, *L. carniolicus* (Bergström and Löfqvist, 1970), *Pheidole lamellidens* (Hayashi *et al*., 1973a), *Acanthomyops claviger* (Regnier and Wilson, 1968), *Camponotus ligniperda* (Bergström and Löfqvist, 1971), *C. japonicus* (Hayashi *et al*., 1973a), *C. americanus*, *C. pennsylvanicus* (Ayre and Blum, 1971), *C. herculeanus* (Ayre and Blum,	

(*continued*)

Table 2.2 (Continued)

	1971; Bergström and Löfqvist, 1972), *C. intrepidus* (Brophy *et al.*, 1973), *Anoplolepis custodiens* (Schreuder and Brand, 1972), *Oecophylla longinoda* (Bradshaw *et al.*, 1973), *Monacis bispinosa* (Blum and Wheeler, 1974), *Novomessor cockerelli* (Vick *et al.*, 1969), *Myrmica rubra* (Morgan and Wadhams, 1972), *Pogonomyrmex rugosus*, and *P. barbatus* (Regnier *et al.*, 1973).		
3-Methyldodecane $C_{13}H_{28}$ MW 184	Hymenoptera Formicidae: *Camponotus intrepidus* (Brophy *et al.*, 1973), *Pogonomyrmex rugosus*, and *P. barbatus* (Regnier *et al.*, 1973).	Dufour's gland of workers.	It is accompanied by 5-methyldodecane, an equally trace constituent.
4-Methyldodecane $C_{13}H_{28}$ MW 184	Hymenoptera Formicidae: *Formica nigricans* (Bergström and Löfqvist, 1973)	Dufour's gland of workers.	It is a trace component which is not detectable in the secretions of other *Formica* species.

5-Methyldodecane $C_{13}H_{28}$ MW 184	Hymenoptera Formicidae: *Camponotus intrepidus* (Brophy *et al.*, 1973).	Dufour's gland of workers.	It is a trace component.
6-Methyldodecane $C_{13}H_{28}$ MW 184	Hymenoptera Formicidae: *Pogonomyrmex rugosus* and *P. barbatus* (Regnier *et al.*, 1973).	Dufour's gland of workers.	It is in admixture with 3-methyldodecane.
Tetradecene $C_{14}H_{28}$ MW 196	Opiliones Phalangiidae: *Phalangium opilio* (Blum *et al.*, 1973e). Hymenoptera Formicidae: *Formica nigricans*, *F. rufa*, *F. polyctena* (Bergström and Löfqvist, 1973), *Anoplolepis custodiens* (Schreuder and Brand, 1972), *Camponotus ligniperda* (Bergström and Löfqvist, 1972), and *Novomessor cockerelli* (Vick *et al.*, 1969).	Phalangiidae: Paired glands located on the flanks of the prosoma between the first and second pair of coxae. Formicidae: Dufour's gland of workers.	It is one of an extensive series of trace alkenes in the glandular exudates. The location of the double bond has not been determined for the hydrocarbons produced by any of these species.

(continued)

Table 2.2 (Continued)

n-Tetradecane $C_{14}H_{30}$ MW 198	Opiliones Phalangiidae: *Phalangium opilio* (Blum *et al*., 1973e).	Phalangiidae: Paired glands located on the flanks of the prosoma between the first and second pair of coxae.	It is not produced by adults of *D. intermedius* and it is not detectable in the secretion of the third abdominal gland of the larvae.
	Hemiptera Pyrrhocoridae: *Dysdercus intermedius* (Calam and Youdeowei, 1968; Youdeowei and Calam, 1969).	Pyrrhocoridae: First and second dorsal abdominal glands of larvae.	Its presence in the secretions of formicine spp. and a primitive myrmiciine sp. may indicate that this alkane has a widespread distribution in the Formicidae.
	Hymenoptera Formicidae: *Formica nigricans*, *F. rufa*, *F. polyctena* (Bergström and Löfqvist, 1973), *Pheidole lamellidens* (Hayashi *et al*., 1973a), *Oecophylla longinoda* (Bradshaw *et al*., 1973), *Camponotus ligniperda* (Bergström and Löfqvist, 1971), *C. herculeanus* (Bergström and Löfqvist, 1972), *C. intrepidus* (Brophy *et al*., 1973), *Anoplolepis custodiens* (Schreuder and Brand, 1972), *Iridomyrmex humilis* (Cavill and Houghton, 1973), *Myrmecia gulosa* (Cavill and Williams, 1967), *Pogonomyrmex rugosus*, and *P. barbatus* (Regnier *et al*., 1973).	Formicidae: Dufour's gland of workers.	

3-Methyltridecane  $C_{14}H_{30}$ MW 198	Hymenoptera Formicidae: *Formica nigricans*, *F. rufa*, *F. polyctena* (Bergström and Löfqvist, 1973), *Camponotus ligniperda* (Bergström and Löfqvist, 1971), *C. herculeanus* (Bergström and Löfqvist, 1972), *C. intrepidus* (Brophy *et al.*, 1973), and *Iridomyrmex humilis* (Cavill and Houghton, 1973, 1974b).	Dufour's gland of workers.	It is a trace component which appears to be an uncommon formicine exocrine product.
5-Methyltridecane  $C_{14}H_{30}$ MW 198	Hymenoptera Formicidae: *Formica nigricans*, *F. rufa*, *F. polyctena* (Bergström and Löfqvist, 1973), *Camponotus ligniperda*, *C. herculeanus* (Bergström and Löfqvist, 1972), *C. intrepidus* (Brophy *et al.*, 1973), and *Pogonomyrmex barbatus* (Regnier *et al.*, 1973).	Dufour's gland of workers.	It is a trace component which constitutes one of three branched hydrocarbons in these secretions.
3,5-Dimethyldodecane  $C_{14}H_{30}$ MW 198	Hymenoptera Formicidae: *Pogonomyrmex rugosus* and *P. barbatus* (Regnier *et al.*, 1973).	Dufour's gland of workers.	It is a minor constituent.

(continued)

Table 2.2 (Continued)

α-Farnesene $C_{15}H_{24}$ MW 204	Hymenoptera Andrenidae: *Andrena bicolor*, *A. denticulata*, *A. nigroaenea*, *A. carbonaria*, *A. helvola*, and *A. haemorrhoa* (Bergström and Tengö, 1974 Formicidae: *Aphaenogaster longiceps* (Cavill *et al*., 1967), *Formica sanguinea*, *F. fusca*, *Polyergus rufescens* (Bergström and Löfqvist, 1968), *Camponotus ligniperda* (Bergström and Löfqvist, 1971), *C. herculeanus* (Bergström and Löfqvist, 1972), and *Myrmica rubra* (Morgan and Wadhams, 1972).	Andrenidae: Dufour's gland of females. Formicidae: Dufour's gland of workers.	Four different farnesenes are present in the andrenid secretions. It comprises 90% of the volatiles present in the secretion of the formicid *A. longiceps* and is the major constituent in the glandular exudate of the slave-raiding species *P. rufescens*. The stereochemistry of the farnesenes in these secretions has not been established.
trans-β-Farnesene $C_{15}H_{24}$ MW 204	Homoptera Aphididae: *Aphis gossypii*, *Schizaphis graminum*, *Macrosiphum rosae*, (Bowers *et al*., 1972), *M. avenae* (Wientjens *et al*., 1973), *Acyrthosiphon pisum* (Bowers *et al*., 1972), *A. solani* (Nault and Bowers, 1974), *Myzus persicae* (Edwards *et al*., 1973), *Rhopalosiphum padi*, *Metopolophium dirhodum* (Wientjens *et al*., 1973), *Hyadaphis erysimi*, and *Sipha flava* (Nault and Bowers, 1974).	Fat cells which extend into the cornicles or subcornicle area.	It is evacuated through the paired cornicles near the tip of the abdomen and functions as a defensive compound and a dispersing (alarm) pheromone. It is pheromonally active against six other aphid species which may also produce this compound. The exact site of its synthesis has not been established.

β-Selinene $C_{15}H_{24}$ MW 204	Lepidoptera Papilionidae: *Battus polydamas* (Eisner *et al.*, 1971b).	Larval osmeterium located in neck membrane.	It comprises about 80% of the secretion.
Pentadecadiene $C_{15}H_{28}$ MW 208	Hymenoptera Formicidae: *Formica nigricans*, *F. rufa*, and *F. polyctena* (Bergström and Löfqvist, 1973).	Dufour's gland of workers.	It is one of four dienes present as trace constituents in the secretions of *Formica* spp.
1-Pentadecene $C_{15}H_{30}$ MW 210	Opiliones Phalangiidae: *Phalangium opilio* (Blum *et al.*, 1973e).	Phalangiidae: Paired glands located on the flanks of the prosoma between the first and second pair of coxae.	The location of the double bond in the phalangiid alkene has not been established. It is only a major constituent in secretions of tenebrionids in the tribe Phaleriini.
	Coleoptera Tenebrionidae: *Eleodes* spp., *Embaphion* spp., *Parastizopus* spp., *Phaleria* spp., *Platydema* spp., *Uloma* spp., *Psorodes* spp., *Meracantha* spp., *Pyanisia* spp., *Gonopus* spp., *Melanopterus* spp., *Opatrinus* spp., *Cibdelis* spp., *Zadenos* spp. (Tschinkel, 1975b), *Tribolium confusum* (von Endt and Wheeler, 1971) and *Tribolium* spp. (Tschinkel, 1975b).	Tenebrionidae: Paired abdominal sternal glands.	

(continued)

Table 2.2 (Continued)

7-Pentadecene $C_{15}H_{30}$ MW 210	Hymenoptera Formicidae: *Formica rufibarbis* (Bergström and Löfqvist, 1968), *F. nigricans*, *F. rufa*, *F. polyctena* (Bergström and Löfqvist, 1973), *Camponotus ligniperda* (Bergström and Löfqvist, 1971), *C. herculeanus*, (Bergström and Löfqvist, 1972), *C. intrepidus* (Brophy *et al*., 1973), *C. americanus* (Ayre and Blum, 1971), *Anoplolepis custodiens* (Schreuder and Brand, 1972), *Iridomyrmex humilis* (Cavill and Houghton, 1973), and *Myrmica rubra* (Morgan and Wadhams, 1972).	Dufour's gland of workers.	The location of the double bond has only been established for the alkene in *I. humilis*.
n-Pentadecane $C_{15}H_{32}$ MW 212	Opiliones Phalangiidae: *Phalangium opilio* (Blum *et al*., 1973e).	Phalangiidae: Paired glands located on the flanks of the prosoma between the first and second pair of coxae.	It is not produced by the two anterior abdominal glands of larval *D. intermedius* and is absent from the metasternal scent gland secretion of the adults.
	Hemiptera Pyrrhocoridae: *Dysdercus intermedius* (Calam and Youdeowei, 1968).	Pyrrhocoridae: Posterior (third) dorsal abdominal gland of larva.	It is a widespread and major constituent in the Dufour's gland of ants in four subfamilies. In the secretion of *A. custodiens* it increases by about 40-fold during the fall and winter.
	Pentatomidae: *Musgraveia sulciventris* and *Biprorulus bibax* (MacLeod *et al*., 1975).	Pentatomidae: Metasternal scent gland of adults.	
	Coleoptera Staphylinidae: *Hesperus semirufus* and *Philonthus politus* (Bellas *et al*., 1974).	Staphylinidae: Pygidial glands.	

Hymenoptera		
Apidae: *Ceratina cucurbitina* (Wheeler *et al.*, 1977a).	Apidae: Mandibular glands of adults.	It is the only alkene in the secretion of male *C. cucurbitina* which possesses a chain length of less than C_{23}.
Formicidae: *Myrmecia gulosa* (Cavill and Williams, 1967), *Formica fusca*, *F. rufibarbis* (Bergström and Löfqvist, 1968), *F. nigricans*, *F. rufa*, *F. polyctena* (Bergström and Löfqvist, 1973), *F. japonica* (Hayashi *et al.*, 1973a), *Oecophylla longinoda* (Bradshaw *et al.*, 1973), *Lasius fuliginosus* (Bernardi *et al.*, 1967), *L. alienus* (Bergström and Löfqvist, 1970), *Camponotus ligniperda* (Bergström and Löfqvist, 1971), *C. japonicus* (Hayashi *et al.*, 1973a), *C. herculeanus* (Bergström and Löfqvist, 1972), *C. americanus*, *C. pennsylvanicus* (Ayre and Blum, 1971), *C. intrepidus* (Brophy *et al.*, 1973), *Anoplolepis custodiens* (Schreuder and Brand, 1972), *Monacis bispinosa*, *Azteca* spp. (Blum and Wheeler, 1974), *Pheidole lamellidens* (Hayashi *et al.*, 1973a), *Myrmica rubra* (Morgan and Wadhams, 1972), *Novomessor cockerelli* (Vick *et al.*, 1969), *Pogonomyrmex rugosus*, *P. barbatus* (Regnier	Formicidae: Dufour's gland of workers.	

(continued)

Table 2.2 (Continued)

	et al., 1973), and *Iridomyrmex humilis* (Cavill and Houghton, 1973, 1974b).		
6-Methyltetradecane  $C_{15}H_{32}$ MW 212	Hymenoptera Formicidae: *Pogonomyrmex rugosus* and *P. barbatus* (Regnier *et al*., 1973).	Dufour's gland of workers.	It is a major constituent in the secretion.
3,4-Dimethyltridecane  $C_{15}H_{32}$ MW 212	Hymenoptera Formicidae: *Pogonomyrmex rugosus* and *P. barbatus* (Regnier *et al*., 1973).	Dufour's gland of workers.	It is in admixture with another dimethylalkane.
Homofarnesene $C_{16}H_{26}$ MW 218	Hymenoptera Formicidae: *Myrmica rubra* (Morgan and Wadhams, 1972).	Dufour's gland of workers.	The stereochemistry of this trace constituent has not been established.

Hexadecene $C_{16}H_{32}$ MW 224	Opiliones Phalangiidae: *Phalangium opilio* (Blum *et al.*, 1973e). Hymenoptera Formicidae: *Formica nigricans* (Bergström and Löfqvist, 1973), *Anoplolepis custodiens* (Schreuder and Brand, 1972), and *Myrmica rubra* (Morgan and Wadhams, 1972).	Phalangiidae: Paired glands located on the flanks of the prosoma between the first and second pair of coxae. Formicidae: Dufour's gland of workers.	The location of the double bond has not been established for this minor component which does not appear to be widespread in formicine secretions.
n-Hexadecane $C_{16}H_{34}$ MW 226	Opiliones Phalangiidae: *Phalangium opilio* (Blum *et al.*, 1973e). Coleoptera Tenebrionidae: *Tribolium confusum* (Keville and Kannowski, 1975). Hymenoptera Formicidae: *Myrmecia gulosa* (Cavill and Williams, 1967), *Formica nigricans*, *F. rufa*, *F. polyctena* (Bergström and Löfqvist, 1973), *Camponotus ligniperda* (Bergström and Löfqvist, 1971), *C. herculeanus* (Bergström and Löfqvist, 1972), *C. intrepidus* (Brophy *et al.*, 1973), *Anoplolepis custodiens* (Schreuder and Brand, 1972), *Pheidole lamel-*	Phalangiidae: Paired glands located on the flanks of the prosoma between the first and second pair of coxae. Tenebrionidae: Paired abdominal sternal glands. Formicidae: Dufour's gland of workers.	It is a minor or trace constituent in all secretions.

(continued)

Table 2.2 (Continued)

Compound	Source	Gland	Remarks
	lidens (Hayashi et al., 1973a), Myrmica rubra (Morgan and Wadhams, 1972), Novomessor cockerelli (Vick et al., 1969), and Iridomyrmex humilis (Cavill and Houghton, 1973, 1974b).		
3-Methylpentadecane $C_{16}H_{34}$ MW 226	Hymenoptera Formicidae: Camponotus intrepidus (Brophy et al., 1973) and Iridomyrmex humilis (Cavill and Houghton, 1973, 1974b).	Dufour's gland of workers.	It is a minor constituent.
5-Methylpentadecane $C_{16}H_{34}$ MW 226	Hymenoptera Formicidae: Formica nigricans (Bergström and Löfqvist, 1973), Camponotus intrepidus (Brophy et al., 1973), and Iridomyrmex humilis (Cavill and Houghton, 1973, 1974b).	Dufour's gland of workers.	It accompanies the 3-methyl isomer as a minor constituent in the secretion of C. intrepidus but is not detectable in those of other Camponotus species.
Bishomofarnesene $C_{17}H_{28}$ MW 232	Hymenoptera Formicidae: Myrmica rubra (Morgan and Wadhams, 1972).	Dufour's gland of workers.	It is a trace constituent which constitutes the largest terpene hydrocarbon detected in arthropod exocrine secretions.

Heptadecadiene $C_{17}H_{32}$ MW 236	Coleoptera Tenebrionidae: *Tribolium confusum* (Keville and Kannowski, 1975). Hymenoptera Formicidae: *Formica nigricans*, *F. rufa*, *F. polyctena* (Bergström and Löfqvist, 1973), *Camponotus ligniperda* (Bergström and Löfqvist, 1971), *C. herculeanus* (Bergström and Löfqvist, 1972), *Myrmica rubra* (Morgan and Wadhams, 1972), and *Oecophylla longinoda* (Bradshaw *et al*., 1973).	Tenebrionidae: Paired abdominal sternal glands. Formicidae: Dufour's gland of workers.	The locations of the double bonds have not been established for this minor glandular product.
1-Heptadecene $C_{17}H_{34}$ MW 238	Opiliones Phalangiidae: *Phalangium opilio* (Blum *et al*., 1973e). Coleoptera Tenebrionidae: a few *Eleodes* spp., *Psorodes* spp., *Meracantha* spp., *Eurynotus* spp., *Gonopus* spp., *Toxicum* spp., *Neatus* spp., *Merinus* spp., *Zadenos* spp., *Cratidus* spp., *Lariversius* spp., *Parastizopus* spp., *Eleates* spp., and *Phaleria* spp. (Tschinkel, 1975b).	Phalangiidae: Paired glands located on the flanks of the prosoma between the first and second pair of coxae. Tenebrionidae: Paired abdominal sternal glands.	It is usually a minor component but it is a major constituent in *Lariversius*, *Psorodes*, and *Merinus* spp. secretions.

(continued)

Table 2.2 (Continued)

7-Heptadecene $C_{17}H_{34}$ MW 238	Hymenoptera Formicidae: *Iridomyrmex humilis* (Cavill and Houghton, 1973, 1974b).	Dufour's gland of workers.	It is one of two heptadecenes in the secretion.
cis-8-Heptadecene $C_{17}H_{34}$ MW 238	Hymenoptera Apidae: *Ceratina cucurbitina* (Wheeler *et al*., 1977a). Formicidae: *Myrmecia gulosa* (Cavill and Williams, 1967), *Myrmica rubra* (Morgan and Wadhams, 1972), *Monacis bispinosa*, *Azteca* spp. (Blum and Wheeler, 1974), *Formica rufibarbis* (Bergström and Löfqvist, 1968), *F. nigricans*, *F. rufa*, *F. polyctena* (Bergström and Löfqvist, 1973), *Lasius niger*, *L. alienus* (Bergström and Löfqvist, 1970), *Oecophylla longinoda* (Bradshaw *et al*., 1973), *Camponotus ligniperda* (Bergström and Löfqvist, 1971), *C. herculeanus* (Bergström and Löfqvist, 1972), *Anoplolepis custodiens* (Schreuder and Brand, 1972), and *Iridomyrmex humilis* (Cavill and Houghton, 1973, 1974b).	Apidae: Mandibular glands of females. Formicidae: Dufour's gland of workers.	It comprises more than 50% of the volatiles in the secretions of *M. gulosa* and *M. rubra*. The location of the double bond has not been determined for the alkenes produced by the other species. It is one of six alkenes in the secretion of *C. cucurbitina*.

Compound	Source	Gland	Remarks
n-Heptadecane $C_{17}H_{36}$ MW 240	Opiliones Phalangiidae: *Phalangium opilio* (Blum *et al*., 1973e).	Phalangiidae: Paired glands located on the flanks of the prosoma between the first and second pair of coxae.	
	Hymenoptera Apidae: *Ceratina cucurbitina* (Wheeler *et al*., 1977a).	Apidae: Mandibular glands of females.	It is one of eight hydrocarbons in the secretion of *C. cucurbitina*, all of which are odd numbered.
	Formicidae: *Polyergus rufescens*, *Formica fusca*, *F. rufibarbis* (Bergström and Löfqvist, 1968), *F. nigricans*, *F. rufa*, *F. polyctena* (Bergström and Löfqvist, 1973), *F. japonica* (Hayashi *et al*., 1973a), *Lasius niger*, *L. alienus* (Bergström and Löfqvist, 1970), *Camponotus ligniperda* (Bergström and Löfqvist, 1971), *C. herculeanus* (Bergström and Löfqvist, 1972), *C. intrepidus* (Brophy *et al*., 1973), *C. japonicus*, *Pheidole lamellidens* (Hayashi *et al*., 1973a), *Oecophylla longinoda* (Bradshaw *et al*., 1973), *Anoplolepis custodiens* (Schreuder and Brand, 1972), *Monacis bispinosa*, *Azteca* spp. (Blum and Wheeler, 1974), *Iridomyrmex humilis* (Cavill and Houghton, 1973), *Myrmica rubra* (Morgan and Wadhams, 1972), *Myrmecia gulosa* (Cavill and Williams,	Formicidae: Dufour's gland of workers.	It is a trace constituent in the secretions of the formicine species whereas in the dolichoderines *M. bispinosa* and *Azteca* spp., it is a major exocrine product.

(*continued*)

Table 2.2 (Continued)

	1967), _Novomessor cockerelli_ (Vick _et al_., 1969), _Solenopsis invicta_, _S_. _richteri_, and _S_. _geminata_ (Brand _et al_., 1972).		
3-Methylhexadecane $C_{17}H_{36}$ MW 240	Hymenoptera Formicidae: _Camponotus intrepidus_ (Brophy _et al_., 1973).	Dufour's gland of workers.	It comprises about 7% of the secretion.
4-Methylhexadecane $C_{17}H_{36}$ MW 240	Hymenoptera Formicidae: _Iridomyrmex humilis_ (Cavill and Houghton, 1973, 1974b).	Dufour's gland of workers.	It is a minor constituent.
5-Methylhexadecane $C_{17}H_{36}$ MW 240	Hymenoptera Formicidae: _Camponotus intrepidus_ (Brophy _et al_., 1973).	Dufour's gland of workers.	It is one of 10 monomethyl-branched alkenes in this secretion. All of these hydrocarbons are present as 3- and 5-methyl-branched pairs.

9-Octadecene $C_{18}H_{36}$ MW 252	Hymenoptera Formicidae: *Formica nigricans*, *F. rufa*, *F. polyctena* (Bergström and Löfqvist, 1973), *Anoplolepis custodiens* (Schreuder and Brand, 1972), *Myrmica rubra* (Morgan and Wadhams, 1972), and *Iridomyrmex humilis* (Cavill and Houghton, 1973, 1974b).	Dufour's gland of workers.	It is a trace constituent. The location of the double bond has been established only for the *I. humilis* alkene.
4-Methylheptadecene $C_{18}H_{36}$ MW 252	Hymenoptera Formicidae: *Iridomyrmex humilis* (Cavill and Houghton, 1974b).	Dufour's gland of workers.	It is a trace constituent. The location of the double bond has not been determined.
3-Methylheptadecane $C_{18}H_{38}$ MW 254	Hymenoptera Formicidae: *Iridomyrmex humilis* (Cavill and Houghton, 1973, 1974b).	Dufour's gland of workers.	It is one of two methyl-branched heptadecanes in the secretion.

(continued)

Table 2.2 (Continued)

5-Methylheptadecane $C_{18}H_{38}$ MW 254	Hymenoptera Formicidae: *Iridomyrmex humilis* (Cavill and Houghton, 1973, 1974b).	Dufour's gland of workers.	It is one of seven methyl-branched alkanes in the secretion.
n-Octadecane $C_{18}H_{38}$ MW 254	Hymenoptera Apidae: *Bombus derhamellus* (Calam, 1969). Formicidae: *Formica nigricans*, *F. rufa*, *F. polyctena* (Bergström and Löfqvist, 1973), *F. japonica*, *Pheidole lamellidens* (Hayashi *et al*., 1973a), *Camponotus ligniperda* (Bergström and Löfqvist, 1971), *C. herculeanus* (Bergström and Löfqvist, 1972), *Anoplolepis custodiens* (Schreuder and Brand, 1972), *Myrmica rubra* (Morgan and Wadhams, 1972), *Novomessor cockerelli* (Vick *et al*., 1969), and *Iridomyrmex humilis* (Cavill and Houghton, 1973, 1974b).	Apidae: Labial glands of males. Formicidae: Dufour's gland of workers.	It is a trace constituent of the glandular exudates.

Nonadecadiene $C_{19}H_{36}$ MW 264	Hymenoptera Formicidae: Formica nigricans, F. rufa, F. polyctena (Bergström and Löfqvist, 1973), and Azteca spp. (Blum and Wheeler, 1974).	Dufour's gland of workers.	It is a trace constituent.
1-Nonadecene $C_{19}H_{38}$ MW 266	Coleoptera Tenebrionidae: Cratidus osculans and Lariversius spp. (Tschinkel, 1975b).	Paired abdominal sternal glands.	It is one of the two long-chain alkenes which fortify the quinone-rich secretions of Cratidus and Lariversius spp.
9-Nonadecene $C_{19}H_{38}$ MW 266	Hymenoptera Apidae: Bombus hortorum (Kullenberg et al., 1970) and Ceratina cucurbitina (Wheeler et al., 1977a). Formicidae: Myrmica rubra (Morgan and Wadhams, 1972), Monacis bispinosa, Azteca spp. (Blum and Wheeler, 1974), Lasius alienus (Bergström and Löfqvist, 1970), Formica nigricans, F. rufa, F. polyctena (Bergström and Löfqvist, 1973), Camponotus ligniperda (Bergström and Löfqvist, 1971), Anoplolepis custodiens (Schreuder and Brand, 1972), and Iridomyrmex	Apidae: Labial glands of Bombus and mandibular glands of C. cucurbitina. Formicidae: Dufour's gland of workers.	Although it is the major component in the secretion of B. hortorum, it could not be detected in 13 other species of Bombus. A quantitatively important alkene in the glandular exudate of M. rubra, I. humilis, and several Azteca spp. The position of the double bond has only been determined for the alkenes produced by M. gulosa, M. rubra, and I. humilis.

(continued)

Table 2.2 (Continued)

Compound	Species	Gland	Remarks
	humilis (Cavill and Houghton, 1973, 1974b).		
n-Nonadecane $C_{19}H_{40}$ MW 268	Hymenoptera Andrenidae: *Andrena bicolor* and *A. nigroaenea* (Bergström and Tengö, 1974). Apidae: *Bombus derhamellus* (Calam, 1969). Formicidae: *Formica rufibarbis* (Bergström and Löfqvist, 1968), *F. nigricans*, *F. rufa*, *F. polyctena* (Bergström and Löfqvist, 1973), *F. japonica*, *Pheidole lamellidens* (Hayashi *et al*., 1973a), *Lasius niger* (Bergström and Löfqvist, 1970), *Camponotus ligniperda* (Bergström and Löfqvist, 1971), *C. herculeanus* (Bergström and Löfqvist, 1972), *C. japonicus* (Hayashi *et al*., 1973a), *Anoplolepis custodiens* (Schreuder and Brand, 1972), *Myrmica rubra* (Morgan and Wadhams, 1972), *Novomessor cockerelli* (Vick *et al*., 1969), and *Iridomyrmex humilis* (Cavill and Houghton, 1973, 1974b).	Andrenidae: Dufour's gland of females. Apidae: Labial glands of males. Formicidae: Dufour's gland of workers.	It accounts for nearly 30% of the volatiles in the secretion of *A. custodiens* whereas in those of *M. rubra* and the formicine species, it is a trace constituent.

Eicosene $C_{20}H_{40}$ MW 280	Hymenoptera Apidae: Bombus derhamellus (Calam, 1969). Formicidae: Formica rufa (Bergström and Löfqvist, 1973) and Anaplolepis custodiens (Schreuder and Brand, 1972).	Apidae: Mandibular glands of males. Formicidae; Dufour's gland of workers.	It is a minor constituent which does not appear to be characteristic of Formica spp. The position of the double bond is not known.
3-Methylnonadecane $C_{20}H_{42}$ MW 282	Hymenoptera Formicidae: Iridomyrmex humilis (Cavill and Houghton, 1973, 1974b).	Dufour's gland of workers.	It is the longest-chain branched alkane in the secretion.
n-Eicosane $C_{20}H_{42}$ MW 282	Hymenoptera Apidae: Bombus derhamellus (Calam, 1969). Formicidae: Formica nigricans, F. rufa, F. polyctena (Bergström and Löfqvist, 1973), F. japonica, Pheidole lamellidens (Hayashi et al., 1973a), Anoplolepis custodiens (Schreuder and Brand, 1972), and Iridomyrmex humilis (Cavill and Houghton, 1973, 1974b).	Apidae: Labial gland of males. Formicidae: Dufour's gland of workers.	It is a minor constituent in this hydrocarbon-rich secretion.

(continued)

Table 2.2 (Continued)

Heneicosene $C_{21}H_{42}$ MW 294	Hymenoptera Apidae: Bombus derhamellus (Calam, 1969) and Ceratina cucurbitina (Wheeler et al., 1977a). Formicidae: Anoplolepis custodiens (Schreuder and Brand, 1972), Formica nigricans, F. rufa, and F. polyctena (Bergström and Löfqvist, 1973).	Apidae: Labial gland of Bombus males and mandibular glands of C. cucurbitina. Formicidae: Dufour's gland of workers.	It is a trace constituent. The position of the double bond has not been established.
n-Heneicosane $C_{21}H_{44}$ MW 296	Hymenoptera Andrenidae: Andrena bicolor, A. denticulata, A. nigroaenea, A. carbonaria, A. helvola, and A. haemorrhoa (Bergström and Tengö, 1974). Colletidae: Colletes cunicularius (Bergström, 1974). Apidae: Bombus lucorum, B. derhamellus (Calam, 1969; Bergström et al., 1973), and B. lapponicus (Bergström and Svensson, 1973). Formicidae: Formica nigricans F. rufa, (Bergström and Löfqvist, 1973), F. japonica (Hayashi et al., 1973a), Lasius alienus (Bergström and Löfqvist, 1970), Anoplolepis custodiens (Schreuder and Brand, 1972), and Iridomyrmex humilis (Cavill and Houghton, 1973).	Andrenidae: Dufour's gland of females. Colletidae: Dufour's gland of females. Apidae: Labial gland of males. Formicidae: Dufour's gland of workers.	It is a major constituent in the andrenid secretions and that of A. custodiens.

Docosene $C_{22}H_{44}$ MW 308	Hymenoptera Apidae: *Bombus derhamellus* (Calam, 1969).	Labial glands of males.	It is a minor constituent. The location of the double bond has not been determined.
n-Docosane $C_{22}H_{46}$ MW 310	Dictyoptera Rhinotermitidae: *Coptotermes lacteus* (Moore, 1968). Hymenoptera Apidae: *Bombus derhamellus* (Calam, 1969). Formicidae: *Formica nigricans*, *F. rufa* (Bergström and Löfqvist, 1973), *F. japonica*, *Camponotus japonicus*, *Pheidole lamellidens* (Hayashi *et al.*, 1973a), and *Iridomyrmex humilis* (Cavill and Houghton, 1973, 1974b).	Rhinotermitidae: Frontal gland of soldiers. Apidae: Labial glands of males. Formicidae: Dufour's gland of workers.	It is one of a series of long-chain paraffins ejected in admixture with an aqueous phase in the termite secretion.
Tricosene $C_{23}H_{46}$ MW 322	Hymenoptera Apidae: *Bombus derhamellus* (Calam, 1969) and *Ceratina cucurbitina* (Wheeler *et al.*, 1977a). Formicidae: *Formica nigricans*, *F. rufa*, and *F. polyctena* (Bergström and Löfqvist, 1973).	Apidae: Labial glands of *Bombus* males and mandibular glands of *Ceratina* females. Formicidae: Dufour's gland of workers.	The location of the double bond has not been established for this major glandular constituent in the apid secretions.

(continued)

Table 2.2 (Continued)

n-Tricosane  $C_{23}H_{48}$ MW 324	Dictyoptera Rhinotermitidae: *Coptotermes lacteus* (Moore, 1968).	Rhinotermitidae: Frontal gland of soldiers.	It is a trace exocrine constituent of *C. ligniperda* but a major constituent in the secretion of *B. derhamellus*.
	Hymenoptera Halictidae: *Halictus calceatus* (Bergström, 1974).	Halictidae: Dufour's gland of females.	
	Apidae: *Bombus lucorum* (Calam, 1969; Bergström *et al.*, 1973), *B. derhamellus* (Calam, 1969), *B. lapponicus*, *B. jonellus* (Bergström and Svensson, 1973), *Ceratina cucurbitina* (Wheeler *et al.*,1977a).	Apidae: Labial glands of *Bombus* males and Mandibular glands of *C. cucurbitina*.	
	Formicidae: *Camponotus ligniperda* (Bergström and Löfqvist, 1971), *C. japonicus*, *Formica japonica*, *Pheidole lamellidens* (Hayashi *et al.*, 1973a), and *Iridomyrmex humilis* (Cavill and Houghton, 1973, 1974b).	Formicidae: Dufour's gland of workers.	
	Andrenidae: *Andrena bicolor*, *A. denticulata*, *A. nigroaenea*, *A. carbonaria*, *A. helvola*, and *A. haemorrhoa* (Bergström and Tengö, 1974).	Andrenidae: Dufour's gland of females.	

Tetracosene $C_{24}H_{48}$ MW 336	Hymenoptera Apidae: *Bombus derhamellus* (Calam, 1969).	Labial glands of males.	It is a minor constituent. The location of the double bond is unknown.
n-Tetracosane  $C_{24}H_{50}$ MW 338	Dictyoptera Rhinotermitidae: *Coptotermes lacteus* (Moore, 1968). Hymenoptera Apidae: *Bombus derhamellus* (Calam, 1969). Formicidae: *Iridomyrmex humilis* (Cavill and Houghton, 1973), *Pheidole lamellidens*, and *Formica japonica* (Hayashi *et al.*, 1973a).	Rhinotermitidae: Frontal gland of soldiers. Apidae: Labial glands of males. Formicidae: Dufour's gland of workers.	It is the major alkane in the glandular discharge of *C. lacteus*.
Pentacosene $C_{25}H_{50}$ MW 350	Hymenoptera Apidae: *Bombus derhamellus* (Calam, 1969) and *Ceratina cucurbitina* (Wheeler *et al.*, 1977a).	Labial glands of *Bombus* males and mandibular glands of *Ceratina* females.	It is a major constituent in the secretion of *B. derhamellus*. The position of the double bond is unknown.

(*continued*)

Table 2.2 (Continued)

n-Pentacosane $C_{25}H_{52}$ MW 352	Dictyoptera Rhinotermitidae: *Coptotermes lacteus* (Moore, 1968). Hymenoptera Halictidae: *Halictus calceatus* (Bergström, 1974). Apidae: *Bombus lucorum* (Calam, 1969; Bergström *et al.*, 1973), *B. derhamellus* (Calam, 1969), *B. lapponicus* (Bergström and Svensson, 1973), and *Ceratina cucurbitina* (Wheeler *et al.*, 1977a). Formicidae: *Iridomyrmex humilis* (Cavill and Houghton, 1973), *Formica japonica*, *Camponotus japonicus*, and *Pheidole lamellidens* (Hayashi *et al.*, 1973a).	Rhinotermitidae: Frontal gland of soldiers. Halictidae: Dufour's gland of females. Apidae: Labial glands of *Bombus* males and mandibular glands of *Ceratina* males. Formicidae: Dufour's gland of workers.	It is the longest-chain alkane in the exudates of *B. lucorum* and *B. derhamellus*.
n-Hexacosane $C_{26}H_{54}$ MW 366	Dictyoptera Rhinotermitidae: *Coptotermes lacteus* (Moore, 1968). Hymenoptera Formicidae: *Iridomyrmex humilis* (Cavill and Houghton, 1973), *Formica japonica*, and *Pheidole lamellidens* (Hayashi *et al.*, 1973a).	Rhinotermitidae: Frontal gland of soldiers. Formicidae: Dufour's gland of workers.	It is a minor constituent.

Heptacosene $C_{27}H_{54}$ MW 378	Hymenoptera Apidae: *Ceratina cucurbitina* (Wheeler *et al.*, 1977a).	Mandibular glands of females.	It is one of the six alkenes which dominate the secretion.
n-Heptacosane $C_{27}H_{56}$ MW 380	Dictyoptera Rhinotermitidae: *Coptotermes lacteus* (Moore, 1968). Hymenoptera Apidae: *Ceratina cucurbitina* (Wheeler *et al.*, 1977a). Formicidae: *Iridomyrmex humilis* (Cavill and Houghton, 1973), *Formica japonica*, and *Pheidole lamellidens* (Hayashi *et al.*, 1973a).	Rhinotermitidae: Frontal gland of soldiers. Apidae: Mandibular glands of males. Formicidae: Dufour's gland of workers.	It is the longest-chain hydrocarbon in the rhinotermitid defensive. secretion which is dominated by alkanes and a mucopolysaccharide.
n-Octacosane $C_{28}H_{58}$ MW 394	Hymenoptera Formicidae: *Pheidole lamellidens* (Hayashi *et al.*, 1973a).	Unknown.	It is a trace constituent.

(*E*)-β-Farnesene and germacrene A have only been detected as products of aphids (Bowers *et al.*, 1972, 1977) and germacrene B has only been encountered in papilionid defensive exudates (Honda, 1980). Early instar papilionid larvae also produce the sesquiterpenes β-elemene and caryophyllene as well as a series of monoterpene hydrocarbons, all of which are absent from the secretions of last instar larvae. Monoterpene hydrocarbons, which are a hallmark of termitid soldiers in many genera (Table 2.2), are also produced by rhopalid species (Aldrich *et al.*, 1979) which appear to be exceptional among the Hemiptera in producing this class of compounds. The same is true of the monoterpene-rich poison gland secretion of the ant *Myrmicaria natalensis* (Brand *et al.*, 1974) which is radically different from poison gland secretions of ants in other genera.

In addition to a large variety of monoterpenes and distinctive sesquiterpenes such as α-selinene (Evans *et al.*, 1978) and biflora-4,10(19),15-triene (Wiemer *et al.*, 1980), termitid soldiers are distinctive in producing the only diterpene sesquiterpenes characterized as arthropod natural products. Termites produce a dazzling variety of diterpenes, and compounds such as cubitene (Prestwich *et al.*, 1978), cembrene A, and (3Z)-cembrene A are representative of the C_{20} compounds that have been identified as termitid defensive compounds.

In contrast to the terpenoid versatility of the termites, hymenopterans are very conservative synthesizers of terpene hydrocarbons. Only one terpene, α-farnesene, has been identified as a bee natural product; and ants, while they synthesize 10 terpenes (Table 2.2), produce relatively simple isoprenoids. Most of the monoterpenes identified as formicid defensive compounds are produced by one species, *Myrmicaria natalensis* (Brand *et al.*, 1974), and α-farnesene constitutes the only sesquiterpene hydrocarbon characterized from the exudates of these insects. On the other hand, the incredible variety of nonterpenoid hydrocarbons synthesized in the Dufour's gland of ants demonstrates that they have the ability to facilely generate alkanes and alkenes. Possibly, this hydrocarbon proficiency is derived from metabolic pathways that are well established for the production of epicuticlar hydrocarbons. Whatever the *raison d'être* of ants' biosynthetic paramountcy as hydrocarbon producers, it is evident that, from a qualitative standpoint, these insects utilize these compounds as defensive allomones with a frequency that is unparalleled among the Arthropoda.

Chapter 3

Alcohols

The virtuosity of arthropods as biosynthesizers of alcohols is demonstrated by the fact that about 60 of these compounds (see also Chapter 18) are present in the defensive exudates of species in the crustacean order Isopoda, the order Opiliones, and the insect orders Dictyoptera, Hemiptera, Coleoptera, Lepidoptera and Hymenoptera (Table 3.1). However, at least 50% of the alcohols identified as products of arthropod exocrine glands are produced by hymenopterous species, and these compounds often constitute only minor constituents in the secretions of members of the other six orders. Although it is possible that carbinols are characteristic defensive substances of isopods, these compounds are not usually emphasized as defensive constituents by cockroaches, termites, beetles or lepidopterous larvae.

AN OVERVIEW

The great diversity of alcohols found in arthropod defensive secretions should not obscure the fact that these compounds have a very restricted distribution in these invertebrates. Most of the identified carbinols occur in species in one or two families, and about three-fourths of these alcohols are limited to the exocrine products of species in a single family; only six alcohols have been identified in more than two families. The unequal distribution of these compounds in the Arthropoda is further emphasized by the fact that, although alcohols have been detected in the defensive exudates of species in 18 families (Table 3.1), in more than a third of these cases a family is identified with only one alcohol. For example,

Table 3.1

Distribution of Alcohols in the Defensive Secretions of the Arthropoda[a]

Class	Order	Family
Crustacea	Isopoda	Armadillididae
		Oniscidae
Arachnida	Opiliones (Phalangida)	Phalangiidae
Insecta	Dictyoptera	
	Blattaria	Blattidae
	Isoptera	Termitidae
	Hemiptera	Alydidae
		Coreidae
		Pentatomidae
		Hyocephalidae
	Coleoptera	Tenebrionidae
		Cerambycidae
		Gyrinidae
		Staphylinidae
	Lepidoptera	Cossidae
		Papilionidae
	Hymenoptera	Andrenidae
		Apidae
		Formicidae

[a] Also see Chapter 18.

the alcoholic nature of the defensive exudates of the Armadillididae, Hyocephalidae, Gyrinidae, Cerambycidae, Staphylinidae, and Papilionidae is predicated on the presence of one particular alcohol in each family, which has, with one exception (Staphylinidae), been identified in only one species in the entire family.

Selin-11-en-4-β-ol, one of the osmeterial components produced by *Battus polydamus* (Eisner *et al.*, 1971c), has not been identified in the secretions of many species in the subfamily Papilioninae. All other species which have been analyzed in this subfamily discharge a defensive exudate fortified with isobutyric and 2-methylbutyric acids (Eisner *et al.*, 1970). The novel compound phoracanthol, a major exocrine component of *Phoracantha semipunctata* (Moore and Brown, 1972), may be characteristic of species in this genus, but it has not been identified in the defensive products elaborated by species in other cerambycid genera. Similarly, isopiperitenol, the cyclic alcohol produced by *Stenus bipunctatus* and *S.*

comma (Schildknecht, 1970; Schildknecht *et al.*, 1976), does not appear to be a typical staphylinid defensive product. Even the polyenols produced by larvae of *Cossus cossus* (Trave *et al.*, 1966), while they may be characteristic of cossid species, can hardly be regarded as typical defensive compounds of lepidopterous larvae.

In some cases, alcohols are minor defensive products which are almost certainly related to the main glandular components. 1-Hexanol, a characteristic constituent produced in the metasternal scent glands of species in the Coreidae and Alydidae, accompanies hexanal, the major exocrine product synthesized by these hemipterans. Very likely, the carbinol and aldehyde are metabolically related. Similarly, four of the five alcohols identified in the abdominal defensive glands of blattids are minor constituents in secretions that are thoroughly dominated by their corresponding carbonyl compounds. For example, 2-hexen-1-ol is present in the defensive discharges of *Polyzosteria* species, but their secretions contain more than 90% 2-hexenal (Wallbank and Waterhouse, 1970). This is equally true of 2-methylene-1-butanol vis-à-vis 2-methylenebutanal (Waterhouse and Wallbank, 1967). It seems clear that in a number of cases alcoholic constituents may actually constitute precursors of the dominant compounds identified with the defensive "punch" of the secretion (see Chapter 16, Section II).

Arthropods have not particularly stressed very short-chain alcohols as defensive compounds, and nearly two-thirds of the carbinols identified as exocrine products are in the range C_{10}–C_{20} (Table 3.2). More than 50% of these compounds are primary alcohols, and only one aromatic alcohol (2-phenylethyl alcohol) has so far been detected in arthropod defensive secretions. About one-fourth of these compounds are terpenes which are primarily synthesized in the exocrine glands of ants and bees. However, whereas ants emphasize monoterpene alcohols, bees appear to favor sesqui- and diterpene alcohols as defensive products. Some species of termitid soldiers, on the other hand, produce defensive exudates that contain monoterpene hydrocarbons and distinctive diterpene alcohols, the trinervitenes (Prestwich *et al.*, 1976a; Vrkoč *et al.*, 1977).

Hymenopterous species have a virtual monopoly on the utilization of alcohols as defensive constituents. About two-thirds of the carbinols which have been identified occur in either the mandibular or Dufour's gland secretions of the Formicidae or Apidae (Table 3.2), and most of the alcohols identified as products of hymenopterous exocrine glands are unique arthropod defensive products. However, the natural product

Table **3.2** Alcohols in arthropod defensive secretions

Name and Formula	Occurrence	Glandular Source	Comments
2-Methylene-1-butanol CH_2OH $C_5H_{10}O$ MW 86	Dictyoptera Blattidae: *Platyzosteria jungii*, *P. castanea*, *P. morosa*, and *P. ruficeps* (Waterhouse and Wallbank, 1967).	Ventral abdominal gland of adults opening between the sixth and seventh sternites.	It constitutes ca. 1.5% of the secretion and is believed to contribute significantly to the effectiveness of the exudate by acting as a spreading agent.
2-Pentanol OH $C_5H_{12}O$ MW 88	Dictyoptera Blattidae: *Platyzosteria armata* (Wallbank and Waterhouse, 1970).	Ventral abdominal gland of adults opening between the sixth and seventh sternites.	It is one of the main constituents in this secretion which is very atypical of those produced by polyzosteriine species.
2-Methyl-1-butanol CH_2OH $C_5H_{12}O$ MW 88	Dictyoptera Blattidae: *Platyzosteria jungii*, *P. castanea*, *P. morosa*, and *P. ruficeps* (Waterhouse and Wallbank, 1967).	Ventral abdominal gland of adults opening between the sixth and seventh sternites.	It is present at a concentration of 0.1-0.2% in the secretions and is probably metabolically related to 2-methylbutanal which also occurs in these exudates.

3-Methyl-1-butanol 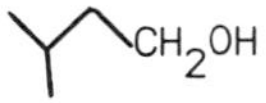 $C_5H_{12}O$ MW 88	Coleoptera Gyrinidae: *Gyrinus natator* (Schildknecht *et al*., 1972c).	Pygidial glands.	The paired glands contain 0.2 μg of this alcohol. The presence of the corresponding aldehyde in the secretion indicates it and the carbinol are metabolically related.
trans-2-Hexen-1-ol 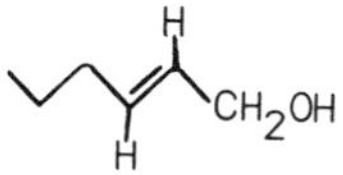 $C_6H_{12}O$ MW 100	Dictyoptera Blattidae: *Polyzosteria limbata*, *P. viridissima*, *P. oculata*, *P. cuprea*, and *P. pulchra* (Wallbank and Waterhouse, 1970).	Blattidae: Ventral abdominal gland of adults opening between the sixth and seventh sternites.	The aldehyde-rich secretions of the polyzosteriine cockroaches contain about 1% of it. Both *Crematogaster* species belong to the subgenus *Atopogyne*, a taxon whose species are distinguished by the production of *trans*-2-hexenal. The presence of both the α,β-unsaturated aldehyde and alcohol in these secretions points to their metabolic relationship.
	Hemiptera Alydidae: *Alydus eurinus* (Aldrich and Yonke, 1975). Coreidae: *Euthochtha galeator* (Aldrich and Yonke, 1975).	Alydidae: Adult metasternal scent gland. Coreidae: Adult metasternal scent gland.	
	Hymenoptera Formicidae: *Crematogaster africana* and *C. buchneri* (Crewe *et al*., 1972).	Formicidae: Mandibular glands of workers.	

(continued)

Table 3.2 (Continued)

1-Hexanol $CH_3(CH_2)_4CH_2OH$ $C_6H_{14}O$ MW 102	Hemiptera Alydidae: *Alydus eurinus* and *A. pilosulus* (Aldrich and Yonke, 1975). Coreidae: *Amorbus rubiginosus*, *A. alternatus*, *A. rhombifer*, *Mictis profana*, *M. caja*, *Aulacosternum nigrorubrum*, *Pachycolpura manca*, *Agriopocoris frogatti* (Waterhouse and Gilby, 1964), *Pternistria bispina* (Baker and Kemball, 1967), and *Amblypelta nitida* (Baker *et al.*, 1972). Hyocephalidae: *Hyocephalus* sp. (Waterhouse and Gilby, 1964).	Alydidae: Adult metasternal scent gland. Coreidae: Adult metasternal scent gland. Also dorsal abdominal glands of larvae of *A. nitida*. Hyocephalidae: Adult metasternal scent gland.	It constitutes about 4% of the coreid secretions. Its presence in the hyocephalid secretion is taken as support for the placement of this family in the coreoid complex.
	Coleoptera Tenebrionidae: *Eleodes beameri* (Tschinkel, 1975a).	Tenebrionidae: Paired abdominal sternal glands.	It is a very atypical product of tenebrionid secretions.
	Hymenoptera Formicidae: *Oecophylla longinoda* (Bradshaw *et al.*, 1973).	Formicidae: Mandibular glands of workers.	
2-Heptanol $CH_3(CH_2)_4CH(OH)CH_3$ $C_7H_{16}O$ MW 116	Dictyoptera Blattidae: *Platyzosteria armata* (Wallbank and Waterhouse, 1970).	Blattidae: Ventral abdominal gland of adults opening between the sixth and seventh sternites.	It is accompanied by 2-heptanone in the secretion of *P. armata*. This is the only *Platyzosteria* secretion which is not dominated by α,β-unsaturated aldehydes.
	Hymenoptera Apidae: *Trigona postica*, *T. depilis*, *T. tubiba*, *T. xan-*	Apidae: Mandibular glands of workers.	It constitutes a major

	thotricha, *T. bipunctata* (Blum *et al*., 1973b), *T. mexicana*, *T. pectoralis* (Luby *et al*., 1973), and *T. spinipes* (Kerr *et al*., 1973). Formicidae: *Atta texana* (Riley *et al*., 1974) and *Azteca* spp. (Blum and Wheeler, 1974).	Formicidae: Mandibular glands of workers (*A. texana*) and anal glands of workers (*Azteca* spp.)	constituent in all the bee secretions and thoroughly dominates the exudate of *T. spinipes*. 2-Heptanone accompanies it in both of the *Azteca* secretions. The secretion of *T. spinipes* is the only 2-heptanol-containing exudate that lacks 2-heptanone.
3-Heptanol $C_7H_{16}O$ MW 116	Hymenoptera Formicidae: *Atta texana* (Riley *et al*., 1974).	Mandibular glands of workers.	It is present as the (+)-isomer. Each worker contains about 0.01 μg of it.
2-Phenylethanol $C_8H_{10}O$ MW 122	Hymenoptera Formicidae: *Camponotus clarithorax* (Lloyd *et al*., 1975).	Mandibular glands of males.	It constitutes about 15% of the volatiles detected in this alcohol-rich secretion.

(continued)

Table 3.2 (Continued)

(5-Ethylcyclopent-1-enyl) methanol 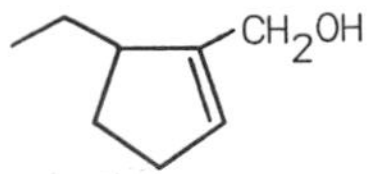 CH_2OH $C_8H_{14}O$ MW 126	Coleoptera Cerambycidae: *Phoracantha semipunctata* (Moore and Brown, 1972).	Adult metasternal scent gland.	It is the second most abundant component in the secretion and has been assigned the name phoracanthol. It is accompanied by the corresponding aldehyde and is believed to be synthesized by the head-to-tail union of four acetate units, possibly in the immature stages.
2-Octen-1-ol  CH_2OH $C_8H_{16}O$ MW 128	Hemiptera Coreidae: *Euthochtha galeator* (Aldrich and Yonke, 1975).	Metathoracic glands of adults and larval dorsal abdominal glands.	It is a minor component in the secretion.
6-Methyl-5-hepten-2-ol  OH $C_8H_{16}O$ MW 128	Hymenoptera Formicidae: *Iridomyrmex* nr. *pruinosus* (Crewe and Blum, 1971).	Anal glands of workers.	The major component in the secretion is 6-methyl-5-hepten-2-one.

Compound	Occurrence	Source	Remarks
1-Octanol CH_2OH $C_8H_{18}O$ MW 130	Isopoda Armadillididae: *Armadillidium* sp. (Cavill *et al*., 1966). Hymenoptera Formicidae: *Lasius niger* (Bergström and Löfqvist, 1970).	Armadillididae: Unknown. Formicidae: Mandibular glands of workers.	It constitutes about 0.17% of the volatiles in the isopod extract. Defensive glands are well known in isopods and it is almost certain that it constitutes an element in the deterrent arsenal of this isopod. It is one of two alcohols in the *Lasius* secretion. Other species in this genus characteristically produce terpenes in their mandibular glands.
3-Octanol OH $C_8H_{18}O$ MW 130	Hymenoptera Formicidae: *Myrmica punctiventris*, *M. fracticornis*, *M. americana*, *M. ruginodis*, *M. rubra*, *A. sabuleti*, *M. scabrinodis* (Crewe and Blum, 1970a), *M. brevinodis* (Crewe and Blum, 1970a,b), *Crematogaster* spp. (Crewe *et al*., 1970; Crewe *et al*., 1972; Schlunneger and Leuthold, 1972), *Cyphomyrmex rimosus*, *Trachymyrmex septentrionalis*, *T. seminole*, *Acromyrmex octospinosus* (Crewe *et al*., 1972), *Atta texana*, and *A. cephalotes* (Riley *et al*., 1974).	Mandibular glands of workers of all species. Also mandibular glands of females and males of *Crematogaster* spp.	The secretion of only one species (*C. rimosus*) did not contain 3-octanone as a concomitant. It is a major constituent in the secretions of many species but is usually present at a lower concentration than 3-octanone. The ratio of 3-octanone:3-octanol varied from 1.2:1 (*M. rubra*) to 128:1 (*A. octospinosus*). It has been detected in 30 *Crematogaster* spp. in 6 subgenera. In secre-

(*continued*)

			tions produced by species in the subgenus _Atopogyne_ which are dominated by 2-hexenal, it is a very minor component.
4-Methyl-3-heptanol 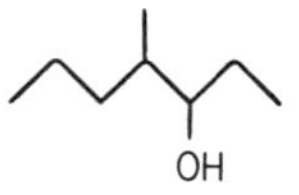 $C_8H_{18}O$ MW 130	Hymenoptera Formicidae: _Pogonomyrmex badius_, _P. barbatus_, _P. californicus_, _P. desertorum_, _P. occidentalis_, _P. rugosus_ (McGurk _et al._, 1966), _Trachymyrmex septentrionalis_ (Crewe and Blum, 1972), _Atta texana_, and _A. cephalotes_ (Riley _et al._, 1974).	Mandibular glands of workers.	It is accompanied by 4-methyl-3-heptanone in all the secretions and is probably metabolically related to this ketone.
3-Nonen-1-ol  $C_9H_{18}O$ MW 142	Isopoda Oniscidae: _Porcellio scaber_ (Cavill _et al._, 1966).	Unknown.	About 5% of the volatiles present in extracts of this isopod have been identified as the cis and trans isomers of this enol.
2,6-Dimethyl-5-hepten-1-ol 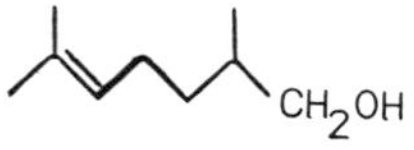 $C_9H_{18}O$ MW 142	Hymenoptera Formicidae: _Acanthomyops claviger_ (Law _et al._, 1965; Regnier and Wilson, 1968), _Lasius neoniger_ (Law _et al._, 1965), _L. alienus_ (Regnier and Wilson, 1969), and _Camponotus clarithorax_ (Lloyd _et al._, 1975).	Mandibular glands of workers of _Acanthomyops_ and _Lasius_ spp. Mandibular glands of males of _C. clarithorax_.	It is a major component in the secretions of male _Acanthomyops_ but is a minor constituent in those of the workers and the _Lasius_ spp. It constitutes about 5% of the _Camponotus_ secretion.

1-Nonanol CH_2OH $C_9H_{20}O$ MW 144	Isopoda Oniscidae: *Porcellio scaber* (Cavill *et al.*, 1966). Hymenoptera Formicidae: *Lasius niger* (Bergström and Löfqvist, 1970).	Oniscidae: Unknown. Formicidae: Mandibular glands of workers.	It accounts for about 1% of slaterol, the alcohol-rich secretion isolated from this isopod. It is a trace component in the *Lasius* secretion.
2-Nonanol OH $C_9H_{20}O$ MW 144	Hymenoptera Apidae: *Trigona benjoim* (Blum *et al.*, 1973b), *T. pectoralis*, *T. mexicana* (Luby *et al.*, 1973), and *T. spinipes* (Kerr *et al.*, 1973).	Mandibular glands of workers.	It is the only alcohol present in the secretion of *T. benjoim* and is a minor component in the other *Trigona* exudates.
6-Methyl-3-octanol OH $C_9H_{20}O$ MW 144	Hymenoptera Formicidae: *Myrmica brevinodis*, *M. rubra*, *M. punctiventris*, *M. fracticornis*, *M. americana*, *M. ruginodis*, *M. sabuleti*, *M. scabrinodis* (Crewe and Blum, 1970a), and *Crematogaster* spp. (Crewe *et al.*, 1972).	Mandibular glands of workers.	Both it and 6-methyl-3-octanone are minor constituents in nearly all of these secretions. However, these two compounds are major components in the secretion of *M. scabrinodis*, raising the possibility that these two compounds are metabolically related.

(continued)

Table 3.2 (Continued)

Compound	Source	Gland	Remarks
Isopiperitenol 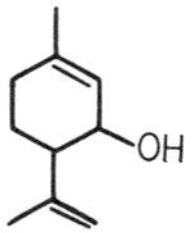 $C_{10}H_{16}O$ MW 152	Coleoptera Staphylinidae: *Stenus bipunctatus* (Schildknecht, 1970) and *S. comma* (Schildknecht *et al.*, 1976).	Pygidial glands.	It is one of three surface-active terpenes utilized to propel disturbed beetles very rapidly through the water.
Geraniol 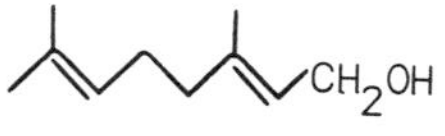 $C_{10}H_{18}O$ MW 154	Hymenoptera Andrenidae: *Andrena denticulata* and *A. nigroaenea* (Bergström and Tengö, 1974). Apidae: *Bombus pratorum* and *B. cullumanus* (Kullenberg *et al.*, 1970). Formicidae: *Atta sexdens* (Blum *et al.*, 1968a).	Andrenidae: Dufour's gland of workers. Apidae: Labial glands of males. Formicidae: Mandibular glands of workers.	It is a minor component in all of the secretions. It is only present in the secretions of two out of 20 apid species and is accompanied by geranyl acetate in both cases. It has been detected in the secretion of only one out of seven species of *Atta*.
2-Decen-1-ol  $C_{10}H_{20}O$ MW 156	Hemiptera Pentatomidae: *Biprorulus bibax* (MacLeod *et al.*, 1975).	Adult metasternal scent gland.	It constitutes less than 1% of the volatiles in the secretion.
3-Decen-1-ol 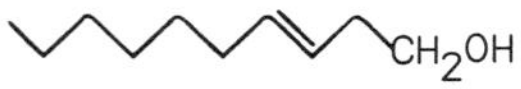 $C_{10}H_{20}O$ MW 156	Isopoda Oniscidae: *Porcellio scaber* (Cavill *et al.*, 1966).	Unknown.	It accounts for about 80% of slaterol, the volatile components isolated from this isopod. It is present as both the cis and trans isomers.

Citronellol CH_2OH $C_{10}H_{20}O$ MW 156	Hymenoptera Apidae: *Bombus pratorum* and *Psithyrus bohemicus* (Kullenberg *et al*., 1970). Formicidae: *Acanthomyops claviger*, *Lasius neoniger* (Law *et al*., 1965), *L. umbratus* (Blum *et al*., 1968b), *L. alienus* (Regnier and Wilson, 1969), *L. speculiventris* (Wilson and Regnier, 1971), *Atta laevigata*, and *A. capiguara* (Blum *et al*., 1968a).	Apidae: Labial glands of males. Formicidae: Mandibular glands of males of *A. claviger* and *L. neoniger*. Mandibular glands of workers of all other species.	It is produced by only two of 20 bee species that were studied. It is quantitatively important in the secretions of males of *A. claviger* and workers of *L. umbratus*. It is a minor component in the exudates of the other species of ants and has been detected in only two out of seven *Atta* spp. surveyed.
1-Decanol CH_2OH $C_{10}H_{22}O$ MW 158	Hymenoptera Formicidae: *Formica sanguinea* (Bergström and Löfqvist, 1968).	Dufour's gland of workers.	It is a minor component in a secretion containing a series of alkanols.
3-Ethyl-2-octanol OH $C_{10}H_{22}O$ MW 158	Dictyoptera Termitidae: *Trinervitermes bettonianus* (Prestwich, 1975).	Frontal gland of soldiers.	It is accompanied by several monoterpene hydrocarbons.
1-Undecanol CH_2OH $C_{11}H_{24}O$ MW 172	Hymenoptera Formicidae: *Formica sanguinea* (Bergström and Löfqvist, 1968) and *Lasius niger* (Bergström and Löfqvist, 1970).	Dufour's gland of workers.	It is one of several alcohols present in the secretion of *F. sanguinea*.

(continued)

Table 3.2 (Continued)

2-Undecanol  $C_{11}H_{24}O$ MW 172	Hymenoptera Apidae: *Trigona depilis*, *T. postica* (Blum *et al.*, 1973b), *T. pectoralis*, and *T. mexicana* (Luby *et al.*, 1973).	Mandibular glands of workers.	It is a minor concomitant of 2-undecanone in all secretions.
1-Dodecanol  $C_{12}H_{26}O$ MW 186	Hymenoptera Formicidae: *Formica sanguinea* (Bergström and Löfqvist, 1968) and *Lasius niger* (Bergström and Löfqvist, 1970).	Dufour's gland of workers.	It is a minor constituent in both glandular exudates.
1-Tridecanol  $C_{13}H_{28}O$ MW 200	Hymenoptera Formicidae: *Camponotus ligniperda* (Bergstrom and Löfqvist, 1971).	Dufour's gland of workers.	It is one of seven aliphatic alcohols present in the secretion of this formicine species.
2-Tridecanol  $C_{13}H_{28}O$ MW 200	Hymenoptera Apidae: *Trigona depilis*, *T. xanthotricha* (Blum *et al.*, 1973b), *T. pectoralis*, and *T. mexicana* (Luby *et al.*, 1973).	Mandibular glands of workers.	It is a major component in the secretion of *T. pectoralis* but is quantitatively less important in those of the other species. It is only known from species in the subgenus *Scaptotrigona*.

3,5,13-Tetradeca-trien-1-ol CH_2OH $C_{14}H_{24}O$ MW 208	Lepidoptera Cossidae: Cossus cossus (Trave et al., 1966).	Mandibular glands of larvae.	It is one of two tri-unsaturated alcohols in the secretion and accounts for 14% of the volatiles.
4,6,13-Tetradeca-trien-1-ol CH_2OH $C_{14}H_{24}O$ MW 208	Lepidoptera Cossidae: Cossus cossus (Trave et al., 1966).	Mandibular glands of larvae.	It is one of two tetra-decatrienols which have been named cossine 2.
5,13-Tetradecadien-1-ol CH_2OH $C_{14}H_{26}O$ MW 210	Lepidoptera Cossidae: Cossus cossus (Trave et al., 1966).	Mandibular glands of larvae.	It has been assigned the trivial name cossine 2 and accounts for 12% of the exudate.
Tetradecenol $C_{14}H_{28}O$ MW 212	Hymenoptera Formicidae: Formica nigricans and F. polyctena (Bergström and Löfqvist, 1973).	Dufour's gland of workers.	It is a trace constituent in both secretions. The location of the double bond has not been determined.

(continued)

Table 3.2 (Continued)

1-Tetradecanol  CH_2OH $C_{14}H_{30}O$ MW 214	Hymenoptera Apidae: *Bombus agrorum*, *B. lapidarius*, *Psithyrus rupestris*, and *P. silvestris* (Kullenberg *et al.*, 1970).	Labial glands of males.	It is a minor component in all secretions.
trans-Farnesol  CH_2OH $C_{15}H_{26}O$ MW 222	Hymenoptera Andrenidae: *Andrena bicolor*, *A. denticulata*, *A. nigroaenea*, and *A. haemorrhoa* (Bergström and Tengo, 1974). Apidae: *Psithyrus barbutellus*, *Bombus hortorum* (Kullenberg *et al.*, 1970), and *B. pratorum* (Calam, 1969; Kullenberg *et al.*, 1970).	Andrenidae: Dufour's gland of workers. Apidae: Labial glands of males.	It constitutes a major component in all the apid secretions and is qualitatively important in the andrenid exudates.
Selin-11-en-4-ol 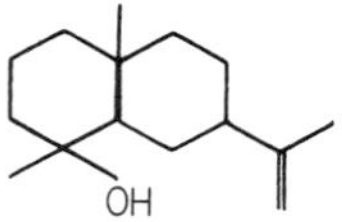 OH $C_{15}H_{26}O$ MW 222	Lepidoptera Papilionidae: *Battus polydamus* (Eisner *et al.*, 1971b).	The osmeterium, a gland which opens on the mid-dorsal line as a two-pronged invagination of the neck membrane in larvae.	It is one of two eudesmane sesquiterpenes in the secretion and constitutes 25% of the mixture.

2,3-Dihydro-6-*trans*-farnesol  CH_2OH $C_{15}H_{28}O$ MW 224	Hymenoptera Apidae: *Bombus terrestris* (Kullenberg *et al.*, 1970) and *B. jonellus* (Bergström and Svensson, 1973).	Labial glands of males.	It is the major constituent in the secretion of *B. terrestris* and is produced as the (-)-isomer (Ställberg-Stenhagen, 1970).
1-Pentadecanol  CH_2OH $C_{15}H_{32}O$ MW 228	Hymenoptera Formicidae: *Camponotus ligniperda* (Bergström and Löfqvist, 1971) and *C. herculeanus* (Bergström and Löfqvist, 1972).	Dufour's gland of workers.	It is a trace constituent in both secretions.
2-Pentadecanol  OH $C_{15}H_{32}O$ MW 228	Hymenoptera Apidae: *Trigona postica*, *T. depilis* (Blum *et al.*, 1973b), *T. mexicana*, and *T. pectoralis* (Luby *et al.*, 1973).	Mandibular glands of workers.	It is only a major constituent in the secretion of *T. pectoralis*.
7-Hexadecen-1-ol  CH_2OH $C_{16}H_{32}O$ MW 240	Hymenoptera Apidae: *Psithyrus silvestris*, *Bombus hypnorum* (Kullenberg *et al.*, 1970), and *B. agrorum* (Calam 1969; Kullenberg *et al.*, 1970). Formicidae: *Formica nigricans* and *F. polyctena* (Bergström and Löfqvist, 1973).	Apidae: Labial glands of males. Formicidae: Dufour's gland of workers.	The position of the double bond has only been established for the enol produced by *B. agrorum*. It is a major glandular constituent in the secretion of this apid.

(*continued*)

Table 3.2 (Continued)

Compound	Occurrence	Source	Remarks
9-Hexadecen-1-ol CH_2OH $C_{16}H_{32}O$ MW 240	Hymenoptera Apidae: *Bombus lapidarius* (Calam, 1969; Kullenberg *et al*., 1970).	Labial glands of males.	It completely dominates this secretion and was not detected in the exudates of 19 other species of bees.
11-Hexadecen-1-ol CH_2OH $C_{16}H_{32}O$ MW 240	Hymenoptera Apidae: *Psithyrus bohemicus* (Kullenberg *et al*., 1970).	Labial glands of males.	It thoroughly dominates this secretion and is one of three isomeric hexadecenols identified as bumblebee natural products.
1-Hexadecanol CH_2OH $C_{16}H_{34}O$ MW 242	Hymenoptera Apidae: *Bombus terrestris*, *B. lapidarius*, *B. pratorum*, *B. hypnorum*, *Psithyrus rupestris*, and *P. globosus* (Kullenberg *et al*., 1970). Formicidae: *Lasius niger* (Bergström and Löfqvist, 1970), *Camponotus ligniperda* (Bergström and Löfqvist, 1971), and *C. herculeanus* (Bergström and Löfqvist, 1972).	Apidae: Labial glands of males. Formicidae: Dufour's gland of workers.	It is a minor component in all secretions.

2-Heptadecanol  $C_{17}H_{36}O$ MW 256	Hymenoptera Apidae: *Trigona depilis*, *T. postica*, *T. bipunctata* (Blum *et al*., 1973b), *T. pectoralis*, and *T. mexicana* (Luby *et al*., 1973).	Mandibular glands of workers.	It is a major component in the secretion of *T. mexicana* and is quantitatively important in the other exudates.
9-Octadecen-1-ol  $C_{18}H_{36}O$ MW 268	Hymenoptera Apidae: *Bombus muscorum*, *B. hortorum*, *Psithyrus campestris*, and *P. globosus* (Kullenberg *et al*., 1970).	Labial glands of males.	It is only a major constituent in the secretion of *B. muscorum*. The position of the double bond has not been established for the octadecenol produced by the other species.
Geranylgeraniol  $C_{20}H_{34}O$ MW 290	Hymenoptera Apidae: *Bombus subterraneus*, *B. hortorum*, *B. lucorum*, and *B. pratorum* (Kullenberg *et al*., 1970). Formicidae: *Formica nigricans*, *F. fusca*, and *F. polyctena* (Bergström and Löfqvist, 1973).	Apidae: Labial glands of males. Formicidae: Dufour's gland of workers.	Among the *Bombus* secretions it is a major constituent in only that of *B. subterraneus*. It is present as the all-*trans*-isomer in the *Formica* secretions and constitutes 0.2–1.2% of the volatiles.
Geranylcitronellol  $C_{20}H_{36}O$ MW 292	Hymenoptera Apidae: *Bombus terrestris*, *B. hypnorum*, and *Psithyrus rupestris* (Kullenberg *et al*., 1970).	Labial glands of males.	It is a major constituent in the secretion of *P. rupestris*. The laevorotary isomer has been shown to be identical to the compound produced by *B. terrestris* (Ahlquist *et al*., 1971).

(continued)

Table 3.2 (Continued)

Eicosen-1-ol $C_{20}H_{40}O$ MW 296	Hymenoptera Apidae: *Psithyrus campestris* (Kullenberg *et al.*, 1970).	Labial glands of males.	It is the major constituent in the secretion and is lacking in those of 19 other species of bumblebees. The location of the double bond has not been determined.

chemistry of bees and ants is often quite distinctive, and even though they both produce large numbers of alcohols, they only share about six common compounds (e.g., citronellol, geraniol).

Ant species in the subfamilies Myrmicinae, Dolichoderinae, and Formicinae synthesize 28 of the identified carbinols (46%), and 15 of these compounds (25%) are not known to occur in the defensive secretions of arthropods in other families. Most of these alcohols have been detected in the Dufour's gland secretions of formicine species (~60%), whereas the myrmicine compounds are products of the mandibular glands. About 75% of the ant-derived alcohols are in the range C_6–C_{12} and these include a few isoprenoids (e.g., 6-methyl-5-hepten-2-ol, citronellol) that are shared between species in two subfamilies. Bees, on the other hand, generally produce longer chain alcohols and display a considerably greater versatility in terms of biosynthesizing terpene alcohols.

About 33% of the alcoholic defensive substances are produced in the mandibular glands of bees in the family Apidae. Apid species produce 12 alcohols (20%) which are unique arthropod natural products, and all these carbinols have been detected in four genera (Table 3.2). Males of the genus *Bombus* are the sole source of about one-sixth of the alcohols identified as defensive products and the closely related genus *Psithyrus* is nearly as biosynthetically versatile (Kullenberg *et al.*, 1970). Workers of several species in the genus *Trigona* (*Scaptotrigona*) are virtually the sole source of a large series of secondary alcohols (Blum *et al.*, 1973b; Luby *et al.*, 1973). Although most of the ant-derived carbinols have a carbon content of less than C_{12}, nearly 75% of those produced by bees are in the range C_{11}–C_{20}. Bees generally favor the synthesis of sesqui- and diterpene alcohols, some of which have also been identified as exocrine products of ants (e.g., geranylgeraniol).

Alcohols emerge as major defensive substances only in the Hymenoptera and possibly the Isopoda. In the latter order, two species in the families Armadillididae and Oniscidae have yielded four alcohols, two of which are unique defensive products (Table 3.2). A survey of additional isopod species will be necessary before it can be ascertained if alcohols are indeed typical exocrine products of these crustaceans. On the other hand, it will not be surprising if the defensive glands of bees and ants continue to yield a wealth of important alcoholic constituents in their exocrine secretions.

Chapter 4

Aldehydes

Although aldehydic allomones have been characterized from the defensive secretions of arthropods in three classes (Table 4.1), these compounds are mainly products of insect species. Whereas more than 60 aldehydes have been identified in these exudates, only four of these compounds have been detected as products of opilionids and diplopods (Table 4.2 and Chapter 18). Indeed, only one aldehyde (benzaldehyde) can really be considered as a typical allomone of polydesmoid millipedes. Insects, on the other hand, produce a dazzling variety of aldehydes, many of which are restricted in their arthropod distribution to species in a single family. For example, about 20% of the aldehydes that have been detected in arthropod defensive exudates are unique compounds limited to the Hemiptera. Nearly 20% of these allomones are idiosyncratic beetle products, as is also the case for the aldehydes synthesized by hymenopterans.

Most of the arthropod-derived aldehydes are aliphatic compounds, only five aromatic constituents having been detected in their secretions (Table 4.2). There is a marked tendency for insects, particularly hemipterans, to produce conjugated aldehydes, and nearly 50% of all the detected aldehydes are α,β-conjugated compounds (Table 4.2). More than a fourth of the aldehydic allomones are terpenoid constituents which include mono-, sesqui-, and diterpenes, particularly as products of beetles and ants. The virtuosity of beetles and hymenopterans as aldehydic chemists is further demonstrated by the fact that they produce 42 and 34%, respectively, of all aldehydes identified as arthropod natural products. As aldehydic synthesizers the species in these orders are only rivaled by hemipterans, true bugs producing nearly 40% of these arthropod-derived carbonyl compounds. Among other insect orders, only species in the Dictyoptera, which produce nearly one-fifth of these compounds, are a significant source of defensive aldehydic allomones.

Table 4.1

Distribution of Aldehydes in the Defensive Secretions of the Arthropoda[a]

Class	Order	Family
Arachnida	Opiliones (Phalangida)	Phalangiidae
Diplopoda	Polydesmida	Xystodesmidae
		Gomphodesmidae
		Polydesmidae
		Euryuridae
		Paradoxosomatidae
		Chelodesmidae
		Nearctodesmidae
	Spirobolida	Rhinocricidae
Insecta	Dictyoptera	
	Blattaria	Blattidae
	Isoptera	Termitidae
	Orthoptera	Phasmidae
	Hemiptera	Alydidae
		Cimicidae
		Coreidae
		Corixidae
		Cydnidae
		Gelastocoridae
		Hyocephalidae
		Lygaeidae
		Miridae
		Naucoridae
		Notonectidae
		Pentatomidae
		Pyrrhocoridae
		Rhopalidae
		Scutelleridae
	Coleoptera	Carabidae
		Cerambycidae
		Chrysomelidae
		Dytiscidae
		Gyrinidae
		Staphylinidae
		Tenebrionidae
	Hymenoptera	Oxaeidae
		Andrenidae
		Colletidae
		Anthophoridae
		Apidae
		Formicidae

[a] Also see Chapter 18.

Table **4.2** Aldehydes in arthropod defensive secretions

Name and Formula	Occurrence	Glandular Source	Comments
Acetaldehyde CH_3CHO C_2H_4O MW 44	Hemiptera Cimicidae: *Cimex lectularius* (Collins, 1968). Miridae: *Leptoterna dolabrata* (Collins and Drake, 1965). Pyrrhocoridae: *Dysdercus intermedius* (Calam and Scott, 1969).	Cimicidae: Adult metasternal scent gland. Miridae: Adult metasternal scent gland. Pyrrhocoridae: Adult metasternal scent gland.	It is a minor component in all the secretions. It was not detected in the larval secretion of *D. intermedius*.
2-Propenal C_3H_4O MW 56	Hemiptera Pentatomidae: *Nezara viridula* (Gilby and Waterhouse, 1965) and *Chrysocoris stolli* (Choudhuri and Das, 1968). Cydnidae: *Scaptocoris divergens* (Roth, 1961).	Pentatomidae: Adult metasternal scent gland. Cydnidae: Adult metasternal scent gland.	It constitutes one of the major components in the secretion of *C. stolli* whereas it is a minor product in the secretion of *N. viridula*.
n-Propanal CHO (structure) C_3H_6O MW 58	Hemiptera Coreidae: *Libyaspis angolensis* (Cmelik, 1969). Pentatomidae: *Chrysocoris stolli* (Choudhuri and Das, 1968). Cydnidae: *Scaptocoris divergens* (Roth, 1961).	Coreidae: Adult metasternal scent gland and dorsal abdominal glands of larvae. Pentatomidae: Adult metasternal scent gland. Cynidae: Adult metasternal scent gland.	It is a major constituent in the larval scrretion of *L. angolensis*, which contains another saturated aldehyda.

2-Butenal CHO C_4H_6O MW 70	Hemiptera Pentatomidae: Nezara viridula (Gilby and Waterhouse, 1965) and Chrysocoris stolli (Choudhuri and Das, 1968). Cydnidae: Scaptocoris divergens (Roth, 1961).	Pentatomidae: Adult metasternal scent gland. Cydnidae: Adult metasternal scent gland.	It is a major constituent in the secretion of C. stolli which also contains five other enals.
2-Methylene propanal CHO C_4H_6O MW 70	Dictyoptera Blattidae: Platyzosteria castanea, P. jungii, P. morosa, and P. ruficeps (Waterhouse and Wallbank, 1967).	Ventral abdominal gland of adults opening between the sixth and seventh sternites.	It is one of four aldehydes in the secretions and accounts for ca. 0.2% of the volatiles.
Butanal CHO C_4H_8O MW 72	Hemiptera Coreidae: Pternistria bispina (Baker and Kemball, 1967), Riptortus clavatus (Tsuyuki et al., 1965), Amorbus rhombifer (Waterhouse and Gilby, 1964), and Merocoris distinctus (Aldrich and Yonke, 1975).	Coreidae: Adult metasternal scent gland. Pentatomidae: Adult metasternal scent gland and dorsal abdominal glands of larvae.	It is a minor constituent in the secretions of all species and was not detected in the larval exudate of P. bispina.
Isobutanal CHO C_4H_8O MW 72	Hemiptera Alydidae: Megalotomus quinquespinosus (Aldrich and Yonke, 1975).	Adult metasternal scent gland.	It is a minor constituent in a secretion dominated by aliphatic fatty acids.

(continued)

Table 4.2 (Continued)

2-Pentenal CHO C_5H_8O MW 84	Hemiptera Cydnidae: *Scaptocoris divergens* (Roth, 1961).	Adult metasternal scent gland.	It is one of a large series of α,β-unsaturated aldehydes produced by this species.
2-Methylene butanal (2-Ethyl acrolein) CHO C_5H_8O MW 84	Dictyoptera Blattidae: *Platyzosteria jungii*, *P. castanea*, *P. morosa*, *P. ruficeps* (Waterhouse and Wallbank, 1967), *P.* nr. *montana*, *P. occidentalis*, and *Methana convexa* (Wallbank and Waterhouse, 1970).	Ventral abdominal gland of adults opening between the sixth and seventh sternites.	It constitutes about 95% of the volatiles in the secretions of four *Platyzosteria* species and is a major component (30%) in the exudated of the other species.
2-Methyl butanal CHO $C_5H_{10}O$ MW 86	Dictyoptera Blattidae: *Platyzosteria castanea*, *P. jungii*, *P. morosa*, and *P. ruficeps* (Waterhouse and Wallbank, 1967). Hemiptera Alydidae: *Megalotomus quinquespinosus* (Aldrich and Yonke, 1975).	Blattidae: Ventral abdominal gland of adults opening between the sixth and seventh sternites.	It is a trace constituent in the blattid secretions but is one of the major nonacidic components in the alydid exudate. It is accompanied by 2-methyl-1-butanol to which it may be metabolically related.

Compound	Source	Gland	Comments
3-Methyl butanal (Isovaleraldehyde) $C_5H_{10}O$ MW 86	Coleoptera Gyrinidae: *Gyrinus natator* (Schildknecht *et al*., 1972c). Staphylinidae: *Thyreocephalus lorquini* and *Eulissus orthodoxus* (Bellas *et al*., 1974). Carabidae: *Anthia thoracica* and *Thermophilum homoplatum* (Scott *et al*., 1975).	Gyrinidae: Pygidial glands. Staphylinidae: Pygidial glands. Carabidae: Pygidial glands.	Each pair of pygidial glands of *G. natator* contains 2.5 μg of this aldehyde. It is a major constituent in the staphylinid secretions.
4-Oxo-*trans*-2-hexenal $C_6H_8O_2$ MW 112	Hemiptera Alydidae: *Megalotomus quinquespinosus*, *Alydus eurinus*, and *A. pilosulus* (Aldrich and Yonke, 1975). Coreidae: *Pternistria bispina* (Baker and Kemball, 1967), *Euthochtha galeator*, *Archimerus alternatus*, *Leptoglossus oppositus*, and *L. clypealis* (Aldrich and Yonke, 1975). Corixidae: *Sigara falleni* (Pinder and Staddon, 1965a,b) and *Corixa dentipes* (Pinder and Staddon, 1965b). Cydnidae: *Macrocystus* sp. (Baggini *et al*., 1966). Gelastocoridae: *Gelastocoris oculatus* (Staddon, 1973). Lygaeidae: *Oncopeltus fasciatus* (Games and Staddon, 1973b). Pentatomidae: *Nezara viridula* (Gilby and Waterhouse, 1965), *Tessaratoma aethiops* (Baggini	Alydidae: Dorsal abdominal glands of larvae. Coreidae: Dorsal abdominal glands of larvae of all species. Corixidae: Adult metasternal scent gland. Cydnidae: Adult metasternal scent gland. Gelastocoridae: Adult metasternal scent gland. Lygaeidae: Posterior dorsal abdominal scent gland of larvae. Pentatomidae: Adult metasternal scent gland of all species and dorsal abdominal glands of	This keto-aldehyde appears to be a rather typical product of the dorsal abdominal glands of larval hemipterans. It is often lacking in the secretions of adult bugs. It is one (25-50%) of the three unsaturated aldehydes that make up the larval secretion of *P. bispina* which shares no common compounds with the adult exudate. It constitutes 7% of the secretion of both larvae and adults of the pentatomid *T. aethiops* and is a minor component in the larval secretion of *O. fasciatus*. It was only detected in the secretion of the posterior dorsal abdominal gland

(*continued*)

Table 4.2 (Continued)

	et al., 1966), *Biprorulus bibax*, *Musgraveia sulciventris* (MacLeod *et al*., 1975), and *Libyaspis angolensis* (Cmelik, 1969). Pyrrhocoridae: *Dysdercus intermedius* (Calam and Youdeowei, 1968).	larvae of *T. aethiops* and *L. angolensis*. Pyrrhocoridae: Dorsal abdominal glands of larvae.	of *O. fasciatus* larvae. It is the only aldehyde in the exudate of the cydnid *Macrocystis* sp. The corixid secretions contain about 95% of this aldehyde which appears to be the main defensive compound produced by species in this family.
trans-2-Hexenal $C_6H_{10}O$ MW 98	Dictyoptera Blattidae: *Eurycotis floridana* (Roth *et al*., 1956; Dateo and Roth, 1967a), *E. decipiens*, *E. bioelleyi* (Dateo and Roth, 1967a,b), *Pelmatosilpha coriacea* (Blum, 1964), *Euzosteria nobilis* (Dateo and Roth, 1967b; Wallbank and Waterhouse, 1970), *Platyzosteria novaeseelandiae* (Roth and Willis, 1960), *P.* (=*Cutilia*) *soror* (Chadha *et al*., 1961b), *P. coolgardiensis*, *P. scabra*, *P. scabrella*, *P. stradbrokensis*, *P. nitidella*, *P.* sp. (Wallbank and Waterhouse, 1970), *Zonioploca pallida*, *Z. bicolor*, *Megazosteria patula*, *Desmozosteria scripta*, *Drymaplaneta semivitta*, *D. shelfordi*, *D. communis* (Wallbank and Waterhouse, 1970), *Polyzosteria limbata*, *P. viri-*	Blattidae: Ventral abdominal gland of adults opening between the sixth and seventh sternites. Alydidae: Dorsal abdominal glands of larvae. Coreidae: Dorsal abdominal glands of larvae of *L. angolensis*, *E. galeator*, *A. alternatus*, *L. oppositus*, and *L. clypealis*. Adult metasternal scent gland of *L. angolensis*, *E. galeator*, *A. alternatus*, and all other species. Cydnidae: Adult metasternal scent gland.	It is a common constituent in the secretions of blattids in the tribes Eurycotini, Polyzosteriini, and Methanini, three taxa in the subfamily Polyzosteriinae. This enal was not detected in the secretions of some *Platyzosteria* and *Drymaplaneta* spp. as well as polyzosteriine species in four other genera. It constitutes at least 90% of the volatiles in the secretions of more than three-fourths of the species in the tribe Polyzosteriini whereas the exudates of species in the tribe Methanini (*Drymaplaneta*) contain 10-47%

dissima, P. oculata, P. cuprea, P. pulchra, and P. mitchelli (Wallbank and Waterhouse, 1970). Hemiptera Alydidae: Megalotomus quinquespinosus, Alydus eurinus, and A. pilosulus (Aldrich and Yonke, 1975). Cimicidae: Cimex lectularius (Schildknecht et al., 1964; Collins, 1968). Coreidae: Acanthocephala femorata (Blum et al., 1961), A. declivis, A. granulosa (McCullough, 1966a, 1967b), Acanthocoris sordidus (Tsuyuki et al., 1965), Libyaspis angolensis (Cmelik, 1969), Leptoglossus oppositus, L. clypealis, Euthochtha galeator, and Archimerus alternatus (Aldrich and Yonke, 1975). Cydnidae: Scaptocoris divergens (Roth, 1961). Gelastocoridae: Gelastocoris oculatus (Staddon, 1973). Lygaeidae: Oncopeltus fasciatus (Games and Staddon, 1973a,b) Pentatomidae: Brochymena quadripustulata (Blum, 1961),	Gelastocoridae: Adult metasternal scent gland. Lygaeidae: Anterior and posterior dorsal abdominal glands of larvae. Pentatomidae: Adult metasternal scent gland of all species and dorsal abdominal glands of larvae of L. angolensis. Pyrrhocoridae: Dorsal abdominal glands of larvae and adult metasternal scent gland. Tenebrionidae: Paired abdominal sternal glands. Formicidae: Mandibular glands of workers.	of this enal. The secretions of the Polyzosteria spp. contain a minor amount of trans-2-hexenol which may be related biosynthetically to 2-hexenal. It is one of the major glandular constituents produced by C. lectularius. The secretion of larvae of the pyrrhocorid D. intermedius contains from 3-23% of this enal whereas it is the main constituent in adult exudates. In the gelastocorid secretion it is accompanied by three other unsaturated aldehydes and is the least concentrated (3%) aldehyde present. It is a minor component in the secretions of both the anterior and posterior glands of larvae of O. fasciatus and is more concentrated in the exudate of the former gland. It is one of three α,β-unsaturated aldehydes produced by the coreid P. bispina and varies

(continued)

Table 4.2 (Continued)

	Dolycoris baccarum, (Schildknecht, *et al*., 1962), *Biprorulus bibax* (Park and Sutherland, 1962; MacLeod *et al*., 1975), *Poecilometis strigatus*, *Nezara viridula* (Waterhouse *et al*., 1961; Gilby and Waterhouse, 1965), *Musgraveia sulciventris* (Waterhouse *et* al., 1961; MacLeod *et al*., 1975), *Piezodorus teretipes* (Gilchrist *et al*., 1966), *Scotinophora lurida* (Tsuyuki *et al*., 1965), *Chrysocoris stolli* (Choudhuri and Das, 1968), *Tessaratoma aethiops* (Baggini *et al*., 1966), *Libyaspis angolensis* (Cmelik, 1969), and *Palomena viridissima* (Schildknecht *et al*., 1964). Scutelleridae: *Eurygaster* sp. (Schildknecht and Weis, 1962b).	in concentration from 0.4-19%. It is the only compound detectable in the secretions of the pentatomids *B. quadripustulata* and *P. teretipes*. It constitutes 50% of the defensive exudate of *D. baccarum* and is a major glandular constituent in almost all of the other pentatomid secretions. In the two-phase secretion of *N. viridula* it represents 1% of the volatiles in the whole scent. The denser minor phase of the secretion contains 2.6% of this enal whereas the lighter phase contains only 0. 0.6%. In *N. viridula* it is believed to be synthesized in the storage reservoir of the scent gland rather than in the lateral glands. Since the lateral glands are less isolated from the hemolymph than the storage reservoir, synthesis of this toxic enal in the latter would result
	Coleoptera Tenebrionidae: *Eleodes beameri* (Tschinkel, 1975a).	
	Hymenoptera Formicidae: *Crematogaster africana* (Bevan *et al*., 1961; Crewe *et al*., 1972), *C. buchneri*, *C. depressa*, *C. jullieni*, *C. luciae* (Crewe *et al*., 1972), and *C.* spp. (Blum *et al*., 1969a; Crewe *et al*., 1972).	

			in it being isolated in a relatively impermeable sac. It is a minor component (2%) in the secretion of the tenebrionid *E. beameri* and is one of seven aldehydes present in this atypical secretion.
2-Methylene pentanal $C_6H_{10}O$ MW 98	Dictyoptera Blattidae: *Platyzosteria castanea*, *P. jungii*, *P. morosa*, *P. ruficeps* (Waterhouse and Wallbank, 1967), and *Methana convexa* (Wallbank and Waterhouse, 1970).	Ventral abdominal gland of adults opening between the sixth and seventh sternites.	It is a trace constituent in the *Platyzosteria* secretions and is one of three methylene alkanals produced by all species. It constitutes a major constituent (60%) in the secretion of *M. convexa*.
Hexanal $C_6H_{12}O$ MW 100	Dictyoptera Blattidae: *Polyzosteria limbata*, *P. viridissima*, *P. oculata*, *P. cuprea*, and *P. pulchra* (Wallbank and Waterhouse, 1970). Hemiptera Alydidae: *Alydus eurinus* and *A. pilosulus* (Aldrich and Yonke, 1975). Coreidae: *Mictis profana*, *M. caja*, *Amorbus rubiginosus*, *A. alternatus*, *A. rhombifer*,	Blattidae: Ventral abdominal gland of adults opening between the sixth and seventh sternites. Alydidae: Adult metasternal scent gland. Coreidae: Adult metasternal scent gland of all species and dorsal abdominal glands of larvae of *A. nitida*.	It is a trace constituent in all the blattid secretions. It is the only aldehyde present in the alydid secretions and is accompanied by 1-hexanol to which it may be metabolically related. It is a rather characteristic component in adult coreid secretions and has not been detected in the exudates of

(*continued*)

Table 4.2 (Continued)

	Aulacosternum nigrorubrum, *Pachycolpura manca*, *Agriopocoris frogatti* (Waterhouse and Gilby, 1964), *Leptocoris apicalis*, (Baggini *et al*., 1966), *Acanthocoris sordidus* (Tsuyuki *et al*., 1965), *Hygia opaca*, *Plinachtus bicoloripes* (Tsuyuki *et al*., 1965), *Pternistria bispina* (Baker and Kemball, 1967), *Leptoglossus oppositus* (McCullough, 1968; Aldrich and Yonke, 1975), *L. clypeatus* (McCullough, 1969), *L. clypealis* (Aldrich and Yonke, 1975), *Amblypelta nitida* (Baker *et al*., 1972), *Mozena obtusa* (McCullough, 1973), *M. lunata* (McCullough, 1974a), and *Chelinidea vittiger* (McCullough, 1974b). Hyocephalidae: *Hyocephalus* sp. (Waterhouse and Gilby, 1964). Pentatomidae: *Graphosoma rubrolineatum* (Tsuyuki *et al*., 1965) and *Chrysocoris stolli* (Choudhuri and Das, 1968). Pyrrhocoridae: *Dysdercus intermedius* (Calam and Youdeowei, 1968). Coleoptera Tenebrionidae: *Eleodes beameri* (Tschinkel, 1975a).	Hyocephalidae: Adult metasternal scent gland. Pentatomidae: Adult metasternal scent gland. Pyrrhocoridae: Posterior dorsal abdominal gland of larvae. Tenebrionidae: Paired abdominal sternal glands. Formicidae: Mandibular glands of workers.	species in only three genera (*Acanthocephala*, *Riptortus*, and *Libyaspis*). Concentrations of this alkanal range from 5% (*Aulacosternum*) to 74% (*Mictis*) with ca. 35% being an average value. 1-Hexanol accompanies it as a minor constituent in most of the coreid secretions. A sample from aged bugs (*A. frogatti*) contained pure *n*-hexanal whereas this compound accounted for about 50% of the volatiles in the secretion of young bugs. About 19% by weight of the secretion of *L. clypeatus* is made up by this alkanal. The secretion of the *Hyocephalus* sp. contains 98% of it. This fact is consistent with the placement of this family in the coreoid complex. It constitutes only about 0.47% of the pyrrhocorid secretion. It is the major saturated aldehyde (ca. 16%) in the

	Hymenoptera Formicidae: *Oecophylla longinoda* (Bradshaw *et al*., 1975).		the secretion of *E*. *beameri* and is accompanied by a trace of 1-hexanol.
Benzaldehyde CHO (benzene ring) C_7H_6O MW 106	Polydesmida Paradoxosomatidae: *Oxidus gracilis* (Guldensteeden-Egeling, 1882; H. E. Eisner *et al*., 1963; Blum *et al*., 1973a) and *Orthomorpha coarctata* (Monteiro, 1961; Duffield *et al*., 1974).	Paradoxsomatidae: Lateral paired glands opening dorsolaterally on the anterior half of most diplosegments.	In the polydesmid secretions it is generated by a cyanogenic precursor and is always accompanied by HCN. It constitutes about 10-20% of the secretion of *A*. *corrugata* and is the major constituent in all secretions which have been analyzed. In some secretions it is accompanied by trace amounts of phenolic compounds (1-2%) or mandelonitrile benzoate and benzoyl cyanide. It is synthesized from aromatic amino acids (e.g., phenylalanine).
	Gomphodesmidae: *Gomphodesmus pavani* (Barbetta *et al*., 1966) and *Astrodesmus laxus* (H. E. Eisner *et al*., 1975).	Gomphodesmidae: Lateral paired glands opening dorsolaterally on the anterior half of most diplosegments.	
	Euryuridae: *Euryurus australis* and *E*. *leachii* (Duffield *et al*., 1974).	Euryuridae: Lateral paired glands opening dorsolaterally on the anterior half of most diplosegments.	
	Polydesmidae: *Pachydesmus crassicutis* (Blum and Woodring, 1962), *Pseudopolydesmus serratus* (H. E. Eisner *et al*., 1963), *P*. *branneri* (H. E. Eisner *et al*., 1975), *P*. *erasus* (Duffey *et al*., 1977), *Polydesmus collaris* (Casnati *et al*., 1963), and *P*. *virginiensis* (W. M. Wheeler, 1890).	Polydesmidae: Lateral paired glands opening dorsolaterally on the anterior half of most diplosegments.	One μl of secretion was collected from each larva of *P*. *atomaria*. The secretion consisted of about 7% of an organic phase and about 93% aqueous material. The presence of HCN in the chrysomelid secretions indicates that the
	Xystodesmidae: *Apheloria corrugata* (H. E. Eisner *et al*., 1963), *A*. *coriacea* (Weatherston and Gardiner, 1973), *A*. *trimaculata*, *A*. *kleinpeteri*	Xystodesmidae: Lateral paired glands opening dorsolaterally on the anterior half of most diplosegments.	

(*continued*)

Table 4.2 (Continued)

	(H. E. Eisner *et al*., 1975), *Cherokia georgiana* (H. E. Eisner *et al*., 1963), *Cleptoria rileyi*, *Brachoria* sp., *Sigiria* sp., *Stelgipus agrestis*, *Paimoikia* sp., *Motyxia tiemanni*, *M. sequoiae* (Duffey *et al*., 1976), and *Harpaphe haydeniana* (Duffey *et al*., 1974). Chelodesmidae: *Caraibodesmus* sp. (Duffey *et al*., 1976). Nearctodesmidae: *Nearctodesmus cerasinus* (Duffey *et al*., 1977).	Chelodesmidae: Lateral paired glands opening dorsolaterally on the anterior half of most diplosegments. Nearctodesmidae: Lateral paired glands opening dorsolaterally on the anterior half of most diplosegments.	aldehyde is synthesized from cyanogenic precursors in much the same way as it is produced by millipedes.
	Coleoptera		
	Carabidae: *Megacephala australis* (Moore and Brown, 1971a).	Carabidae: Pygidial glands.	Its presence in the secretion of the carabid *M. australis* indicates that it may be a character state for species in the subfamily Cicindelinae.
	Chrysomelidae: *Paropsis atomaria*, *Chrysophtharta variicollis*, and *C. amoena* (Moore, 1967).	Chrysomelidae: Paired glands opening on the eighth dorsal tergite of larvae.	
	Hymenoptera		
	Apidae: *Trigona postica*, *T. tubiba*, *T. depilis*, *T. xanthotricha* (Blum *et al*., 1973b), *T. pectoralis*, and *T. mexicana* (Luby *et al*., 1973).	Apidae: Mandibular glands of workers.	It is a major constituent in the *Trigona* secretions which are otherwise dominated by aliphatic ketones and alcohols.
	Formicidae: *Veromessor pergandei* (Blum *et al*., 1969b) and *Azteca* spp. (Blum and Wheeler, 1974).	Formicidae: Mandibular glands of workers of *V. pergandei* and anal glands of workers of *Azteca* spp.	It does not appear to be an especially characteristic component in *Veromessor* secretions since it could not be detected in the exudates of three other species in this genus. It is a major constituent in one of two *Azteca* spp. but was not detected as an anal gland product of a large number of other species in this genus.

Compound	Source	Gland	Remarks
Salicylaldehyde (structure: benzene ring with CHO and OH) $C_7H_6O_2$ MW 122	Coleoptera Carabidae: Calosoma prominens (Eisner et al., 1963c), C. sycophanta (Casnati et al., 1965), C. affine, C. alternans, C. macrum, C. parvicollis (McCullough, 1966b), C. externum, C. marginalis (McCullough and Weinheimer, 1966), C. schayeri C. oceanicum (Moore and Wallbank, 1968), Callisthenes luxatus (McCullough, 1972a), Bembidion quadriguttatum, Asphidion flavipes (Schildknecht et al., 1968a,b), Loxandrus longiformis, and L. sp. (Moore and Wallbank, 1968).	Carabidae: Pygidial glands.	It is frequently accompanied by aliphatic acids (e.g., methacrylic) in the carabid secretions. It varies from 3-80% of the Calosoma secretions. The secretion of Callisthenes luxatus contains only about 8% of it, the exudate being dominated by methacrylic acid, as is that of C. oceanicum. It is lacking from the secretion of Calosoma scrutator. Although it is produced by Bembidion quadriguttatum, it was not detected in the secretions of three other species in this genus. Loxandrus was the only one of thirteen pterostichine genera whose pygidial gland exudate contained it.
	Chrysomelidae: Melasoma populi (Hollande, 1909; Pavan, 1953), Plagiodera sp. (Hollande, 1909), Phyllodecta vittelinae (Wain, 1943), and Chrysomela scripta (Wallace and Blum, 1969).	Chrysomelidae: Paired eversible glands opening dorsally on the meso- and metathorax of first instar larvae and on the first seven abdominal segments of later larval instars.	The secretions of chrysomelid larvae can be conserved by being sucked back into the glandular reservoirs.
	Hymenoptera Anthophoridae: Pithitis smaragdula (Hefetz et al., 1979).	Anthophoridae: Mandibular glands.	

(continued)

Table 4.2 (Continued)

p-Hydroxybenzaldehyde 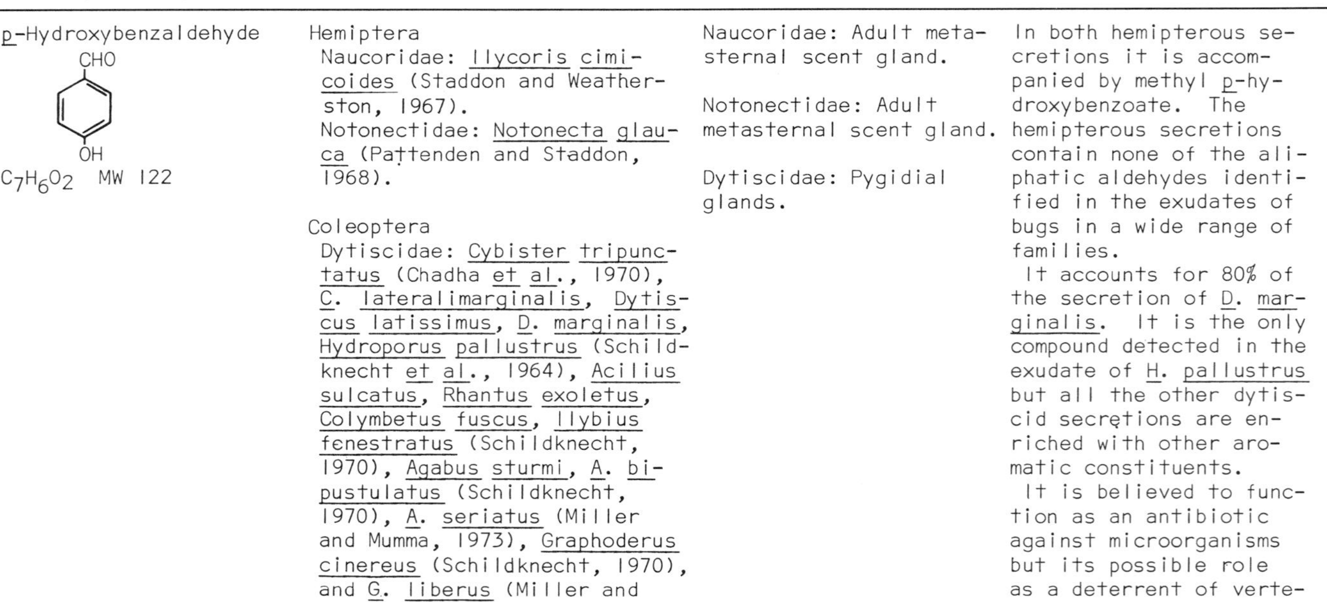 $C_7H_6O_2$ MW 122	Hemiptera Naucoridae: *Ilycoris cimicoides* (Staddon and Weatherston, 1967). Notonectidae: *Notonecta glauca* (Pattenden and Staddon, 1968). Coleoptera Dytiscidae: *Cybister tripunctatus* (Chadha *et al*., 1970), *C. lateralimarginalis*, *Dytiscus latissimus*, *D. marginalis*, *Hydroporus pallustrus* (Schildknecht *et al*., 1964), *Acilius sulcatus*, *Rhantus exoletus*, *Colymbetus fuscus*, *Ilybius fenestratus* (Schildknecht, 1970), *Agabus sturmi*, *A. bipustulatus* (Schildknecht, 1970), *A. seriatus* (Miller and Mumma, 1973), *Graphoderus cinereus* (Schildknecht, 1970), and *G. liberus* (Miller and Mumma, 1973).	Naucoridae: Adult metasternal scent gland. Notonectidae: Adult metasternal scent gland. Dytiscidae: Pygidial glands.	In both hemipterous secretions it is accompanied by methyl p-hydroxybenzoate. The hemipterous secretions contain none of the aliphatic aldehydes identified in the exudates of bugs in a wide range of families. It accounts for 80% of the secretion of *D. marginalis*. It is the only compound detected in the exudate of *H. pallustrus* but all the other dytiscid secretions are enriched with other aromatic constituents. It is believed to function as an antibiotic against microorganisms but its possible role as a deterrent of vertebrate predators has not been evaluated in any detail.

Compound	Order, family, species	Gland	Remarks
2-Heptenal  $C_7H_{12}O$ MW 112	Hemiptera Lygaeidae: *Oncopeltus fasciatus* (Games and Staddon, 1973a,b). Cydnidae: *Scaptocoris divergens* (Roth, 1961). Pentatomidae: *Oebalus pugnax* (Blum *et al.*, 1960). Coleoptera Tenebrionidae: *Eleodes beameri* (Tschinkel, 1975a).	Lygaeidae: Anterior and posterior dorsal abdominal glands of larvae. Cydnidae: Adult metasternal scent gland. Pentatomidae: Adult metasternal scent gland. Tenebrionidae: Paired abdominal sternal glands.	It is a minor constituent (0.5-1%) in the exudates of both the anterior and posterior glands of *O. fasciatus*. The *trans*-isomer is produced by *O. pugnax* adults. It is the least concentrated α,β-unsaturated aldehyde produced by *E. beameri*.
trans-4-Heptenal  $C_7H_{12}O$ MW 112	Hemiptera Pentatomidae: Unidentified sp. (Murkerji and Sharma, 1966).	Adult metasternal scent gland.	It is the only aldehydic constituent identified from a total extract of bugs.
Heptanal  $C_7H_{14}O$ MW 114	Coleoptera Tenebrionidae: *Eleodes beameri* (Tschinkel, 1975a).	Paired abdominal sternal glands.	It is the least concentrated of the three saturated aldehydes in the secretion.

(continued)

Table 4.2 (Continued)

6-Methyl salicylaldehyde 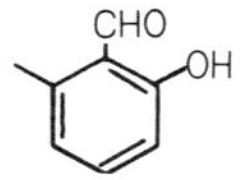 $C_8H_8O_2$ MW 136	Coleoptera Cerambycidae: *Phoracantha semipunctata* (Moore and Brown, 1972).	Adult metasternal scent glands opening to the surface through small pores near the distal borders of the metasternum.	It is the major constituent in the secretion and the only aromatic compound produced.
5-Ethylcyclopent-1-enecarbaldehyde 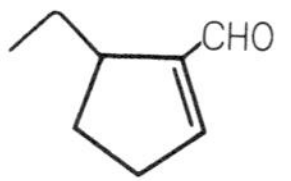 $C_8H_{12}O$ MW 124	Coleoptera Cerambycidae: *Phoracantha semipunctata* (Moore and Brown, 1972).	Adult metasternal scent glands opening to the surface through small pores near the distal borders of the metasternum.	It is the main alicyclic aldehyde present in the secretion and has been assigned the trivial epithet phoracanthal.
4-Oxo-*trans*-2-octenal 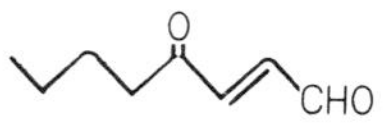 $C_8H_{12}O_2$ MW 140	Hemiptera Lygaeidae: *Oncopeltus fasciatus* (Games and Staddon, 1973b). Pentatomidae: *Nezara viridula* (Gilby and Waterhouse, 1965) and *Musgraveia sulciventris* (MacLeod *et al*., 1975). Pyrrhocoridae: *Dysdercus intermedius* (Calam and Youdeowei, 1968).	Lygaeidae: Anterior and posterior dorsal abdominal glands of larvae. Pentatomidae: Adult metasternal scent gland. Pyrrhocoridae: Posterior dorsal abdominal gland of larvae.	It constitutes 81% of the secretion from the posterior gland of *O. fasciatus* but only 7% of the exudate from the anterior gland. It is a minor constituent in the pyrrhocorid secretion (ca. 5%) and that of *N. viridula* (0.5%).

2-Ethylcyclopentane-carbaldehyde $C_8H_{14}O$ MW 126	Coleoptera Cerambycidae: *Phoracantha semipunctata* (Moore and Brown, 1972).	Adult metasternal scent glands opening to the surface through small pores near the distal borders of the metasternum.	Both the cis and trans isomers are present in the secretion as minor constituents.
trans-2-Octenal $C_8H_{14}O$ MW 126	Dictyoptera Blattidae: *Polyzosteria cuprea*, *P. pulchra*, *Platyzosteria nitidella*, and *P.* spp. (Wallbank and Waterhouse, 1970). Hemiptera Alydidae: *Megalotomus quinquespinosus* and *Alydus eurinus* (Aldrich and Yonke, 1975). Cimicidae: *Cimex lectularius* (Schildknecht *et al.*, 1964; Collins, 1968). Coreidae: *Leptocoris apicalis* (Baggini *et al.*, 1966), *Leptocorisa varicornis* (Choudhuri and Das, 1968), *Pternistria bispina* (Baker and Jones, 1969), *Euthochtha galeator*, and *Archimerus alternatus* (Aldrich and Yonke, 1975). Cydnidae: *Scaptocoris divergens* (Roth, 1961).	Blattidae: Ventral abdominal gland of adults opening between the sixth and seventh sternites. Alydidae: Dorsal abdominal glands of larvae. Cimicidae: Adult metasternal scent gland. Coreidae: Adult metasternal scent gland of *L. apicalis* and *L. varicornis*. Dorsal abdominal glands of larvae of *P. bispina*, *E. galeator*, and *A. alternatus*. Cydnidae: Adult metasternal scent gland. Gelastocoridae: Adult metasternal scent gland.	It is a minor component (2%) in the *Polyzosteria* secretions but accounts for 26-60% of the volatiles in the *Platyzosteria* exudates. It is a relatively minor component in the alydid secretions but accounts for 30% of the exudate of *C. lectularius*. It constitutes about 40% of the secretion of the coreid *L. apicalis* and is accompanied by 2-decenal as a minor concomitant. This enal accounts for about 52% of the secretion of larval *P. bispina* and is moderately concentrated in the exudates of the other larval coreids.

(continued)

Table 4.2 (Continued)

	Gelastocoridae: *Gelastocoris oculatus* (Staddon, 1973). Lygaeidae: *Oncopeltus fasciatus* (Games and Staddon, 1973a,b). Miridae: *Leptoterna dolabrata* (Collins and Drake, 1965). Pentatomidae: *Nezara viridula* (Gilby and Waterhouse, 1965), *Dolycoris baccarum* (Schildknecht *et al*., 1962, 1964), *Biprorulus bibax* (Park and Sutherland, 1962), *Tessaratoma aethiops* (Baggini *et al*., 1966), *Chrysocoris stolli* (Choudhuri and Das, 1968), *Scotinophora lurida*, *Aelia fieberi* (Tsuyuki *et al*., 1965), *Poecilometes strigatus* (Waterhouse *et al*., 1961), *Musgraveia sulciventris* (Waterhouse *et al*., 1961; Gilby and Waterhouse, 1967; MacLeod *et al*., 1975), and *Pallomena viridissima* (Schildknecht *et al*., 1964). Pyrrhocoridae: *Dysdercus intermedius* (Calam and Youdeowei, 1968; Calam and Scott, 1969). Scutelleridae: *Eurygaster* sp. (Schildknecht and Weis, 1962b).	Lygaeidae: Anterior and posterior dorsal abdominal glands of larvae. Miridae: Adult metasternal scent gland. Pentatomidae: Adult metasternal scent gland of all species. Dorsal abdominal glands of larvae of *T. aethiops* and *M. sulciventris*. Pyrrhocoridae: Adult metasternal scent gland and posterior dorsal abdominal gland of larvae. Scutelleridae: Adult metasternal scent gland.	It constitutes about 25% of the gelastocorid secretion and accounts for nearly 90% of the volatiles present in the exudate of the anterior gland of *O. fasciatus*. The posterior gland secretion contains only 15% of this enal. Its concentration in the pentatomid secretions is highly variable. It accounts for only 0.5% of the volatiles in the exudate of *N. viridula* whereas that of *D. baccarum* contains 20% of this enal. The secretion of larvae of *T. aethiops* contains nearly twice as much of it as the adult secretion (27% vs. 15%). It is shared by both adults and larvae of *M. sulciventris*, as are all other compounds. It is a major component in the secretions of both larvae and adults of *D. intermedius*.

Compound	Order, family, species (reference)	Gland	Remarks
Octanal  $C_8H_{16}O$ MW 128	Dictyoptera Blattidae: *Polyzosteria pulchra* and *P. cuprea* (Wallbank and Waterhouse, 1970).	Blattidae: Ventral abdominal gland of adults opening between the sixth and seventh sternites.	It is a minor (2%) and atypical constituent in polyzosteriine secretions.
	Hemiptera Coreidae: *Plinachtus bicoloripes* (Tsuyuki *et al.*, 1965). Pyrrhocoridae: *Dysdercus intermedius* (Calam and Youdeowei, 1968)	Coreidae: Adult metasternal scent gland. Pyrrhocoridae: Adult metasternal scent gland.	It accounts for only a small percentage of the volatiles in both the hemipterous and coleopterous secretions.
	Coleoptera Gyrinidae: *Dineutes discolor* (Wheeler, 1975). Tenebrionidae: *Eleodes beameri* (Tschinkel, 1975a).	Gyrinidae: Pygidial glands. Tenebrionidae: Paired abdominal sternal glands.	It is one of seven aldehydes in the exudate of *E. beameri.*
2-Nonenal  $C_9H_{16}O$ MW 140	Hemiptera Coreidae: *Leptocorisa varicornis* (Choundhuri and Das, 1968).	Coreidae: Adult metasternal scent gland.	It is a major constituent in the secretion of *E. beameri* accounting for nearly 30% of the volatiles.
	Coleoptera Tenebrionidae: *Eleodes beameri* (Tschinkel, 1975a).	Tenebrionidea: Paired abdominal sternal glands.	
2,6-Dimethyl-5-heptenal  $C_9H_{16}O$ MW 140	Hymenoptera Formicidae: *Acanthomyops claviger* (Regnier and Wilson, 1968), *Lasius alienus* (Regnier and Wilson, 1969), and *L. carniolicus* (Bergström and Löfqvist, 1970).	Mandibular glands of workers.	It is a trace constituent in all secretions and is accompanied by 2,6-dimethyl-5-heptenol-l in two of the exudates.

(continued)

Table 4.2 (Continued)

Cuminaldehyde CHO $C_{10}H_{12}O$ MW 148	Polydesmida Xystodesmidae: *Rhysodesmus vicinus* (Pallares, 1946).	Lateral paired glands opening dorsolaterally on the anterior half of most diplosegments.	It is stored as the glucoside of *p*-isopropylmandelonitrile and is generated, along with HCN, during the secretory act.
Dolichodial CHO CHO $C_{10}H_{14}O_2$ MW 166	Orthoptera Phasmidae: *Anisomorpha buprestoides* (Meinwald *et al*., 1962). Hymenoptera Formicidae: *Dolichoderus clarki*, *D. dentata*, *D. scabridus*, *Iridomyrmex rufoniger*, *I. myrmecodiae* (Cavill and Hinterberger, 1960, 1962), and *I. humilis* (Cavill and Houghton, 1974b).	Phasmidae: Paired dorsal prothoracic glands of larvae and adults. Formicidae: Anal glands of workers.	The compound produced by the phasmid has been named anisomorphal. It polymerizes rapidly on exposure to air and was detected in extracts of *I. humilis* with great difficulty. Ants in some colonies of *I. rufoniger* and *D. scabridus* are not reported to produce dolichodial but rather, other oxygenated monoterpenes.
Citral CHO Geranial Neral CHO $C_{10}H_{16}O$ MW 152	Coleoptera Staphylinidae: *Bledius mandibularis* and *B. spectabilis* (Wheeler *et al*., 1972a), *Thyreocephalus lorquini*, and *Eulissus orthodoxus* (Bellas *et al*., 1974). Hymenoptera Oxaeidae: *Oxaea flavescens*	Staphylinidae: Pygidial glands. Oxaeidae: Mandibular glands. Colletidae: Mandibular glands. Apidae: Mandibular	Both the cis (neral) and trans (geranial) isomers are present in nearly all the secretions. It is a relatively minor component in all staphylinid secretions except that of *B. mandibularis* (ca. 20%).

(Blum *et al*., 1974b). Colletidae: *Hylaeus cressoni* (Blum and Bohart, 1972), *H*. (=*Prosopis*) *communis*, *H*. *confusa*, *H*. *hyalinata*, and *H*. *pectoralis* (Bergström and Tengö, 1973). Apidae: *Lestrimellita limao* (Blum, 1966) and *Trigona subterranea* (Blum *et al*., 1970). Formicidae: *Atta sexdens* (Butenandt *et al*., 1959; Blum *et al*., 1968a), *Acanthomyops claviger* (Chadha *et al*., 1962; Regnier and Wilson, 1968), *A*. *subglaber*, *A*. *latipes* (Wilson and Regnier, 1971), and *Lasius fuliginosus* (Bernardi *et al*., 1967).	glands. Formicidae: Mandibular glands.	Only the trans isomer is present in the secretions of *T*. *lorquini* and *E*. *orthodoxus*. Geranial is the main isomer present in all the bee secretions and it is possible that the cis isomer predominates in general in the exudates of bees. Geranial accounts for about 66% of the volatiles in the secretion of *T*. *subterranea*. Citral completely dominates (ca. 95%) the mandibular gland secretion of *A*. *sexdens* but is not a characteristic exocrine product of this genus. It appears to be a typical minor product of the mandibular glands of *Acanthomyops* spp. but has only been detected in the exudates of one out of thirteen species of *Lasius*.

(continued)

Table 4.2 (Continued)

2-Methylene butanal dimer 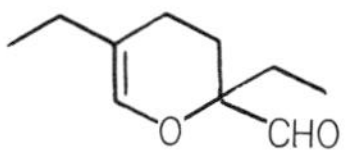 $C_{10}H_{16}O_2$ MW 168	Dictyoptera Blattidae: *Platyzosteria occidentalis*, *P. jungii*, *P. morosa*, *P. ruficeps* (Waterhouse and Wallbank, 1967), and *P.* nr. *montana* (Wallbank and Waterhouse, 1970).	Ventral abdominal gland of adults opening between the sixth and seventh sternites.	It accompanies the main component, 2-methylene butanal, in all secretions and accounts for 1-7% of the volatiles.
Iridodial $C_{10}H_{16}O_2$ MW 168	Coleoptera Staphylinidae: *Staphylinus olens* (Abou-Donia *et al.*, 1971), *Thyreocephalus lorquini*, and *Eulissus orthodoxus* (Bellas *et al.*, 1974). Cerambycidae: *Aromia moschata* (Vidari *et al.*, 1973). Hymenoptera Formicidae: *Iridomyrmex detectus*, *I. conifer* (Cavill *et al.*, 1956a,b), *I. nitideceps*, *I. rufoniger* (Cavill and Hinterberger, 1960, 1962), *I. pruinosus* (McGurk *et al.*, 1968), *Tapinoma nigerrimum* (Trave and Pavan, 1956), *T. sessile* (McGurk *et al.*, 1968), *Conomyrma pyramicus* (McGurk *et al.*, 1968), and *Azteca* nr. *velox* (Wheeler *et al.*, 1975).	Staphylinidae: Pygidial glands. Cerambycidae: Paired metasternal scentglands. Formicidae: Anal glands of workers.	It constitutes 90% of the secretion of *S. olens* and is the major component in the exudates of the other two staphylinid exudates. Both the γ and Δ isomers are present in the secretion of *A. moschata*. Its name is derived from the dolichoderine ant genus *Iridomyrmex*, from which it was first isolated. It was not detected in some colonies of *I. rufoniger* and *D. scabridus*, having been replaced by other monoterpenes. The anal glands of *I. pruinosus*, *T. sessile*, and *C. pyramicus* produce four isomers of it but in each case one isomer consists of 80% of the isomeric iridodials.

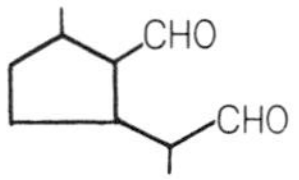

trans-2-Decenal $C_{10}H_{18}O$ MW 154	Dictyoptera Blattidae: Platyzosteria sp. (Wallbank and Waterhouse, 1970).	Blattidae: Ventral abdominal gland of adults opening between the sixth and seventh sternites.	A trace of the cis isomer has been detected in the secretion of N. viridula. It is present as a trace constituent in the blattid secretion and is accompanied by two other enals. It does not have a widespread distribution in the secretions of hemipterous families and has not been detected in any larval exudates. It is one of two α,β-unsaturated aldehydes in the secretion of the coreid L. apicalis and accounts for 12% of the volatiles. It appears to be an especially characteristic component in pentatomid defensive secretions. It is a minor constituent in the secretions of D. baccarum (5%) and B. bibax (3%) whereas it is quantitatively important in those of N. viridula and P. viridissima (25%).
	Hemiptera Coreidae: Leptocoris apicalis (Baggini et al., 1966), Leptocorisa varicornis (Choudhuri and Das, 1968), and Libyaspis angolensis (Cmelik, 1969).	Coreidae: Adult metasternal scent gland.	
	Pentatomidae: Nezara viridula (Waterhouse et al., 1961; Gilby and Waterhouse, 1965; Tsuyuki et al., 1965), N. antennata (Tsuyuki et al., 1965), Dolycoris baccarum, Palomena viridissima (Schildknecht et al., 1962, 1964), Chrysocoris stolli (Choudhuri and Das, 1968), Aelia fieberi, Graphosoma rubrolineatum, Scotinophora lurida, Menida scotti (Tsuyuki et al., 1965), Commius elegans (Gilby and Waterhouse, 1967), Musgraveia sulciventris, and Biprorulus bibax (MacLeod et al., 1975).	Pentatomidae: Adult metasternal scent gland.	
	Coleoptera Tenebrionidae: Eleodes beameri (Tschinkel, 1975a).	Tenebrionidae: Paired abdominal sternal glands.	

(continued)

Table 4.2 (Continued)

Citronellal CHO $C_{10}H_{18}O$ MW 154	Coleoptera Staphylinidae: *Thyreocephalus lorquini* and *Eulissus orthodoxus* (Bellas *et al*., 1974). Hymenoptera Formicidae: *Acanthomyops claviger* (Chadha *et al*., 1962), *A. latipes*, *A. subglaber* (Wilson and Regnier, 1971), *Lasius umbratus* (Blum *et al*., 1968b), *L. alienus* (Regnier and Wilson, 1969), *L. flavus*, *L. carniolicus* (Bergström and Löfqvist, 1970), *L. spathepus* (Kistner and Blum, 1971), and *L. speculiventris* (Wilson and Regnier, 1971).	Staphylinidae: Pygidial glands. Formicidae: Mandibular glands of workers.	It constitutes a major component in both staphylinid secretions and is accompanied by four compounds. It accounts for more than 90% of the volatiles in the mandibular gland exudate of *A. claviger*. It is a major constituent in the secretions of *L. carniolicus* and *L. spathepus* but is a minor component in the exudates of the other *Lasius* species. Its known distribution in ants is limited to these two genera in the subfamily Formicinae.
trans-2-Dodecenal CHO $C_{12}H_{22}O$ MW 182	Spirobolida Rhinocricidae: *Rhinocricus insulatus* (Wheeler *et al*., 1964).	Lateral paired glands opening dorsolaterally on the anterior half of most diplosegments.	It is the dominant component in a secretion which also contains *p*-toluquinone.

2-*n*-Butyl-2-octenal 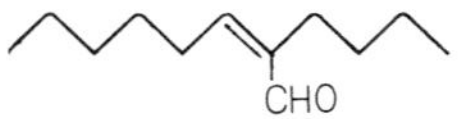 CHO $C_{12}H_{22}O$ MW 182	Hemiptera Coreidae: *Amblypelta nitida* (Baker *et al.*, 1972). Hymenoptera Formicidae: *Oecophylla longinoda* (Bradshaw *et al.*, 1975).	Coreidae: Dorsal abdominal glands of larvae and adult metasternal scent gland. Formicidae: Mandibular glands of workers.	It represents the aldol condensation-dehydration product of *n*-hexanal, a major glandular constituent in both secretions.
Dodecanal  CHO $C_{12}H_{24}O$ MW 184	Coleoptera Staphylinidae: *Drusilla canaliculata* (Brand *et al.*, 1973b).	The tergal gland, a structure opening between the sixth and seventh abdominal tergites.	It is the major aldehydic constituent present in the secretion and is accompanied by three other aliphatic aldehydes.
Gyrinidal  O O CHO $C_{14}H_{18}O_3$ MW 234	Coleoptera Gyrinidae: *Gyrinus ventralis* (Meinwald *et al.*, 1972), *G. natator* (Schildknecht *et al.*, 1972b), *Dineutes hornii*, *D. serrulatus* (Meinwald *et al.*, 1972), *D. assimilis*, and *D. nigrior* (Miller *et al.*, 1975).	Pygidial glands.	It constitutes about 50% of the volatiles in the secretions of *D. assimilis* and *D. nigrior* and is accompanied by three other norsesquiterpenes. It is the presumed precursor for the other norsesquiterpenes present.
Isogyrinidal  O O CHO $C_{14}H_{18}O_3$ MW 234	Coleoptera Gyrinidae: *Dineutes assimilis* and *D. nigrior* (Miller *et al.*, 1975).	Pygidial glands.	It is a minor component (6%) in both secretions and may be produced from gyrinidal by isomerization.

(*continued*)

Table 4.2 (Continued)

5,8-Tetradecadienal  CHO $C_{14}H_{24}O$ MW 208	Coleoptera Staphylinidae: *Drusilla canaliculata* (Brand *et al*., 1973b).	The tergal gland, a structure opening between the sixth and seventh abdominal tergites.	It is a minor constituent in the secretion which is accompanied by the monounsaturated and saturated C_{14} aldehydes.
5-Tetradecenal  CHO $C_{14}H_{26}O$ MW 210	Coleoptera Staphylinidae: *Drusilla canaliculata* (Brand *et al*., 1973b).	The tergal gland, a structure opening between the sixth and seventh abdominal tergites.	It constitutes a major constituent in the secretion and is probably biosynthetically related to its concomitant, 5,8-tetradecadienal.
Tetradecanal  CHO $C_{14}H_{28}O$ MW 212	Coleoptera Staphylinidae: *Drusilla canaliculata* (Brand *et al*., 1973b). Hymenoptera Apidae: *Bombus sporadicus*, *Psithyrus bohemicus*, *P. silvestris*, and *P. globosus* (Kullenberg *et al*., 1970).	Staphylinidae: The tergal gland, a structure opening between the sixth and seventh abdominal tergites. Apidae: Labial glands of males.	It is the least concentrated of the four aldehydic constituents present in the staphylinid secretion. It is a minor component in all the bee exudates.
Farnesal  CHO $C_{15}H_{24}O$ MW 220	Hymenoptera Formicidae: *Lasius fuliginosus* (Bernardi *et al*., 1967).	Mandibular glands of workers.	It accounts for 7% of the volatiles in a secretion that is completely dominated by three sesquiterpenes.

2,3-Dihydro-6-*trans*-farnesal  CHO $C_{15}H_{26}O$ MW 222	Hymenoptera Apidae: *Bombus jonellus* (Bergström and Svensson, 1973). Formicidae: *Lasius alienus* and *L. flavus* (Bergström and Löfqvist, 1970).	Apidae: Labial glands of males. Formicidae: Mandibular glands of workers.	The secretion of *B. jonellus* is dominated by 2,3-dihydro-6-*trans*-farnesol and the aldehyde is a minor constituent. It is not quantitatively important in the *Lasius* secretions.
Hexadecanal  CHO $C_{16}H_{38}O$ MW 238	Hymenoptera Apidae: *Bombus agrorum* and *Psithyrus bohemicus* (Kullenberg *et al*., 1970).	Labial glands of males.	Both secretions are dominated by hexadecenol-I to which it may be metabolically related.
Geranylgeranial  CHO $C_{20}H_{32}O$ MW 288	Hymenoptera Formicidae: *Lasius carniolicus* (Bergström and Löfqvist, 1970).	Mandibular glands of workers.	It is a minor constituent in the exudate which is distinctive in containing one of the few diterpenes synthesized by arthropods.
Geranylcitronellal  CHO $C_{20}H_{34}O$ MW 290	Hymenoptera Formicidae: *Lasius carniolicus* (Bergström and Löfqvist, 1970).	Mandibular glands of workers.	It is one of the major compounds in a secretion which is dominated by mono- and diterpenes.

AN OVERVIEW

In general, the identified aldehydes do not have a widespread occurrence in arthropod secretions, three-fourths of these compounds being limited to the defensive exudates of species in single orders (Table 4.2 and Chapter 18). Only two aldehydes, (*E*)-2-hexenal and hexanal, have been detected in the exudates of species in four orders, and even in these cases they are unusual allomones for species in some of these taxa. Sometimes these carbonyl compounds are atypical of the classes of defensive constituents produced by the members of an order, as is the case for (*E*),(*E*)-2,4-dimethyl-2,4-dienal (Jones *et al.*, 1977), a product of the phalangiid *Leiobunum nigripalpi*, a species in a genus that emphasizes ketones as allomones. The same can be said for (*E*)-2-dodecenal, a defensive product of a millipede (Wheeler *et al.*, 1964) that is a member of a quinone-producing taxon. On the other hand, polydesmoid millipedes characteristically produce an aldehydic allomone, benzaldehyde (Table 4.2), and one species is distinctive in producing cinnimaldehyde as a surprising surrogate for benzaldehyde.

Polyzosteriine cockroaches are distinctive in producing 10 aldehydes in the restricted range C_4–C_{10}, half of which are α,β-unsaturated compounds; four of these aldehydes are novel arthropod natural products. Conjugated aldehydes are also typical of hemipterans, species of which produce a remarkable diversity of aldehydes essentially in the range C_2–C_{10}. True bugs produce all the α,β-unsaturated aldehydes in the range C_3–C_{10} as well as most of the corresponding saturated aldehydes (Table 4.2). The aldehydic hemipterans synthesize distinctive compounds such as 4-oxo-(*E*)-2-hexenal and 4-oxo-(*E*)-2-octenal that are among the most idiosyncratic allomones produced by species in a single order. Some of these insects also generate an aromatic allomone, *p*-hydroxybenzaldehyde, in their exocrine glands. The aldehydic versatility of hemipterans is further illustrated by the detection of at least seven of these compounds in the secretions of single species (Gilby and Waterhouse, 1965; Roth, 1961).

Beetles produce more aldehydes than members of any other order, these compounds occurring in the defensive exudates of species in seven families (Table 4.1). As is the case with other classes of compounds, the aldehydes synthesized by beetles include a variety of aromatic constituents. Four of the five aromatic aldehydes detected as arthropod natural products occur in the secretions of beetles, and compounds such as 6-methylsalicylaldehyde are restricted to the exudates of species in a single coleopterous genus (Moore and Brown, 1972, 1976). In the case of a few of these secretions, aldehydic virtuosity is unexpected. An example is the tenebrionid *Eleodes beameri* whose abdominal exudate contains seven aliphatic aldehydes (Tschinkel, 1975a), several of which are identical to hemipterous natural products (Table 4.2). However, whereas the aldehydic richness of the secretion of one species, *E. beameri*, is exclusively responsible for the family Tenebrionidae being classified as a producer of this class of compounds, the same cannot be said for species in other coleopterous families.

Staphylinids produce nine aldehydes, resulting in these beetles being the richest source of these compounds in the Coleoptera. Included among these compounds are very idiosyncratic allomones such as 5-ethylcyclopent-1-enecarbaldehyde and 2-ethylcyclopentanecarbaldehyde (Moore and Brown, 1972). The presence of the monoterpenes citral, citronellal, and iridodial in staphylinid secretions illustrates the ability of beetles to produce terpenoid allomones, eight of which are produced by coleopterans. The characteristic monoterpenes chrysomelidial, epichrysomelidial, and plagiodial produced by chrysomelid larvae (Sugawara *et al.*, 1979b; Matsuda and Sugawara, 1980) further document the ability of beetles to synthesize a wide variety of monoterpenes. Gyrinid species, on the other hand, produce distinctive norsesquiterpenes, gyrinidal and iosgyrinidal (Table 4.2), compounds that appear to be particularly idiosyncratic glandular products of species in this family.

Hymenopterans also produce a large variety of aldehydes, primarily because of the biosynthetic versatility of the exocrine glands of ants. Except for (*E*)-2-hexenal and hexanal, both of which have a very limited distribution in the Formicidae (Table 4.2), the aldehydic allomones of ants are in the range C_9–C_{20}. Ants produce 18 of the 21 aldehydes identified as hymenopterous natural products, and nine of these compounds are limited in their arthropod distribution to ant secretions. Like beetles, these formicids are outstanding terpene chemists, synthesizing mono-, sesqui-, and diterpenes which account for half of the terpene aldehydes identified in arthropod defensive secretions (Table 4.2). Bees share with ants the ability to synthesize a few monoterpene aldehydes (e.g., citral, citronellal), but diterpene aldehydes have only been detected in formicid defensive secretions. Distinctive compounds such as 2,6-dimethyl-5-heptenal, dolichodial, and farnesal represent the terpenoid variation that characterizes these formicid natural products, which are often produced in relatively large quantities. It will prove no surprise if additional aldehydic compounds are detected when their paramount biosynthesizers, the ants, are subjected to further analytical scrutiny.

Chapter 5

Ketones

Although ketones (also see Chapter 18) have been identified as defensive products of arthropod species in seven orders (Table 5.1), this class of compounds is not especially typical of the allomones produced by members of several of these taxa. For example, although five ketones are present in the exudates of species in three hemipterous families (Table 5.1), these compounds are generally trace constituents accompanying major aldehydic compounds. Similarly, of the six ketones identified in the secretions of beetles in three families, only two of these compounds, gyrinidone and gyrinidione, appear to be characteristic products of species in one family (Gyrinidae). 2-Undecanone and 2-tridecanone, the two ketonic concomitants of formic acid in notodontid secretions (Eisner *et al.*, 1972; Weatherston *et al.*, 1979), are not at all representative of the defensive products identified in other lepidopterous secretions. Finally, it is not unlikely that the four ketones characterized in an acridid secretion (Table 5.2) constitute products that have been sequestered or metabolized from ingested plant compounds.

Ketonic allomones appear to be characteristic of harvestmen in the family Phalangiidae, some species of rhinotermitid and termitid termites, and ants and bees (Table 5.1). Although only six ketones have been identified in opilionid secretions, these compounds appear to have been emphasized as defensive products in a wide variety of phalangiid species. In the Dictyoptera, whereas only one ketone (2-heptanone) has been identified in cockroach exudates, 16 compounds representing 30% of the ketonic allomones characterized in arthropod secretions have been detected as frontal gland products of soldier termites in two families. But it is in the Hymenoptera, especially the bees and ants, that virtuosity as

Table 5.1

Distribution of Ketones in the Defensive Secretions of the Arthropoda[a]

Class	Order	Family
Arachnida	Opiliones (Phalangida)	Phalangiidae
Insecta	Dictyoptera	
	Blattaria	Blattidae
	Isoptera	Rhinotermitidae
		Termitidae
	Orthoptera	Acrididae
	Hemiptera	Cimicidae
		Pentatomidae
		Rhopalidae
	Coleoptera	Staphylinidae
		Gyrinidae
		Tenebrionidae
	Lepidoptera	Notodontidae
	Hymenoptera	Mutillidae
		Oxaeidae
		Andrenidae
		Apidae
		Formicidae

[a] Also see Chapter 18.

ketonic synthesizers is most highly evolved. Nearly 60% of the ketones identified as arthropod natural products are produced by hymenopterans, and almost 50% of these ketonic allomones are produced by ants. When it comes to ketones, ants are in a class by themselves.

AN OVERVIEW

Arthropods produce more than 50 ketones in the range C_4–C_{22}, but relatively few of these compounds have a widespread distribution among their secretions. Two-thirds of these compounds are limited in their arthropod occurrence to species in a single family, and when ketones are present in the secretions of members of several families, these taxa are usually in a single order (e.g., Hymenoptera). Two-thirds of the identified ketones are idiosyncratic familial compounds, with about 30% of these novel allomones occurring in hymenopterous secretions. Indeed, ketonic allomones have a more restricted familial distribution than the defensive

Table **5.2** Ketones in arthropod defensive secretions

Name and Formula	Occurrence	Glandular Source	Comments
2-Butanone 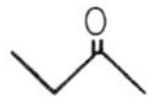 C_4H_8O MW 72	Hemiptera Cimicidae: *Cimex lectularius* (Collins, 1968). Pentatomidae: *Nezara viridula* (Gilby and Waterhouse, 1965) and *Chrysocoris stolli* (Choudhuri and Das, 1968).	Cimicidae: Adult metasternal scent gland. Pentatomidae: Adult metasternal scent gland.	It constitutes a minor or trace component in all the secretions.
2-Pentanone 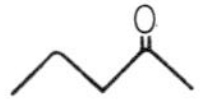 $C_5H_{10}O$ MW 86	Hymenoptera Formicidae: *Azteca* sp. and *Monacis bispinosa* (Blum and Wheeler, 1978).	Anal glands of workers.	It is a trace constituent in both secretions. It is not a typical anal gland product of *Azteca* spp.
2-Hexen-4-one 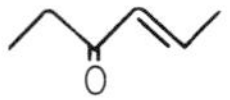 $C_6H_{10}O$ MW 98	Hemiptera Pentatomidae: *Nezara viridula* (Gilby and Waterhouse, 1965).	Adult metasternal scent gland.	It represents a trace constituent and is accompanied by several other short-chain ketones.
2-Methylcyclopentanone 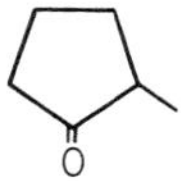 $C_6H_{10}O$ MW 98	Hymenoptera Formicidae: *Azteca* spp. (Wheeler *et al*., 1975).	Anal glands of workers.	It has been identified as an anal gland product of five species but is lacking in the secretions of several other *Azteca* spp. It is a minor constituent in all secretions.

Compound	Source	Gland	Remarks
3-Hexanone 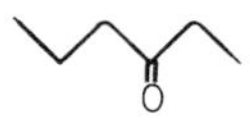 $C_6H_{12}O$ MW 100	Hemiptera Pentatomidae: *Nezara viridula* (Gilby and Waterhouse, 1965).	Adult metasternal scent gland.	It is a trace constituent in the secretion.
2-Heptanone $C_7H_{14}O$ MW 114	Dictyoptera Blattidae: *Platyzosteria armata* (Wallbank and Waterhouse, 1970).	Blattidae: Ventral abdominal gland of adults opening between the sixth and seventh sternites.	It is accompanied by 2-heptanol in the blattid secretion but is very atypical of *Platyzosteria* exudates, most of which contain α,β-unsaturated aldehydes.
	Hymenoptera Oxaeidae: *Oxaea flavescens* (Blum *et al.*, 1974b).	Oxaeidae: Mandibular glands of males.	It is replaced by 6-methyl-5-hepten-2-one in the secretion of females of *O. flavescens*.
	Apidae: *Trigona postica*, *T. depilis*, *T. tubiba*, *T. xanthotricha* (Blum *et al.*, 1973b), *T. mexicana*, *T. pectoralis* (Luby *et al.*, 1973), and *Apis mellifera* (Shearer and Boch, 1965).	Apidae: Mandibular glands of workers.	This alkanone has only been detected in *Trigona* spp. in the subgenus *Scaptotrigona*. It is a major compound in all secretions and is also utilized as a pheromone in different behavioral contexts. Mature workers of *A. mellifera* produce up to 20 μg of this ketone but its synthesis is supressed if the bees are caged and not al-
	Formicidae: *Iridomyrmex pruinosus* (Blum *et al.*, 1963), *Conomyrma pyramicus* (Blum and Warter, 1966; McGurk *et al.*, 1968), *Azteca* spp. (Wheeler *et al.*, 1975), *Monacis bispinosa* (Blum and Wheeler, 1978), *Crematogaster jehovae* (Crewe *et al.*, 1972), *Atta texana* (Moser *et al.*, 1968), *A. robusta*, *A. capiguara*, *A.*	Formicidae: Anal glands of workers of *Iridomyrmex*, *Conomyrma*, *Azteca*, and *Monacis* spp. Mandibular glands of workers of *Atta* spp., females of *A. texana*, and *Crematogaster* workers.	

(continued)

Table 5.2 (Continued)

	laevigata, A. sexdens, A. bisphaerica, A. columbica (Blum et al., 1968a), and A. cephalotes (Riley et al., 1974).		lowed to forage. It is not produced by three other species of Apis and its occurrence in the secretion of A. mellifera appears to constitute a valid chemotaxonomic character.
4-Methyl-2-hexanone 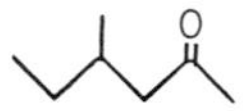 $C_7H_{14}O$ MW 114	Hymenoptera Formicidae: Dolichoderus clarki (Cavill and Hinterberger, 1960, 1962).	Anal glands of workers.	It constitutes a minor anal gland product and is not typical of Iridomyrmex secretions.
4-Methyl-3-hexanone 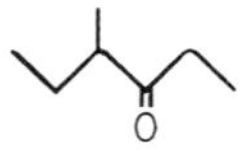 $C_7H_{14}O$ MW 114	Hymenoptera Formicidae: Manica mutica and M. hunteri (Fales et al., 1972).	Mandibular glands of workers.	It is a minor component in a secretion which contains five ethyl ketones.
o-Aminoacetophenone 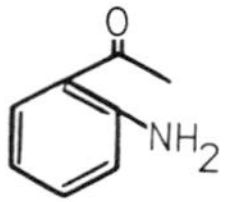 C_8H_9NO MW 135	Hymenoptera Formicidae: Mycocepurus goeldii (Blum et al., 1974a).	Mandibular glands of workers.	It constitutes the only aromatic compound identified as a mandibular gland product of an attine species.

Compound	Source	Gland	Comments
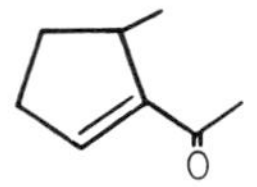 2-Acetyl-3-methylcyclopentene $C_8H_{12}O$ MW 124	Hymenoptera Formicidae: *Azteca* nr. *nigriventris* and *A.* nr. *instabilis* (Wheeler *et al.*, 1975).	Anal glands of workers.	It is a major constituent in the secretions of these two species but was not detected as an anal gland product of several other *Azteca* species.
6-Methyl-5-hepten-2-one  $C_8H_{14}O$ MW 126	Coleoptera Staphylinidae: *Stenus bipunctatus* (Schildknecht *et al.*, 1970) and *S. comma* (Schildknecht *et al.*, 1976). Hymenoptera Oxaeidae: *Oxaea flavescens* (Blum *et al.*, 1974b). Formicidae: *Liometopum microcephalum* (Casnati *et al.*, 1964), *Tapinoma nigerrimum* (Trave and Pavan, 1956), *Dolichoderus scabridus* (Cavill and Hinterberger, 1960, 1962), *Iridomyrmex detectus*, *I. conifer* (Cavill *et al.*, 1956a,b), *I. rufoniger*, *I. nitideceps*, (Cavill and Hinterberger, 1960, 1962), *I.* nr. *pruinosus* (Crewe and Blum, 1971), *Conomyrma pyramicus* (McGurk *et al.*, 1968), *Azteca* spp., *Monacis bispinosa* (Blum and	Staphylinidae: Smaller pair of pygidial glands. Oxaeidae: Mandibular glands of females. Formicidae: Mandibular glands of workers of *L. carniolicus* and *L. fuliginosus*. Anal glands of all other species.	It is one of three terpenes secreted into the water by adults of *S. bipunctatus* when disturbed. These compounds function to propel the beetle along the surface by depressing the surface tension of the water (*Entspannungsschwimmen*). It is a typical defensive product of ant species in the subfamily Dolichoderinae. It is accompanied by 2-heptanone in the anal gland exudates of some *Azteca* spp. and *M. bispinosa*, demonstrating that both terpenoid and nonterpenoid ketones can be synthesized by these organs. It has only been found in two of seven

(*continued*)

Table 5.2 (Continued)

	Wheeler, 1978), *Lasius carniolicus* (Bergström and Löfqvist, 1970), and *L. fuliginosus* (Bernardi *et al*., 1967).		species of *Lasius* and is not an especially typical exocrine product of this formicine genus.
cis-1-Acetyl-2-methylcyclopentane $C_8H_{14}O$ MW 126	Hymenoptera Formicidae: *Azteca* nr. *velox*, *A.* nr. *nigriventris*, and *A.* nr. *instabilis* (Wheeler *et al*., 1975).	Anal glands of workers.	It is a major constituent in the anal gland secretions of two *Azteca* spp. but is not detectable in the secretions of several other species.
2-Octanone $C_8H_{16}O$ MW 128	Hymenoptera Formicidae: *Pseudomyrmex* spp. (Blum and Wheeler, 1974).	Mandibular glands of workers.	It is a minor constituent in all secretions which is accompanied by three other alkanones.

Compound	Species	Gland source	Comments
3-Octanone $C_8H_{16}O$ MW 128	Hymenoptera Formicidae: Myrmica brevinodis (Crewe and Blum, 1970a,b), M. rubra, M. punctiventris, M. fracticornis, M. americana, M. ruginodis, M. sabuleti, M. scabrinodis (Crewe and Blum, 1970b), Crematogaster spp. (Crewe et al., 1969; Crewe et al., 1970; Crewe et al., 1972; Schlunnegger and Leuthold, 1972), Manica mutica, M. hunteri (Fales et al., 1972), Trachymyrmex septentrionalis, T. seminole, Acromyrmex octospinosus (Crewe and Blum, 1972), Atta texana, A. cephalotes (Riley et al., 1974), and Camponotus schaeferri (Duffield and Blum, 1975b).	Mandibular glands of workers, females, and males of Crematogaster spp. Mandibular glands of workers of all other species.	It is an especially characteristic component in the mandibular gland secretions of a wide range of genera in the subfamily Myrmicinae. It is usually accompanied by 3-octanol to which it may be biogenetically related. In almost all the myrmicine secretions it is accompanied by other ethyl ketones or methyl ketones (Atta spp.). This alkanone is not typical of formicine species. Its occurrence in the mandibular gland secretion of C. schaeferri is consistent with the conclusion that the systematic position of this species is equivocal.
4-Methyl-3-heptanone $C_8H_{16}O$ MW 128	Opiliones Phalangiidae: Leiobunum formosum, L. speciosum (Blum and Edgar, 1971), and L. vittatum (Meinwald et al., 1971).	Phalangiidae: Paired glands located on the flanks of the prosoma between the first and second pair of coxae.	It is the major constituent in the phalangiid secretions and is accompanied by an unusual C_{10} ketone in the secretion of L. vittatum.
	Hymenoptera Mutillidae: Dasymutilla occidentalis and D. mutata	Mutillidae: Mandibular glands of males and females.	It is the major ketonic constituent in the exudates of most of the

(continued)

Table 5.2 (Continued)

	(Schmidt and Blum, 1977). Formicidae: Pogonomyrmex badius, P. barbatus, P. californicus, P. desertorum, P. rugosus (McGurk et al., 1966), Atta robusta, A. capiguara, A. laevigata, A. sexdens, A. bisphaerica, A. columbica (Blum et al., 1968a), A. texana (Moser et al., 1968; Riley et al., 1974), A. cephalotes (Riley et al., 1974), Trachmyrmex seminole (Crewe and Blum, 1972), Manica mutica, M. hunteri (Fales et al., 1972), Neoponera villosa (Duffield and Blum, 1973), and Camponotus abdominalis (Duffield and Blum, 1975c).	Formicidae: Mandibular glands of workers, males, and females of Atta spp.. Mandibular glands of workers of all other species.	Atta species which also contain 2-heptanone. It is the only alkanone produced by Pogonomyrmex, Neoponera, and Camponotus spp. Manica spp. synthesize five ketones in their mandibular gland tissues and it is the second most concentrated alkanone produced.
2-Methyl-4-heptanone $C_8H_{18}O$ MW 128	Hymenoptera Formicidae: Tapinoma nigerrimum (Trave and Pavan, 1956).	Anal glands of workers.	It is the only 4-alkanone identified in ant defensive secretions and is accompanied by 6-methyl-5-hepten-2-one.

2,6,6-Trimethyl-cyclohex-2-ene-1,4-dione $C_9H_{12}O_2$ MW 152	Orthoptera Acrididae: *Romalea microptera* (Eisner *et al.*, 1971a).	Paired glands evacuating through the mesothoracic spiracles.	It is a trace constituent in the secretion and may be derived from ingested food.
Isophorone (3,5,5-Trimethyl-2-cyclohexen-1-one) $C_9H_{14}O$ MW 138	Orthoptera Acrididae: *Romalea microptera* (Eisner *et al.*, 1971a).	Paired glands evacuating through the mesothoracic spiracles.	It is a concomitant of isophorone in the secretion and may originate in the discharge by sequestration from ingested food.
1-Nonen-3-one $C_9H_{16}O$ MW 140	Coleoptera Tenebrionidae: *Eleodes beameri* (Tschinkel, 1975a).	Paired abdominal sternal glands.	It comprises 1% of a secretion containing at least 10 carbonyl compounds.
2-Nonanone $C_9H_{18}O$ MW 142	Hemiptera Pentatomidae: *Nezara viridula* (Gilby and Waterhouse, 1965). Hymenoptera Apidae: *Trigona postica*, *T.*	Pentatomidae: Adult metasternal scent gland. Apidae: Mandibular glands of workers.	It constitutes a trace component in the pentatomid secretion. It is a major constituent in the *Trigona* secretions and is accom-

(continued)

Table 5.2 (Continued)

	tubiba, *T. depilis*, *T. xanthotricha* (Blum *et al*., 1973b), *T. mexicana*, and *T. pectoralis* (Luby *et al*., 1973). Formicidae: *Azteca* spp. (Blum and Wheeler, 1974).	Formicidae: Anal glands of workers.	panied by 2-nonanol to which it may be metabolically related. This methyl ketone has only been detected in *Trigona* species in the subgenus *Scaptotrigona*.
3-Nonanone $C_9H_{18}O$ MW 142	Coleoptera Tenebrionidae: *Eleodes beameri* (Tschinkel, 1975a). Hymenoptera Formicidae: *Myrmica rubra*, *M. brevinodis*, *M. punctiventris*, *M. fracticornis*, *M. americana*, *M. ruginodis*, *M. sabuleti*, *M. scabrinodis* (Crewe and Blum, 1970b), and *Crematogaster* spp. (Crewe *et al*., 1972).	Tenebrionidae: Paired abdominal sternal glands. Formicidae: Mandibular glands of workers.	It is a minor component in all secretions and is one of three ethyl ketones which are produced by ant species in both genera. It is a very atypical constituent in tenebrionid secretions.
6-Methyl-3-octanone $C_9H_{18}O$ MW 142	Hymenoptera Formicidae: *Myrmica rubra*, *M. brevinodis*, *M. punctiventris*, *M. fracticornis*, *M. americana*, *M. ruginodis*, *M. sabuleti*, *M. scabrinodis* (Crewe and Blum, 1970b), and *Crematogaster* spp. (Crewe *et al*., 1972).	Mandibular glands of *Myrmica* spp. Mandibular glands of workers, females, and males of *Crematogaster* spp.	In all secretions it represents a trace constituent. It is always accompanied by 6-methyl-3-octanol and may be biogenetically related to this carbinol.

Verbenone (2,6,6-Trimethylbicyclo [3.1.1]-2-hepten-3-one) $C_{10}H_{14}O$ MW 150	Orthoptera Acrididae: _Romalea microptera_ (Eisner _et al._, 1971a).	Paired glands evacuating through the mesothoracic spiracles.	It is one of eight compounds identified in the secretion and is present at a concentration of 50 ppm in the froth.
4,6-Dimethyl-4-octen-3-one (Manicone) $C_{10}H_{18}O$ MW 154	Hymenoptera Formicidae: _Manica mutica_ and _M. hunteri_ (Fales _et al._, 1972).	Mandibular glands of workers.	Manicone is the major component in the secretion and is accompanied by four other ethyl ketones.
(*E*)-4,6-Dimethyl-6-octen-3-one (Leiobunone) $C_{10}H_{18}O$ MW 154	Opiliones Phalangiidae: _Leiobunum vittatum_ (Meinwald _et al._, 1971).	Paired glands located on the flanks of the prosoma between the first and second pair of coxae.	The secretion of _L. vittatum_ is dominated by leiobunone and 4-methyl-3-heptanone.

(*continued*)

Table 5.2 (Continued)

Compound	Order / Family: Species (Reference)	Gland	Comments
3-Decanone $C_{10}H_{20}O$ MW 156	Hymenoptera Formicidae: *Manica mutica* and *M. hunteri* (Fales *et al.*, 1972).	Mandibular glands of workers.	It is the longest-chain ethyl ketone produced in myrmicine secretions and is accompanied by unbranched C_8 and C_9 ethyl ketones.
2-Undecanone $C_{11}H_{22}O$ MW 170	Lepidoptera Notodontidae: *Heterocampa manteo* (Eisner *et al.*, 1972). Hymenoptera Apidae: *Trigona postica*, *T. tubiba*, *T. depilis*, *T. xanthotricha* (Blum *et al.*, 1973b), *T. mexicana*, and *T. pectoralis* (Luby *et al.*, 1973). Formicidae: *Azteca alfari* (Blum and Wheeler, 1974).	Notodontidae: Cervical gland opening on the underside of the neck in larvae. Apidae: Mandibular glands of workers. Formicidae: Anal glands of workers.	It is one of the two minor ketonic constituents in the notodontid secretion which is dominated by formic acid. The ketones are reported to increase considerably the deterrent effectiveness of formic acid. In the *Trigona* secretions this methyl ketone is often a major constituent and is usually accompanied by 2-undecanol. It is a major constituent in the secretion of *A. alfari*.
Romallenone $C_{13}H_{20}O_3$ MW 224	Orthoptera Acrididae: *Romalea microptera* (Eisner *et al.*, 1971a).	Paired glands evacuating through the mesothoracic spiracles.	This complex allenic sesquiterpenoid is not repellent to arthropod predators. It probably represents a degradation product of an allenic plant pigment which is incorporated into the secretion by tracheal

Compound	Taxa	Gland	Remarks
			gland cells. Romallenone is the main constituent in this defensive exudate.
2-Tridecanone  $C_{13}H_{26}O$ MW 198	Dictyoptera Rhinotermitidae: Schedorhinotermes putorius (Quennedey et al., 1973).	Rhinotermitidae: Frontal gland of soldiers.	This methyl ketone constitutes one of five ketones detected in this unusual termite secretion.
	Lepidoptera Notodontidae: Heterocampa manteo (Eisner et al., 1972).	Notodontidae: Cervical gland opening on the underside of the neck in larvae.	The secretion of H. manteo contains only about 0.35% 2-tridecanone as well as a trace of 2-undecanone.
	Hymenoptera Apidae: Trigona postica, T. depilis, T. tubiba, T. xanthotricha (Blum et al.,1973b), T. pectoralis, and T. mexicana (Luby et al., 1973).	Apidae: Mandibular glands of workers.	It represents a major component in the Trigona secretions and is accompanied by the corresponding carbinol, to which it may be biosynthetically related. It has only been found in bees in the subfamily Scaptotrigona.
	Formicidae: Lasius umbratus (Quilico et al., 1957a; Blum et al., 1968b), L. carniolicus (Bergström and Löfqvist, 1970), L. alienus (Regnier and Wilson, 1969; Bergström and Löfqvist, 1970), L. neoniger, L. flavus, L. nearcticus, L. speculiventris (Wilson and Regnier, 1971), Acanthomyops claviger (Regnier and Wilson, 1968), A. latipes, A. subglaber (Wilson and Regnier, 1971), Formica rufibarbis (Bergström and Löfqvist, 1968), F. subsericea, F.	Formicidae: Dufour's gland of workers.	It has only been encountered in Dufour's gland secretions of ant species in the subfamily Formicinae. It is a major and typical constituent in Lasius and Acanthomyops secretions but is lacking from the secretions of several Formica species.

(continued)

Table 5.2 (Continued)

	neogagates, F. rubicunda (Wilson and Regnier, 1971), Camponotus ligniperda (Bergström and Löfqvist, 1971), and C. herculeanus (Bergström and Löfqvist, 1972).		
Gyrinidone 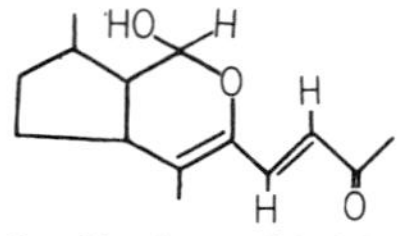 HO H O H H O $C_{14}H_{20}O_3$ MW 236	Coleoptera Gyrinidae: Dineutes discolor (Wheeler et al., 1972b), D. assimilis, and D. nigrior (Miller et al., 1975).	Pygidial glands.	It is the major constituent in the secretion of D. discolor whereas it is a minor constituent in the exudates of the other two species. Gyrinidal accompanies it in the secretion of D. assimilis and D. nigrior and may be its precursor.
Gyrinidione 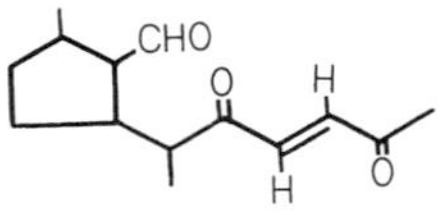 CHO O H H O $C_{14}H_{20}O_3$ MW 236	Coleoptera Gyrinidae: Dineutes assimilis and D. nigrior (Miller et al., 1975).	Pygidial glands.	It constitutes 36% of the secretions of both species and is accompanied by three other norsesquiterpene carbonyl compounds.
1-Tetradecen-3-one  O $C_{14}H_{26}O$ MW 210	Dictyoptera Rhinotermitidae: Schedorhinotermes putorius (Quennedey et al., 1973) and S. lamanianus (Prestwich, 1975).	Frontal gland of soldiers.	It constitutes at least 70% of the ketonic secretions of both species.

3-Tetradecanone $C_{14}H_{28}O$ MW 212	Rhinotermitidae: *Schedorhinotermes putorius* (Quennedey *et al.*, 1973) and *S. lamanianus* (Prestwich, 1975). Hymenoptera Formicidae: *Lasius carniolicus* (Bergström and Löfqvist, 1970).	Rhinotermitidae: Frontal gland of soldiers. Formicidae: Dufour's gland of workers.	It is a trace constituent in the rhinotermitid secretions and is accompanied by 1-tetradecen-3-one to which it may be metabolically related. It is a trace component in the secretion of *L. carniolicus* and is not a typical exocrine product of species in this genus.
2-Pentadecanone $C_{15}H_{30}O$ MW 226	Dictyoptera Rhinotermitidae: *Schedorhinotermes putorius* (Quennedey *et al.*, 1973). Hymenoptera Apidae: *Trigona postica*, *T. tubiba*, *T. depilis*, *T. xanthotricha* (Blum *et al.*, 1973b), *T. pectoralis*, and *T. mexicana* (Luby *et al.*, 1973). Formicidae: *Lasius fuliginosus* (Bernardi *et al.*, 1967), *L. alienus* (Regnier and Wilson, 1969; Bergström and Löfqvist, 1970), *L. umbratus*, *L. nearcticus*, *L. speculiventris* (Wilson and Regnier, 1971), *Acanthomyops claviger* (Regnier and Wilson, 1968), *A. latipes*, *A. subglaber* (Wilson and Regnier, 1971), and *Camponotus*	Rhinotermitidae: Frontal gland of soldiers. Apidae: Mandibular glands of workers. Formicidae: Dufour's gland of workers.	It is one of two trace ketonic constituents in the rhinotermitid secretion. It is one of six methyl ketones in the *Trigona* secretions and is always accompanied by 2-pentadecanol to which it may be metabolically related. Among ants it has only been detected as a Dufour's gland product of species in the subfamily Formicinae. It is a major constituent in a few *Lasius* spp. but is generally a minor exocrine product in the secretions of the other formicines.

(*continued*)

Table 5.2 (Continued)

Compound	Order, family, species (reference)	Source	Remarks
	ligniperda (Bergström and Löfqvist, 1971).		
1,15-Hexadecadien-3-one  $C_{16}H_{28}O$ MW 236	Dictyoptera Rhinotermitidae: *Schedorhinotermes lamanianus* (Prestwich, 1975).	Frontal gland of soldiers.	It is one of three diunsaturated ketones in the secretions.
1-Hexadecen-3-one  $C_{16}H_{30}O$ MW 238	Dictyoptera Rhinotermitidae: *Schedorhinotermes putorius* (Quennedey *et al.*, 1973) and *S. lamanianus* (Prestwich, 1975).	Frontal gland of soldiers.	It constitutes 6-15% of these ketone-rich secretions.
3-Hexadecanone  $C_{16}H_{32}O$ MW 240	Dictyoptera Rhinotermitidae: *Schedorhinotermes lamanianus* (Prestwich, 1975). Hymenoptera Formicidae: *Lasius carniolicus* (Bergström and Löfqvist, 1970).	Rhinotermitidae: Frontal gland of soldiers. Formicidae: Dufour's gland of workers.	It is a minor constituent in the termite secretion. It is a trace constituent which is accompanied by 3-tetradecanone in the ant secretion. These are the only two ethyl ketones identified as formicine exocrine products.

2-Heptadecanone $C_{17}H_{34}O$ MW 254	Hymenoptera Apidae: *Trigona postica*, *T. tubiba*, *T. depilis*, *T. xanthotricha* (Blum *et al.*, 1973b), *T. pectoralis*, and *T. mexicana* (Luby *et al.*, 1973). Formicidae: *Lasius fuliginosus* (Bernardi *et al.*, 1967) and *L. carniolicus* (Bergström and Löfqvist, 1970).	Apidae: Mandibular glands of workers. Formicidae: Dufour's gland of workers.	It is accompanied by 2-heptanol in the *Trigona* secretions. It is a trace constituent in both of the *Lasius* secretions.
2-Nonadecanone $C_{19}H_{38}O$ MW 282	Hymenoptera Formicidae: *Lasius alienus* (Bergström and Löfqvist, 1970).	Dufour's gland of workers.	It is one of four methyl ketones in this formicine secretion.

compounds of any other class of exocrine products. Only a few ketones (e.g., 4-methyl-3-hexanone, 2-tridecanone) have been detected in the secretions of species in more than two orders (Table 5.2), and in these cases the compounds appear to be exceptional products for members of two of the orders.

Phalangiids are distinctive in producing a variety of C_7–C_{11} mono- and dimethyl-branched ethyl ketones, four of which are only known from the secretions of these arachnids. These exudates are unusual in containing both saturated ketones and their unsaturated counterparts (e.g., 4-methyl-3-heptanone and 4-methyl-4-hepten-3-one) (Jones *et al.*, 1976b, 1977). On the other hand, whereas unsaturated dimethyl alkenones such as (*E*)-4,6-dimethyl-6-octen-3-one are sometimes found in phalangiid secretions (Meinwald *et al.*, 1971), the corresponding saturated ketones have only been detected in mutillid exudates (Fales *et al.*, 1980).

Although ketones are not typical defensive products of hemipterans, four of the five identified compounds constitute novel defensive compounds. For example, short-chain compounds such as 2-butanone and 3-hexanone are limited in their arthropod distribution to cimicid and pentatomid exudates (Table 5.2). Furthermore, the recent identification of piperitone as a natural product of a rhopalid species (Aldrich *et al.*, 1979) demonstrates that hemipterans possess the ability to synthesize novel terpenoid ketones as well.

Ketonic allomones may be exceptional natural products of termites even though 16 have already been characterized in termite secretions. Although the frontal gland secretions of many species of termite soldiers have been analyzed, ketonic constituents have been limited to the exudates of four species. Six of the eight novel ketonic natural products characterized in termite secretions occur as allomones of two *Schedorhinotermes* species (Quennedey *et al.*, 1973; Prestwich, 1975) and include several C_{14} and C_{15} 3-alkenones (e.g., 1-tetradecen-3-one). On the other hand, an *Amitermes* species produces the C_{11}–C_{17} methyl ketones (Meinwald *et al.*, 1978) that are typical exocrine products of hymenopterans. The two novel ketonic diterpenes isolated from *Nasutitermes octopilis* (Prestwich *et al.*, 1979) are representative of the extensive diterpene theme that characterizes the glandular products of many species of termitid soldiers.

Bees in three families produce 20% of the ketones identified as arthropod natural products, and these compounds mostly consist of 2-alkanones in the range C_5–C_{17} (Table 5.2 and Chapter 18). These compounds appear to be characteristic of *Bombus* and some *Trigona* species and are frequently accompanied by their corresponding 2-alkanols. Among bees, species of *Andrena* are the most versatile ketonic synthesizers, and three novel compounds (e.g., 7-methyl-4-octanone) have been identified as glandular products of members of this genus (Tengö and Bergström, 1976).

Ants are the ketonic biochemists *par excellence*, nearly half of the arthropod-derived ketones being identified as their cephalic and abdominal products. An incredible variety of methyl and ethyl ketones have been characterized in their exudates, and 20% of the ketonic natural products identified as arthropod defensive compounds are derived from formicid exudates. Compounds such as 4-methyl-2-hexanone, 6-methyl-3-octanone, and 4,6-dimethyl-4-octen-3-one typify the ability of these invertebrates to produce distinctive methyl-branched ketones in their exocrine glands (Table 5.2). *o*-Aminoacetophenone, the only aromatic ketone identified as an arthropod defensive allomone, is also a formicid natural product (Blum *et al.*, 1974c). All odd-numbered methyl ketones in the range C_5–C_{19} have been characterized as glandular products of ants, particularly species in the subfamilies Dolichoderinae and Formicinae. Ants also produce an abundance of ethyl ketones, nine of which have been identified as exocrine products. Species in the myrmicine genus *Manica* are noteworthy in synthesizing five of these compounds in their mandibular glands; three of these compounds have not been identified in the secretions of any other ant species (Fales *et al.*, 1972).

Novel anal gland products of dolichoderine ants in the genus *Azteca* include distinctive alicyclic ketones such as 2-acetyl-3-methylcylopentene (Wheeler *et al.*, 1975). The presence of such compounds in the defensive secretions of ants further emphasizes the remarkable ketonic themes that characterize the exocrine tissues of their mandibular, anal, and Dufour's glands. It would be no exaggeration to describe these invertebrates as the ketonic virtuosos of the arthropod world.

Chapter 6

Carboxylic Acids

Arthropods in the Arachnida, Diplopoda, Chilopoda, and Insecta utilize carboxylic acids (see also Chapter 18) as defensive agents, but these compounds are typical exocrine products in a relatively limited number of families. More than 40 acids have been identified in the secretions of species of arachnids, millipedes, centipedes, and the insect orders Dictoyoptera, Hemiptera, Coleoptera, Lepidoptera, and Hymenoptera (Table 6.1). However, only about a fifth of these compounds are widely distributed in the families in which they occur, and the remaining carboxylic acids are either present in relatively few species or are trace constituents in defensive exudates. For example, in the order Lepidoptera, eight acids have been identified as exocrine products, but only four of these compounds are major constituents and the remaining acids are trace components which have been identified in no more than one species in a family (Table 6.2).

The only insect orders that appear to have emphasized carboxylic acids as defensive compounds are the Coleoptera and Hymenoptera. About one-half of the acids identified in exocrine discharges is produced by beetles, particularly species in the families Carabidae and Dytiscidae (Table 6.2 and Chapter 18). The pygidial glands of coleopterans are eminently capable of synthesizing a wide variety of short-chain and aromatic acids and, in one investigation alone, more than 25% of the carboxylic acids characterized in arthropod defensive exudates were identified (Moore and Wallbank, 1968). Recent studies on dytiscid natural products demonstrate that these aquatic coleopterans produce a diversity of distinctive acids that include 3-indoleacetic acid and sulfur-containing acids (Dettner and Schwinger, 1977; Dettner, 1979).

Table 6.1

Distribution of Carboxylic Acids in the Defensive Secretions of the Arthropoda[a]

Class	Order	Family
Arachnida	Uropygi	Thelyphonidae
Diplopoda	Polydesmida	Paradoxsomatidae
		Gomphodesmidae
		Xystodesmidae
		Polydesmidae
Chilopoda	Geophilomorpha	Geophilidae
Insecta	Dictyoptera	
	Blattaria	Blattidae
	Hemiptera	Alydidae
		Coreidae
		Hyocephalidae
		Reduviidae
	Coleoptera	Cantharidae
		Carabidae
		Dytiscidae
		Tenebrionidae
		Staphylinidae
	Lepidoptera	Notodontidae
		Papilionidae
	Hymenoptera	Apidae
		Formicidae

[a] Also see Chapter 18.

In general, the carboxylic acids identified as arthropod exocrine products have been regarded as deterrents for either invertebrate or vertebrate predators. On the other hand, several of these compounds are considered to function as defensive agents against bacteria and fungi and, as a consequence, these natural products are believed to constitute selective antibiotic agents (Schildknecht and Koob, 1970; Schildknecht, 1971). The putative role that these acids play is predicated on *in vitro* studies in which the pure acids have been demonstrated to inhibit the growth of microorganisms. However, such tests do not establish whether these compounds function in this capacity under normal circumstances, especially since evaluations of the growth-inhibiting properties of natural products may not be especially meaningful unless the relevant microorganisms are employed for testing. Many of the defensive compounds present in insect secretions possess antibacterial properties (see review by Valcurone and Baggini, 1957), but there are no strong grounds for con-

Table **6.2** Acids in arthropod defensive secretions

Name and Formula	Occurrence	Glandular Source	Comments
Formic acid HCOOH CH_2O_2 MW 46	Polydesmida Polydesmidae: *Polydesmus collaris* (Casnati *et al.*, 1963).	Polydesmidae: Lateral paired glands opening dorsolaterally on the anterior half of most diplosegments.	It is a trace component in the polydesmid secretion.
	Lepidoptera Notodontidae: *Dicranura vinula* (Poulton, 1888; Schildknecht and Schmidt, 1963), *Schizura leptinoides* (Monro *et al.*, 1962), and *Heterocampa manteo* (Eisner *et al.*, 1972).	Notodontidae: Cervical gland opening on the underside of the neck in larvae.	The notodontid secretions, which contain up to 25% aqueous formic acid, are fortified with minor ketonic constituents that facilitate the penetration of this highly ionized cytotoxin through the arthropod cuticle.
	Coleoptera Carabidae: Species in the following genera: *Siagonyx*, *Dicrochile*, *Acinopus*, *Diaphoromerus*, *Amblytelus*, *Abacomorphus*, *Sphodrosomus*, *Notonomus*, *Sarticus*, *Loxodactylus*, *Philophloeus*, *Eudalia*, *Helluo*, *Sphallomorpha* (Moore and Wallbank, 1968), *Agonum*, *Platynus*, *Ophonus*, *Diachromus*, *Pseudophonus*, *Anisodactylus*, *Dichirotrichus*, *Stenolophus*, *Badister*, *Licinus*, *Lebia*, *Odacantha*, *Drypta*, *Polystichus* (Schildknecht *et al.*, 1968a), *Harpalus* (Moore and Wallbank,	Carabidae: Pygidial glands. Formicidae: Poison gland of worker and female.	It is produced by carabids in at least 10 subfamilies and is rarely accompanied by other acids. Hydrocarbons, which may facilitate its penetration through the lipophilic cuticle of arthropods, are frequently concomitants. In *Dicaelus*, the secretion contains about a 75% solution of acid. It is an ubiquitous poison gland product in the ant subfamily Formicinae but is not found

	1968; Schildknecht *et al.*, 1968a; McCullough, 1966c), *Calathus* (Moore and Wallbank, 1968; Schildknecht *et al.*, 1968a), *Dicaelus* (McCullough, 1967a), *Progaleritina* (McCullough, 1971), *Anthia*, and *Thermophilium* (Scott *et al.*, 1975).		
	Hymenoptera Formicidae: Species in the following genera: *Formica* (Melander and Brues, 1906; Stumper, 1922, 1952; O'Rourke, 1950; Osman and Brander, 1961), *Lasius* (Stumper, 1922, 1952; O'Rourke, 1950), *Camponotus* (Melander and Brues, 1906; Stumper, 1922, 1952; Hermann and Blum, 1968), *Cataglyphis* (Stumper, 1922,1952), *Plagiolepis* (Stumper, 1952), *Polyergus* (Stumper, 1951, 1952), and *Acanthomyops* (Regnier and Wilson, 1968).		in the secretions of species in other subfamilies. The venom may contain solutions of at least 60% formic acid and amounts of 2 mg/ant have been reported. It is the only volatile organic compound detected in the venom and its penetration through the arthropod cuticle is facilitated by the Dufour's gland products which are discharged in admixture with the venom.

(continued)

Table 6.2 (Continued)

Acetic acid CH_3COOH $C_2H_4O_2$ MW 60	Uropygi Thelyphonidae: *Mastigoproctus giganteus* (Eisner *et al.*, 1961).	Thelyphonidae: Paired glands in the opisthosoma opening on the tip of the postabdomen.	The thelyphonid secretion contains an 84% solution of acetic acid which is accompanied by caprylic acid, a cuticular wetting agent.
	Polydesmida Polydesmidae: *Polydesmus collaris* (Casnati *et al.*, 1963).	Polydesmidae: Lateral paired glands opening dorsolaterally on the anterior half of most diplosegments.	It is a minor constituent in the benzaldehyde-rich polydesmid secretion.
	Hemiptera Coreidae: *Amorbus rubiginosus*, *A. alternatus*, *A. rhombifer*, *Mictis profana*, *M. caja*, *Aulacosternum nigrorubrum*, *Pachycolpura manca*, *Agricoporis frogatti* (Waterhouse and Gilby, 1964), *Acanthocephala granulosa* (McCullough, 1966a), *Leptoglossus oppositus* (McCullough, 1968), *L. clypeatus* (McCullough, 1969), *L. clypealis*, *Euthochtha galeator*, *Archimerus alternatus* (Aldrich and Yonke, 1975), *Amblypelta nitida* (Baker *et al.*, 1972), *Mozena obtusa* (McCullough, 1973), *M. lunata* (McCullough, 1974a), and *Chelinidea vittiger* (McCullough, 1974b). Hyocephalidae: *Hyocephalus* sp. (Waterhouse and Gilby, 1964).	Coreidae: Adult metasternal scent gland. Hyocephalidae: Adult metasternal scent gland.	Coreid secretions typically contain about 5% of this acid as well as acetate esters. Its concentration may increase with age since overwintered bugs of *L. clypeatus* are reported to contain more acid than first-generation summer individuals. Its presence in the hyocephalid secretion is consistent with the placement of this family in the coreoid complex.
	Coleoptera Carabidae: *Castelnaudia*	Carabidae: Pygidial glands. Papilionidae: The osmeterium, a gland which opens on the mid-dorsal line as a two-pronged invagination of the neck membrane in larvae. Formicidae: Obtained from extracts of whole ants but probably derived from the anal glands of *L. microcephalum*.	It is a trace constituent in the papilionid secretion along with several other acids which are atypical constituents in osmeterial exudates. Its occurrence in the secretions of a few

	superba (Moore and Wallbank, 1968), *Anthia thoracica*, *Thermophilum homoplatum*, and *T. burchelli* (Scott *et al.*, 1975).		carabid species appears to be unusual since ground beetles typically produce formic or C_4-C_6 acids in their pygidial glands.
	Lepidoptera Papilionidae: *Papilio anchisiades* (López and Quesnel, 1970).		
	Hymenoptera Formicidae: *Myrmicaria natalensis* (Quilico *et al.*, 1962) and *Liometopum microcephalum* (Casnati *et al.*, 1964).		
n-Propionic acid 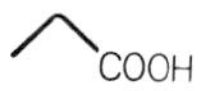 $C_3H_6O_2$ MW 74	Lepidoptera Papilionidae: *Papilio anchisiades* (López and Quesnel, 1970).	Papilionidae: The osmeterium, a gland which opens on the mid-dorsal line as a two-pronged invagination of the neck membrane in larvae.	In the papilionid secretion it is one of three fatty acids which are present as minor concomitants of the main glandular products.
	Hymenoptera Formicidae: *Myrmicaria natalensis* (Quilico *et al.*, 1962).	Formicidae: Obtained from extracts of whole ants.	
Methacrylic acid 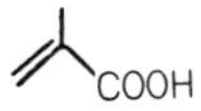 $C_4H_6O_2$ MW 86	Coleoptera Carabidae: *Calosoma sycophanta* (Casnati *et al.*, 1965), *C. marginalis*, *C. scrutator* (McCullough and Weinheimer, 1966), *C. schayeri*, *C. ocean-*	Pygidial glands.	This acid is a typical constituent in the defensive exudates of about 27 carabid genera in the subfamilies Carabinae, Broscinae,

(continued)

Table 6.2 (Continued)

	icum (Moore and Wallbank, 1968), *Callisthenes luxatus* (McCullough, 1972a), *Pamborus alternans*, *P. viridis*, *P. guerini*, *P. pradieri*, *Carenum bonellii*, *C. interruptum*, *C. tinctillatum*, *Laccopterum foveigerum*, *Philoscaphus tuberculatus*, *Eurylychnus blagravei*, *E. olliffi*, *Cratoferonia phylarchus*, *Trichosternus nudipes*, *Castelnaudia superba*, *Cratogaster melas*, *Prosopogmus harpaloides*, *Pseudoceneus iridescens*, *Rhytisternus laevilaterus*, *Abacomorphus asperulus* (Moore and Wallbank, 1968), *Carabus taedatus* (Benn *et al*., 1973), *C. auratus*, *C. granulatus*, *C. problematicus*, *Leistus ferrugineus*, *Nebria livida*, *Poecilus cupreus*, *Pterostichus niger*, *P. macer*, *P. vulgaris*, *P. melas*, *P. metallicus*, *Abax ater*, *A. parallelus*, *A. ovalis*, *Molops elatus*, *Amara similata*, *A. familiaris*, *Cychrus rostratus* (Schildknecht *et al*., 1968a), *Pasimachus elongatus*, (McCullough, 1972b), *Scaphonotus andrewsi*, *S. vidirus*, and *S. webbi* (Wheeler *et al*., 1970).	Scaritinae, Pterostichinae, Amarinae, and Nebriinae. The concentration in the glandular reservoirs often reaches 75-90% methacrylic acid. It is almost invariably accompanied by other acids although at least two species, *C. rostratus* and *P. harpaloides*, secrete pure methacrylic acid. Defensive exudates containing only acidic constituents consist of one-phase systems whereas if aldehydic compounds are present (e.g., *C. schayeri*), two-phase systems are present.

Crotonic acid 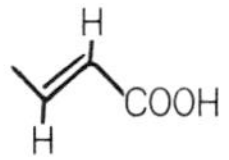 $C_4H_6O_2$ MW 86	Coleoptera Carabidae: Laccopterum foveigerum and Philoscaphus tuberculatus (Moore and Wallbank, 1968).	Pygidial glands.	It is an atypical and minor constituent in the secretions of two scaritine carabids which is accompanied by several other aliphatic acids.
Isocrotonic acid 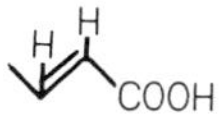 $C_4H_6O_2$ MW 86	Coleoptera Carabidae: Carenum bonellii, C. interruptum, C. tinctallatum, Laccopterum foveigerum, and Philoscaphus tuberculatus (Moore and Wallbank, 1968).	Pygidial glands.	The secretions of the Carenum spp., a taxon in the Scaritinae, contain about 43% of this acid whereas in the exudates of the other two species it is quantitatively less important.
n-Butyric acid  $C_4H_8O_2$ MW 88	Hemiptera Coreidae: Amorbus rhombifer (Waterhouse and Gilby, 1964) and Pternistria bispina (Baker and Kemball, 1967). Alydidae: Megalotomus quinquespinosus, Alydus eurinus, and A. pilosulus (Aldrich and Yonke, 1975).	Coreidae: Adult metasternal scent gland. Alydidae: Adult metasternal scent gland.	The defensive exudates of the coreid and alydid species contain minor amounts of this acid.
	Coleoptera Carabidae: Promecoderus sp. (Moore and Wallbank, 1968).	Carabidae: Pygidial glands.	It is an especially atypical product of carabid pygidial glands and has not been detected in the secretions of any other species in the Broscinae.
	Lepidoptera Papilionidae: Papilio anchisiades (López and Quesnel,	Papilionidae: The osmeterium, a gland which opens on the mid-dorsal line as a two-pronged invagination of the neck membrane in larvae.	The papilionid secretion contains trace amounts of this acid and it has been reported to occur occasionally as a trace constituent

(continued)

Table 6.2 (Continued)

	1970).		
	Hymenoptera Formicidae: *Liometopum microcephalum* (Casnati *et al*., 1964).	Formicidae: Obtained from extracts of whole ants but probably derived from the anal glands.	in two other papilionid secretions.
Isobutyric acid (structure: $(CH_3)_2CH$–COOH) $C_4H_8O_2$ MW 88	Hemiptera Reduviidae: *Rhodnius prolixus* (Pattenden and Staddon, 1972), *Triatoma phyllosoma*, and *Panstrongylus mictis* (Games *et al*., 1974).	Reduviidae: Brindley's glands, paired structures opening dorsolaterally in the region between the thorax and abdomen.	It appears to be a characteristic defensive product of reduviid species.
	Alydidae: *Megalotomus quinquespinosus* (Aldrich and Yonke, 1975).	Alydidae: Adult metasternal scent gland.	
	Coleoptera Carabidae: *Omophron limbatum*, *Notiophilus biguttatus*, *Elaphrus riparius*, *Loricera pilicornis*, *Broscus cephalotes*, *Bembidion lampros*, and *B*. *andreae* (Schildknecht *et al*., 1968a).	Carabidae: Pygidial glands.	The secretions of species in six carabid subfamilies are enriched with this acid but its absence in the exudates of most species in the Bembidiinae and Broscinae indicates that it is not an especially typical product in these subfamilies.
	Lepidoptera Papilionidae: *Baronia brevicornis*, *Eurytides marcellus* (Eisner *et al*., 1970), *Graphium carpedon* (Crossley and Waterhouse, 1969), *Papilio machaon* (Eisner and Meinwald, 1965), *P*. *cresphontes*, *P*.	Papilionidae: The osmeterium, a gland which opens on the middorsal line as a two-pronged invagination of the neck membrane in larvae.	Isobutyric acid constitutes about 50% of the osmeterial secretions of papilionid larvae in the Papilioniinae and Baroniinae. Its concentration is little affected by larval diet. The osmeterial secretion consists of a two-phase exudate and the denser aqueous phase contains
		Formicidae: Obtained from extracts of whole ants.	

	demodocus, P. glaucus, P. palamedes, P. polyxenes, P. troilus (Eisner et al., 1970), P. aegus, P. anactus, P. demoleus, (Crossley and Waterhouse, 1969), P. thaos, and P. anchisiades (López and Quesnel, 1970). Hymenoptera Formicidae: Myrmicaria natalensis (Quilico et al., 1962).		an appreciable quantity of this acid.
β-Hydroxy-n-butyric acid $C_4H_8O_3$ MW 104	Lepidoptera Papilionidae: Papilio aegus (Seligman and Doy, 1972).	The osmeterium, a gland which opens on the mid-dorsal line as a two-pronged invagination of the neck membrane in larvae.	It is present in the secretions of the early larval instars but could not be detected in that of the last larval instar. 2-Methylbutyric acid and isobutyric acid are the only acids detectable in the exudate of the ultimate larval instar.
2-Methylenebutyric acid $C_5H_8O_2$ MW 100	Dictyoptera Blattidae: Platyzosteria castanea, P. jungii, P. morosa, and P. ruficeps (Waterhouse and Wallbank, 1967).	Blattidae: A bilobed gland which opens on the midventral line between the sixth and seventh abdominal sternites.	The secretions of the Platyzosteria spp. contain about 0.4% of this acid which may represent an oxidation product of the main defensive constituent, 2-methylene butanal.

(continued)

Table 6.2 (Continued)

Compound	Source	Gland	Remarks
	Coleoptera Carabidae: *Pamborus alternans*, *P. viridis*, *P. guerini*, *P. pradieri*, *Laccopterum foveigerum* (Moore and Wallbank, 1968), and *Carabus taedatus* (Benn *et al.*, 1973).	Carabidae: Pygidial glands.	The defensive exudates of *Pamborus* spp. contain about 26% of this compound whereas it is a minor component in the secretion of the scaritine *L. foveigerum*. Its occurrence in the secretion of the carabine *C. taedatus* demonstrates that it is not restricted to the genus *Pamborus* in this subfamily.
Angelic acid 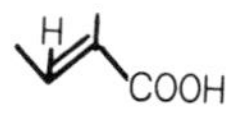 $C_5H_8O_2$ MW 100	Coleoptera Carabidae: *Carenum bonellii*, *C. interruptum*, *C. tinctillatum*, *Abacomorphus asperulus* (Moore and Wallbank, 1968), *Anthia thoracica*, *Thermophilum homoplatum*, and *T. burchelli* (Scott *et al.*, 1975).	Pygidial glands.	This compound appears to be a typical and major constituent in the defensive exudates of scaritine species in the genus *Carenum* in which it is always accompanied by isocrotonic and methacrylic acids. Although it is found in the secretion of the pterostichine *A. asperulus*, it is not typical of the exudates of species in this subfamily.

Tiglic acid (structure: $CH_3CH{=}C(CH_3)COOH$, H shown) $C_5H_8O_2$ MW 100	Coleoptera Carabidae: *Calosoma sycophanta* (Casnati *et al*., 1965), *Carenum tinctillatum*, *Laccopterum foveigerum*, *Philoscaphus tuberculatus*, *Eurylychnus blagravei*, *E. olliffi*, *Cratoferonia phylarchus*, *Trichosternus nudipes*, *Castelnaudia superba*, *Cratogaster melas*, *Pseudoceneus iridescens*, *Rhytisternus laevilaterus*, *Abacomorphus asperulus* (Moore and Wallbank, 1968), *Carabus auratus*, *C. granulatus*, *C. problematicus*, *Leistus ferrugineus*, *Nebria livida*, *Poecilus cupreus*, *Pterostichus niger*, *P. macer*, *P. vulgaris*, *P. melas*, *P. metallicus*, *Abax ater*, *A. parallelus*, *A. ovalis*, *Molops elatus*, *Amara similata*, *A. familiaris* (Schildknecht *et al*., 1968a), *Scaphonotus andrewsi*, and *S. webbi* (Wheeler *et al*., 1970).	Pygidial glands.	It is invariably accompanied by methacrylic acid in the pygidial gland reservoir and is never present in a greater concentration than the latter. It has been detected in species in 22 genera in six subfamilies and virtually never consists of more than 25% of the exudate. Ratios of methacrylic acid: tiglic acid range from 2:1 to 12:1. These two acids sometimes occur in the absence of other constituents (e.g., *Carabus*, *Abax*, *Nebria*) but they are frequently accompanied by alkanes in the secretions of *Pterostichus* and *Amara* species.

(*continued*)

Table 6.2 (Continued)

n-Pentanoic acid (Valeric acid)  $C_5H_{10}O_2$ MW 102	Coleoptera Carabidae: *Bembidion quadriguttatum* and *Asphidion flavipes* (Schildknecht *et al*., 1968a).	Pygidial glands.	The defensive exudates of these two bembidiine species are also enriched with alkanes and an aromatic aldehyde and probably exist as a two-phase system.
2-Methylbutyric acid 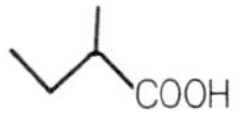 $C_5H_{10}O_2$ MW 102	Hemiptera Alydidae: *Megalotomus quinquespinosus* (Aldrich and Yonke, 1975). Lepidoptera Papilionidae: *Baronia brevicornis*, *Eurytides marcellus* (Eisner *et al*., 1970), *Graphium carpedon* (Crossley and Waterhouse, 1969), *Papilio machaon* (Eisner and Meinwald, 1965), *P. cresphontes*, *P. demodocus*, *P. glaucus*, *P. palamedes*, *P. polyxenes*, *P. troilus* (Eisner *et al*., 1970), *P. aegus*, *P. anactus*, *P. demoleus* (Crossley and Waterhouse, 1969), *P. thaos*, and *P. anchisiades* (López and Quesnel, 1970).	Alydidae: Adult metasternal gland. Papilionidae: The osmeterium, a gland which opens on the mid-dorsal line as a two-pronged invagination of the neck membrane in larvae.	Along with isobutyric acid, it constitutes one of the major constituents in the osmeterial secretions of larvae in the papilionid subfamilies Papilioninae and Baroniinae. Its concentration in the aqueous portion of the two-phase secretion is considerably lower than the more water-soluble isobutyric acid.

3-Methylbutyric acid (Isovaleric acid) $C_5H_{10}O_2$ MW 102	Polydesmida Polydesmidae: *Polydesmus collaris* (Casnati *et al*., 1963).	Polydesmidae: Lateral paired glands opening dorsolaterally on the anterior half of most diplosegments.	It is the major fatty acid present in the polydesmid secretion.
	Coleoptera Carabidae: *Promecoderus* sp., *Eurylychnus olliffi* (Moore and Wallbank, 1968), *Omophron limbatum*, *Notiophilus biguttatus*, *Elaphrus riparius*, *Loricera pilicornis*, *Broscus cephalotes*, *Bembidion lampros*, and *B. andreae* (Schildknecht *et al*., 1968a).	Carabidae: Pygidial glands.	In the carabid secretions it occurs as a concomitant of isobutyric acid in species in six subfamilies. However, two species in one of these subfamilies (Broscinae) eject secretions which contain this acid as a minor constituent in the absence of isobutyric acid.
	Lepidoptera Papilionidae: *Papilio demoleus* (Crossley and Waterhouse, 1969).	Papilionidae: The osmeterium, a gland which opens on the mid-dorsal line as a two-pronged invagination of the neck membrane in larvae.	It is a trace constituent in the papilionid secretions which are dominated by two other methyl-branched acids.
	Hymenoptera Formicidae: *Myrmicaria natalensis* (Quilico *et al*., 1962), *Liometopum microcephalum* (Casnati *et al*., 1964), and *Iridomyrmex nitidus* (Cavill, 1970).	Formicidae: The anal glands of workers of *I. nitidus*. Obtained from extracts of whole ants of *L. microcephalum* and *M. natalensis* and probably derived from the anal glands of the former.	

(*continued*)

Table 6.2 (Continued)

trans-2-Hexenoic acid $C_6H_{10}O_2$ MW 114	Dictyoptera Blattidae: *Polyzosteria limbata*, *P. oculata*, *P. cuprea*, and *P. pulchra* (Wallbank and Waterhouse, 1970).	Blattidae: A bilobed gland which opens on the midventral line between the sixth and seventh abdominal sternites.	It is a minor component in the polyzosteriine secretions and may represent an oxidation product of the main defensive compound, *trans*-2-hexenal.
	Coleoptera Carabidae: *Carenum bonellii* and *Laccopterum foveigerum* (Moore and Wallbank, 1968).	Carabidae: Pygidial glands.	It is a major constituent in the secretion of the scaritine carabid *L. foveigerum*. Although the position of the double bond has not been established for the carabid compound, it probably contains α,β-unsaturation, as do the other carabid alkenoic acids.

n-Hexanoic acid (Caproic acid) COOH $C_6H_{12}O_2$ MW 116	Hemiptera Coreidae: *Pternistria bispina* (Baker and Kemball, 1967) and *Amblypelta nitida* (Baker *et al*., 1972). Alydidae: *Alydus eurinus* and *A. pilosulus* (Aldrich and Yonke, 1975). Coleoptera Carabidae: *Calosoma schayeri*, *C. oceanicum*, *Laccopterum foveigerum*, and *Promecoderus* sp. (Moore and Wallbank, 1968).	Coreidae: Adult metasternal scent gland of *P. bispina* and *A. nitida* and dorsal abdominal glands of fifth instar larvae of *A. nitida*. Carabidae: Pygidial glands.	It is a minor constituent in the hemipterous secretions. In the secretions of *L. foveigerum* and *Promecoderus* sp., it constitutes 80% of the secretion.
4-Methylpentanoic acid (Isocaproic acid) COOH $C_6H_{12}O_2$ MW 116	Coleoptera Carabidae: *Philoscaphus tuberculatus* (Moore and Wallbank, 1968).	Pygidial glands.	It is a minor constituent in a secretion dominated by four alkenoic acids.
β-Hydroxyhexanoic acid COOH OH $C_6H_{12}O_3$ MW 132	Hymenoptera Formicidae: *Atta sexdens* (Schildknecht and Koob, 1971).	Paired metapleural glands of adult workers.	The secretion of this myrmicine species is dominated by two other β-hydroxy acids. It is present at a concentration of less than 0.1 μg/gland but is absent from the secretions of other myrmicine species.

(continued)

Table 6.2 (Continued)

Gluconic acid $HOH_2C-CH(OH)-CH(OH)-CH(OH)-CH(OH)-COOH$ $C_6H_{12}O_7$ MW 196	Dictyoptera Blattidae: Eurycotis decipiens, E. floridana, E. biolleyi (Dateo and Roth, 1967a,b), Polyzosteria limbata, and P. viridissima (Wallbank and Waterhouse, 1970).	A bilobed gland which opens on the midventral line between the sixth and seventh abdominal sternites.	This acid is present in the aqueous layer of a two-phase secretion and constitutes about a 6.5% solution. It is in equilibrium with its γ- and Δ-lactones.
Benzoic acid C_6H_5COOH $C_7H_6O_2$ MW 122	Polydesmida Paradoxsomatidae: Oxidus gracilis (Duffey et al., 1977) and Orthomorpha coarctata (Monteiro, 1961).	Paradoxsomatidae: Lateral paired glands opening dorsolaterally on the anterior half of most diplosegments.	It is a minor constituent in the secretion of O. gracilis but is one of three major aromatic compounds in the exudates of C. georgiana and C. rileyi. It is accompanied by benzaldehyde and benzoyl cyanide in the xystodesmid secretions.
	Gomphodesmidae: Gomphodesmus pavani (Barbetta et al., 1966).	Gomphodesmidae: Lateral paired glands opening dorsolaterally on the anterior half of most diplosegments.	
	Xystodesmidae: Cherokia georgiana, Sigmoria nantalahae, Cleptoria rileyi, Brachoria sp., and Motyxia tiemanni (Duffey et al., 1977).	Xystodesmidae: Lateral paired glands opening dorsolaterally on the anterior half of most diplosegments.	
	Polydesmidae: Polydesmus collaris (Barbetta et al., 1966).	Polydesmidae: Lateral paired glands opening dorsolaterally on the anterior half of most diplosegments.	
	Coleoptera Dytiscidae: Dytiscus marginalis (Schildknecht and Weis, 1962a), D. latissimus, Acilius sulcatus, Graphoderus cinereus, Rhantus exoletus, Colymbetes fuscus, Agabas sturmi, Ilybius fenestratus, Cybister lateralimarginalis (Schildknecht, 1971), and C. tripunctatus (Chadha et al., 1970).	Dytiscidae: Pygidial glands.	The pygidial gland secretion of D. marginalis contains 11% of this acid which has a widespread distribution in dytiscid exudates.

p-Hydroxybenzoic acid 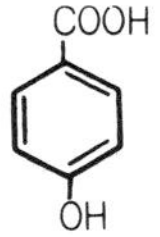 $C_7H_6O_3$ MW 138	Coleoptera Dytiscidae: Graphoderus cinereus, Rhantus exsoletus, Colymbetus fuscus, and Ilybius fenestratus (Schildknecht, 1971).	Pygidial glands.	It is a minor and not especially typical component of dytiscid secretions and was not detected as a pygidial gland product of 11 other species. It is reported to play a role as an antibacterial agent but its role as a defensive product per se has not been evaluated.
Phenylacetic acid 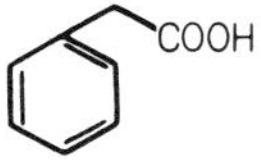 $C_8H_8O_2$ MW 136	Hymenoptera Formicidae: Atta sexdens (Schildknecht and Koob, 1970), Manica (=Myrmica) rubida, Myrmecina graminicola, Tetramorium caespitum, Harpoxenus sublaevis (Maschwitz et al., 1970), Myrmica rubra (=laevinodis), and Messor barbarus (Schildknecht and Koob, 1971).	Paired metapleural glands of adult workers.	It has a wide distribution among species in the subfamily Myrmicinae. It is presumed to be a bactericide based on in vitro tests but its exact role in the colonial milieu has not been established.
2,4-Dimethyl-2-hexenoic acid 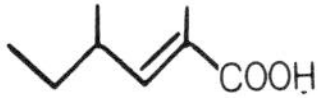 $C_8H_{14}O_2$ MW 142	Hymenoptera Formicidae: Camponotus nearcticus and C. rasilis (Brand et al., 1973d).	Mandibular glands of adult males.	This compound is only known as a natural product because of its production by these two formicine species. It probably functions as a sex-specific phermone as well as a defensive compound.

(continued)

Table 6.2 (Continued)

n-Octanoic acid (Caprylic acid) $C_8H_{16}O_2$ MW 144	Uropygi Thelyphonidae: *Mastigoproctus giganteas* (Eisner *et al.*, 1961).	Thelyphonidae: Paired glands on the opisthosoma opening on the tip of the postabdomen.	The thelyphonid secretion contains 5% caprylic acid and 84% acetic acid. The former compound acts as a cuticular wetting agent enabling the more hydrophilic C_2 acid to penetrate rapidly.
	Coleoptera Tenebrionidae: *Eleodes longicollis* (Y. (Meinwald and Eisner, 1964; Tschinkel, 1975b), *E. acuticauda*, *E. ubscura*, *E. grandicollis*, *E. femorata*, *E. gracilis*, *E. laticollis*, *E. hispilabris*, *E. sponsa*, *E. gigantea*, *E. dissimilis*, *E. metablapylis*, *Embaphion muracatum*, *E.* spp., and *Neobaphion* spp. (Tschinkel, 1975b).	Tenebrionidae: Paired abdominal sternal glands.	In the tenebrionid secretions this acid is limited to a few genera in the tribe Eleodini. It is a minor constituent in the secretions of *Eleodes* spp. but constitutes about 20% of the discharges of *Embaphion* and *Neobaphion* spp. It is often lacking in the secretions of *Eleodes* spp. in which it had been previously identified.
	Hymenoptera Formicidae: *Camponotus clarithorax* (Lloyd *et al.*, 1975).	Formicidae: Mandibular glands of adult males.	
β-Hydroxyoctanoic acid $C_8H_{16}O_3$ MW 160	Hymenoptera Formicidae: *Atta sexdens* (Schildknecht and Koob, 1971).	Paired metapleural glands of adult workers.	It is present at a concentration of 0.5-5.0 μg/gland but is lacking in the secretions of other myrmicine species.

n-Nonanoic acid  COOH $C_9H_{18}O_2$ MW 158	Hymenoptera Formicidae: *Camponotus clarithorax* (Lloyd *et al.*, 1975).	Mandibular glands of adult males.	It is one of four acids in this multicomponent secretion.
Dihydromatricaria acid  COOH $C_{10}H_{10}O_2$ MW 162	Coleoptera Cantharidae: *Chauliognathus lecontei* (Meinwald *et al.*, 1968b) and *C. pulchellus* (Moore and Brown, 1978).	Paired glands opening on the lateral margins of the prothorax and first eight abdominal segments of adults.	This compound is the first acetylenic acid to be identified as an animal natural product. Since the adults feed on plants in the family Compositae, a known source of methyl esters of acetylenic acids, it is possible that this compound is derived from an ingested constituent.
Geranic acid  COOH $C_{10}H_{16}O_2$ MW 168	Hymenoptera Formicidae: *Camponotus clarithorax* (Lloyd *et al.*, 1975).	Mandibular glands of adult males.	The secretion also contains two geranoate esters.

(continued)

Table 6.2 (Continued)

Citronellic acid  $C_{10}H_{18}O_2$ MW 170	Hymenoptera Formicidae: *Camponotus clarithorax* (Lloyd *et al*., 1975).	Mandibular glands of adult males.	It is one of two terpene acids in the secretion.
10-Hydroxy-*trans*-2-decenoic acid  $C_{10}H_{18}O_3$ MW 186	Hymenoptera Apidae: *Apis mellifera* (Butenandt and Rembold, 1957).	Mandibular glands of adult workers.	This compound is a major constituent in royal jelly and has been reported to be an effective inhibitor of bacterial and fungal growth (Blum *et al*., 1959).
β-Hydroxydecanoic acid  $C_{10}H_{20}O_3$ MW 188	Hymenoptera Formicidae: *Atta sexdens*, *Messor barbarus*, *Myrmica rubra* (=*laevinodis*), and *Acromyrmex* sp. (Schildknecht and Koob, 1971).	Paired metapleural glands of adult workers.	This acid is the major constituent present in the secretion and has been assigned the trivial epithet myrmicacin. Workers of *A. sexdens* produce 2.0–3.5 μg/gland of the levorotatory D-acid. It is believed to be continuously secreted into the nest where it inhibits the growth of foreign fungi. This compound has been recently identified in the fungus garden of *A. sexdens* but at physiological concen-

β-Hydroxydecanoic acid (con't)			trations it appears to function as a fungal growth regulator rather than a fungicide as previously suggested.
10-Methyldodecanoic acid  $C_{13}H_{26}O_2$ MW 214	Hymenoptera Formicidae: *Camponotus ligniperda*, *C. herculeanus*, *C. pennsylvanicus*, and *C. noveboracensis* (Brand *et al.*, 1973e).	Mandibular glands of adult males.	This compound constitutes a major constituent in the secretions of *C. herculeanus* and *C. pennsylvanicus*, two closely related species.
4-Hydroxyoctadec-9-enoic acid  $C_{18}H_{34}O_3$ MW 298	Hymenoptera Formicidae: *Lasius flavus* (Bergström and Löfqvist, 1970).	Dufour's gland of workers.	It is a trace constituent in the secretion which is accompanied by its corresponding γ-lactone.

cluding that these substances are, in fact, normally utilized as antibiotic agents, or for that matter, are not effective deterrents for animal predators (Hepburn *et al.*, 1973). Thus, although several of the carboxylic acids that are described will be treated as defensive substances because of their reported activities against microorganisms, it is not meant to imply that they have been evolved to function primarily, if at all, as antibiotic agents.

It is not unlikely that some of the carboxylic acids identified as allomones may be produced autoxidatively from aldehydic compounds. For example, benzoic acid, a common defensive compound of polydesmid millipedes, probably constitutes an autoxidation product of benzaldehyde, a major exocrine product of these arthropods. The frequent occurrence in arthropod defensive secretions of minor quantities of acids corresponding to the major aldehydic products suggests that the former may be autoxidatively generated from the latter.

AN OVERVIEW

At a glance it is evident that, when it comes to carboxylic acids, arthropods have evolved relatively short-chain compounds to function as defensive allomones. The virtuosity of these invertebrates as synthesizers of short-chain acids is illustrated by the fact that most of the more than 40 identified acids are in the restricted range C_1–C_{10}. For the most part, these acids are quite volatile and presumably the majority of them should constitute potent olfactants that will be highly stimulatory when they impinge on the chemoreceptors of potential predators. However, notwithstanding both their widespread distribution in arthropod defensive exudates and their proved effectiveness as predator deterrents in a few cases, these compounds have been emphasized by species in relatively few arthropod orders.

The distribution of carboxylic acids in the pygidial gland secretions of carabid beetles illustrates the scattered distribution of these compounds within species in a family. Of the 15 acids identified in carabid secretions, only three compounds, formic, methacrylic, and tiglic acids, appear to have fairly widespread distributions (Table 6.2). The remaining acids either constitute idiosyncratic products of selected species or represent minor or trace constituents in their defensive exudates. Similarly, in the Polydesmida, only one (i.e., benzoic) of the eight identified acids identified in this order (Table 6.2 and Chapter 18) is a rather widespread and major constituent. The same is true of the allomones characterized in lepidopterous secretions. Although formic acid may be a hallmark of notodontid secretions, only two of the eight acids identified in papilionid osmeterial secretions, isobutyric and 2-methylbutyric acids, occur as relatively widespread and major glandular products.

At this juncture, ants constitute the most prolific acid producers among the arthropods although, as is the case for other groups of these invertebrates, only one acid (formic) is a major and relatively common allomonal product. Nearly half of the acids identified in arthropod secretions is produced in the exocrine glands of ants. Furthermore, more than 50% of these ant-derived compounds are limited in their distribution to the exudates of formicids (Table 6.2) and, in many cases, these allomones are only known as products of a single species. Ants are further distinguished by producing the only monoterpenoid acids (e.g., neric, geranic) identified in arthropod defensive secretions as additional testimony for their well-developed ability to biosynthesize terpenes in their mandibular glands. That the synthesis of distinctive acidic allomones is not restricted to the mandibular glands of these insects is illustrated by the identification of β-hydroxy acids in the metapleural gland secretions of myrmicine ants (Schildknecht and Koob, 1971).

Although aromatic acids are produced by arthropod species in three orders, Polydesmida, Geophilomorpha, and Insecta, these compounds constitute consistently major products of beetles only in the family Dytiscidae (Table 6.2 and Chapter 18). Indeed, all six aromatic acids identified as arthropod natural products have been characterized as pygidial gland products of dytiscids, and four of these compounds are limited in their distribution to the exudates of these beetles. Benzoic acid, which is a characteristic product of these beetles, is also produced by millipedes and centipedes as part of a defensive secretion containing metabolically related compounds such as benzaldehyde. On the other hand, with the exception of phenylacetic acid, a metapleural glandular product of myrmicine ants, aromatic acids have not been detected as defensive compounds of any insects except dytiscid beetles. The prowess of these beetles as producers of aromatic acids is not necessarily unexpected, since among the insects coleopterans are preeminent in the synthesis of aromatic allomones.

Some of the carboxylic acids have such limited arthropod distributions that they may possess some value as potential character states for selected taxa. For example, dihydromatricaria acid, the only acetylenic acid identified in arthropod defensive exudates, has been identified as a product of only a few cantharid beetles (Meinwald *et al.*, 1968b; Moore and Brown,

1978). A few species of scaritine carabids produce acids that appear to have a very limited distribution in the Arthropoda. Crotonic, isocrotonic, and angelic acids, which are restricted in their arthropod distribution to the exudates of scaritine species (Moore and Wallbank, 1968), constitute very distinctive allomonal products for their producers. These novel acidic allomones would seem to constitute natural product trademarks of the selected coleopterans that generate them.

Chapter 7

1,4-Quinones and Hydroquinones

Excluding hydroquinones, fifteen 1,4-quinones have been identified in arthropod defensive secretions in the classes Arachnida, Diplopoda, and Insecta (Table 7.1) (see also Chapter 18). Although these compounds are produced by a large variety of species belonging to these three classes, from a qualitative standpoint insects are clearly preeminent. More than 70% of the 1,4-quinones are produced by insects, and six of these compounds are limited in their arthropod distribution to the exudates of insect species. By contrast, four of the 1,4-quinones are restricted to arachnid (opilionid) exudates and only one compound, 2,3-dimethoxy-1,4-benzoquinone, is peculiar to millipede defensive secretions (Table 7.2). Opilionids share two of their 1,4-quinones, 2,3-dimethyl-1,4-benzoquinone and 6-methyl-1,4-naphthoquinone, with a few insect species, whereas millipedes share four of their five quinones with insects. No 1,4-quinone is common to all three arthropod classes.

Insects owe their dominant role as producers of quinones primarily to the biosynthetic versatility of beetles. The quinonoid repertoire of earwigs, cockroaches, termites, and grasshoppers appears to be very limited, whereas that of the beetles encompasses all 1,4-quinones known to be produced by insects (Table 7.2). The superiority of coleopterans as quinone biochemists is further illustrated by the fact these insects are the source of all of these compounds that are limited in their arthropod distribution to insect secretions. Furthermore, unlike harvestmen or mil-

Table 7.1

Distribution of 1,4-Quinones in the Arthropoda[a]

Class	Order	Family
Arachnida	Opiliones (Phalangida)	Gonyleptidae
		Cosmetidae
		Phalangiidae
Diplopoda	Julida	Nemasomidae
		Julidae
		Parajulidae
	Spirobolida	Atopetholidae
		Pachybolidae
		Rhinocricidae
		Spirobolidae
		Floridobolidae
	Spirostreptida	Cambalidae[b]
		Spirostreptidae
		Harpagophoridae
		Odontopygidae
		Choctellidae
Insecta	Dermaptera	Forficulidae
	Dictyoptera	
	Isoptera	Mastotermitidae
		Termitidae
	Blattaria	Blaberidae
		Blattidae[c]
	Orthoptera	Acrididae
	Coleoptera	Staphylinidae
		Carabidae
		Tenebrionidae[d]

[a] Also see Chapter 18.
[b] Formerly the order Cambalida.
[c] Quinones have been detected in the secretion of a *Deropeltis* sp. (Eisner, 1970).
[d] Includes the former family Alleculidae (Doyen, 1972).

lipedes, beetles produce quinones in a variety of glands and in the case of species in one family (Staphylinidae), at least two nonhomologous glands are the sites of syntheses of these compounds (Table 7.2). And only in the Coleoptera have some species (Carabidae) evolved an elegant crepitation mechanism to explosively discharge their hot 1,4-quinones. Indeed, to a large extent the quinonoid virtuosity of arthropods reflects the morphological and biochemical idiosyncrasies of beetles.

AN OVERVIEW

Although 1,4-quinones have been evolved as defensive compounds by arthropods in three different classes, it is evident that the species in each of the taxa possess characteristic quinonoid emphases. Opilionids in the families Cosmetidae and Gonyleptidae typically synthesize di- and trimethyl-1,4-benzoquinones, and only one of these compounds, 2,3-dimethyl-1,4-benzoquinone, has been infrequently detected in nonarachnid secretions (Table 7.2). While this quinone has been identified in a few tenebrionid (Eisner *et al.*, 1974a) and carabid (Eisner *et al.*, 1977) exudates, it is clearly not a typical defensive allomone of species in either family.

Opilionids also produce naphthoquinones, as evidenced by the detection of two of these compounds in the secretion of *Phalangium opilio* (Wiemer *et al.*, 1978), a species in a different suborder than the species producing 1,4-benzoquinones. However, the secretion of this phalangiid appears to be rather exceptional, since species in this family typically produce aliphatic ketones and aldehydes in their defensive glands (Jones *et al.*, 1977). 6-Methyl-1,4-naphthoquinone, one of the two quinones synthesized by *P. opilio,* has also been detected in the abdominal secretions of tenebrionid beetles in the genus *Argoporis* (Tschinkel, 1969), the species of which are unusual in producing bicyclic quinones in addition to the typical 1,4-benzoquinones of insects.

Millipedes characteristically generate defensive exudates that contain a combination of toluquinone and 2-methoxy-3-methyl-1,4-benzoquinone, often with 1,4-benzoquinone also being present as a trace constituent (Table 7.2). The latter compound is clearly only a typical allomone of arthropods in the millipede orders Spirobolida, Spirostreptida, and Julida, three closely related taxa. On the other hand, 2,3-dimethoxy-1,4-benzoquinone, a product of a few parajulid and nemasomid species (Weatherston and Percy, 1969; Weatherston and Cheesman, 1975), is a very uncharacteristic quinonoid product of millipedes or, for that matter, arthropods in general.

The quinonoid emphasis of spiroboloid millipedes should not obscure the fact that these invertebrates can also produce completely unrelated defensive allomones. The secretion of the spirobolid *Rhinocricus insulatus* contains (*E*)-2-dodecenal (Wheeler *et al.*, 1964) in addition to toluquinone, whereas the quinone-rich exudate of the parajulid *Oriulus delus* is fortified with *o*-cresol (Kluge and Eisner, 1971). These results

Table **7.2** *p*-Quinones and hydroquinones in arthropod defensive secretions

Name and Formula	Occurrence	Glandular Source	Comments
1,4-Benzoquinone (structure: para-benzoquinone ring, O=C₆H₄=O) $C_6H_4O_2$ MW 108	Julida Julidae: *Julus terrestris* (Béhal and Phisalix, 1901) and *Ophyiulus hewetti* (Duffey *et al*., 1976).	Julidae: Lateral paired glands opening dorsolaterally on the anterior half of most diplosegments.	It was originally called "la quinone" after its isolation from the secretion of *J. terrestris*. It is not an especially characteristic exocrine product of millipedes and usually constitutes a minor constituent. However, it is the only quinonoid compound present in the secretion of *S. castaneus*. In insect secretions this compound generally constitutes a minor constituent although it is the only quinone reported to be produced by the tenebrionid *T. obscurus* and the termites *O. badius* and *H. obscuriceps*. In the mastotermitid and termitid secretions the quinones accompany proteinaceous constituents and viscous, polymeric products are formed after the quinone-protein mixtures
	Parajulidae: *Uroblaniulus canadensis* (Weatherston and Percy, 1969), *Uroblaniulus* sp., *Bollmaniulus* sp., *Ptyoilus* sp., *Saiulus* sp., and *Tuniulus hewetti* (Duffey *et al*., 1976).	Parajulidae: Lateral paired glands opening dorsolaterally on the anterior half of most diplosegments.	
	Spirobolida Atopetholidae: *Atopetholus* sp. (Duffey *et al*., 1976).	Atopetholidae: Lateral paired glands opening dorsolaterally on the anterior half of most diplosegments.	
	Pachybolidae: *Trigoniulus lumbricinus* (Monro *et al*., 1962) and *Leptogoniulus naresi* (Duffey *et al*., 1976).	Pachybolidae: Lateral paired glands opening dorsolaterally on the anterior half of most diplosegments.	
	Rhinocricidae: *Eurhinocricus* spp. and *Rhinocricus monilicornis* (Duffey *et al*., 1976). Spirobolidae: *Narceus annularis* (Monro *et al*., 1962) and *N. americanus* (Duffey *et al*., 1976).	Rhinocricidae: Lateral paired glands opening dorsolaterally on the anterior half of most diplosegments.	
	Spirostreptida Cambalidae: *Cambala hubrichti* (Eisner *et al*., 1965) and *C.*		

annulata (Duffey *et al.*, 1976). Choctellidae: *Choctella cumminsi* (Duffey *et al.*, 1976). Spirostreptidae: *Orthoporus ornatus* (Duffey *et al.*, 1976) and *Spirostreptus castaneus* (Barbier and Lederer, 1957).	Spirobolidae: Lateral paired glands opening dorsolaterally on the anterior half of most diplosegments. Cambalidae: Lateral paired glands opening dorsolaterally on the anterior half of most diplosegments.	are exposed to air. Quinones have not been detected in some termitid species in the genera *Odontotermes* and *Macrotermes*.
Dictyoptera Mastotermitidae: *Mastotermes darwiniensis* (Moore, 1968). Termitidae: *Macrotermes carbonarius* (Maschwitz *et al.*, 1972), *Hypotermes obscuriceps* (Maschwitz and Tho, 1974), and *Odontotermes badius* (Wood *et al.*, 1975). Blaberidae: *Diploptera punctata* (Roth and Stay, 1958) and *Blaberus giganteus* (Weatherston and Gardiner, 1975).	Choctellidae: Lateral paired glands opening dorsolaterally on the anterior half of most diplosegments. Spirostreptidae: Lateral paired glands opening dorsolaterally on the anterior half of most diplosegments.	The secretion of the acridid *R. microptera* is not a typical quinonoid exudate of insects since it contains a diversity of compounds unrelated to 1,4-quinones. It is discharged from reactor glands as a hot secretion in carabid species in the subfamilies Brachininae (*Brachinus* spp.) and Ozaeninae (*Mystropomus* and *Arthropterus* spp.).
Orthoptera Acrididae: *Romalea microptera* (Eisner *et al.*, 1971a).	Acrididae: Paired glands evacuating through the mesothoracic spiracles.	It is not a typical product of species in the carabid genus *Chlaenius*, having been detected in the secretion of only one of five species.
Coleoptera Tenebrionidae: Species in the following genera: *Tenebrio* (Schildknecht and Weis, 1960a), *Blaps* (Schildknecht and Weis, 1960a; Schildknecht *et al.*, 1964; Ikan *et al.*, 1970;	Mastotermitidae: Cephalic gland of soldiers. Termitidae: Labial glands of soldiers.	It is generally a minor constituent in tenebrionid secretions although in those of *Eleodes longicollis*, *Alobates pennsylvanica*, *Zophobas ru-*

(*continued*)

Table 7.2 (Continued)

	Tschinkel, 1975b), *Zophobas* (Tschinkel, 1969, 1975b), *Scaurus* (Schildknecht *et al*., 1964), *Eleodes* (Chadha *et al*., 1961a; Tschinkel, 1975b), *Cratidus*, *Meracantha*, *Iphthimus*, *Alobates*, *Nyctobates*, *Neatus*, *Merinus*, *Polopinus* (Tschinkel, 1975b), and *Leichenum* (Happ, 1967).	Blaberidae: Tracheal glands leading to the second abdominal spiracles of larvae and adults. Tenebrionidae: Paired abdominal sternal glands. Also paired prothoracic glands of species in the genera *Diaperis* and *Tribolium*.	*gipes*, and *Iphthimus* spp., it constitutes more than 5% of the secretion. It is almost invariably present in the defensive exudates of tenebrionid species which secrete only quinones (*Iphthimus*, *Alobates*, *Polopinus*, *Zophobas*, and *Tenebrio* spp.).
	Carabidae: *Pheropsophus catoirei* (Schildknecht and Holoubek, 1961), *P. verticalis* (Moore and Wallbank, 1968), *Chlaenius vistitus*, *Callistus lunatus*, *Clivina fossor*, (Schildknecht *et al*., 1968a), *C. basalis*, *Mystropomus regularis* (Moore and Wallbank, 1968), *Brachinus crepitans* (Schildknecht, 1957), *B. explodens*, and *B. sclopeta* (Schildknecht and Holoubek, 1961).	Carabidae: Pygidial glands.	
	Staphylinidae: *Lomechusa strumosa* (Blum *et al*., 1971) and *Drusilla canaliculata* (Brand *et al*., 1973b).	Staphylinidae: The tergal gland, a dorsal structure opening between the sixth and seventh abdominal tergites.	

Hydroquinone $C_6H_6O_2$ MW 110	Spirobolida Rhinocricidae: *Rhinocricus holomelanus* (Duffey *et al.*, 1976). Orthoptera Acrididae: *Romalea microptera* (Eisner *et al.*, 1971a). Coleoptera Carabidae: *Brachinus crepitans* (Schildknecht and Holoubek, 1961) and *Blaps sulcata* (Ikan *et al.*, 1970). Staphylinidae: *Drusilla canaliculata* (Brand *et al.*, 1973b).	Rhinocricidae: Lateral paired glands opening dorsolaterally on the anterior half of most diplosegments. Acrididae: Paired glands evacuating through the mesothoracic spiracles. Carabidae: Pygidial glands. Staphylinidae: The tergal gland, a dorsal structure opening between the sixth and seventh abdominal tergites.	It has a widespread occurrence in arthropod secretions and serves as a precursor for 1,4-benzoquinone which it accompanies in these exudates. It is almost certainly present in all secretions in which 1,4-benzoquinone has been identified.
2-Methyl-1,4-benzoquinone (Toluquinone) $C_7H_6O_2$ MW 122	Julida Julidae: *Archiulus sabulosus* (Trave *et al.*, 1959), *Brachyiulus unilineatus*, *Cylindroiulus teutonicus* (Schildknecht and Weis, 1961), and *Ophyiulus hewitti* (Duffey *et al.*, 1976). Parajulidae: *Oriulus delus* (Kluge and Eisner, 1971), *Uroblaniulus* sp., *Bollmaniulus* sp., *Ptyoilus* sp., *Saiulus* sp., and *Tuniulus hewitti*	Julidae: Lateral paired glands opening dorsolaterally on the anterior half of most diplosegments. Parajulidae: Lateral paired glands opening dorsolaterally on the anterior half of most diplosegments.	It is generally a major constituent in the defensive secretions of virtually all diplopods and insects in which it has been detected. It is the only quinone present in the secretions of *P. laminatus* & *R. insulatus* although it is accompanied by an α, β-unsaturated aldehyde in the secretion of the

(continued)

Table 7.2 (Continued)

	(Duffey *et al*., 1976).		latter species. It is only lacking from the exudates of a few species of millipedes in the order Julida and has been detected in the secretions of numerous diplopods in the orders Julida, Spirobolida, and Spirostreptida. In the mastotermitid *M. darwiniensis* it is a trace constituent whereas it is generally a quantitatively important component of termitid secretions. It is the only quinone present in the secretions of the termitids *M. globicola* and two *Odontotermes* species. *M. carbonarius* is the only termitid species for which both 1,4-benzoquinone and 2-methyl-1,4-benzoquinone have been detected in a frontal gland secretion. No other quinone is reported to be present in the secretions of the tenebrionids *Morisia planta*, *Pimelia confusa*,
	Nemasomidae: *Blaniulus guttulatus* (Weatherston, 1971) and *Nopoilus minutus* (Weatherston and Cheesman, 1975).	Nemasomidae: Lateral paired glands opening dorsolaterally on the anterior half of most diplosegments.	
	Spirobolida		
	Atopetholidae: *Atopetholus* sp. (Duffey *et al*., 1976).	Atopetholidae: Lateral paired glands opening dorsolaterally on the anterior half of most diplosegments.	
	Floridobolidae: *Floridobolus penneri* (Monro *et al*., 1962).	Floridobolidae: Lateral paired glands opening dorsolaterally on the anterior half of most diplosegments.	
	Pachybolidae: *Pachybolus laminatus* (Barbier and Lederer, 1957), *Trigoniulus lumbricinus* (Monro *et al*., 1962), and *Leptogoniulus naresi* (Duffey *et al*., 1976).	Pachybolidae: Lateral paired glands opening dorsolaterally on the anterior half of most diplosegments.	
	Rhinocricidae: *Rhinocricus insulatus* (Wheeler *et al*., 1964), *R. varians* (Moussatché *et al*., 1969), *R. holomelanus*, *R. monilicornis*, and *Eurhinocricus* spp. (Duffey *et al*., 1976).	Rhinocricidae: Lateral paired glands opening dorsolaterally on the anterior half of most diplosegments.	
	Spirobolidae: *Chicobolus spinigerus*, *Narceus annularis*, *N. gordanus* (Monro *et al*., 1962), and *N. americanus* (Duffey *et al*., 1976).	Spirobolidae: Lateral paired glands opening dorsolaterally on the anterior half of most diplosegments.	
	Spirostreptida		
	Cambalidae: *Cambala hubrichti* (Eisner *et al*., 1965) and *C.*		

annulata (Duffey *et al*., 1976).
Choctellidae: *Choctella cumminsi* (Duffey *et al*., 1976).
Spirostreptidae: *Aulonopygus aculeatus* (Barbier, 1959), *Rhapidostreptus virgator*, *Spirostreptus castaneus* (Barbier and Lederer, 1957), *S. multisulcatus*, *Peridontopyge vachoni*, *P. aberrans* (Barbier, 1959), *P. conani*, *P. rubescens* (Smolanoff *et al*., 1975a), *Orthoporus flavior*, *O. punctilliger* (Eisner *et al*., 1965), *O. ornatus* (Duffey *et al*., 1976), *Doratogonus annulipes* (Eisner *et al*., 1965), *Collostreptus fulvus* (Perissé and Salles, 1970), *Archispirostreptus gigas* (Wood, 1974), and *A. tumuliporus* (Smolanoff *et al*., 1975a).
Odontopygidae: *Prionopetalum frundsbergi* and *P. tricuspis* (Wood, 1974).
Harpagophoridae: Unidentified sp. (Eisner *et al*., 1965).

Dermaptera
Forficulidae: *Forficula auricularia* (Schildknecht and Weis, 1960b).

Cambalidae: Lateral paired glands opening dorsolaterally on the anterior half of most diplosegments.

Spirostreptidae: Lateral paired glands opening dorsolaterally on the anterior half of most diplosegments.

Odontopygidae: Lateral paired glands opening dorsolaterally on the anterior half of most diplosegments.

Harpagophoridae: Lateral paired glands opening dorsolaterally on the anterior half of most diplosegments.

Forficulidae: Two pairs of glands opening dorsally on the posterior margins of the third and fourth abdominal tergites.

Mastotermitidae: Cephalic glands of soldiers.

and *Eleodes beameri*. The secretion of *E. beameri* is further distinguished by the presence of a variety of carbonyl compounds typical of those found in hemipterous secretions. Toluquinone and benzoquinone are present in almost all tenebrionid secretions and the former predominates in only about 11% of the species studied.

It is often the major volatile constituent present in the carabid secretions, which appear to lack nonquinonoid constituents, for the most part. Although it is produced by two species of *Clivina* in the carabid subfamily Scaritinae, species in other scaritine genera synthesize alkenoic acids in their pygidial glands. It is discharged as part of a hot secretion generated in the reactor gland of species in the

(continued)

Table 7.2 (Continued)

	Dictyoptera Mastotermitidae: *Mastotermes darwiniensis* (Moore, 1968). Termitidae: *Macrotermes carbonarius* (Maschwitz *et al*., 1972), *Microtermes globicola*, *Odontotermes redemanni*, and *O. praevalens* (Maschwitz and Tho, 1974).	Termitidae: Labial glands of soldiers.	carabid subfamilies Brachininae and Ozaeninae.
	Blaberidae: *Diploptera punctata* (Roth and Stay, 1958) and *Blaberus giganteus* (Weatherston and Gardiner, 1975).	Blaberidae: Tracheal glands leading to the second abdominal spiracles of larvae and adults.	The secretion of the staphylinid *L. strumosa* is dominated by it but those of species in two other genera in this family are not.
	Coleoptera Tenebrionidae: Species in the following genera: *Tenebrio* (Schildknecht and Weis, 1960a; Schildknecht *et al*., 1964; Tschinkel, 1975b), *Blaps* (Schildknecht and Weis, 1960a; Ikan *et al*., 1970; Tschinkel, 1975b), *Prionychus* (Schildknecht *et al*., 1964), *Gnaptor*, *Morisia*, *Pimelia* (Schildknecht and Weis, 1960a), *Diaperis* (Roth and Stay, 1958; Schildknecht *et al*., 1964; Tschinkel, 1975b), *Alphitobius* (Tseng *et al*., 1971; Tschinkel, 1975b), *Eleodes* (Blum and Crain, 1961; Chadha *et al*., 1961a; Tschinkel, 1975b), *Tribolium* (Alexander	Tenebrionidae: Paired abdominal sternal glands. Also paired prothoracic glands of species in the genera *Diaperis* and *Tribolium*. Carabidae: Pygidial glands. Staphylinidae: In *L. strumosa* and *D. canaliculata*, the tergal gland, a dorsal structure opening between the sixth and seventh abdominal tergites. In *Bledius* spp., the pygidial glands.	The tergal gland exudate secreted by *D. canaliculata* contains, in addition to *p*-quinones, large amounts of aliphatic aldehydes and hydrocarbons. The pygidial gland secretions of the *Bledius* spp. are dominated by a γ-lactone and the quinone is a relatively minor constituent produced by both species.

and Barton, 1943; Hackman et al., 1948; Loconti and Roth, 1953; Tschinkel, 1975b), Leichenum (Happ, 1967), Argoporis (Tschinkel, 1972, 1975b), Zophobas (Tschinkel, 1969, 1975b), Helops, Opatroides, (Schildknecht et al., 1964), Scaurus (Schildknecht et al., 1964; Tschinkel, 1975b), Gnathocerus, Uloma, Psorodes, Meracantha, Pyanisia, Schelodontes, Eurynotus, Gonopus, Trigonopus, Anomalipus, Melanopterus, Iphthimus, Alobates, Toxicum, Coelocnemis, Centronopus, Nyctobates, Opatrinus, Cibdelis, Neatus, Merinus, Polopinus, Zadenos, Eulabis, Apsena, Epantius, Amphidora, Cratidus, Trogloderus, Embaphion, Neobaphion, Lariversius, Notibius, Conibius, Blapstinus, Gonocephalum, Parastizopus, Bolitotherus, Eleates, Phaleria, Platydema, Metaclisia, and Neomida (Tschinkel, 1975b).
Carabidae: Brachinus crepitans (Schildknecht, 1957), B. explodens, B. sclopeta, Pheropsophus catoirei (Schildknecht and Holoubek, 1961), P. verticalis (Moore and Wallbank, 1968), Chlae-

(continued)

Table 7.2 (Continued)

	nius *vistitus*, *Callistus lunatus*, *Clivina fossor* (Schildknecht *et al.*, 1968a), *C. basalis*, *Mystropomus regularis*, and *Arthropterus* sp. (Moore and Wallbank, 1968). Staphylinidae: *Lomechusa strumosa* (Blum *et al.*, 1971), *Drusilla canaliculata* (Brand *et al.*, 1973b), *Bledius mandibularis*, and *B. spectabilis* (Wheeler *et al.*, 1972a).		
2-Methoxy-1,4-benzoquinone $C_7H_6O_3$ MW 138	Coleoptera Tenebrionidae: *Tribolium castaneum* (Loconti and Roth, 1953).	Paired abdominal sternal glands and paired prothoracic glands.	It is a trace constituent (ca. 0.001%) in the secretion and unlike the alkylquinones in the exudate, is not repellent to adult beetles.
2-Methylhydroquinone $C_7H_8O_2$ MW 124	Julida Julidae: *Archiulus sabulosus* (Schildknecht and Krämer, 1962). Rhinocricidae: *Rhinocricus holomelanus* (Duffey *et al.*, 1976). Dermaptera Forficulidae: *Forficula auricularia* (Schildknecht and Krämer, 1962).	Julidae: Lateral paired glands opening dorsolaterally on the anterior half of most diplosegments. Rhinocricidae: Lateral paired glands opening dorsolaterally on the anterior half of most diplosegments.	It is a precursor of 2-methyl-1,4-benzoquinone and probably accompanies this quinone in all secretions. In combination with hydroquinone, it constitutes a 10% aqueous solution which is stored in a reservoir until released into a reaction chamber where it is

Compound	Source	Gland	Remarks
	Coleoptera Tenebrionidae: *Tenebrio molitor* (Schildknecht and Krämer, 1962) and *Blaps sulcata* (Ikan *et al*., 1970). Carabidae: *Brachinus crepitans* (Schildknecht and Holoubek, 1961). Staphylinidae: *Drusilla canaliculata* (Brand *et al*., 1973b).	Forficulidae: Two pairs of glands opening dorsally on the posterior margins of the third and fourth abdominal tergites. Tenebrionidae: Paired abdominal sternal glands. Carabidae: Pygidial glands. Staphylinidae: The tergal gland, a dorsal structure opening between the sixth and seventh abdominal tergites.	oxidized by peroxidase in the presence of oxygen (*Brachinus* spp.). The oxidized quinols are then forcibly discharged under the pressure of the free oxygen that was enzymatically (catalase) generated from hydrogen peroxide which was present as a 25% solution along with the quinols.
2-Ethyl-1,4-benzoquinone 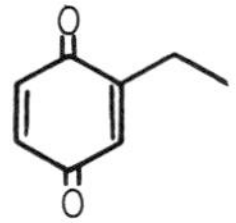 $C_8H_8O_2$ MW 136	Spirobolida Rhinocricidae: *Rhinocricus varians* (Moussatché *et al*., 1969). Dermaptera Forficulidae: *Forficula auricularia* (Schildknecht and Weis, 1960b). Dictyoptera Blaberidae: *Diploptera punctata* (Roth and Stay, 1958) and *Blaberus giganteus* (Wea-	Rhinocricidae: Lateral paired glands opening dorsolaterally on the anterior half of most diplosegments. Forficulidae: Two pairs of glands opening dorsally on the posterior margins of the third and fourth abdominal tergites.	Its identification in the millipede secretion was based on the R_f value of the quinol obtained after reduction of the presumed quinone with alkaline sodium dithionite. It does not predominate in the quinonoid secretions of the forficulid and blaberid species. In the coleopterous secretions it is generally

(continued)

Table 7.2 (Continued)

	therston and Gardiner, 1975).	Blaberidae: Tracheal glands leading to the second abdominal spiracles of larvae and adults.	either a trace or minor constituent in the defensive exudates. In tenebrionid species it is virtually always accompanied by *p*-toluquinone and an analysis of the ratios of these two quinones demonstrated that it predominated in the secretions of about 90% of the species. In several genera in the tribe Scaurini (e.g., *Eulabis*) it constitutes nearly 90% of the quinone mixture along with *p*-toluquinone. The secretions of species in the tribe Amarygmini appear to be unusual in that it is a distinctly minor component in comparison to *p*-toluquinone.
	Coleoptera		
	Tenebrionidae: Species in the following genera: *Tenebrio* (Schildknecht and Weis, 1960a; Schildknecht *et al*., 1964; Tschinkel, 1975b), *Blaps* (Schildknecht and Weis, 1960a; Schildknecht *et al*., 1964; Ikan *et al*., 1970; Tschinkel, 1975b), *Prionychus*, *Gnaptor* (Schildknecht *et al*., 1964), *Diaperis* (Roth and Stay, 1958; Schildknecht *et al*., 1964; Tschinkel, 1975b), *Alphitobius* (Tseng *et al*., 1971; Tschinkel, 1975b), *Eleodes* (Blum and Crain, 1961; Chadha *et al*., 1961a; Tschinkel, 1975b), *Tribolium* (Alexander and Barton, 1943; Hackman *et al*., 1948; Loconti and Roth, 1953; Tschinkel, 1975b), *Leichenum* (Happ,1967), *Argoporis* (Tschinkel, 1972, 1975b), *Helops*, *Opatroides* (Schildknecht *et al*., 1964), *Zophobas* (Tschinkel, 1969, 1975b), *Scaurus* (Schildknecht *et al*., 1964; Tschinkel, 1975b), *Gnathocerus*, *Uloma*, *Psorodes*, *Meracantha*, *Pyanisia*, *Schelodontes* *Eurynotus*, *Gono-*	Tenebrionidae: Paired abdominal sternal glands. Also paired prothoracic glands of species in the genera *Diaperis* and *Tribolium*.	
		Carabidae: Pygidial glands.	
		Staphylinidae: The tergal gland, a dorsal structure opening between the sixth and seventh abdominal tergites.	

Compound	Source	Gland	Remarks
	pus, *Trigonopus*, *Anomalipus*, *Melanopterus*, *Iphthimus*, *Alobates*, *Toxicum*, *Coelocnemis*, *Centronopus*, *Nyctobates*, *Opatrinus*, *Cibdelis*, *Neatus*, *Merinus*, *Polopinus*, *Zadenos*, *Eulabis*, *Apsena*, *Epantius*, *Amphidora*, *Cratidus*, *Troglo-derus*, *Embaphion*, *Neobaphion*, *Lariversius*, *Notibius*, *Coni-bius*, *Blapstinus*, *Gonoce-phalum*, *Parastizopus*, *Bolito-therus*, *Eleates*, *Phaleria*, *Platydema*, *Metaclisa*, and *Neomida* (Tschinkel, 1975b). Carabidae: *Mystropomus regularis* and *Arthropterus* sp. (Moore and Wallbank, 1968). Staphylinidae: *Lomechusa strumosa* (Blum *et al.*, 1971).		
2,3-Dimethyl-1,4-benzoquinone $C_8H_8O_2$ MW 136	Opiliones Gonyleptidae: *Heteropachyloidellus robustus* (Fieser and Ardao, 1956). Cosmetidae: *Vonones sayi* (Eisner *et al.*, 1971b). Spirobolida Rhinocricidae: *Rhinocricus varians* (Moussatche *ét al.*, 1969). Coleoptera Tenebrionidae: *Adelium perca-*	Gonyleptidae: Paired glands opening on the flanks of the prosoma between the first and second pair of coxae. Cosmetidae: Paired glands opening on the flanks of the prosoma between the first and second pair of coxae.	It is the major quinonoid constituent in the opilionid secretions and appears to be a characteristic exocrine product of harvestmen. It is accompanied by another dimethylquinone in the secretion of *H. robustus*. Its identification in the diplopod secretion was based on the R_f value of the quinol

(*continued*)

Table 7.2 (Continued)

Compound	Taxon (reference)	Gland	Remarks
	tum and *A. pustolosum* (Eisner *et al.*, 1974a).	Rhinocricidae: Lateral paired glands opening dorsolaterally on the anterior half of most diplosegments. Tenebrionidae: Paired abdominal sternal glands.	obtained by reduction of the free quinone. It is a particularly atypical constituent in tenebrionid secretions.
2,5-Dimethyl-1,4-benzoquinone $C_8H_8O_2$ MW 136	Opiliones Gonyleptidae: *Heteropachyloidellus robustus* (Fieser and Ardao, 1956).	Paired glands opening on the flanks of the prosoma between the first and second pair of coxae.	It constitutes about 14% of a quinonoid secretion dominated by the 2,3-dimethylquinone.
2-Methoxy-3-methyl-1,4-benzoquinone $C_8H_8O_3$ MW 152	Julida Julidae: *Archiulus sabulosus* (Trave *et al.*, 1959), *Brachyiulus unilineatus*, *Cylindroiulus teutonicus* (Schildknecht and Weis, 1961), and *Ophyiulus hewetti* (Duffey *et al.*, 1976). Parajulidae: *Uroblaniulus* sp., *Bollmaniulus* sp., *Ptyoilus* sp., *Saiulus* sp., and *Tuniulus hewetti* (Duffey *et al.*, 1976). Nemasomidae: *Blaniulus guttulatus* (Weatherston, 1971).	Julidae: Lateral paired glands opening dorsolaterally on the anterior half of most diplosegments. Parajulidae: Lateral paired glands opening dorsolaterally on the anterior half of most diplosegments. Atopetholidae: Lateral paired glands opening dorsolaterally on the anterior half of most diplosegments.	It is an especially characteristic constituent in the secretions of millipedes and has been reported to be absent from the exudates of only two rhinocricid species. In *Rhinocricus insulatus* it has been replaced by *trans*-2-dodecenal. In some millipede species it constitutes nearly 80% of a quinonoid mixture in combination with *p*-toluquinone. It has been suggested

Spirobolida		that its low melting point (29°C) is lowered by the presence of *p*-toluquinone, enabling it to function as a "solvent" in the defensive exudates. It has also been suggested that it may be biosynthesized from *p*-toluquinone in at least one millipede species. 6-Methylsalicylic acid has been demonstrated to be an excellent precursor for it in at least one species of millipede. Among tenebrionid species, it has only been unequivocally identified in species in the subgenus *Blapylis* of *Eleodes*. It is generally present in lower concentration than either *p*-toluquinone or 2-ethyl-1,4-benzoquinone in the secretions of *E*. (*Blapylis*) spp. It may also be present in the secretions of species in the tenebrionid genus *Amphidora*. It may not be typical
Atopetholidae: *Atopetholus* sp. (Duffey *et al*., 1976).		
Floridobolidae: *Floridobolus penneri* (Monro *et al*., 1962).	Floridobolidae: Lateral paired glands opening dorsolaterally on the anterior half of most diplosegments.	
Pachybolidae: *Trigoniulus lumbricinus* (Monro *et al*., 1962) and *Leptogoniulus naresi* (Duffey *et al*., 1976).	Pachybolidae: Lateral paired glands opening dorsolaterally on the anterior half of most diplosegments.	
Rhinocricidae: *Rhinocricus holomelanus*, *R*. *monilicornis*, and *Eurhinocricus* spp. (Duffey *et al*., 1976).	Rhinocricidae: Lateral paired glands opening dorsolaterally on the anterior half of most diplosegments.	
Spirobolidae: *Chicobolus spinigerus*, *Narceus annularis*, *N*. *gordanus* (Monro *et al*., 1962), and *N*. *americanus* (Duffey *et al*., 1976).	Spirobolidae: Lateral paired glands opening dorsolaterally on the anterior half of most diplosegments.	
Spirostreptida		
Cambalidae: *Cambala hubrichti* (Eisner *et al*., 1965) and *C*. *annulata* (Duffey *et al*., 1976).	Cambalidae: Lateral paired glands opening dorsolaterally on the anterior half of most diplosegments.	
Choctellidae: *Choctella cumminsi* (Duffey *et al*., 1976).	Choctellidae: Lateral paired glands opening dorsolaterally on the anterior half of most diplosegments.	
Spirostreptidae: *Doratogonus annulipes*, *Orthoporus conifer*, *O*. *flavior* (Eisner *et al*., 1969), *O*. *ornatus* (Duffey *et al*., 1976), *Archispirostreptus gigas* (Wood, 1974),		

(*continued*)

Table 7.2 (Continued)

	A. tumuliporus (Smolanoff *et al.*, 1975a), and *Collostreptus fulvus* (Perissé and Salles, 1970).	Spirostreptidae: Lateral paired glands opening dorsolaterally on the anterior half of most diplosegments.	of carabid species in the genus *Clivina* since it was not detected in another species in this taxon.
	Odontopygidae: *Prionopetalum frundsbergi* and *P. tricuspis* (Wood, 1974).	Odontopygidae: Lateral paired glands opening dorsolaterally on the anterior half of most diplosegments.	It constitutes the major quinone present in the tergal gland exudate of *D. canaliculata*.
	Harpagophoridae: Unidentified sp. (Eisner *et al.*, 1965).		
	Coleoptera		
	Tenebrionidae: *Eleodes blanchardi* and *E.* spp. (Tschinkel, 1975b).	Harpagophoridae: Lateral paired glands opening dorsolaterally on the anterior half of most diplosegments.	
	Carabidae: *Clivina fossor* (Schildknecht *et al.*, 1968a).	Tenebrionidae: Paired abdominal sternal glands.	
	Staphylinidae: *Drusilla canaliculata* (Brand *et al.*, 1973b).	Carabidae: Pygidial glands.	
		Staphylinidae: The tergal gland, a dorsal structure opening between the sixth and seventh abdominal tergites.	

2,3-Dimethoxy-1,4-benzoquinone 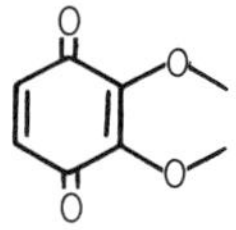 $C_8H_8O_4$ MW 168	Julida Parajulidae: *Uroblaniulus canadensis* (Weatherston and Percy, 1969). Nemasomidae: *Nopoilus minutus* (Weatherston and Cheesman, 1975).	Parajulidae: Lateral paired glands opening dorsolaterally on the anterior half of most diplosegments. Nemasomidae: Lateral paired glands opening dorsolaterally on the anterior half of most diplosegments.	It is accompanied by *p*-benzoquinone in the secretion of *U. canadensis* and *p*-toluquinone in the exudate of *N. minutus*.
2-Ethylhydroquinone 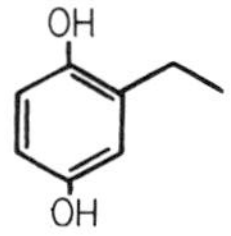 $C_8H_{10}O_2$ MW 138	Dermaptera Forficulidae: *Forficula auricularia* (Schildknecht and Krämer, 1962). Coleoptera Tenebrionidae: *Blaps sulcata* (Ikan *et al.*, 1970).	Forficulidae: Two pairs of glands opening dorsally on the posterior margins of the third and fourth abdominal tergites. Tenebrionidae: Paired abdominal sternal glands.	It represents the precursor of 2-ethyl-1,4-benzoquinone and is probably present in all secretions containing this 1,4-quinone.
2-Methoxy-3-methylhydroquinone 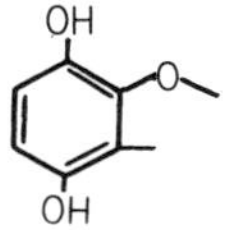 $C_8H_{10}O_3$ MW 154	Julida Julidae: *Archiulus sabulosus* (Schildknecht and Krämer, 1962). Coleoptera Staphylinidae: *Drusilla canaliculata* (Brand *et al.*, 1973b).	Julidae: Lateral paired glands opening dorsolaterally on the anterior half of most diplosegments. Staphylinidae: The tergal gland, a dorsal structure opening between the sixth and seventh abdominal tergites.	It is the precursor of 2-methoxy-3-methylbenzoquinone in these and other arthropod defensive secretions.

(continued)

Table 7.2 (Continued)

2-*n*-Propyl-1,4-benzoquinone 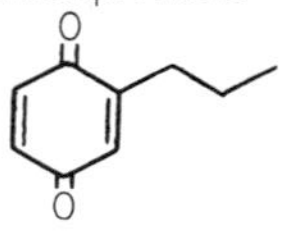 $C_9H_{10}O_2$ MW 150	Coleoptera Tenebrionidae: *Blapstinus* spp., *Conibius* spp., and *Notibius* spp. (Tschinkel, 1975b).	Paired abdominal sternal glands.	It has only been detected in genera in the tribe Pedinini. Species in the genus *Notibius* contain about 85% of this quinone and the normally present *p*-toluquinone is either absent or present as a trace constituent.
2,3,5-Trimethyl-1,4-benzoquinone 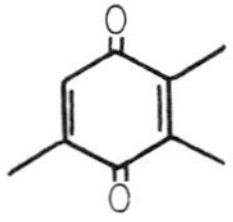 $C_9H_{10}O_2$ MW 150	Opiliones Gonyleptidae: *Heteropachyloidellus robustus* (Fieser and Ardao, 1956). Cosmetidae: *Vonones sayi* (Eisner *et al*., 1971b).	Gonyleptidae: Paired glands opening on the flanks of the prosoma between the first and second pair of coxae. Cosmetidae: Paired glands opening on the flanks of the prosoma between the first and second pair of coxae.	In both secretions it is accompanied by 2,3-dimethyl-1,4-benzoquinone. In the secretion of *H. robustus* it represents about 20% of the available quinones whereas in the exudate of *V. sayi* it varies from 25-50% of the available quinones.
6-Methyl-1,4-naphthoquinone 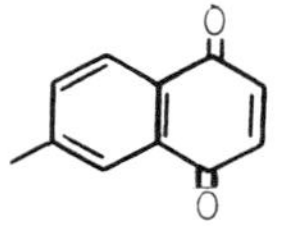 $C_{11}H_8O_2$ MW 172	Coleoptera Tenebrionidae: *Argoporis alutacea*, *A. costipennis*, *A. bicolor*, and *A. rufipes* (Tschinkel, 1969, 1975b).	Paired abdominal sternal glands.	It constitutes about 66% of a mixture of four naphthoquinones.

6-Ethyl-1,4-naphthoquinone 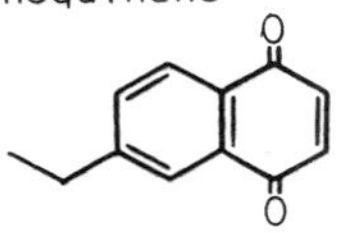 $C_{12}H_{10}O_2$ MW 186	Coleoptera Tenebrionidae: *Argoporis alutacea*, *A. costipennis*, *A. bicolor*, and *A. rufipes* (Tschinkel, 1969, 1975b).	Paired abdominal sternal glands.	It is the second-most concentrated naphthoquinone in the secretion (ca. 25%).
6-*n*-Propyl-1,4-naphthoquinone 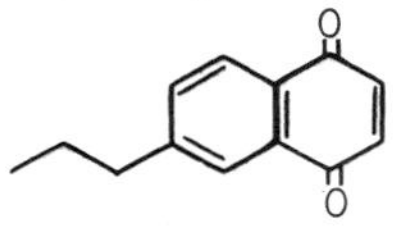 $C_{13}H_{12}O_2$ MW 200	Coleoptera Tenebrionidae: *Argoporis alutacea*, *A. costipennis*, *A. bicolor*, and *A. rufipes* (Tschinkel, 1969, 1975b).	Paired abdominal sternal glands.	It is a minor constituent in the secretion and is also accompanied by the usual tenebrionid 1,4-benzoquinones.
6-*n*-Butyl-1,4-naphthoquinone 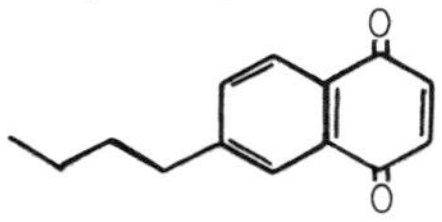 $C_{14}H_{14}O_2$ MW 214	Coleoptera Tenebrionidae: *Argoporis alutacea*, *A. costipennis*, *A. bicolor*, and *A. rufipes* (Tschinkel, 1969, 1975b).	Paired abdominal sternal glands.	Along with three other alkylnaphthoquinones, it is limited in its arthropod distribution to one genus in the tribe Scaurini.

stress the possibility that other species of millipedes in typical quinone-producing taxa may also produce natural product surprises.

The qualitative preeminence of insects as quinonoid producers is rather misleading, because most of the novel 1,4-quinones synthesized by species in the Insecta are products of one family, the Tenebrionidae. Earwigs, cockroaches, termites, grasshoppers, staphylinids and carabids typically synthesize a mixture of toluquinone–ethyl-1,4-benzoquinone in their exocrine glands, with 1,4-benzoquinone sometimes occurring as a trace concomitant (Table 7.2). Toluquinone has the most widespread distribution of any quinone in arthropod defensive secretions, with ethyl-1,4-benzoquinone, a typical insect product, being a close second. 2,3-Dimethyl-1,4-benzoquinone and 2-methoxy-3-methyl-1,4-benzoquinone are rarely present in carabid secretions and clearly represent atypical quinonoid allomones of species in this family. Tenebrionids, on the other hand, synthesize all the 1,4-quinones identified as insect natural products, and species in selected genera exhibit a quinonoid richness that is unmatched in the animal kingdom.

The excellent comparative investigation of Tschinkel (1975b) clearly documented the biosynthetic expertise of tenebrionids when it comes to 1,4-quinones. In addition to the typical 1,4-quinones produced by these beetles, such compounds as 2-propyl-1,4-benzoquinone and a variety of alkylated naphthoquinones are generated in the exocrine glands of certain species. Other species produce large amounts of 2-methoxy-3-methyl-1,4-benzoquinone or 2,3-dimethyl-1,4-benzoquinone (Table 7.2), typical allomones of millipedes and some opilionids, respectively. Qualitatively, the champion 1,4-quinone producer among arthropods may be tenebrionids in the genus *Argoporis*, species of which produce six quinones including two 1,4-benzoquinones and four 1,4-naphthoquinones (Tschinkel, 1969). It is worth noting that the identification of one of the novel tenebrionid quinones, 2-methoxy-1,4-benzoquinone (Loconti and Roth, 1953), has been questioned (Tschinkel, 1975b).

Although no defensive exudates of tenebrionid species have ever been reported to be quinone-free, these secretions sometimes contain considerable amounts of nonquinonoid constituents. Alkenes and octanoic acid are not infrequent quinonoid concomitants (Tschinkel, 1975b), and one species, *Eleodes beameri*, produces an incredible variety of carbonyl compounds that are more typical of a hemipterous secretion than a coleopterous one (Tschinkel, 1975a), while another produces novel lactones (Lloyd *et al.*, 1978). As is the case for some millipede secretions, nonquinonoid allomones may also be produced by tenebrionids, and their presence may possess considerable ecological significance, if the biology of the species is thoroughly illuminated, as stressed by Lloyd *et al.* (1978a).

The quinonoid conservatism of the opilionids and millipedes vis-à-vis the insects' expansiveness in producing these compounds may be explicable in terms of the biosynthetic pathways available to the species in these taxa. S. S. Duffey (personal communication, 1974) has demonstrated that the quinones synthesized by a spirobolid millipede are derived only from preformed aromatic precursors, a metabolic development that severely limits the quinonoid structural variation available to these arthropods. On the other hand, while at least some insects synthesize unsubstituted 1,4-benzoquinone from aromatic amino acids, alkylated 1,4-quinones are produced by the acetate–malonate pathway (J. Meinwald *et al.*, 1966a). The ability of insects to biosynthesize substituted 1,4-quinones from simple precursors could provide these invertebrates with the ability to produce a large variety of alkylated quinones provided that the requisite metabolic pathways had been evolved. If both arachnids and millipedes lack this biosynthetic flexibility, their quinonoid repertoires would be quite limited relative to those of insects. Therefore, the demonstrated preeminence of insects, and tenebrionids in particular, as synthesizers of 1,4-quinones may have its roots in a metabolic plasticity not available to arthropods in other taxa.

Chapter 8

Esters

The identification of esters (see also Chapter 18) in the defensive secretions of arthropod species in six orders (Table 8.1) demonstrates that this class of exocrine compounds has a very widespread distribution. The importance of these allomones in the chemical defenses of arthropods is illustrated by the fact that nearly 90 esters, in the range of C_5–C_{26}, have been characterized as exocrine products. About 70% of these compounds are in the range C_8–C_{18} and very short-chain compounds (C_5–C_7) occur infrequently in defensive exudates; most of the compounds above C_{18} are termitid diterpene esters. However, although these compounds have been detected in the glandular exudates of arthropods in 27 families (Table 8.1), the individual esters often have such a limited occurrence as to appear to be idiosyncratic defensive allomones. Indeed, the ordinal and familial distributions of these esters indicate that arthropods synthesize them with great selectivity.

Hymenopterans, particularly bees and ants, produce more than 40% of the esters that have been identified as defensive compounds. Esters are also important allomonal constituents of both the Hemiptera and Coleoptera, with more than 20% of these compounds being produced by true bugs and about 17% by adult and larval beetles. However, in general, species in each of these orders generate characteristic types of esters that reflect the biosynthetic peculiarities of their glandular tissues. The restricted distributions of esters in arthropod secretions is documented by the occurrence of 90% of the identified esters as products of species in only one order (Table 8.2). Furthermore, more than two-thirds of these allomones are limited to arthropods in single families, and more than a third of the esters are products restricted to a single species. Since only

Table 8.1

Distribution of Esters in the Defensive Secretions of the Arthropoda[a]

Class	Order	Family
Diplopoda	Julida	Parajulidae
	Polydesmida	Polydesmidae
		Paradoxsomatidae
		Gomphodesmidae
		Xystodesmidae
Insecta	Dictyoptera	
	Isoptera	Termitidae
	Hemiptera	Alydidae
		Belastomatidae
		Coreidae
		Cydnidae
		Hyocephalidae
		Naucoridae
		Notonectidae
		Pentatomidae
		Lygaeidae
	Coleoptera	Cerambycidae
		Carabidae
		Chrysomelidae
		Dytiscidae
	Lepidoptera	Notodontidae
		Cossidae
		Papilionidae
	Hymenoptera	Sphecidae
		Anthophoridae
		Andrenidae
		Apidae
		Formicidae

[a] Also see Chapter 18.

eight compounds are shared by two orders and only two compounds have been identified in more than two orders, it appears that allomonal esters have been evolved very selectively in the Arthropoda.

The nature of the acidic moieties of the esters further illustrates the distributional peculiarities that characterize these compounds in arthropod glandular exudates. Although nearly 50% of these esters are acetates (Table 8.2), species in a diversity of arthropod taxa have utilized a wide, and often selective, variety of acids to synthesize characteristic esters. For example, benzoates have only been identified in coleopterous

Table **8.2** Esters in arthropod defensive secretions

Name and Formula	Occurrence	Glandular Source	Comments
Methyl _p_-hydroxybenzoate 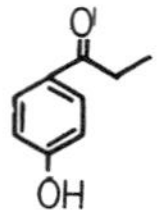 $C_8H_8O_3$ MW 152	Hemiptera Naucoridae: _Ilyocoris cimicoides_ (Staddon and Weatherston, 1967). Notonectidae: _Notonecta glauca_ (Pattenden and Staddon, 1968). Coleoptera Dytiscidae: _Dytiscus marginalis_ (Schildknecht and Weis, 1962a), _D. latissimus_, _Cybister lateralimarginalis_ (Schildknecht _et al._, 1964), _C. tripunctatus_, _Acilius sulcatus_, _Graphoderus cinereus_, _Rhantus exoletus_, _Colymbetes fuscus_, _Copelatus ruficollis_, _Agabus sturmi_, _A. bipustulatus_, and _Ilybius fenestratus_ (Schildknecht, 1970).	Naucoridae: Adult metasternal scent gland. Notonectidae: Adult metasternal scent gland. Dytiscidae: Pygidial glands.	It is accompanied by _p_-hydroxybenzaldehyde in the secretions of _I. cimicoides_, _N. glauca_, and 11 of 12 dytiscid species. In the two hemipterous secretions it comprises 30% of the volatiles whereas in _D. marginalis_ each beetle produces 0.18 μg of this ester or ca. 7% of the exudate. Schildknecht (1970) considers this compound to function as an antimicrobial agent although its role as a deterrent of invertebrate predators has not been evaluated. It is absent from the pygidial gland secretion of the dytiscid _Hydroporus palustis_ (Schildknecht _et al._, 1964).

Methyl salicylate 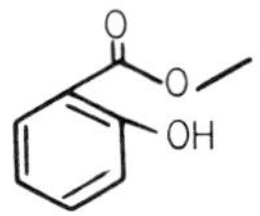 $C_8H_8O_3$ MW 152	Coleoptera Carabidae: *Idiochroma dorsalis* (Schildknecht *et al*., 1968c).	Pygidial glands.	It is an atypical compound in carabid defensive secretions which is accompanied by formic acid and hydrocarbons. The presence of this ester probably results in a two-phase secretion.
Methyl 3,4-dihydroxybenzoate (Methyl protocatechuate) 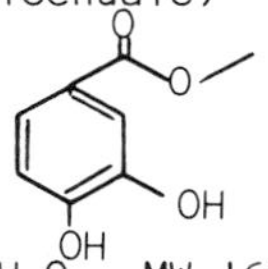 $C_8H_8O_4$ MW 168	Coleoptera Dytiscidae: *Cybister lateralimarginalis*, *Colymbetes fuscus*, and *Agabus bipustulatus* (Schildknecht, 1970).	Pygidial glands.	It is present in three of 11 dytiscid species, which are in the tribes Dytiscinae and Colymbetinae. The free acid has been isolated from the cuticle of some insect species and is a degradation product of tyrosine which is utilized in the sclerotization of the cuticle.
Methyl anthranilate 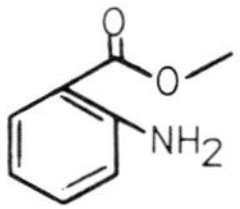 $C_8H_9NO_2$ MW 151	Hymenoptera Formicidae: *Camponotus nearcticus*, *C. rasilis* (Brand *et al*., 1973d), and *C. ligniperda* (Brand *et al*., 1973e).	Mandibular glands of males.	It is a major component in the secretions of these formicine species which functions as a pheromone as well as a defensive substance.

(*continued*)

Table 8.2 (Continued)

trans-2-Hexenyl acetate  $C_8H_{14}O_2$ MW 142	Hemiptera Belastomatidae: Lethocerus indicus (Butenandt and Tam, 1957) and L. cordofanus (Pattenden and Staddon, 1970). Coreidae: Euthochtha galeator and Archimerus alternatus (Aldrich and Yonke, 1975). Pentatomidae: Nezara viridula (Gilby and Waterhouse, 1965). Lygaeidae: Oncopeltus fasciatus (Games and Staddon, 1973b).	Belastomatidae: Adult metasternal scent gland. Coreidae: Adult metasternal scent gland. Pentatomidae: Adult metasternal scent gland. Lygaeidae: Adult metasternal scent gland.	It had been reported that only males of L. indicus possessed scent glands but it has been now demonstrated that the females also possess these glands although they are greatly reduced (Pattenden and Staddon, 1970). The glands of L. indicus males contain ca. 50 µl of secretion vs. ca. 2 µl for the females. The secretion of L. cordofanus contain 98% of this alkenyl acetate. It is a minor component in the secretion of N. viridula and is not commonly produced by pentatomids. This compound is probably produced in the lateral scent glands rather than by the epidermal cells of the reservoir (Gilby and Waterhouse, 1967).
Isobutyl isobutyrate 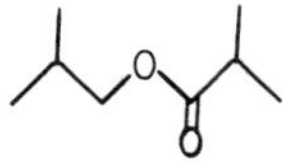 $C_8H_{16}O_2$ MW 144	Hemiptera Alydidae: Megalotomus quinquespinosus (Aldrich and Yonke, 1975).	Adult metasternal scent glands.	It is one of six esters in this secretion.

Compound	Occurrence	Source	Remarks
n-Butyl butyrate 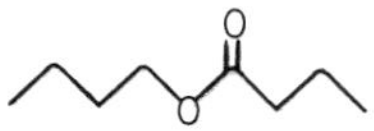 $C_8H_{16}O_2$ MW 144	Hemiptera Coreidae: Amorbus rhombifer and Mictis caja (Waterhouse and Gilby, 1964). Alydidae: Megalotomus quinquespinosus, Alydus eurinus, and A. pilosulus (Aldrich and Yonke, 1975).	Coreidae: Adult metasternal scent gland. Alydidae: Adult metasternal scent gland.	It is an uncommon coreid defensive product detected in only two out of about 20 species. Although only detected in one of 3 species of Amorbus, it is the major constituent (53%) in A. rhombifer. It is a trace component in M. caja which emphasizes a different ester in its exudate.
n-Hexyl acetate  $C_8H_{16}O_2$ MW 144	Hemiptera Alydidae: Alydus eurinus and A. pilosulus (Aldrich and Yonke, 1975). Coreidae: Amorbus rubiginosus, A. alternatus, A. rhombifer, Mictis profana, M. caja, Aulacosternum nigrorubrum, Pachycolpura manca, Agripocoris frogatti (Waterhouse and Gilby, 1964), Pternistria bispina (Baker and Kemball, 1967), Amblypelta nitida (Baker et al., 1972), Mozena obtusa (McCullough, 1973), M. lunata (McCullough, 1974a), Leptoglosus oppositus, L. clypealis (Aldrich and Yonke, 1975), and Chelinidea vittiger (McCullough, 1974b). Hyocephalidae: Hyocephalus sp. (Waterhouse and Gilby, 1964).	Alydidae: Adult metasternal scent gland. Coreidae: Adult metasternal scent gland. Hyocephalidae: Adult metasternal scent gland.	It is a rather typical defensive component of coreids which usually accompanies n-hexanal and acetic acid in the secretion. It is usually a major component and in 9 species averages 45% of the volatiles in the secretion. It is not produced by larvae of P. bispina and is only detected in last instar larvae and adults of A. nitida. It is a trace constituent in the Hyocephalus secretion.

(continued)

Table 8.2 (Continued)

Methyl 6-methylsalicylate $C_9H_{10}O_3$ MW 166	Hymenoptera Formicidae: *Camponotus nearcticus*, *C. pennsylvanicus*, *C. noveboracensis*, *C. subbarbatus* (Brand *et al*., 1973d), *C. herculeanus*, *C. ligniperda* (Brand *et al*., 1973e), and *Gnamptogenys pleurodon* (Duffield and Blum, 1975a).	Mandibular glands of *Camponotus* males and workers of *G. pleurodon*.	It is one of three major components in the secretion of *C. nearcticus* but is the major compound produced by *C. pennsylvanicus* and *C. noveborancensis*. This ester probably provides these male ants with some degree of protection against predators in addition to functioning as a sex-specific pheromone.

Methyl 2,5-dihydroxyphenyl acetate (Methyl homogentisate) 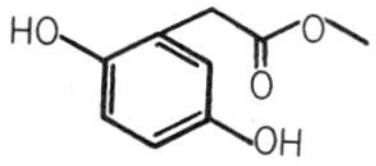 $C_9H_{10}O_4$ MW 182	Coleoptera Dytiscidae: *Dytiscus marginalis* (Schildknecht, 1970).	Pygidial glands.	It is only present in one of 11 dytiscid species and is the most minor constituent in the secretion of *D. marginalis* comprising ca. 0.5% of the secretion or 14 µg/beetle. This methyl ester of a hydroquinone is probably derived from tyrosine as it is in mammals suffering from a biochemical lesion.
Ethyl 3,4-dihydroxybenzoate (Ethyl protocatechuate) 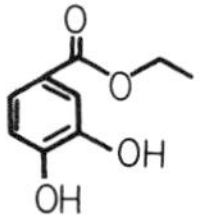 $C_9H_{10}O_4$ MW 182	Coleoptera Dytiscidae: *Cybister laterallimarginalis* (Schildknecht, 1970).	Pygidial glands.	It is an uncommon exocrine product of dytiscids which is absent from the secretions of ten other species including another member of the genus *Cybister*. It is probably derived from tyrosine which is readily converted to protocatechuic acid by some insect species.

(continued)

Table 8.2 (Continued)

2-Methylbutyl butyrate 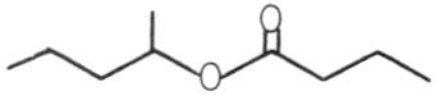 $C_9H_{18}O_2$ MW 158	Hemiptera Alydidae: Megalotomus quinquespinosus (Aldrich and Yonke, 1975).	Adult metasternal scent gland.	It is one of three 2-methylbutyl esters in the secretion.
2-Methylbutyl isobutyrate 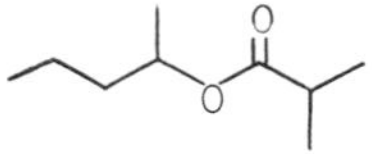 $C_9H_{18}O_2$ MW 158	Hemiptera Alydidae: Megalotomus quinquespinosus (Aldrich and Yonke, 1975).	Adult metasternal scent gland.	It is one of four butyl esters in the secretion.
trans-2-Hexenyl butyrate  $C_{10}H_{18}O_2$ MW 170	Hemiptera Belastomatidae: Lethocerus indicus (Devakul and Maarse, 1964).	Adult metasternal scent gland.	It is one of eight minor constituents in a secretion which is completely dominated by trans-2-hexenyl acetate.
trans-2-Octenyl acetate 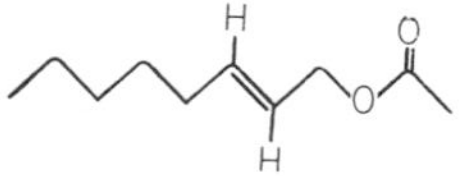 $C_{10}H_{18}O_2$ MW 170	Hemiptera Coreidae: Euthochtha galeator, Archimerus alternatus, Leptoglossus oppositus, and L. cylpealis (Aldrich and Yonke, 1975). Pentatomidae: Musgraveia	Coreidae: Adult metasternal scent gland. Pentatomidae: Adult metasternal scent gland. Cydnidae: Adult meta-	In the storage reservoir of N. viridula it comprises only 0.5% of the secretion and is barely detectable in the exudate of M. sulciventris. It is apparently syn-

	(=*Rhoecocoris*) *sulciventris* (Park and Sutherland, 1962), *Nezara viridula* (Gilby and Waterhouse, 1965), and *Tessaratoma aethiops* (Baggini *et al.*, 1966). Cydnidae: *Macrocystus* sp. (Baggini *et al.*, 1966). Lygaeidae: *Oncopeltus fasciatus* (Games and Staddon, 1973b).	sternal scent gland. Lygaeidae: Adult metasternal scent gland.	thesized in the lateral glands of the scent gland complex (Gilby and Waterhouse, 1967). This alkenyl acetate could not be detected in the abdominal secretion of larvae of *T. aethiops*. It also appears to be a minor component in cydnid secretions accounting for only 4% of the steam volatiles.
n-Butyl hexanoate $C_{10}H_{20}O_2$ MW 172	Hemiptera Alydidae: *Megalotomus quinquespinosus* (Aldrich and Yonke, 1975). Coreidae: *Amblypelta nitida* (Baker *et al.*, 1972).	Alydidae: Adult metasternal scent gland. Coreidae: Adult metasternal scent gland.	It is not produced in the larval abdominal glands and is a minor component in the secretion of *A. nitida*.
n-Hexyl butyrate $C_{10}H_{20}O_2$ MW 172	Hemiptera Alydidae: *Alydus eurinus* and *A. pilosulus* (Aldrich and Yonke, 1975). Coreidae: *Amblypelta nitida* (Baker *et al.*, 1972) and *Pternistria bispina* (Baker and Kemball, 1967).	Alydidae: Adult metasternal scent gland. Coreidae: Adult metasternal scent gland.	This ester is a minor component in the secretion of *A. nitida* and is not produced in the larval defensive glands. Similarly, it is only produced by adults of *P. bispina* but in the secretion of this species this ester is a major constituent and the concentration of *n*-hexyl

(*continued*)

Table 8.2 (Continued)

			acetate, a typical ester in coreid secretions, is low.
n-Octyl acetate $C_{10}H_{20}O_2$ MW 172	Hemiptera Coreidae: *Leptocoris apicalis* (Baggini *et al.*, 1966) and *Amblypelta nitida* (Baker *et al.*, 1972).	Larval dorsal abdominal glands (*A. nitida*) and adult metasternal scent gland.	Although this alkyl acetate was identified in a total extract of adults of *L. apicalis*, it is almost certainly a component of the metasternal scent gland secretion. It accounts for 40% of the volatiles detected in the secretion of this species whereas in that of *A. nitida* it is not a major constituent. It is only detected in the secretion of the ultimate larval instar of *A. nitida*.
n-Nonyl formate $C_{10}H_{20}O_2$ MW 172	Coleoptera Carabidae: *Helluo costatus* (Moore and Wallbank, 1968).	Pygidial glands.	Nonyl esters appear to be characteristic of species in the subfamily Helluoninae. Although this ester is a minor component in the secretion (ca. 8%), it probably has the important role of acting as a wetting agent and promoting the penetration of the major defensive component, formic acid.

Methyl 8-hydroxyquinoline-2-carboxylate 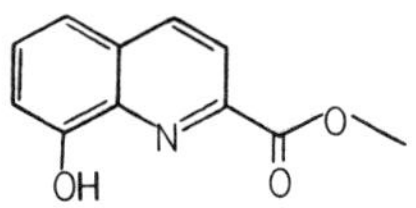 $C_{11}H_9NO_3$ MW 203	Coleoptera Dytiscidae: *Ilybius fenestratus* (Schildknecht *et al.*, 1969a).	Paired prothoracic glands of adults.	This alkaloid, which is the major glandular constituent (350 μg) and is responsible for the yellow color of the secretion, is not toxic to amphibians and fish. However, it produces clonic spasms in mammals such as mice, predators which this amphibious beetle may encounter.
2-Methylbutyl hexanoate 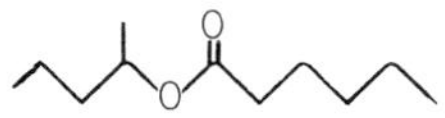 $C_{11}H_{22}O_2$ MW 186	Hemiptera Alydidae: *Megalotomus quinquespinosus* (Aldrich and Yonke, 1975).	Adult metasternal scent gland.	It is a trace constituent in the secretion.
n-Nonyl acetate  $C_{11}H_{22}O_2$ MW 186	Coleoptera Carabidae: *Helluo costatus* (Moore and Wallbank, 1968), *Helluomorphoides ferrugineus* and *H. latitarsus* (Eisner *et al.*, 1968). Hymenoptera Formicidae: *Formica sanguinea* (Bergström and Löfqvist, 1968).	Carabidae: Pygidial glands. Formicidae: Dufour's gland of the female.	This ester has been demonstrated to facilitate the penetration of the major defensive substance, formic acid, through the insect cuticle and to be much more effective in inducing cleansing reflexes in insects than the acid alone. It is not detected in the Dufour's gland secretion of workers of *F. sanguinea*.

(continued)

Table 8.2 (Continued)

2-Phenylethyl isobutyrate 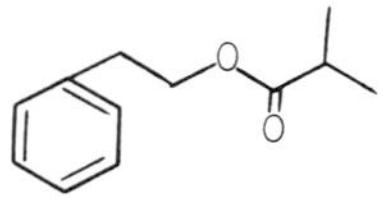 $C_{12}H_{16}O_2$ MW 192	Coleoptera Chrysomelidae: _Chrysomela interrupta_ (Blum _et al._, 1972).	Dorsally paired thoracic and abdominal glands of larvae.	It comprises ca. 20% of the glandular volatiles.
trans-2-Decenyl acetate  $C_{12}H_{22}O_2$ MW 198	Hemiptera Pentatomidae: _Nezara viridula_ (Gilby and Waterhouse, 1965), _Commius elegans_ (Gilby and Waterhouse, 1967), and _Biprorulus bibax_ (MacLeod _et al._, 1975). Cydnidae: _Macrocystus_ sp. (Baggini _et al._, 1966).	Pentatomidae: Adult metasternal scent gland. Cydnidae: Adult metasternal scent gland.	It is a minor component (ca. 3%) in the secretions of these species which accompanies _trans_-2-decenal and is synthesized in the lateral glands of the scent gland complex. It may be the precursor for the more toxic aldehyde which is apparently formed in the reservoir of the scent gland.
Citronellyl acetate 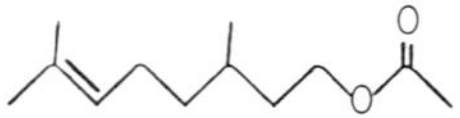 $C_{12}H_{22}O_2$ MW 198	Hymenoptera Apidae: _Bombus pratorum_ (Kullenberg _et al._, 1970).	Labial glands of males.	It is present as a minor component in the exudates of only a few species of _Bombus_.

Ethyl decanoate $C_{12}H_{24}O_2$ MW 200	Hymenoptera Apidae: *Bombus terrestris* and *B. lucorum* (Kullenberg *et al.*, 1970).	Labial glands of males.	It is a minor constituent detected in only two of 12 *Bombus* species.
n-Hexyl hexanoate $C_{12}H_{24}O_2$ MW 200	Hemiptera Alydidae: *Alydus eurinus* and *A. pilosulus* (Aldrich and Yonke, 1975). Coreidae: *Pternistria bispina* (Baker and Kemball, 1967) and *Amblypelta nitida* (Baker *et al.*, 1972).	Alydidae: Adult metasternal scent gland. Coreidae: Larval dorsal abdominal glands (*A. nitida*) and adult metasternal scent gland.	It is a very minor constituent (0.6%) in the secretion of *P. bispina* which is not detectable in the larval glandular exudate. It is the main glandular component of *A. nitida* and is the only ester which is produced in the glands of all larval instars.
n-Decyl acetate $C_{12}H_{24}O_2$ MW 200	Hymenoptera Formicidae: *Formica sanguinea*, *F. rufibarbis* (Bergström and Löfqvist, 1968), *Lasius niger* (Bergström and Löfqvist, 1970), *F. pergandei*, *F. subintegra* (Regnier and Wilson, 1971), and *Camponotus ligniperda* (Bergström and Löfqvist, 1971).	Dufour's gland of the workers and females.	It is one of several alkyl acetates which is synthesized in the Dufour's gland of formicine ants. It is a major constituent in all species except *C. ligniperda*. In *F. pergandei* this ester accounts for more than 5% of the body weight of a worker and ca. 400 μg are present in a single gland. This

(continued)

Table 8.2 (Continued)

			and other acetates are reported to be utilized as chemical disarming agents by slave-making species such as *F. pergandei* and *F. subintegra*. However, it is a major glandular component in *F. rufibarbis*, a slave species, and *L. niger*, a formicine which is not a slave raider.
2-Phenylethyl 2-methyl butyrate $C_{13}H_{18}O_2$ MW 206	Coleoptera Chrysomelidae: *Chrysomela interrupta* (Blum *et al.*, 1972).	Paired thoracic and dorsal abdominal glands of larvae.	It accounts for nearly 80% of the volatiles in the exudate of *C. interrupta* and constitutes one of the two phenyl esters identified in the secretions of larval Chrysomelinae.

n-Undecyl acetate  $C_{13}H_{26}O_2$ MW 214	Hymenoptera Formicidae: *Formica sanguinea*, *F. rufibarbis* (Bergström and Löfqvist, 1968), *Lasius niger* (Bergström and Löfqvist, 1970), and *Camponotus ligniperda* (Bergström and Löfqvist, 1972).	Dufour's gland of the workers and females.	It is a minor component in the secretions of all species.
11-Dodecenyl acetate  $C_{14}H_{26}O_2$ MW 226	Lepidoptera Cossidae: *Zeuzera pyrina* (Marchesini *et al.*, 1969).	Mandibular glands of larvae.	It may be characteristic of species in the subfamily Zeuzerinae and has been assigned the trivial name zeuzerina. About 85% of the secretion is comprised of zeuzerina but the geometry of the double bond has not been established.
Ethyl dodecanoate  $C_{14}H_{28}O_2$ MW 228	Hymenoptera Apidae: *Bombus sporadicus*, *B. terrestris*, *B. lucorum* ("dark" and "blonde"), and *B. patagiatus* (Kullenberg *et al.*, 1970).	Labial glands of males.	It is the major component in the discharges of *B. lucorum* ("dark") and *B. patagiatus*.

(*continued*)

Table 8.2 (Continued)

n-Dodecyl acetate  $C_{14}H_{28}O_2$ MW 228	Hymenoptera Formicidae: *Formica sanguinea*, *F. rufibarbis* (Bergström and Löfqvist, 1968), *F. pergandei*, *F. subintegra* (Regnier and Wilson, 1971), and *Lasius niger* (Bergström and Löfqvist, 1970). Lepidoptera Cossidae: *Zeuzera pyrina* (Marchesini *et al*., 1969).	Formicidae: Dufour's gland of workers and females.	It is a major constituent in the secretions of *F. pergandei*, *F. sanguinea*, *F. subintegra*, all of which are slave-making species. It accounts for ca. 3% of the body weight of *F. subintegra* workers and is believed to disrupt the social cohesiveness of raided species.
Mandelonitrile benzoate 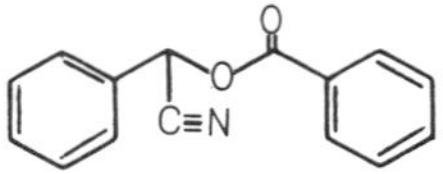 $C_{15}H_{11}NO_2$ MW 237	Polydesmida Polydesmidae: *Polydesmus collaris* (Casnati *et al*., 1963). Xystodesmidae: *Cherokia georgiana*, *Cleptoria rileyi*, *Brachoria* sp., *Sigiria* sp., *Stelgipus agrestis*, *Motyxia sequoiae*, *M. tiemanni*, and *Paimoikia* sp. (Duffey *et al*., 1977). Gomphodesmidae: *Gomphodesmus pavani* (Barbetta *et al*., 1966).	Polydesmidae: Lateral paired glands opening dorsolaterally on the anterior half of most diplosegments. Xystodesmidae: Lateral paired glands opening dorsolaterally on the anterior half of most diplosegments. Gomphodesmidae: Lateral paired glands opening dorsolaterally on the anterior half of most diplosegments.	It constitutes a minor constituent in all secretions and does not appear to represent a storage form of benzaldehyde and HCN, as is mandelonitrile.

n-Tridecyl acetate  $C_{15}H_{30}O_2$ MW 242	Hymenoptera Formicidae: *Camponotus ligniperda* (Bergström and Löfqvist, 1972).	Dufour's gland of workers.	It is a trace constituent.
2-Phenylethyl octanoate 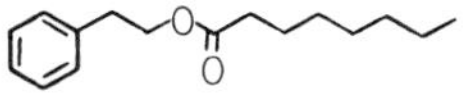 $C_{16}H_{24}O_2$ MW 248	Hymenoptera Formicidae: *Camponotus clarithorax* (Lloyd *et al*., 1975).	Mandibular glands of males.	It is accompanied by the free alcohol and acid.
3,5,13-Tetradecatrienyl acetate  $C_{16}H_{26}O_2$ MW 250	Lepidoptera Cossidae: *Cossus cossus* (Trave *et al*., 1966).	Mandibular glands of larvae.	It is one of the major components in the secretion. It has been named cossina A and occurs along with its alcoholic moiety (cossina 2). In addition, different geometric isomers of this ester are present in this secretion and possibly that of *Zeuzera pyrina*.

(*continued*)

Table 8.2 (Continued)

4,6,13-Tetradecatrienyl acetate  $C_{16}H_{26}O_2$ MW 250	Lepidoptera Cossidae: *Cossus cossus* (Trave *et al.*, 1966).	Mandibular glands of larvae.	It is a major exocrine component which has been assigned the trivial name cossina C. The secretion also contains the free alcohol (cossina 3) as well as a mixture of the same acetates with a different steric configuration.
5,13-Tetradecadienyl acetate  $C_{16}H_{28}O_2$ MW 252	Lepidoptera Cossidae: *Cossus cossus* (Trave *et al.*, 1966) and *Zeuzera pyrina* (Marchesini *et al.*, 1969).	Mandibular glands of larvae.	It is one of the two major constituents in the secretion of *C. cossus* but it is a minor component in the exudate of *Z. pyrina*. This ester, which has been given the trivial epithet cossina A, is accompanied by the free alcohol in the exudate of *C. cossus*. The geometry of the double bonds has not been ascertained. The secretion has excellent contact toxicity against ants.

Ethyl 9-tetradecenoate  $C_{16}H_{30}O_2$ MW 254	Hymenoptera Apidae: _Bombus terrestris_, _B. lucorum_ ("blonde"), _Psithyrus silvestris_, and _P. globosus_ (Calam, 1969; Kullenberg _et al._, 1970).	Labial glands of males.	The position of the double bond has only been determined for this ester in the secretion of _B. lucorum_ in which it is the major constituent. The double bond probably has the cis configuration and these are the first reports of ethyl myristoleate from a natural source. The absence of it from the secretion of the so-called "dark" form of _B. lucorum_ probably indicates that the two forms of this species are really separate species.
Tetradecenyl acetate $C_{16}H_{30}O_2$ MW 254	Hymenoptera Apidae: _Trigona tubiba_, _T. depilis_, and _T. xanthotricha_ (Blum _et al._, 1973b). Formicidae: _Formica nigricans_, _F. rufa_, _F. polyctena_ (Bergström and Löfqvist, 1973), _Camponotus ligniperda_ (Bergström and Löfqvist, 1971), and _C. herculeanus_ (Bergström and Löfqvist, 1972).	Apidae: Mandibular glands of workers. Formicidae: Dufour's gland of workers and females.	It is a major constituent in only the secretion of _T. depilis_. The positions of the double bonds have not been determined.

(continued)

Table 8.2 (Continued)

n-Tetradecyl acetate  $C_{16}H_{32}O_2$ MW 256	Hymenoptera Formicidae: Lasius niger (Bergström and Löfqvist, 1970), Formica pergandei, F. subintegra (Regnier and Wilson, 1971), F. nigricans, F. rufa, F. polyctena (Bergström and Löfqvist, 1973), Camponotus ligniperda (Bergström and Löfqvist, 1971), and C. herculeanus (Bergström and Löfqvist, 1972). Apidae: Bombus sporadicus (Kullenberg et al., 1970).	Formicidae: Dufour's gland of workers and females. Apidae: Labial gland of males.	It is a minor component in the secretions of L. niger, C. herculeanus, and C. ligniperda, but in those of some Formica species it is present in very great amounts. In F. subintegra, it accounts for about 3% of the worker body weight and in both species of Formica it is considered to be one of the chemical disarming agents utilized during slave raids. It is the major glandular compound in B. sporadicus but has not been detected in any other species in this genus.
Geranyl hexanoate 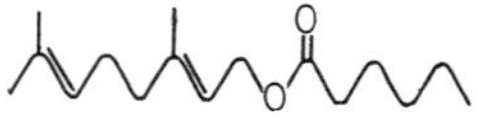 $C_{16}H_{32}O_2$ MW 256	Hymenoptera Andrenidae: Andrena denticulata, A. nigroaenea, and A. carbonaria (Bergström and Tengö, 1974).	Dufour's gland of females.	It is one of three geranyl esters in Andrena secretions and is a major glandular product of A. denticulata and A. nigroaenea.

2-Phenylethyl nonanoate $C_{17}H_{26}O_2$ MW 262	Hymenoptera Formicidae: *Camponotus clarithorax* (Lloyd *et al*., 1975).	Mandibular glands of males.	It is one of at least four esters in the secretion.
Farnesyl acetate $C_{17}H_{28}O_2$ MW 264	Hymenoptera Formicidae: *Lasius niger* (Bergström and Löfqvist, 1970), *Camponotus ligniperda* (Bergström and Löfqvist, 1971), and *C. herculeanus* (Bergström and Löfqvist, 1972). Andrenidae: *Andrena bicolor*, *A. denticulata*, *A. nigroaenea*, and *A. haemorrhoa* (Bergström and Tengö, 1974). Apidae: *Bombus cullumanus*, *B. pratorum*, and *Psithyrus barbutellus* (Kullenberg *et al*., 1970).	Formicidae: Dufour's gland of workers. Andrenidae: Dufour's gland of females. Apidae: Labial glands of males.	This terpene ester is an unusual Dufour's gland product since terpenes are commonly synthesized in the mandibular glands of formicines. It is a trace constituent in the *Camponotus* and *Bombus* secretions which is present in the all-*trans*-configuration. It is one of seven terpene esters produced by *Andrena* spp. and is a major constituent in the secretion of *A. denticulata*.
2,3-Dihydrofarnesyl acetate $C_{17}H_{30}O_2$ MW 266	Hymenoptera Apidae: *Bombus terrestris* (Kullenberg *et al*., 1970).	Labial glands of males.	It is a minor glandular constituent, present in only one of 14 *Bombus* species, which is accompanied by the free alcohol as the major exocrine product.

(*continued*)

Table 8.2 (Continued)

2,6-Dimethyl-5-hepten-1-octanoate 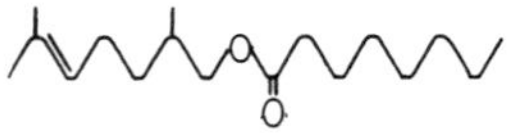 $C_{17}H_{32}O_2$ MW 268	Hymenoptera Formicidae: *Camponotus clarithorax* (Lloyd *et al.*, 1975).	Mandibular glands of males.	Both the free alcohol and acid also fortify the secretion.
Methyl *n*-hexadecanoate  $C_{17}H_{34}O_2$ MW 270	Hymenoptera Formicidae: *Lasius alienus* (Bergström and Löfqvist, 1970).	Dufour's gland of workers.	It is a trace component which has not been detected in the secretions of any other species of *Lasius*.
n-Pentadecyl acetate  $C_{17}H_{34}O_2$ MW 270	Hymenoptera Formicidae: *Camponotus ligniperda* (Bergström and Löfqvist, 1971) and *C. herculeanus* (Bergström and Löfqvist, 1972).	Dufour's gland of workers.	It is a trace component of these two secretions which has not been identified as a Dufour's gland product of any other formicine species.
n-Tridecyl butyrate  $C_{17}H_{34}O_2$ MW 270	Hemiptera Pentatomidae: *Musgraveia sulciventris* (MacLeod *et al.*, 1975).	Adult metasternal scent gland.	It is an atypical pentatomid constituent since unsaturated esters have been identified in all other secretions.

Compound	Taxon	Gland source	Remarks
Ethyl hexadecatrienoate $C_{18}H_{30}O_2$ MW 278	Hymenoptera Apidae: *Bombus lucorum* ("dark") and *B. lucorum* ("blonde") (Kullenberg *et al.*, 1970).	Labial gland of males.	It constitutes a minor constituent in the secretions of both forms for which the double bond positions have not been determined.
Ethyl hexadecadienoate $C_{18}H_{32}O_2$ MW 280	Hymenoptera Apidae: *Bombus lapidarius* (Kullenberg *et al.*, 1970).	Labial gland of males.	The double bond positions have not been determined for this minor exocrine product.
2,6-Dimethyl-5-hepten-l-nonanoate 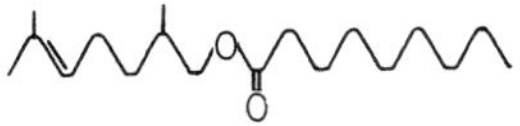 $C_{18}H_{34}O_2$ MW 282	Hymenoptera Formicidae: *Camponotus clarithorax* (Lloyd *et al.*, 1975).	Mandibular glands of males.	It is one of four esters in the secretion.
Ethyl hexadecenoate $C_{18}H_{34}O_2$ MW 282	Hymenoptera Apidae: *Bombus lucorum* ("dark"), *B. lucorum* ("blonde"), and *B. hypnorum* (Kullenberg *et al.*, 1970).	Labial gland of males.	It is a minor exocrine product of all three species. Although the position of the double bond has not been determined, it is probably 9,10 (palmitoleic acid).

(continued)

Table 8.2 (Continued)

9-Hexadecenyl acetate  $C_{18}H_{34}O_2$ MW 282	Julida Parajulidae: Blaniulus guttulatus (Weatherston et al., 1971). Hymenoptera Apidae: Trigona tubiba, T. xanthotricha, and T. depilis (Blum et al., 1973b). Formicidae: Formica nigricans, F. rufa, F. polyctena (Bergström and Löfqvist, 1973), Camponotus ligniperda (Bergström and Löfqvist, 1971), and C. herculeanus (Bergström and Löfqvist, 1972).	Parajulidae: Lateral paired glands opening dorsolaterally on the anterior half of most diplosegments. Apidae: Mandibular glands of workers. Formicidae: Dufour's gland of workers.	It accounts for nearly 25% of the esters in the secretion of B. guttulatus. In T. xanthotricha and T. depilis it is a major exocrine product and in the exudate of T. tubiba there are three isomeric hexadecenyl acetates, two of which are major components. It is a minor glandular constituent of C. ligniperda. The position of the double bonds in these esters detected in hymenopterous secretions has not been established.
Geranyl octanoate 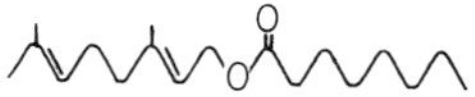 $C_{18}H_{36}O_2$ MW 284	Hymenoptera Andrenidae: Andrena denticulata, A. carbonaria, and A. helvola (Bergström and Tengö, 1974).	Dufour's gland of females.	It is the major constituent in the secretion of A. helvola.
n-Tetradecyl butyrate  $C_{18}H_{36}O_2$ MW 284	Hemiptera Pentatomidae: Musgraveia sulciventris (MacLeod et al., 1975).	Adult metasternal scent gland.	It is one of two butyrate esters in the secretion.

n-Hexadecyl acetate $C_{18}H_{36}O_2$ MW 284	Julida Parajulidae: *Blaniulus guttulatus* (Weatherston *et al.*, 1971). Hymenoptera Apidae: *Bombus sporadicus* (Kullenberg *et al.*, 1970) and *Ceratina cucurbitina* (Wheeler *et al.*, 1977a). Formicidae: *Lasius niger*, *L. alienus* (Bergström and Löfqvist, 1970), *Formica nigricans*, *F. rufa*, *F. polyctena* (Bergström and Löfqvist, 1973), *Camponotus ligniperda* (Bergström and Löfqvist, 1971), and *C. herculéanus* (Bergström and Löfqvist, 1972).	Parajulidae: Lateral paired glands opening dorsolaterally on the anterior half of most diplosegments. Apidae: Mandibular glands of *Ceratina* males and labial glands of *Bombus* males. Formicidae: Dufour's gland of workers.	It comprises ca. 20% of the esters in the secretion of *B. guttulatus* whereas it is a minor exocrine product in the bee and ant exudates. Its exclusive occurrence in the secretions of males of *Ceratina* and *Bombus*, two widely separated genera, may also indicate that it has been developed independently to function as a sex pheromone as well as a defensive compound.
Farnesyl butyrate $C_{19}H_{32}O_2$ MW 292	Hymenoptera Andrenidae: *Andrena bicolor*, *A. denticulata*, and *A. haemorrhoa* (Bergström and Tengö, 1974).	Dufour's gland of females.	It is a minor product in the secretions.

(*continued*)

Table 8.2 (Continued)

Geranyl decanoate  $C_{20}H_{36}O_2$ MW 308	Hymenoptera Andrenidae: Andrena helvola (Bergström and Tengö, 1974).	Dufour's gland of females.	It is the longest-chain geranyl ester identified in Andrena secretions and appears to be a rather atypical Dufour's gland constituent of species in this genus.
9-Octadecenyl acetate  $C_{20}H_{38}O_2$ MW 310	Julida Parajulidae: Blaniulus guttulatus (Weatherston et al., 1971). Hymenoptera Apidae: Bombus muscorum (Kullenberg et al., 1970).	Parajulidae: Lateral paired glands opening dorsolaterally on the anterior half of most diplosegments. Apidae: Labial glands of males.	This ester and 9-octadecenol-1 constitute the major compounds in the secretion of B. muscorum. It comprises 60% of the esters in the millipede exudate.
n-Octadecyl acetate  $C_{20}H_{40}O_2$ MW 312	Hymenoptera Apidae: Psithyrus barbutellus (Kullenberg et al., 1970). Formicidae: Formica nigricans, F. rufa, F. polyctena (Bergström and Löfqvist, 1973), Lasius niger (Bergström and Löfqvist, 1970), and Camponotus ligniperda (Bergström and Löfqvist, 1971).	Apidae: Labial glands of males. Formicidae: Dufour's gland of workers.	It is only present in 1 of 6 species of Psithyrus and 1 of 4 species of Lasius. It is a trace product in all exocrine secretions.

Farnesyl hexanoate  $C_{21}H_{36}O_2$ MW 320	Hymenoptera Andrenidae: *Andrena bicolor*, *A. denticulata*, *A. nigroaenea*, *A. carbonaria*, *A. helvola*, and *A. haemorrhoa* (Bergström and Tengö, 1974).	Dufour's gland of females.	It is the major constituent in the Dufour's gland secretion of all species.
Geranylgeranyl acetate  $C_{22}H_{36}O_2$ MW 332	Hymenoptera Apidae: *Bombus sorocensis*, *B. lucorum* ("dark"), *B. cullumanus*, *B. pratorum*, *B. hypnorum*, and *Psithyrus rupestris* (Kullenberg *et al.*, 1970). Formicidae: *Formica nigricans*, *F. rufa*, and *F. polyctena* (Bergström and Löfqvist, 1973).	Apidae: Labial glands of males. Formicidae: Dufour's gland of workers.	It is a rather typical constituent of *Bombus* species and is a major constituent in *B. sorocensis*, *B. cullumanus*, and *B. hypnorum*.
n-Eicosyl acetate  $C_{22}H_{44}O_2$ MW 340	Hymenoptera Apidae: *Bombus lucorum* ("dark") (Kullenberg *et al.*, 1970).	Labial glands of males.	It is a minor component present in 1 of 14 species of *Bombus*.

(continued)

Table 8.2 (Continued)

Farnesyl octanoate  $C_{23}H_{40}O_2$ MW 348	Hymenoptera Andrenidae: *Andrena bicolor*, *A. denticulata*, *A. nigroaenea*, *A. carbonaria*, *A. helvola*, and *A. haemorrhoa* (Bergström and Tengö, 1974).	Dufour's gland of females.	It is the most widely distributed terpene ester in the *Andrena* secretions and is a major constituent in that of *A. helvola*.
n-Docosyl acetate  $C_{24}H_{48}O_2$ MW 368	Hymenoptera Apidae: *Bombus lucorum* ("dark") (Kullenberg *et al.*, 1970).	Labial glands of males.	It is a minor component in this ester-rich secretion.

and diplopodous secretions, and two compounds, mandelonitrile benzoate and ethyl benzoate, are restricted to the exudates of species in the latter order (Table 8.2). Hemipterous species are distinctive in producing six of the seven *n*-butyrate esters as well as a majority of esters containing hexanoic acid. The only octanoates identified as arthropod defensive compounds are produced by hymenopterans, species of which are also distinctive in producing esters containing the longest chain acids (e.g., C_{12}, C_{14}, C_{16}) found in these allomonal compounds. Indeed, whereas arthropods clearly favor the utilization of acids such as acetic and butyric for the synthesis of these esters, they nevertheless incorporate a wide variety of other acids into these compounds. At least 14 other aliphatic acids have been detected in these compounds, which, with few exceptions (e.g., pentanoate), are even numbered acids (Table 8.2).

Esters are often accompanied by their alcoholic moieties in the glandular reservoirs and in some cases the defensive secretion is fortified with an ester in addition to its acidic and alcoholic components. For example, adult coreids sometimes produce 1-hexanol, acetic acid, and hexyl acetate in their metathoracic glands (Waterhouse and Gilby, 1964), suggesting an obvious biosynthetic relationship for these compounds. That this is the case was demonstrated by Aldrich *et al.* (1979), who reported that coreids contain an esterase in their metathoracic gland that can either synthesize or hydrolyze hexyl acetate from acetic acid and 1-hexanol. Oxidation of the alcohol by a dehydrogenase produces hexanal, a characteristic defensive product of coreids. It will not be surprising if subsequent biosynthetic investigators establish metabolic relationships for other esters and their alcohols co-occurring in defensive secretions.

AN OVERVIEW

The roles of esters as defensive compounds appear to be quite varied, and some of these allomones are believed to function as more than simple repellents or toxicants. Some esters are regarded as ancillary defensive products that act both as wetting agents and facilitators of the penetration through the adversaries' cuticles of the primary irritants in the defensive secretions (Eisner *et al.*, 1968). In other instances esters represent the precursors from which the highly reactive final defensive compounds are generated, as is the case for hexanal which is oxidatively produced from 1-hexanol after hydrolysis of hexyl acetate (Aldrich *et al.*, 1979). Indeed, since these compounds almost invariably accompany other classes of more reactive allomones (e.g., aldehydes, acids), it is not unlikely that esters primarily act as that "extra" ingredient that guarantees that the secretion will possess an inordinately great defensive "punch."

Esters are further distinguished by their selective distribution as glandular products of species in particular arthropod taxa. For example, methyl *p*-hydroxybenzoate is the only one of the 20 esters produced by hemipterans that has been identified as an exocrine product of a nonhemipterous species (Table 8.2). Furthermore, several esters have very restricted distributions in the Hemiptera. Alydid species produce four unique esters as well as sharing four distinctive esters with coreid species (Aldrich and Yonke, 1975). The Coreidae and Alydidae are the ester synthesizers *par excellence* in the Hemiptera, producing eight of the 20 esters identified as exocrine products of species in this order. Pentatomid species are also distinctive in producing all three esters (e.g., (*E*)-2-decenyl acetate) derived from α,β-unsaturated alcohols that have been detected in defensive secretions. Stink bugs are also exceptional in synthesizing tridecyl butyrate (C_{17}) and tetradecyl butyrate (C_{18}) (MacLeod *et al.*, 1975), the only two esters in hemipterous defensive exudates that do not fall in the restricted range C_7–C_{12}.

Among hemipterans, notonectids and naucorids are distinctive in producing methyl *p*-hydroxybenzoate (Staddon and Weatherston, 1967), one of two aromatic allomones identified from species in this order. However, it is chiefly in the secretions of larval and adult beetles that aromatic defensive compounds are encountered. Five aromatic esters have been identified as pygidial gland products of dytiscids (Table 8.2), some of which are esters of acids reported to be utilized for the sclerotization of the insect cuticle (e.g., methyl 3,4-dihydroxybenzoate). The Dytiscidae generate a diverse group of aromatic esters, acids, and aldehydes in their defensive glands, and these compounds are regarded by Schildknecht *et al.* (1964) as strictly antimicrobial in function. Methyl *p*-hydroxybenzoate is invariably accompanied by *p*-hydroxybenzaldehyde in the dytiscid secretions, and this aromatic pair is believed to represent the primary antimicrobial defense of the beetles. However, since the same ester and aldehyde dominate the metasternal gland secretions of certain hemipterous species, there are no substantive grounds for not also regarding these compounds as typical defensive compounds evolved to deter animal predators.

The detection of esters of 2-phenylethanol in larval chrysomelid secretions (Blum *et al.*, 1972) and methyl salicylate in a carabid exudate (Schildknecht *et al.*, 1968c) further documents the virtuosity of beetles as synthesizers of aromatic allomones. Methyl salicylate is an unusual pygidial gland product of carabids, which normally produce secretions dominated by short-chain acids. However, this ester, in common with the few aliphatic esters (e.g., nonyl formate) detected in carabid secretions, probably is an important defensive product because of its ability to act as a wetting and spreading agent that would promote the toxic actions of the concomitant reactive acids.

The presence of two characteristic aromatic esters, mandelonitrile benzoate and ethyl benzoate, in the defensive secretions of millipedes (Table 8.2) can be readily reconciled with the known chemical defenses of these arthropods. Polydesmoid millipedes typically produce benzaldehyde, HCN, and benzoic acid in their defensive glands, and mandelonitrile benzoate may serve as the precursor for these compounds. Ethyl benzoate, which is limited in its known distribution to a single diplopod species (Duffey *et al.*, 1977), may be generated from benzoic acid that is derived from mandelonitrile benzoate.

The presence of four aromatic esters in the mandibular secretions of some male *Camponotus* species seems particularly significant because these are among the first aromatic compounds identified in the Formicidae, a particularly rich source of structurally diverse allomones. No aromatic compounds have been detected as exocrine products of any formicine workers or females and these caste-specific compounds possess pheromonal functions as well (Hölldobler and Maschwitz, 1965). The presence of aromatic compounds in the heads of these large-eyed male formicines could reflect an emphasized metabolism of aromatic amino acids leading to the production of large quantities of eye pigments. Hopefully, subsequent biosynthetic studies will illuminate the metabolic pathways responsible for the generation of these caste-specific esters in male ants.

Male formicines are typical of hymenopterous species in producing sex-specific esters in their defensive glands. Indeed, hymenopterans constitute the major ester synthesizers in the Arthropoda, generating more than 40% of the esters identified as exocrine compounds. In particular, the mandibular, labial, and Dufour's gland secretions of ants and bees contain the only terpene esters identified in these exudates. Male bumblebees, in common with male ants, produce a variety of mono-, sesqui-, and diterpene esters that function both as pheromones and defensive allomones (Kullenberg *et al.*, 1970). A few of these labial gland products (e.g., farnesyl acetate, geranylgeranyl acetate) are also produced in the Dufour's glands of formicine ants (Bergström and Löfqvist, 1973), and further testify to the biosynthetic versatility of hymenopterans as isoprenoid chemists. However, in addition to these terpenoid esters, some hymenopterans produce a large variety of nonterpenoid constituents. For example, ants in the genera *Formica*, *Lasius*, and *Camponotus* produce acetates of all alcohols in the range C_9–C_{18}. It may be taxonomically significant that these esters have only seen detected in the Dufour's gland exudates of species of Formicinae. Although the defensive chemistry of relatively few hymenopterans has been examined, it is obvious that these insects excel in the production of esters as well as compounds belonging to most other chemical classes. The ecological correlates of this exocrine versatility essentially remain to be established.

Chapter 9

Lactones

Lactones (also see Chapter 18) have been characterized as exocrine products of cockroaches, beetles, phasmids, ants, and bees (Table 9.1). In a few cases, the acids corresponding to the lactones have also been identified in the glandular exudates, and the former may be precursors of the latter. Although these cyclic esters do not represent major arthropod defensive allomones, they nevertheless constitute some of the most distinctive natural products isolated from these invertebrates. Indeed, sev-

Table 9.1

Distribution of Lactones in the Defensive Secretions of the Arthropoda[a]

Class	Order	Family
Insecta	Dictyoptera	
	Blattaria	Blattidae
	Orthoptera	Phasmidae
	Coleoptera	Cerambycidae
		Dytiscidae
		Chrysomelidae
		Tenebrionidae
		Staphylinidae
	Hymenoptera	Colletidae
		Halictidae
		Apidae
		Formicidae

[a] Also see Chapter 18.

eral of these allomones are unique compounds (e.g., iridomyrmecin, plagiolactone) whose known distribution is limited to the defensive secretions of the few insect species known to produce them.

AN OVERVIEW

Except for the two gluconolactones present in the secretions of blattid species, all the lactones characterized as defensive allomones are at least C_{10} compounds. Furthermore, since nearly 90% of the 26 identified lactones (Table 9.2 and Chapter 18) are produced by ants, bees, and beetles, this class of compounds must be considered as especially characteristic of the species in these two orders. At this juncture one family of insects, the Formicidae, appears to contain the most versatile biosynthesizers of lactones among the arthropods, producing nearly one-third of the identified compounds. Bees in the families Halictidae, Colletidae, and Apidae also produce a considerable variety of lactones (Table 9.2 and Chapter 18), and the bees and ants together generate more than 50% of the lactones known as arthropod defensive allomones. For all intent and purpose, the lactonic contribution (35%) of coleopterans in five families (Table 9.1) provides the remaining compounds identified as members of this chemical class.

In most cases the insect-derived lactones are in admixture with other classes of compounds, some of which are probably biosynthetically related to the former. For example, iridomyrmecin and isoiridomyrmecin are sometimes anal gland concomitants of iridodial in dolichoderine species (Trave and Pavan, 1956; McGurk *et al.*, 1968), and these iridolactones can be easily derived from the corresponding dials. The same relationship is evident between plagiolactone and chrysomelidial, the major defensive compounds produced by certain chrysomelid larvae (Meinwald *et al.*, 1977). The gluconolactones identified in the exudates of blattid species are in equilibrium with gluconic acid (Dateo and Roth, 1967a) and constitute the aqueous phase. The nonaqueous phase, which is comprised of nearly pure 2-hexenal, contains the major defensive allomones, and the lactones do not appear to be important deterrents for animals.

While the iridolactones iridomyrmecin, isoiridomyrmecin, and isodihydronepetelactone have only been identified as anal gland products of dolichoderine ants, other iridolactones have been characterized in the secretions of nonants. Plagiolactone (Table 9.2) is a particularly distinctive compound produced by chrysomelid larvae in the genus *Plagiodera* (Meinwald *et al.*, 1977), whereas a different iridolactone, gastrolactone,

Table **9.2** Lactones in arthropod defensive secretions

Name and Formula	Occurrence	Glandular Source	Comments
γ-D-Gluconolactone 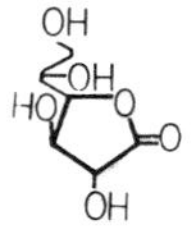 $C_6H_{10}O_6$ MW 178	Dictyoptera Blattidae: _Eurycotis decipiens_, _E. biolleyi_, _E. floridana_ (Dateo and Roth, 1967a), _Polyzosteria limbata_, _P. viridissima_, _P. oculata_, _P. cuprea_, _P. pulchra_, and _P. mitchelli_ (Wallbank and Waterhouse, 1970).	Ventral abdominal gland of adults opening between the sixth and seventh sternites.	It is in equilibrium with gluconic acid and is primarily present in the aqueous phase of the exudates.
δ-D-Gluconolactone 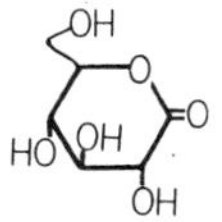 $C_6H_{10}O_6$ MW 178	Dictyoptera Blattidae: _Eurycotis decipiens_, _E. biolleyi_, _E. floridana_ (Dateo and Roth, 1967a), _Polyzosteria limbata_, _P. viridissima_, _P. oculata_, _P. cuprea_, _P. pulchra_, and _P. mitchelli_ (Wallbank and Waterhouse, 1970).	Ventral abdominal gland of adults opening between the sixth and seventh sternites.	In the _Polyzosteria_ spp. it is present as about a 6.5% aqueous solution in equilibrium with the γ-lactone and gluconic acid.
Mellein (3,4-Dihydro-8-hydroxy-3-methylisocoumarin) 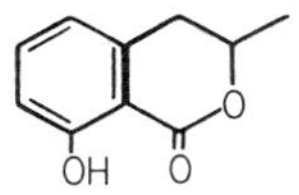 $C_{10}H_{10}O_3$ MW 178	Hymenoptera Formicidae: _Camponotus ligniperda_, _C. herculeanus_, _C. pennsylvanicus_, and _C. noveboracensis_ (Brand _et al._, 1973e).	Mandibular glands of males.	It is also produced by two species of fungi in the genus _Aspergillus_ and bears another trivial epithet, ochracin. It is a major constituent in the glandular secretions of males of the four species of _Camponotus_.

(continued)

Table 9.2 (Continued)

Iridomyrmecin 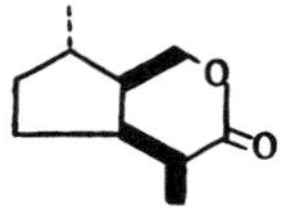 $C_{10}H_{16}O_2$ MW 168	Hymenoptera Formicidae: _Iridomyrmex humilis_ (Pavan, 1959; Fusco _et al._, 1955) and _I. pruinosus_ (McGurk _et al._, 1968).	Anal glands of workers.	It constitutes about 3.3% of the lipid-soluble material and is reported to possess insecticidal properties (Pavan, 1952).
Isoiridomyrmecin 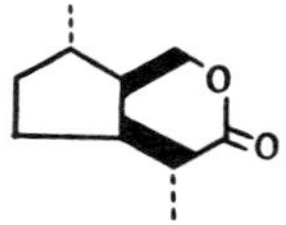 $C_{10}H_{16}O_2$ MW 168	Hymenoptera Formicidae: _Iridomyrmex nitidus_ (Cavill _et al._, 1956a,b), _Dolichoderus scabridus_ (Cavill and Hinterberger, 1960, 1962), and _Tapinoma sessile_ (McGurk _et al._, 1968).	Anal glands of workers.	It is not always reported to be produced by workers of _D. scabridus_. One of two other terpenoid defensive compounds has been identified as an anal gland product of different colonies of this dolichoderine.
Isodihydronepetalactone 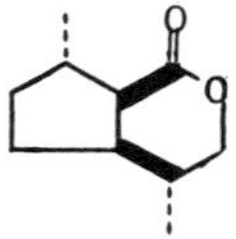 $C_{10}H_{16}O_2$ MW 168	Hymenoptera Formicidae: _Iridomyrmex nitidus_ (Cavill and Clark, 1967).	Anal glands of workers.	It is a concomitant of isoiridomyrmecin and comprises about one third of the iridolactone mixture. The two lactones constitute 8% of the body weight of the whole ants.

Massoilactone $C_{10}H_{16}O_2$ MW 168	Hymenoptera Formicidae: Two *Camponotus* spp. (Cavill *et al.*, 1968).	Mandibular glands of workers.	It is a powerful skin irritant which also has a pronounced effect on vertebrate heart muscle.
γ-Decalactone $C_{10}H_{18}O_2$ MW 170	Hymenoptera Apidae: *Trigona carbonaria* group (Wheeler *et al.*, 1975).	Mandibular glands of workers.	It is one of two decalactones produced by species in this group.
δ-Decalactone $C_{10}H_{18}O_2$ MW 170	Hymenoptera Apidae: *Trigona carbonaria* group (Wheeler *et al.*, 1975).	Mandibular glands of workers.	It is one of two lactones identified in the mandibular gland secretions of stingless bees.
γ-Dodecalactone $C_{12}H_{22}O_2$ MW 198	Coleoptera Staphylinidae: *Bledius mandibularis* and *B. spectabilis* (Wheeler *et al.*, 1972a).	Pygidial glands.	It comprises at least 70% of the secretions of both species and may be metabolically related to 1-undecene, the second most abundant component of the exudates.

(*continued*)

Table 9.2 (Continued)

Marginalin 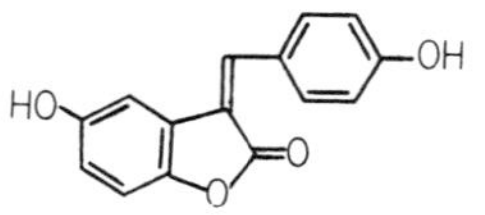 $C_{15}H_{10}O_4$ MW 254	Coleoptera Dytiscidae: *Dytiscus marginalis* (Schildknecht *et al.*, 1970).	Pygidial glands.	Although each beetle contains about 23 µg of this lactone, its function in a phenol-rich secretion is unknown.
4-Hexadec-9-enolide  $C_{16}H_{28}O_2$ MW 252	Hymenoptera Formicidae: *Lasius flavus* (Bergström and Löfqvist, 1970).	Dufour's gland of workers.	It is a trace constituent.
16-Hexadecanolide (Dihydroambrettolide) 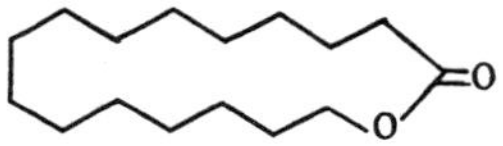 $C_{16}H_{30}O_2$ MW 254	Hymenoptera Halictidae: *Halictus albipes* (Andersson *et al.*, 1967) and *H. calceatus* (Andersson *et al.*, 1967; Bergström, 1974). Colletidae: *Colletes cunicularius* (Bergström, 1974).	Halictidae: Dufour's gland of females. Colletidae: Dufour's gland of females.	It is one of the major constituents in the secretion of *H. calceatus* but it is a trace glandular product of *C. cunicularius*.

18-Octadecenolide $C_{18}H_{32}O_2$ MW 280	Hymenoptera Halictidae: *Halictus albipes* (Andersson *et al.*, 1967) and *H. calceatus* (Andersson *et al.*, 1967; Bergström, 1974). Colletidae: *Colletes cunicularius* (Bergström, 1974).	Halictidae: Dufour's gland of females. Colletidae: Dufour's gland of females.	It is a relatively minor constituent in the section of *C. cunicularius* and a major constituent in those of the halictids. The location of the double bond has not been established.
4-Octadec-9-enolide  $C_{18}H_{32}O_2$ MW 280	Hymenoptera Formicidae: *Lasius flavus* (Bergström and Löfqvist, 1970).	Dufour's gland of workers.	It is a major component which is accompanied by trace amounts of the free acid. It was not detected in the secretions of several other species of *Lasius*.
18-Octadecanolide 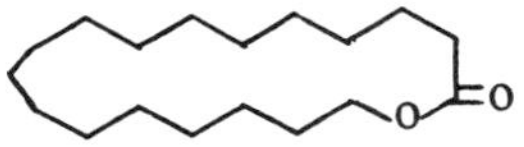 $C_{18}H_{34}O_2$ MW 282	Hymenoptera Halictidae: *Halictus albipes* (Andersson *et al.*, 1967) and *H. calceatus* (Andersson *et al.*, 1967; Bergström, 1974). Colletidae: *Colletes cunicularius* (Bergström, 1974).	Halictidae: Dufour's gland of females. Colletidae: Dufour's gland of females.	It is the major constituent in the secretion of *C. cunicularius* and one of the two major lactones produced by the *Halictus* species.
20-Eicosanolide 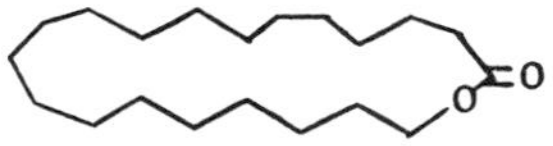 $C_{20}H_{38}O_2$ MW 310	Hymenoptera Colletidae: *Colletes cunicularius* (Bergström, 1974).	Dufour's gland of females.	It is one of four lactones produced by this species and is present as a fairly major constituent.

whose structure has not been clearly established, is synthesized by larvae of a species of *Gastrophysa* (Blum *et al.*, 1978). Recently, nepetalactone was identified as a major constituent in the defensive secretion of a phasmid (Smith *et al.*, 1979).

Qualitatively rich lactonic secretions seem to be characteristic only of bees in the families Colletidae and Halictidae (Bergström, 1974; Hefetz *et al.*, 1978). In virtually all species of these bees that have been examined, at least four macrocyclic lactones are present (Table 9.2), and some of these compounds constitute the largest lactones identified as exocrine products of insects. The defensive secretion of the cerambycid *Phoracantha synonyma* contains three lactones (Moore and Brown, 1976), but these compounds do not appear to be consistent generic products, being absent from another species of *Phoracantha* (Moore and Brown, 1972). For the most part, only one or at best two lactones are glandular products of the other species that have been studied.

While the diverse lactones identified in these secretions are probably effective deterrents for a variety of animal predators, these compounds may possess other roles as well. Bergström and Löfqvist (1970) suggest that the distinctive lactones identified as Dufour's gland constituents of *Lasius flavus* may be important in species recognition vis-à-vis sympatric species of *Lasius*, none of which produces these compounds. The lactonic exudates of colletid and halictid bees may also function as antibiotic agents when applied to the pollen stores on which the larvae must develop. These compounds would then constitute extraordinarily versatile defensive products by being employed against microorganisms in the milieu of the nest, and predators in the foraging territories of the adults.

Notwithstanding the proved defensive roles of some of the lactones (Pavan, 1952; R. W. Kerr, in Cavill *et al.*, 1961), their *raison d'être* in several defensive secretions is obscure at best. Marginalin, the compound responsible for the yellow color of the pygidial gland secretion of *Dytiscus marginalis*, is in admixture with a series of phenolic compounds (Schildknecht *et al.*, 1970) that appear to constitute excellent defensive agents. No role for the lactone is evident. Similarly, the isocoumarins identified as exocrine gland products of *Apsena pubescens* are concomitants of the typical tenebrionid 1,4-benzoquinones (Lloyd *et al.*, 1978a), proved defensive compounds. The factors responsible for the evolution of these lactones as glandular products are unknown, but they surely must have ecological correlates that resulted in the biosynthesis of such idiosyncratic allomones.

Chapter 10

Phenols

Nine phenols (see also Chapter 18) have been identified in the defensive exudates of opilionids, millipedes, and insects (Table 10.1). Neither of the phenols characterized from the opilionid *Cynorta astora* (Eisner *et al.*, 1977), 2,3-dimethylphenol and 2-methyl-5-ethylphenol, have been detected in the secretions of millipedes or insects, although it is possible that the unidentified xylenol produced by a carabid beetle (Moore and Wallbank, 1968) is identical to the dimethylphenol of the harvestman. The presence of phenols in the secretions of millipedes in five families belonging to three orders demonstrates that these compounds have a widespread distribution in members of this arthropod class. In the Insecta, although phenols have been detected in species in four orders (Table 10.1), beetles are clearly preeminent in the biosynthesis of these compounds, producing five of the identified phenolic allomones. The rich variety of phenols characterized in coleopterous defensive exudates further testifies to the well-developed ability of these insects to synthesize aromatic constituents in their exocrine glands.

AN OVERVIEW

The characterized phenolic defensive compounds frequently occur in admixture with constituents belonging to other chemical classes, and in many cases the phenols comprise minor exocrine products in the secretions. This is particularly true for the exudates of polydesmid, euryurid, and paradoxsomatid millipedes, which are dominated by benzaldehyde with the phenolic concomitants phenol and guaiacol consisting

Table 10.1

Distribution of Phenols in the Defensive Secretions of the Arthropoda[a]

Class	Order	Family
Arachnida	Opiliones (Phalangida)	Cosmetidae
Diplopoda	Polydesmida	Euryuridae
		Paradoxosomatidae
		Polydesmidae
	Callipodida	Callipodidae
	Julida	Parajulidae
Insecta	Dictyoptera	
	Blattaria	Blattidae
	Orthoptera	Acrididae
	Trichoptera	Limnephilidae
	Coleoptera	Cerambycidae
		Carabidae
		Tenebrionidae

[a] Also see Chapter 18.

of minor constituents (Blum *et al.*, 1977). On the other hand, the secretions of the callipodid millipedes *Abacion magnum* and *Tetracion jonesi* consist primarily of *p*-cresol (T. Eisner *et al.*, 1963b; Blum *et al.*, 1975) with phenol (Duffield and Blum, 1975c) and nonphenolic compounds being minor concomitants. Among the beetles, the exudates of the tenebrionid and carabid species are essentially phenolic in nature, being dominated by *m*-cresol (Tschinkel, 1969; Moore and Wallbank, 1968); in the cerambycid secretions, on the other hand, toluene is the major constituent (Moore and Brown, 1971b) (Table 10.2).

While even low concentrations of phenolics probably increase the deterrent effectiveness of the exudates, these compounds may possess other significant roles as well. Although the secretions of polydesmid millipedes are fortified with about 95% benzaldehyde, these exudates nevertheless possess a phenolic odor, rather a benzaldehydic one, when first discharged (Monteiro, 1961; Duffield *et al.*, 1974). The initial phenolic note of these secretions may serve as a highly distinctive "early warning system" for predators that have been conditioned to abandon their millipede prey before the latter discharges its copious amounts of benzaldehyde-rich secretion. Thus, the initial phenolic discharge may enable the diplopod to conserve the main defensive products (benzaldehyde-HCN) which are generally only given off after repeated tactile stimulation. The

Table **10.2** Phenols in arthropod defensive secretions

Name and Formula	Occurrence	Glandular Source	Comments
Phenol OH (benzene ring) C_6H_6O MW 94	Polydesmida Paradoxosomatidae: *Oxidus gracilis* (Blum *et al.*, 1973a) and *Orthomorpha coarctata* (Monteiro, 1961; Duffield *et al.*, 1974).	Paradoxosomatidae: Lateral paired glands opening dorsolaterally on the anterior half of most diplosegments.	It is a minor constituent in the secretions of the millipedes and a trace constituent in the exudates of *Z. rugipes* and *P. americana.* It is the major volatile compound present in the secretion of *R. microptera* (ca. 500 ppm)
	Callipodida Callipodidae: *Abacion magnum* (Duffield and Blum, 1975c).	Callipodidae: Lateral paired glands opening dorsolaterally on the anterior half of most diplosegments.	
	Orthoptera Acrididae: *Romalea microptera* (Eisner *et al.*, 1971a).	Acrididae: Paired glands evacuating through the mesothoracic spiracles.	
	Dictyoptera Blattidae: *Periplaneta americana* (Takahashi and Kitamura, 1972).	Blattidae: Ventral glands opening through a bilobed pouch between the sixth and seventh abdominal sternites.	
	Coleoptera Tenebrionidae: *Zophobas rugipes* (Tschinkel, 1969).	Tenebrionidae: Paired prothoracic glands of adults.	

(continued)

Table 10.2 (Continued)

o-Cresol C_7H_8O MW 108	Julida Parajulidae: *Oriulus delus* (Kluge and Eisner, 1971). Orthoptera Acrididae: *Romalea microptera* (Eisner *et al.*, 1971a). Coleoptera Cerambycidae: *Stenocentrus ostricilla* and *Syllitus grammicus* (Moore and Brown, 1971b).	Parajulidae: Lateral paired glands opening dorsolaterally on the anterior half of most diplosegments. Acrididae: Paired glands evacuating through the mesothoracic spiracles. Cerambycidae: Mandibular glands of adults.	It comprises about 20% of the quinone-dominated secretion of *O. delus* and 25% of the toluene-rich exudates of the cerambycids. The beetle exudates are dispensed by an unusual pit-and-tongue organ at the base of each mandible.
m-Cresol C_7H_8O MW 108	Coleoptera Carabidae: *Craspedophorus* sp., *Chlaenius australis*, (Moore and Wallbank, 1968), *C. cordicollis* (Eisner *et al.*, 1963b), *C. tristis*, *C. chrysocephalus*, *C. festivus*, and *Panagaeus bipustulatus* (Schildknecht *et al.*, 1968a). Tenebrionidae: *Zophobas rugipes* (Tschinkel, 1969).	Carabidae: Pygidial glands. Tenebrionidae: Paired prothoracic glands.	It thoroughly dominates the secretions of the carabids and constitutes about 98% of the tenebrionid discharge.
p-Cresol C_7H_8O MW 108	Callipodida Callipodidae: *Abacion magnum* (Eisner *et al.*, 1963b) and *Tetracion jonesi* (Blum *et al.*, 1975). Orthoptera Acrididae: *Romalea microptera* (Eisner *et al.*, 1971a).	Callipodidae: Lateral paired glands opening dorsolaterally on the anterior half of most diplosegments. Acrididae: Paired glands evacuating through the mesothoracic	It is in admixture with a very minor carbonyl constituent(s) and phenol in the millipede secretions whereas the blattid and acridid exudates contain two other phenols as well. It is a minor component

	Dictyoptera Blattidae: Periplaneta americana (Takahashi and Kitamura, 1972).	spiracles. Blattidae: Ventral gland opening through a bilobed pouch between the sixth and seventh abdominal sternites.	which accompanies two indoles in the secretion of P. scabripennis.
	Trichoptera Limnephilidae: Pycnopsyche scabripennis (Duffield et al., 1977).	Limnephilidae: Paired glands opening on the sternites of the seventh abdominal segment of adults.	
Guaiacol (o-Methoxyphenol) [structure: OCH_3, OH] $C_7H_8O_2$ MW 124	Polydesmida Euryuridae: Euryurus australis and E. leachii (Duffield et al., 1974). Paradoxosomatidae: Orthomorpha coarctata (Monteiro, 1961; Duffield et al., 1974).	Euryuridae: Lateral paired glands opening dorsolaterally on the anterior half of most diplosegments. Paradoxosomatidae: Lateral paired glands opening dorsolaterally on the anterior half of most diplosegments.	It is a minor constituent in secretions dominated by benzaldehyde.
m-Ethylphenol [structure: OH] $C_8H_{10}O$ MW 122	Coleoptera Tenebrionidae: Zophobas rugipes (Tschinkel, 1969).	Paired prothoracic glands of adults.	It is a trace constituent.

(continued)

Table 10.2 (Continued)

p-Ethylphenol OH $C_8H_{10}O$ MW 122	Dictyoptera Blattidae: *Periplaneta americana* (Takahashi and Kitamura, 1972).	Ventral glands opening through a bilobed pouch between the sixth and seventh abdominal sternites.	It is present in almost equal concentration with *p*-cresol and presumably is partially responsible for the characteristic "roachy" odor of the species.
Dimethylphenol (Xylenol) $C_8H_{10}O$ MW 122	Coleoptera Carabidae: *Craspedophorus* sp. (Moore and Wallbank, 1968).	Pygidial glands.	It comprises about 10% of a secretion dominated by *m*-cresol. The positions of the methyl groups on the ring have not been determined.

question of how the phenols, which are synthesized in the cyanogenic gland and stored within the aqueous fluid of the reaction chamber (Duffey and Blum, 1977), are released in sufficient quantities to constitute the initial primary odorants in benzaldehyde-dominated exudates remains to be determined.

The occurrence of phenols in the prothoracic glands of Z. *rugipes* may be especially significant since the abdominal glands of this species produce only quinones. Tschinkel (1969) has suggested that the prothoracic glands of this tenebrionid may not possess the oxidative capacity to produce quinones from phenols whereas the necessary enzymes may be present in the abdominal glands. The recent demonstration that phenol is converted to 1,4-benzoquinone in both millipedes and beetles (Duffey and Blum, 1977) is consistent with Tschinkel's speculation regarding phenols as quinonoid precursors. Ultimately it may be demonstrated that quinones are routinely derived from phenolic constituents in many quinone-producing arthropods.

It has also been suggested that for millipedes phenols may play key defensive roles against microorganisms. The pronounced bacteriostatic properties of phenol and related compounds are well known, and phenolics such as phenol and guaiacol within the fluid of the reaction chamber could effectively inhibit soil microorganisms that had penetrated the glandular pore canals (Duffey and Blum, 1977). In addition, these phenolics could provide a vapor pressure in the open pore of the reaction chamber that would effectively limit or prevent the entry of spores or contaminated organic matter (Duffey *et al.*, 1977). In essence, the presence of minor phenolic constituents in the reaction chamber may be highly adaptive to millipedes because of their ability to deter pathogenic bacteria and fungi from utilizing the glandular fluid as a growth medium.

Chapter **11**

Steroids

The prothoracic glands of dytiscid beetles are the source of an impressive group of steroids, many of which are identical to well-known vertebrate hormones (see also Chapter 18). These compounds, which are mostly pregnene and pregnadiene derivatives, must be synthesized from exogenous sterols such as cholesterol (Schildknecht, 1970), since insects do not possess the metabolic capacity to generate the cholestane skeleton *de novo*. The steroidal exudates of dytiscids appear to function primarily as defensive agents against vertebrate predators (Blunck, 1917), and amphibians have been observed to regurgitate living beetles which were encased in a bloody slime (Schildknecht *et al.*, 1967a). Arthropods, on the other hand, regulate their growth and development with nonvertebrate hormones, and they do not exhibit the profound sensitivity to the C_{21} steroids of many vertebrates. However, the pygidial gland secretions of dytiscids may also be utilized to deter predators, notwithstanding the assumption that the products of these exocrine structures function exclusively as antimicrobial agents (Schildknecht, 1970).

Dytiscids are not the only beetles which can biosynthesize steroidal defensive compounds (Table 11.1). Recently, it has been demonstrated that chrysomelid and lampyrid beetles produce cardenolides (Pasteels and Dalooze, 1977) and steroidal pyrones (Eisner *et al.*, 1978), respectively, and some of these compounds constitute unique natural products. A large variety of steroids are produced by beetles in both families (Daloze and Pasteels, 1979; Meinwald *et al.*, 1979; Goetz *et al.*, 1979) and it will not prove surprising if the coleopterans continue to yield new steroidal allomones. Although the chrysomelid-derived compounds constitute true exocrine products, those produced by lampyrids are not se-

Table 11.1

Distribution of Steroids in the Defensive Secretions of the Arthropoda[a]

Class	Order	Family
Insecta	Coleoptera	Dytiscidae Chrysomelidae Lampyridae

[a] Also see Chapter 18.

creted from exocrine glands, but will be discussed here so as to treat all defensive steroids together.

AN OVERVIEW

Species of Dytiscidae, a primitive coleopterous family, possess an extraordinary ability to produce typical vertebrate steroidal hormones as well as several novel steroids (Table 11.2). The syntheses of these compounds, which are produced from cholesterol or related sterols (Schildknecht, 1970), demonstrate that some insects, like vertebrates, possess a well-developed ability to shorten the sterolic side chain while oxidatively modifying the ring system. By contrast, the steroidal hormones which insects utilize as regulators of growth and differentiation, invariably contain an intact side chain which has been hydroxylated (see review by Ritter and Wientjens, 1967). The ability of insects to produce distinctive molting hormones by hydroxylation is in a sense paralleled by the capacity of certain dytiscids to biosynthesize novel steroids in their prothoracic glands. Thus, the 15-hydroxy compounds produced by *Agabus sturmi* (Schildknecht, 1968) and the 12-hydroxy steroids produced by *Cybister* species (Chadha *et al.*, 1970) are distinctive natural products limited in their animal distribution to dytiscids.

Since the prothoracic glands of dytiscids constitute the only known exocrine source of *vertebrate* steroidal hormones in the Arthropoda, these glandular structures apparently represent an unique biosynthetic system in the Insecta. The synthesis of steroids in these exocrine glands probably represents an evolutionary development that made it possible for the dytiscids to exploit their aquatic environments in the presence of abundant vertebrate predators. The narcotic action which these steroids possess for vertebrates (Schildknecht *et al.*, 1967a) demonstrates that the

Table **11.2** Steroids in arthropod defensive secretions

Name and Formula	Occurrence	Glandular Source	Comments
Estrone 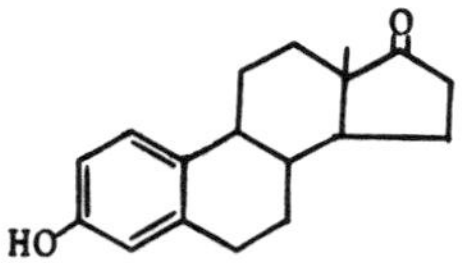 HO O $C_{18}H_{22}O_2$ MW 270	Coleoptera Dytiscidae: *Ilybius fenestratus* (Schildknecht *et al.*, 1967a; Schildknecht and Birringer, 1969b).	Prothoracic glands of adults.	It is the most minor constituent (2 µg/beetle) of an unique glandular exudate containing C_{18}, C_{19}, and C_{21} steroids.
17β-Estradiol 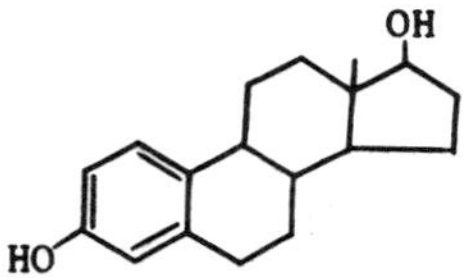 OH HO $C_{18}H_{24}O_2$ MW 272	Coleoptera Dytiscidae: *Ilybius fenestratus* (Schildknecht and Birringer, 1969b).	Prothoracic glands of adults.	It is the major C_{18} steroid (19 µg/beetle) in the exudate.
1,4-Androstadien-17β-ol-3-one 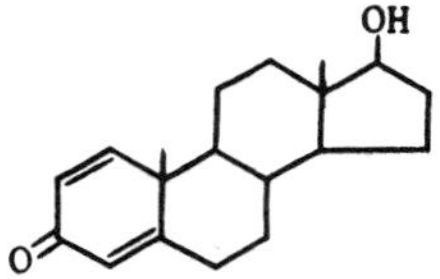 OH O $C_{19}H_{26}O_2$ MW 286	Coleoptera Dytiscidae: *Ilybius fenestratus* (Schildknecht and Birringer, 1969a).	Prothoracic glands of adults.	It is one of two C_{19} steroids in the secretion and each beetle produces about 16 µg.

Testosterone 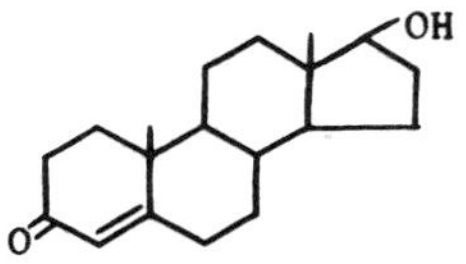 $C_{19}H_{28}O_2$ MW 288	Coleoptera Dytiscidae: *Ilybius fenestratus* and *I. fuliginosus* (Schildknecht *et al.*, 1967a; Schildknecht and Birringer, 1969a,b).	Prothoracic glands of adults.	It is the major steroid (28 μg/beetle) in the secretion and is reported to be a stupefacient for fish.
4,6-Pregnadiene-3,20-dione 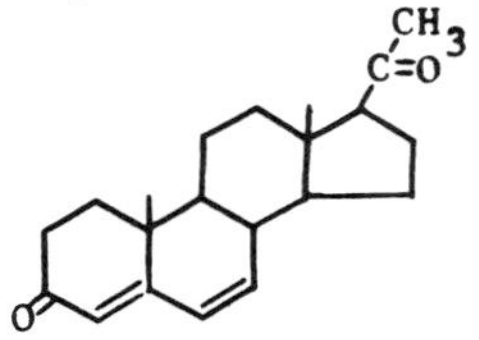 $C_{21}H_{28}O_2$ MW 312	Coleoptera Dytiscidae: *Acilius sulcatus* (Schildknecht *et al.*, 1967b).	Prothoracic glands of adults.	It is a minor constituent (6 μg/beetle) in a secretion which contains at least five C_{21} steroids.
4,6-Pregnadien-12β-ol-3,20-dione 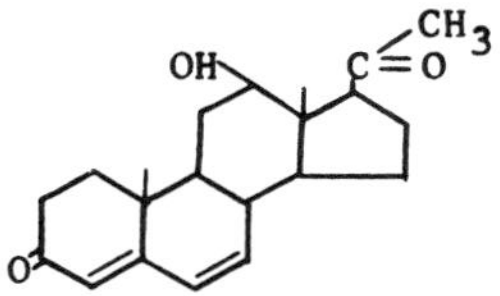 $C_{21}H_{28}O_3$ MW 328	Coleoptera Dytiscidae: *Cybister tripunctatus* (Schildknecht and Körnig, 1968), *C. lateralimarginalis* (Schildknecht, 1971), and *C. limbatus* (Chadha *et al.*, 1970).	Prothoracic glands of adults.	It is a characteristic component in the glandular secretions of *Cybister* spp. which varies in concentration from 1 mg/beetle (*C. tripunctatus*) to 6 μg/beetle (*C. lateralimarginalis*). It has been assigned the trivial name cybisterol.

(*continued*)

Table 11.2 (Continued)

4,6-Pregnadien-15α-ol-3,20-dione 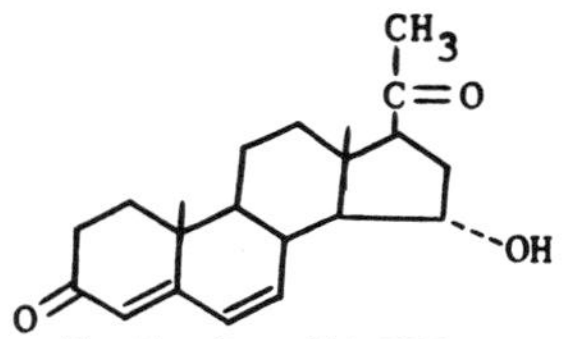 $C_{21}H_{28}O_3$ MW 328	Coleoptera Dytiscidae: *Agabus sturmi* (Schildknecht, 1970).	Prothoracic glands of adults.	It is an uncommon dytiscid steroid which is one of three pregnadienes limited to the genus *Agabus*.
4,6-Pregnadien-21-ol-3,20-dione 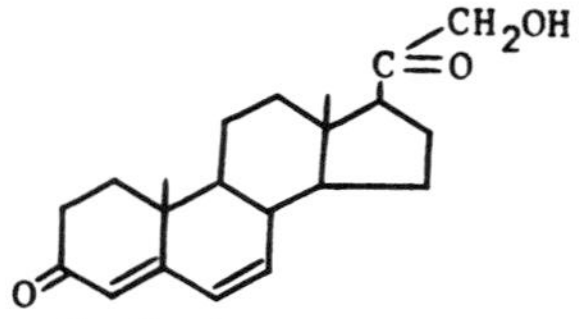 $C_{21}H_{28}O_3$ MW 328	Coleoptera Dytiscidae: *Cybister lateralimarginalis* (Schildknecht, 1968), *C. tripunctatus* (Schildknecht and Körnig, 1968), *C. limbatus* (Sipahimalani *et al.*, 1970), and *Acilius sulcatus* (Schildknecht *et al.*, 1967b).	Prothoracic glands of adults.	Concentrations of this steroid range from 90 μg/beetle (*C. lateralimarginalis*) to 1 μg/beetle (*C. tripunctatus*).
4,6-Pregnadien-20α-ol-3-one 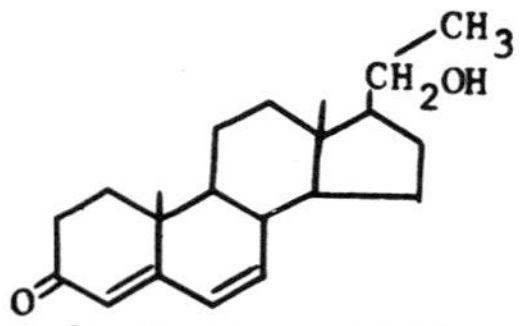 $C_{21}H_{30}O_2$ MW 314	Coleoptera Dytiscidae: *Cybister lateralimarginalis* (Schildknecht *et al.*, 1967d), *C. limbata* (Sipahimalani *et al.*, 1970), *Dytiscus marginalis* (Schildknecht, 1968), and *Acilius sulcatus* (Schildknecht *et al.*, 1967b).	Prothoracic glands of adults.	It is an unique animal steroid which has been assigned the trival name cybisterone. Concentrations range from 140 μg/beetle (*C. lateralimarginalis*) to 7 μg/beetle (*A. sulcatus*).

4-Pregnen-12β-ol-3,20-dione 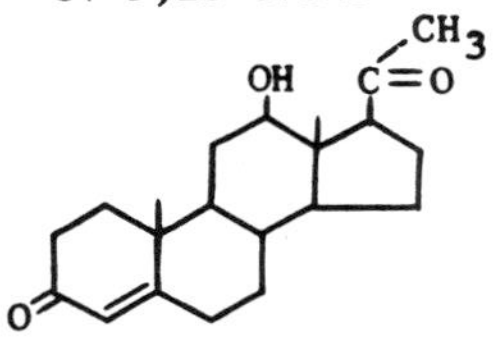 OH, CH_3, C=O $C_{21}H_{30}O_3$ MW 330	Coleoptera Dytiscidae: *Cybister laterali-marginalis* (Schildknecht, 1970) and *C. limbatus* (Chadha *et al*., 1970).	Prothoracic glands of adults.	It is one of several 12-hydroxy steroids which appear to be limited in distribution to the genus *Cybister*.
4-Pregnen-21-ol-3,20-dione (Cortexone) 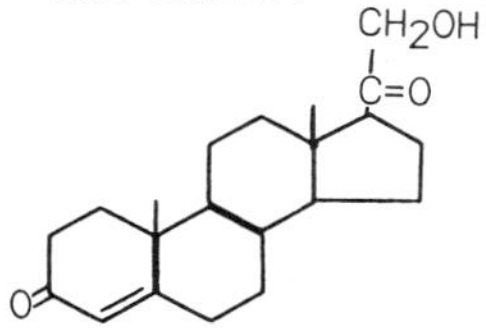 CH_2OH, C=O $C_{21}H_{30}O_3$ MW 330	Coleoptera Dytiscidae: *Dytiscus marginalis* (Schildknecht *et al*., 1966a,b), *Cybister limbatus*, *C. confusus*, *C. tripunctatus*, (Sipahimalani *et al*., 1970; Chadha *et al*., 1970), *C. lateralimarginalis* (Schildknecht, 1971), *Agabus bipustulatus* (Schildknecht, 1968), *A. seriatus* (Miller and Mumma, 1973), *Acilius sulcatus* (Schildknecht *et al*., 1967b), and *Graphoderus liberus* (Miller and Mumma, 1973).	Prothoracic glands of adults.	It is the most widespread steroid identified in the prothoracic secretions of dytiscids. This compound, which is reported to be a powerful narcotic for freshwater fish, ranges in concentration from 1 mg/beetle (*C. limbatus*) to 3 μg/beetle (*C. lateralimarginalis*).

(*continued*)

Table 11.2 (Continued)

4,6-Pregnadiene-12β,20α-diol-3-one 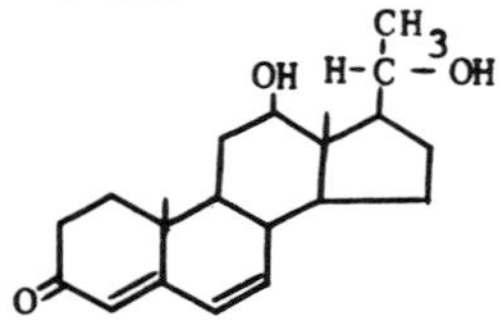 CH3 / OH H-C-OH $C_{21}H_{30}O_3$ MW 330	Coleoptera Dytiscidae: Cybister laterali-marginalis (Schildknecht, 1971).	Prothoracic glands of adults.	It is one of the major compounds (36 μg/beetle) produced in the steroid-rich glands of this species.
4-Pregnen-20α-ol-3-one 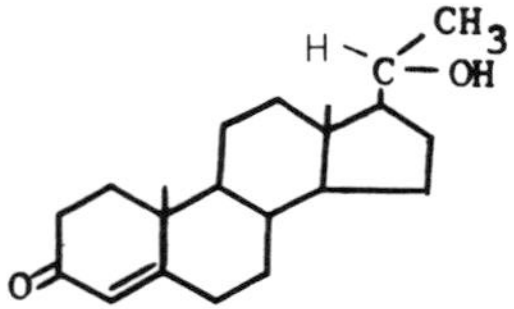 H-C(CH3)-OH $C_{21}H_{32}O_2$ MW 316	Coleoptera Dytiscidae: Dytiscus marginalis (Schildknecht and Hotz, 1967), Acilius sulcatus (Schildknecht et al., 1967b), and Cybister limbata (Sipahimalani et al., 1970; Chadha et al., 1970).	Prothoracic glands of adults.	It is a minor or trace component (1-8 μg/beetle) of all secretions.
4-Pregnen-20β-ol-3-one 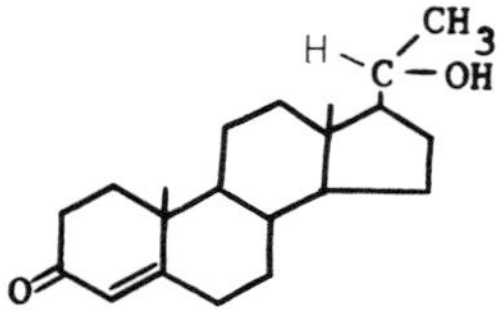 H-C(CH3)-OH $C_{21}H_{32}O_2$ MW 316	Coleoptera Dytiscidae: Ilybius fenestratus (Schildknecht and Birringer, 1969a) and Cybister limbatus (Chadha et al., 1970).	Prothoracic glands of adults.	It is a major glandular product in C. limbatus (100 μg/beetle) and a trace constituent of the secretion of I. fenestratus (1 μg/beetle).

Compound	Order/Family: Species	Source	Remarks
4-Pregnene-20β, 21-diol-3-one 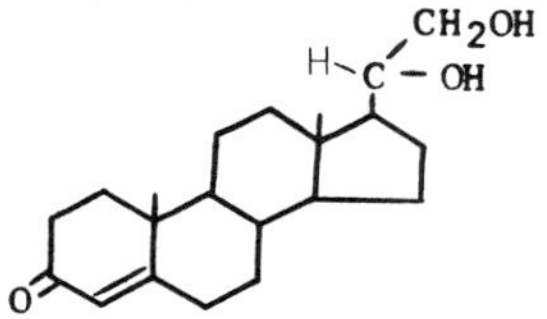 $C_{21}H_{32}O_3$ MW 332	Coleoptera Dytiscidae: *Ilybius fenestratus* (Schildknecht, 1971).	Prothoracic glands of adults.	It is one of two pregnene derivatives in the exudate of this species, both of which constitute minor components.
4-Pregnene-15α, 20β-diol-3-one 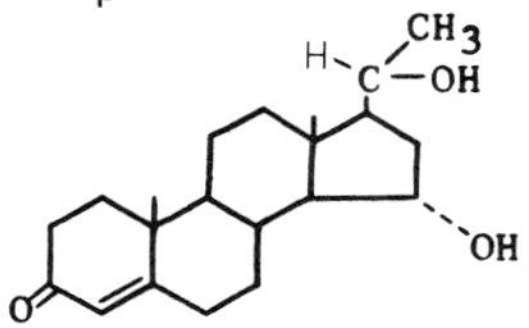 $C_{21}H_{32}O_3$ MW 332	Coleoptera Dytiscidae: *Platambus maculatus* (Schildknecht *et al.*, 1969b) and *Ilybius fenestratus* (Schildknecht, 1971).	Prothoracic glands of adults.	It is the sole steroid in the secretion of *P. maculatus* and is a minor constituent in the exudates of both species (ca. 7 μg/beetle).
4,6-Pregnadien-15α-ol-3,20-dione-isobutyrate 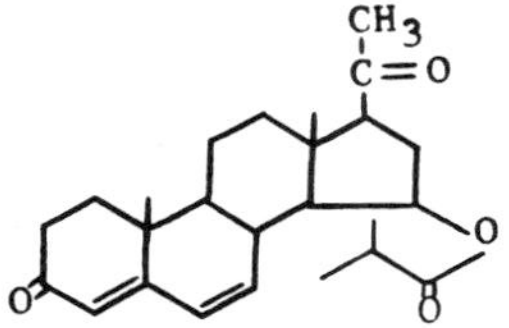 $C_{25}H_{34}O_4$ MW 398	Coleoptera Dytiscidae: *Agabus sturmi* (Schildknecht, 1968).	Prothoracic glands of adults.	It occurs in admixture with the nonesterified steroid.

(continued)

Table 11.2 (Continued)

4,6-Pregnadiene-15 α,20β-diol-3-one-20-isobutyrate 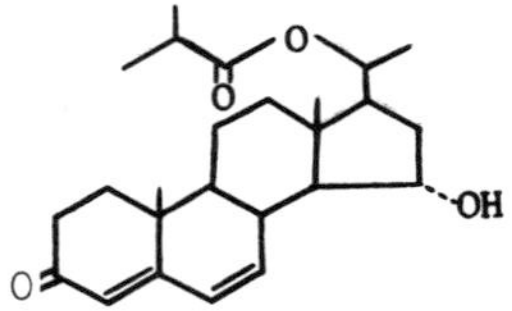 $C_{25}H_{36}O_4$ MW 400	Coleoptera Dytiscidae: *Agabus sturmi* (Schildknecht, 1968).	Prothoracic glands of adults.	It is one of two isobutyrate esters in this secretion.
4-Pregnen-12β-ol-3,20-dione-pentenate-3 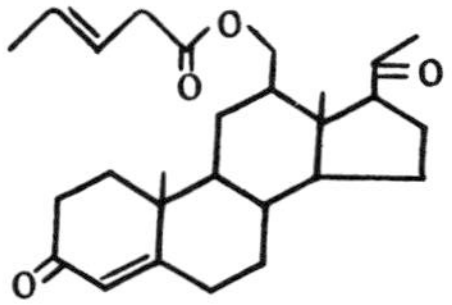 $C_{26}H_{36}O_4$ MW 412	Coleoptera Dytiscidae: *Cybister lateralimarginalis* (Schildknecht, 1971).	Prothoracic glands of adults.	It is a minor constituent (4 μg/beetle) which constitutes the only ester in the secretion of this species.

prothoracic glandular products constitute a potent chemical defensive system against both fish and amphibians, the major available vertebrate predators.

It would be premature to conclude that the functions of the dytiscid prothoracic gland exudates have been fully determined. G. H. Johnson (personal communication, 1973) reported that the secretion of *Cybister fimbriolatus* is enriched with a potent odorant which appears to be a terpenoid constituent. The secretion is highly excitatory to the beetles and causes rapid swimming and dispersion when these dytiscids perceive it. Thus, it is possible that odoriferous constituents in the glandular exudate may cause dispersal of proximate beetles while at the same time enabling vertebrate predators to quickly recognize their prey. In addition, these odoriferous compounds may provide experienced predators with an "early warning" system which enables them to avoid prolonged contact with the potentially toxic beetles. In this way only a small quantity of the exudate may have to be released: the bulk of the steroid-rich secretion would be conserved for subsequent encounters with naive predators. However, the full significance of the prothoracic glandular secretions of the Dytiscidae is apparently still unclear, even if their remarkable chemistry is not.

The cardenolides synthesized by certain species of chrysomelids are particularly distinctive in sometimes containing xylose, a pentose that is not frequently encountered in animal glycosides (Pasteels and Daloze, 1977). Although cardenolides are characteristic of plants in several families, these beetle-derived allomones represent the first examples in this class of compounds produced by animals. Both the aglycones and xylosides are synthesized in elytral and pronotal glands of these aposematic insects (Daloze and Pasteels, 1979) and provide them with an effective defense against a variety of predators. Cardenolides are also present in larvae and eggs of these coleopterans (Pasteels *et al.*, 1979), although the glandular source of these compounds in the former has not been determined. Ultimately, detailed studies on the chemical ecology of chrysomelids and their host plants will be required in order to comprehend the significance of cardenolides being present or absent in species in the same genus.

Lampyrids are unique among arthropods in synthesizing bufodienolides that have been termed lucibufagins (Eisner *et al.*, 1978). The characterization of nine steroidal pyrones (Meinwald *et al.*, 1979; Goetz *et al.*, 1979) in three species of *Photinus* indicates that fireflies will be a rich source of these compounds. Although the lucibufagins are not secreted from exocrine glands, they can be autohemorrhagically discharged from elytral and pronotal pores. In this way the lucibufagins can manifest their distasteful properties before the lampyrid is seriously injured by a predatory vertebrate. Furthermore, the effectiveness of reflex bleeding as a deterrent to small predators (Blum and Sannasi, 1974) is undoubtedly augmented by the presence of the distasteful lucibufagins. Although the pharmacological properties of these steroidal pyrones have not been determined, the production of these natural products appears to be the *sine qua non* that has enabled the fireflies to evolve a nocturnally conspicuous *modus vivendi* in the presence of predatory avian and chiropterous species.

Chapter 12

Miscellaneous Compounds

A real potpourri of natural products including inorganic substances, ethers, alkaloids, furans, sulfides, and nitro compounds have been identified in the defensive secretions of arthropods (Table 12.1). As a matter of convenience, this heterogeneous mixture of allomones is grouped together as miscellaneous compounds (see also Chapter 18). It appears that some of these glandular constituents have a very restricted arthropod distribution and, as a consequence, they may possess value as chemotaxonomic indicators. Beyond that the presence of these diverse and often distinctive natural products in defensive exudates documents the biosynthetic versatility of arthropods and provides strong grounds for anticipating that additional novel compounds will be identified in the multitudinous glandular secretions that have yet to be analyzed.

AN OVERVIEW

The biosynthetic versatility of arthropods is emphasized by the diversity of compounds produced in their exocrine tissues (Table 12.1). A brief examination of these miscellaneous compounds demonstrates that whereas some of these allomones appear to be restricted to species in selected families, others are atypical generic products to the point of being considered idiosyncratic elements of certain species. In short, while generalizations about the chemistry of species in arthropod taxa are obviously useful, they may obscure an appreciation of the exceptions with their exciting biosynthetic and probable ecological correlates.

A case in point is the ant *Lasius fuliginosus*, a formicine that produces a

Table **12.1** Miscellaneous compounds in arthropod secretions

Name and Formula	Occurrence	Glandular Source	Comments
Hydrogen peroxide H-O-O-H H_2O_2 MW 34	Hemiptera Pleidae: *Plea leachi* (Maschwitz, 1971).	Adult metasternal scent gland.	It is secreted as a 10-15% solution and is believed to function as an antibiotic against surface bacteria.
Ammonia NH_3 H_3N MW 17	Coleoptera Silphidae: *Oeceoptoma thorcica*, *Silpha obscura*, and *Phosphuga atrata* (Schildknecht and Weis, 1962b).	Probably the hind gut of the adult.	It is secreted as a 4.5% aqueous solution. It may be derived from the decaying meat fed upon by adults.
Hydrogen cyanide **HC≡N** CHN MW 27	Polydesmida Paradoxosomatidae: *Oxidus gracilis* (Guldensteeden-Egeling, 1882; H.E. Eisner *et al*., 1963), and *Orthomorpha coarctata* (Monteiro, 1961). Gomphodesmidae: *Gomphodesmus pavani* (Barbetta *et al*., 1966) and *Astrodesmus laxus* (H.E. Eisner *et al*., 1975). Euryuridae: *Euryurus australis* and *E. leachii* (Duffield *et al*., 1974).	Paradoxsomatidae: Lateral paired glands opening dorsolaterally on the anterior half of most diplosegments. Gomphodesmidae: Lateral paired glands opening dorsolaterally on the anterior half of most diplosegments.	It appears to be a characteristic defensive product produced by polydesmoid millipedes and along with benzaldehyde, is derived from mandelonitrile. *R. vicinus* is reported to generate HCN from *p*-isopropylmandelonitrile. Xystodesmid millipedes are reported to produce from 16-114 μg of HCN per individual.

Polydesmidae: *Pachydesmus crassicutis* (Blum and Woodring, 1962), *Pseudopolydesmus serratus* (H.E. Eisner *et al*., 1963), *P. branneri* (H.E. Eisner *et al*., 1975), *P. erasus* (Duffey *et al*., 1977), *Polydesmus collaris* (Casnati *et al*., 1963), and *P. virginiensis* (W. M. Wheeler, 1890).	Euryuridae: Lateral paired glands opening dorsolaterally on the anterior half of most diplosegments. Polydesmidae: Lateral paired glands opening dorsolaterally on the anterior half of most diplosegments.	The secretions of the chrysomelids also contain benzaldehyde, a fact which suggests that mandelonitrile is the precursor for the defensive compounds. On the other hand, the centipede secretion is not reported to contain an aromatic aldehyde.
Xystodesmidae: *Apheloria corrugata* (H.E. Eisner *et al*., 1963), *A. coriacea* (Weatherston and Gardiner, 1973), *A. trimaculata*, *A. kleinpeteri* (H.E. Eisner *et al*., 1975), *Cherokia georgiana*, *Nannaria* sp. (H.E. Eisner *et al*., 1963), *Rhysodesmus vicinus* (Pallares, 1946), *Cleptoria rileyi*, *Brachoria* sp., *Sigiria* sp., *Stelgipis agrestis*, *Paimoikia* sp., *Motyxia tiemanni*, (Duffey *et al*., 1977), and *M. sequoiae* (Davenport *et al*., 1952; Duffey *et al*., 1977).	Xystodesmidae: Lateral paired glands opening dorsolaterally on the anterior half of most diplosegments.	
Chelodesmidae: *Caraibodesmus* sp. (Duffey *et al*., 1976).	Chelodesmidae: Lateral paired glands opening dorsolaterally on the anterior half of most diplosegments.	
Nearctodesmidae: *Nearctodesmus cerasinus* (Duffey *et al*., 1976).	Nearctodesmidae: Lateral paired glands opening dorsolaterally on the anterior half of most diplosegments.	
	Geophilidae: Ventral segmentally arranged glands.	

(continued)

Table 12.1 (Continued)

	Geophilomorpha Geophilidae: *Pachymerium ferrugineum* (Schildknecht *et al*., 1968b).	Chrysomelidae: Paired defensive glands at the base of the eighth abdominal tergite.	
	Coleoptera Chrysomelidae: *Paropsis atomaria*, *Chrysophtharta variicolis*, and *C*. *amoena* (Moore, 1967).		
Dimethyldisulfide CH_3-S-S-CH_3 $C_2H_6S_2$ MW 94	Hymenoptera Formicidae: *Paltothyreus tarsatus* (Casnati *et al*., 1967; Crewe and Fletcher, 1974).	Mandibular glands of workers.	It is one of two alkyl sulfides in the secretion and also functions as a releaser of digging behavior.
Dimethyltrisulfide CH_3-S-S-S-CH_3 $C_2H_6S_3$ MW 126	Hymenoptera Formicidae: *Paltothyreus tarsatus* (Casnati *et al*., 1967; Crewe and Fletcher, 1974).	Mandibular glands of workers.	It is the major constituent in the secretion.

Histamine (Also see Nonexocrine defensive compounds of arthropod origin and Nonproteinaceous constituents in arthropod venoms) $C_5H_9N_3$ MW 111	Orthoptera Acrididae: _Poekilocerus bufonius_ (von Euw _et al._, 1967).	A single gland below the first abdominal tergite which discharges through an orifice between the first two tergites.	It constitutes about 1% of the wet weight of the secretion.
Benzoyl cyanide C_8H_5NO MW 131	Polydesmida Xystodesmidae: _Cherokia georgiana_, _Cleptoria rileyi_, _Brachoria_ sp., _Sigiria_ sp., _Stelgipus agrestis_, _Motyxia sequoiae_, _M. tiemanni_, and _Paimokia_ sp. (Duffey _et al._, 1977).	Lateral paired glands opening dorsolaterally on the anterior half of most diplosegments.	It is a major constituent in the secretions of several species and is accompanied by benzaldehyde, HCN, and mandelonitrile benzoate. This compound is a powerful benzoylating agent and its presence undoubtedly increases the defensive efficacy of the secretion considerably.
Indole C_8H_7N MW 117	Trichoptera Limnephilidae: _Pycnopsyche scabripennis_ (Duffield _et al._, 1977).	Paired glands opening on the sternite of the seventh abdominal segment of adults.	It is the major constituent in the secretion and is accompanied by 3-methylindole.

(continued)

Table 12.1 (Continued)

2,6-Dimethyl-3-ethyl-1,4-pyrazine 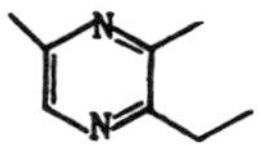 $C_8H_{12}N_2$ MW 136	Hymenoptera Formicidae: Odontomachus brunneus (Wheeler and Blum, 1973).	Mandibular glands of workers.	It is a trace constituent in a secretion containing four pyrazines.
Skatole (3-Methylindole) 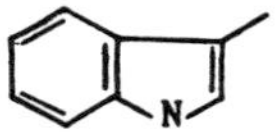 C_9H_9N MW 131	Trichoptera Limnephilidae: Pycnopsyche scabripennis (Duffield et al., 1977). Neuroptera Chrysopidae: Chrysopa septempunctata (Sakan et al., 1970) and C. oculata (Blum et al., 1973d). Hymenoptera Formicidae: Pheidole fallax (Law et al., 1965).	Limnephilidae: Paired glands opening on the sternite of the seventh abdominal segment of adults. Chrysopidae: Paired glands opening on the frontal margin of the adult prothorax. Formicidae: Poison gland of major workers.	It is a minor component in the limnephilid secretion which is dominated by indole. The secretion of C. oculata is dominated by tridecene which probably functions as a solvent for this indole, and facilitates its penetration through the arthropod cuticle.
2,5-Dimethyl-3-n-propyl-1,4-pyrazine 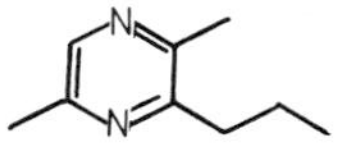 $C_9H_{14}N_2$ MW 150	Hymenoptera Formicidae: Iridomyrmex humilis (Cavill and Houghton, 1974a,b).	Mandibular glands of workers.	It is present at a concentration of only 5 ppm.

2,6-Dimethyl-3-*n*-propyl-1,4-pyrazine 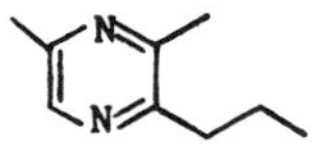 $C_9H_{14}N_2$ MW 150	Hymenoptera Formicidae: *Odontomachus brunneus* (Wheeler and Blum, 1973).	Mandibular glands of workers.	It is one of four pyrazines in the secretion and constitutes about 2% of the mixture.
1,2-Dimethyl-4(3H)-quinazolinone 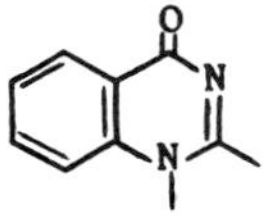 $C_{10}H_{10}N_2O$ MW 174	Glomerida Glomeridae: *Glomeris marginata* (Y.C. Meinwald *et al.*, 1966; Schildknecht *et al.*, 1966c).	Eight glands discharging through pores along the dorsal midline.	It is one of two quinazolines present and comprises about 40% of the mixture. It has been assigned the trivial name of glomerin.
Actinidine 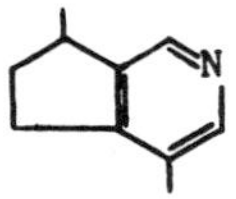 $C_{10}H_{13}N$ MW 147	Coleoptera Staphylinidae: *Hesperus semirufus* and *Philonthus politus* (Bellas *et al.*, 1974).	Pygidial glands.	It is the major constituent in both secretions.
Perillene 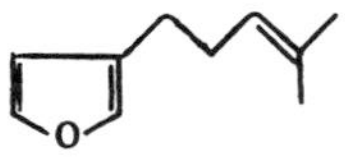 $C_{10}H_{14}O$ MW 150	Hymenoptera Formicidae: *Lasius fuliginosus* (Bernardi *et al.*, 1967), *Tetramorium angulinode*, and *T. sp.* (Longhurst *et al.*, 1979b).	Mandibular glands of workers.	It constitutes about 0.4% of the secretion and is one of two furans present in the *Lasius* exudate. It is the major constituent in the *Tetramorium* secretions.

(continued)

Table 12.1 (Continued)

2,6-Dimethyl-3-*n*-butyl-1,4-pyrazine 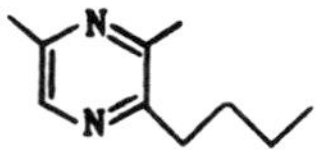 $C_{10}H_{16}N_2$ MW 164	Hymenoptera Formicidae: *Odontomachus brunneus* (Wheeler and Blum, 1973).	Mandibular glands of workers.	It accounts for about 8% of a mixture of four pyrazines.
6,6-Dimethyl-2 azaspiro[4.4]non-1-ene (Polyzonimine)  $C_{10}H_{17}N$ MW 151	Polyzoniida Polyzoniidae: *Polyzonium rosalbum* (Smolanoff *et al*., 1975b).	Lateral paired glands opening dorsolaterally on the anterior half of most diplosegments.	Each millipede produces about 70 μg of this novel nitrogen-containing terpene. Polyzonimine is the first known natural product based on the 2-azaspiro-[4.4]nonane ring system.
1,8-Cineole  $C_{10}H_{18}O$ MW 154	Coleoptera Staphylinidae: *Stenus bipunctatus* (Schildknecht, 1970) and *S. comma* (Schildknecht *et al*., 1976).	Pygidial glands.	It is the main constituent present in the secretion and is one of three surface-active terpenoids which are utilized to propel these beetles rapidly through the water when they are disturbed.

Rose Oxide 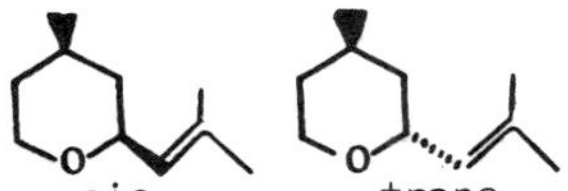 *cis* *trans* $C_{10}H_{18}O$ MW 152	Coleoptera Cerambycidae: *Aromia moschata* (Vidari *et al.*, 1973).	Paired ventral metathoracic glands.	It is present in the secretion as both the cis and trans isomers.
β,β-Dimethylacrylylcholine 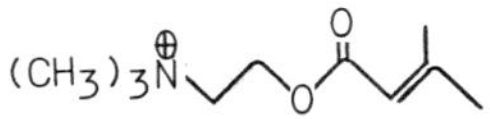 $(CH_3)_3\overset{\oplus}{N}$ $C_{10}H_{20}NO_2$ MW 186	Lepidoptera Arctiidae: *Arctia caja* (Bisset *et al.*, 1960) and *Utetnesia bella* (Rothschild and Haskell, 1966).	Paired prothoracic (cervical) glands of adults.	This pharmacologically active choline ester is present in concentrations equivalent to 1-2 mg of acetylcholine per adult. Other tissues contain large amounts of acetylcholine and in addition, a toxic protein in the abdomen which increases capillary permeability when injected into guinea pigs.
1-Methyl-2-ethyl-4-(3*H*)-quinazolinone 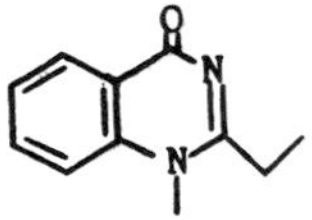 $C_{11}H_{12}N_2O$ MW 188	Glomerida Glomeridae: *Glomeris marginata* (Y. C. Meinwald *et al.*, 1966).	Eight glands discharging through pores along the dorsal midline.	It comprises about 60% of the secretion.

(continued)

Table 12.1 (Continued)

2,6-Dimethyl-3-*n*-pentyl-1,4-pyrazine 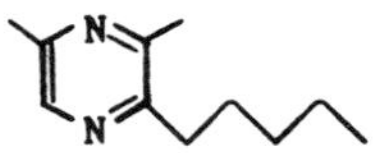 $C_{11}H_{18}N_2$ MW 178	Hymenoptera Formicidae: *Odontomachus brunneus* (Wheeler and Blum, 1973).	Mandibular glands of workers.	It constitutes about 90% of a secretion containing four pyrazines and is also utilized as an alarm pheromone.
2,5-Dimethyl-3-isopentyl-1,4-pyrazine 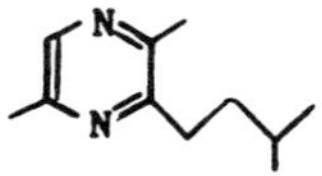 $C_{11}H_{18}N_2$ MW 178	Hymenoptera Formicidae: *Odontomachus hastatus*, *O. clarus* (Wheeler and Blum, 1973), *Iridomyrmex humilis* (Cavill and Houghton, 1974a,b), and *Calomyrmex* sp. (Brown and Moore, 1979).	Mandibular glands of workers.	It is the only volatile compound present in the *Odontomachus* secretions. The presence of different pyrazines in another "subspecies" of *Odontomachus* suggests that the two forms may actually constitute different species.
2,5-Dimethyl-3-(2'-methylbutyl)-pyrazine 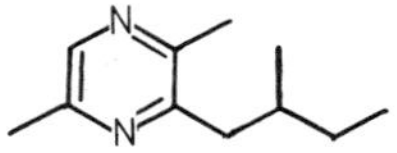 $C_{11}H_{18}N_2$ MW 178	Hymenoptera Formicidae: *Calomyrmex* sp. (Brown and Moore, 1979).	Mandibular glands of workers.	It is accompanied in the secretion by the isomeric isopentylpyrazine.

Nitropolyzonamine 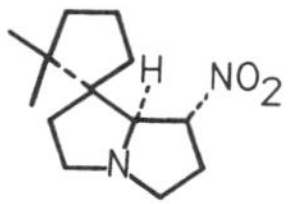 $C_{13}H_{22}N_2O_2$ MW 238	Polyzoniida Polyzoniidae: *Polyzonium rosalbum* (Meinwald *et al.*, 1975).	Lateral paired glands opening dorsolaterally on the anterior half of most diplosegments.	It constitutes 15% of a secretion which is dominated by the bicyclic compound polyzonimine.
2,5-Dimethyl-3-(*E*)- styrylpyrazine 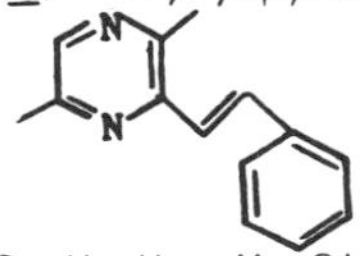 $C_{14}H_{14}N_2$ Mw 210	Hymenoptera Formicidae: *Iridomyrmex humilis* (Cavill and Houghton, 1974a,b).	Mandibular glands of workers.	The (Z)-isomer has also been isolated from the ant but probably constitutes a photoisomeric artifact.
Dendrolasin  $C_{15}H_{22}O$ MW 218	Hymenoptera Formicidae: *Lasius filiginosus* (Quilico *et al.*, 1957b).	Mandibular glands of workers.	It accounts for more than 85% of the glandular secretion and is accompanied by another furan, perilline. Its name is derived from the *Lasius* subgenus *Dendrolasius*.
4,11-Epoxy-*cis*-eudesmane 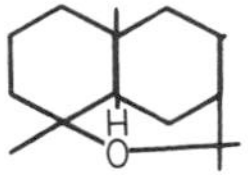 $C_{15}H_{26}O$ MW 222	Dictyoptera Termitidae: *Amitermes evuncifer* (Wadhams *et al.*, 1974).	Frontal gland of soldiers.	It constitutes 90% of the volatiles in the secretion. Only monoterpene hydrocarbons have been identified in the frontal gland exudates of other species of *Amitermes*.

(continued)

Table 12.1 (Continued)

1-Nitro-1-pentadecene  $C_{15}H_{28}NO$ MW 238	Dictyoptera Rhinotermitidae: *Prorhinotermes simplex* (Vrkoč and Ubik, 1974).	Frontal gland of soldiers.	It is present in the secretion as the trans isomer.
Colymbetin	Coleoptera Dytiscidae: *Colymbetes fuscus* (Schildknecht and Tacheci, 1970).	Prothoracic glands of adults.	The secretion is composed of both high and low molecular weight peptides, two of which are reported to possess hypotensive properties when administered intravenously to rats.

rich variety of terpenes in its mandibular glands (Bernardi *et al.*, 1967). Although the occurrence of compounds such as citral and 6-methyl-5-hepten-2-one in the secretion is not particularly surprising, the presence of perillene and dendrolasin is (Table 12.1). These furans are especially atypical products of the genus *Lasius*. Indeed, these compounds may not even be typical of the subgenus in which *L. fuliginosus* is placed since they are lacking in *L. spathepus* (Kistner and Blum, 1971), another species in this taxon. Although perillene has been recently identified as an allomone of some myrmicine species (Longhurst *et al.*, 1979a) (Table 12.1), it and dendrolasin clearly constitute novel and surprising natural products for the genus *Lasius* or for that matter, the Formicinae. The same may be said for actinidine, a compound whose occurrence as an anal gland product of a few *Conomyrma* species (Wheeler *et al.*, 1977b) appears to be quite exceptional for a wide range of ant species in several dolichoderine genera. Ponerine ant species in a variety of genera also produce characteristic allomones. Sulfides are limited as arthropod natural products to species in two ponerine genera (Table 12.1 and Chapter 18) and a host of pyrazines are produced by other ponerine species.

However, the idiosyncratic biochemistry of these ants is more than matched by the natural product versatility of an unrelated group of social insects, the termites. It seems safe to say that the isopterans have often evolved chemical defenses predicated on the availability of highly distinctive allomones. Novel defensive compounds such as 4,11-epoxy-*cis*-eudesmane (Wadhams *et al.*, 1974) and ancistrofuran (Evans *et al.*, 1979) are illustrative of the structural diversity that often characterizes the allomonal products of termitid soldiers. However, the identification of 1-nitro-1-pentadecene as a glandular product of a rhinotermitid species (Table 12.1) demonstrates that among termites, unusual exocrine compounds are not limited to the secretions of the Termitidae. Although this allomone is the only nitro compound identified as an insect exocrine product, a unique compound, nitropolyzonamine, has been characterized in the defensive secretion of a millipede (Meinwald *et al.*, 1975). These results serve to emphasize that chemical surprises may be the order of the day as the natural products of arthropods in a wide variety of higher taxa are subjected to analytical scrutiny.

Chapter 13

Nonexocrine Defensive Compounds of Arthropod Origin

Insect species in the orders Coleoptera and Lepidoptera synthesize a variety of compounds that are stored in specific tissues or are found widely distributed in body organs and blood (Table 13.1 and Chapter 18). While some of these compounds constitute relatively simple substances (e.g., HCN), others represent the most complex nonproteinaceous organic molecules (e.g., pederin) (Table 13.2), which are synthesized by insects as defensive products. In the case of insects that exhibit reflex bleeding (e.g., coccinellids and lampyrids), these compounds may be present in the discharged blood, but they are nevertheless included in this section because they are distributed throughout the body rather than being the products of specific exocrine glands.

AN OVERVIEW

Although relatively few nonexocrine defensive compounds have been characterized so far, a multitude of unrecognized compounds probably are produced by arthropod species. Many aposematic arthropods that have not developed on plants containing toxic plant products are not fed upon readily by predators, notwithstanding their lack of demonstrable exocrine secretions. Since many of these species are distasteful, it seems certain that some of them contain natural products, synthesized *de novo*,

Table 13.1

Distribution of Nonexocrine Compounds in the Defensive Secretions of the Arthropoda[a]

Class	Order	Family
Insecta	Lepidoptera	Zygaenidae
		Arctiidae
		Lymantridae
		Sphingidae
		Saturniidae
	Coleoptera	Coccinellidae
		Meloidae
		Staphylinidae
		Chrysomelidae
		Lampyridae[b]

[a] Also see Chapter 18.
[b] Chemistry described in Chapter 11 on Steroids.

that render them distasteful or potentially toxic and thereby function as defensive agents. Thus, there is good reason to anticipate that a storehouse of novel defensive compounds will be isolated from warningly colored arthropods that lack exocrine glands. This appears to be especially the case for species that exhibit autohemorrhage, the discharged and distasteful blood functioning as a first line of defense against many predators.

Several of the nonexocrine defensive compounds are identical to exocrine products of species in other arthropod taxa. Both histamine and acetylcholine, typical defensive compounds of a variety of lepidopterous species (Table 13.2), are major constituents in the venoms of some vespid wasps (Bhoola *et al.*, 1961). HCN, a nonexocrine product of both larval and adult zygaenid moths (Jones *et al.*, 1962), is a commonly produced glandular product of polydesmid millipedes and some insects as well (Table 12.1). Recently, three of the characteristic nonexocrine alkaloids produced by a variety of coccinellid species have been identified as part of the defensive secretion of a cantharid beetle (Moore and Brown, 1978). In short, what is of selective value is the evolution of effective defensive compounds, whether they be of exocrine or nonexocrine origins. Thus, there are no compelling reasons for believing that compounds currently known only as nonexocrine allomones will not eventually be identified as exocrine products of other species.

Table **13.2** Nonexocrine defensive compounds of arthropod origin

Name and Formula	Occurrence	Source	Comments
Hydrogen cyanide HC≡N CHN MW 27	Lepidoptera Zygaenidae: *Zygaena filipendulae*, *Z. lonicerae*, *Z. trifolii*, and *Procris geryon* (Jones *et al.*, 1962).	It is found in eggs, larvae, and pupae of the *Zygaena* species and adults of *P. geryon*. It is detectable in the hemolymph of adults.	The amount of HCN liberated varies with both the life stage and age of the adult. Older moths appear to be almost cyanide-free whereas freshly emerged adults may produce up to 200 µg Although larvae often develop on plants containing cyanogenic glucosides, HCN is released from crushed larvae reared on acyanogenic plants. The adults,which can liberate HCN-enriched blood from specialized bleeding areas, are highly resistant to this compound.

Histamine (also see Miscellaneous defensive compounds in arthropod secretions and Nonproteinaceous constituents in arthropod venoms) $C_5H_9N_3$ MW 111	Coleoptera Coccinellidae: *Adalia bipunctata* and *Coccinella septempunctata* (Frazer and Rothschild, 1962). Lepidoptera Arctiidae: *Tyria jacobaeae* (Bisset *et al.*, 1960) and *Spilosoma lubricipeda* (Bisset *et al.*, 1960; Morley and Schachter, 1963). Lymantridae: *Euproctis similis* (Frazer and Rothschild, 1962). Sphingidae: *Laothoe populi* and *Smerinthus ocellatus* (Bisset *et al.*, 1960; Morley and Schachter, 1963). Zygaenidae: *Zygaena lonicerae* (Bisset *et al.*, 1960). Saturniidae: *Dirphia* sp. (Valle *et al.*, 1954; Picarelli and Valle, 1971).	Whole bodies of adult insects have been analyzed in general. It is found in the wings of *S. lubricipeda* but is mostly concentrated in the abdomen. The spines of larvae of *Dirphia* sp. contain high concentrations of this amine.	It has either not been detected in the tissues of many other species in these families or is a very minor constituent. Compounds with histamine-like properties are present in a number of additional arctiid species. Concentrations range from 75 μg/gm (*S. ocellatus*) to ca. 700 μg/gm (*H. jacobaeae*). It is 7-8x more concentrated in the body of a female of *S. lubricipeda* than that of a male. The urticating setae of larvae of *Dirphia* sp. contain 86-172 μg of histamine and a larva may contain 100-200 mg/gm.
Acetylcholine (also see Nonproteinaceous constituents in arthropod venoms) $(CH_3)_3\overset{\oplus}{N}$ $C_7H_{16}NO_2$ MW 146	Lepidoptera Arctiidae: *Arctia caja*, *A. villica* and *Tyria jacobaeae* (Morley and Schachter, 1963). Zygaenidae: *Zygaena filipendulae* and *Z. lonicerae* (Morley and Schachter, 1963).	It is concentrated in the ejaculatory duct and accessory glands of males of *A. caja* and the ovaries and eggs of females of this species. The male reproductive system of zygaenid adults is also rich in this ester.	Substances with ACh-like activity have been found in the nonnervous tissues of other lepidopterous species. Additional arctiid species contain low concentrations of ACh in their organs. The ejaculatory duct of males of *A. caja*

(continued)

Table 13.2 (Continued)

			contains 4 mg/gm (dry weight) of this ester. The silk gland of the larvae also contains about 4 mg/gm of ACh.
Cantharidin (Spanish fly) 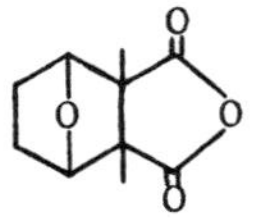 $C_{10}H_{12}O_4$ MW 196	Coleoptera Meloidae: Meloe proscarabeus (Kobert, 1906; Dixon et al., 1963), Mylabris oculata (Colledge, 1910), Macrobasis albida (Viehoever and Capen, 1923), Lytta (=Cantharis) vesicatoria (Blyth and Blyth, 1920), Cissites cephalutes, (=maxillosa) (van Zijp, 1922), Lydus trimaculatus (Kobert, 1906), Epicauta ruficeps (van Zijp, 1917), Decapotama (=Mylabris) lunata (Colledge, 1910), Horia debyi (van Zijp, 1922), Cyaneolytta (=Lytta) gigas (Kobert, 1906), Eletica wahlbergia (Colledge, 1910), and many other species in these genera. (See review by Dixon et al., 1963).	It is found in eggs, larvae, and adults. The main source of this compound in larvae and adults is the hemolymph. It is also concentrated in the male reproductive organs.	It is present in the cis-exo form and occurs in concentrations ranging from 0.3-0.4% (wet weight). Although it is present in adult females, it does not appear to be synthesized by them in this life stage. Adult meloids reflex bleed, especially at the femorotibial joints, and the discharged blood of M. proscarabeus contains 26% cantharidin.

Propyleine (Dehydrococcinellin) $C_{13}H_{21}N$ MW 191	Coleoptera Coccinellidae: *Propylaea quatuordecimpunctata* (Tursch *et al.*, 1972a).	It is concentrated in the hemolymph.	Each beetle contains ca. 25 μg of this compound which constitutes the only alkaloid detectable in the blood.
Precoccinelline (2-Methyl-*cis*, *trans*,-*cis*-perhydro-9*b*-azaphenalene) $C_{13}H_{23}N$ MW 193	Coleoptera Coccinellidae: *Coccinella septempunctata* (Tursch *et al.*, 1971a, 1971b), *C. pentempunctata*, *C. tetradecempunctata* (Pasteels *et al.*, 1973), *Micraspis hexadecempunctata*, and *Cheilomenes propinqua* (Tursch *et al.*, 1975).	It is primarily concentrated in the hemolymph of adults and is also present in eggs and larvae.	It constitutes about 7% of a mixture of two blood-borne alkaloids. Each adult beetle contains about 7 μg of this compound.
Hippodamine (2-Methyl-*trans*, *cis*,-*cis*-perhydro-9*b*-azaphenalene) $C_{13}H_{23}N$ MW 193	Coleoptera Coccinellidae: *Hippodamia convergens* (Tursch *et al.*, 1972b) and *Anisosticta nonadecempunctata* (Pasteels *et al.*, 1973).	The hemolymph of adults is enriched with this azaphenalene.	It is a stereoisomer of precoccinelline and constitutes about 33% of a mixture of two alkaloids present in the blood.

(continued)

Table 13.2 (Continued)

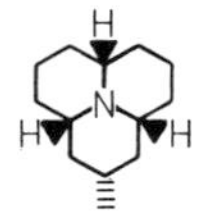 Myrrhine (2-Methyl-*trans*, *trans*,-*trans*-perhydro-9*b*-azaphenalene) $C_{13}H_{23}N$ MW 193	Coleoptera Coccinellidae: *Myrrha octodecimguttata* (Tursch *et al*., 1975).	It is present in the hemolymph of adults.	It is a stereoisomer of both precoccinelline and hippodamine.
Coccinelline 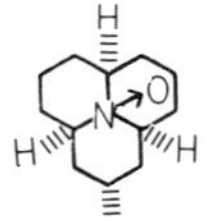 $C_{13}H_{23}NO$ MW 209	Coleoptera Coccinellidae: *Coccinella septempunctata* (Tursch *et al*., 1971a, 1971b), *C. pentempunctata*, *C. undecempunctata*, *C. tetradecempunctata* (Pasteels *et al*., 1973), *C. californica*, and *Cheilomenes propinqua* (Tursch *et al*., 1975).	It is present in the hemolymph of adults.	It is accompanied by the nonoxide alkaloid precoccinelline, which may be its precursor. It is responsible for the bitter taste of the adult beetle, each of which contains ca. 85 μg. It possesses the cis configuration (Karlsson and Losman, 1972).
Convergine 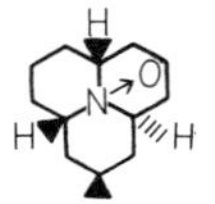 $C_{13}H_{23}NO$ MW 209	Coleoptera Coccinellidae: *Hippodamia convergens* (Tursch *et al*., 1972b).	It is primarily concentrated in the hemolymph.	It is accompanied by a minor constituent, hippodamine.

Adaline 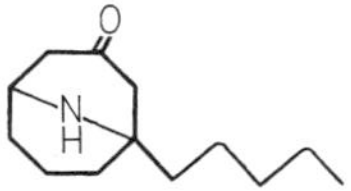 $C_{13}H_{23}NO$ MW 209	Coleoptera Coccinellidae: *Adalia bipunctata* (Tursch *et al.*, 1973) and *A. decempunctata* (Pasteels *et al.*, 1973).	It is concentrated in the hemolymph of adults.	It is the only ketonic alkaloid identified in the Coccinellidae.
Pederone 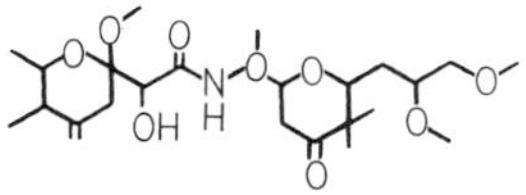 $C_{24}H_{43}NO_9$ MW 489	Coleoptera Staphylinidae: *Paederus fuscipes* (Cardani *et al.*, 1967).	It is present in the hemolymph and organs of adults.	The average yield of this amine is 0.00025–0.005% (wet weight). The position of the carbonyl group at C-4 is inferred from structural studies on the closely related compound pederin (Matsumoto *et al.*, 1968; Furusaki *et al.*, 1968).
Pseudopederin 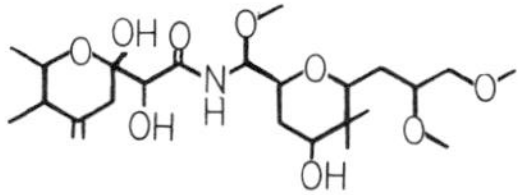 $C_{24}H_{43}NO_9$ MW 489	Coleoptera Staphylinidae: *Paederus fuscipes* (Cardani *et al.*, 1965a, 1966).	It has been isolated from the hemolymph of adults.	It is a trace constituent which accompanies pederin.
Pederin 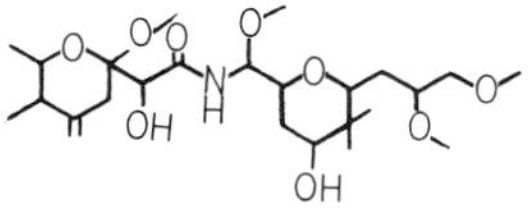 $C_{25}H_{45}NO_9$ MW 503	Coleoptera Staphylinidae: *Paederus fuscipes* (Cardani *et al.*, 1965a, 1966), *P. litoralis*, *P. rubrothoracicus*, and *P. rufocyaneus* (Cardani *et al.*, 1965b).	It is present in hemolymph of adults and is probably found in various tissues as well.	It is the major vesicatory compound produced and is present at a concentration of 1 μg/beetle. Females contain more than males and it may constitute 0.025% of

(continued)

Table 13.2 (Continued)

			the wet weight of the former.
Leptinotarsin MW 50,000	Coleoptera Chrysomelidae: *Leptinotarsa decemlineata* (Hsiao and Fraenkel, 1969; Parker, 1972) and *L. juncta* (Parker, 1972).	It has been detected in eggs, larvae, pupae, and adults. In larvae it is restricted to the hemolymph.	It is an acidic protein that is synthesized independent of diet. Although it is toxic to both insects and vertebrates when injected, it possesses no demonstrable oral toxicity.

In a few cases, nonexocrine defensive compounds can also be considered as exocrine products during a particular life stage. This is clearly the case for the cantharidin that is synthesized in the male accessory glands of meloid beetles (Meyer *et al.*, 1968). Cantharidin is transferred to the female, which does not produce this compound, as part of the seminal ejaculate, and this terpenoid anhydride can clearly be regarded as an exocrine product of the male accessory glands that is subsequently utilized as an allomone by the females. The same may be true for acetylcholine which is synthesized in the ejaculatory duct and accessory glands of males of the moth *Arctia caja* (Morley and Schachter, 1963).

In general, most of the nonexocrine defensive compounds have proven to be very characteristic of the groups that produce them, and these substances appear to possess considerable chemotaxonomic value. At this juncture it is also evident that the evolution of these internal chemical defenses has been emphasized particularly by selected groups of beetles. Cantharidin, a relatively simple bicyclic monoterpene (Table 13.2), is only known to be produced by species in the family Meloidae and has not been detected in members of closely related families. Pederin and its derivatives are only known from a few species in the staphylinid genus *Paederus*, and these amides may constitute valid character states for the species in this taxon. The 13 alkaloids (Table 12.2 and Chapter 18) identified as coccinellid defensive compounds are clearly trademarks of species in many genera in this family, notwithstanding their scattered occurrence in other coleopterous families. Steroidal lucibufagins (see Chapter 11 on Steroids) may also be characteristic nonexocrine compounds of selected lampyrid species (Meinwald *et al.*, 1979; Goetz *et al.*, 1979). Even a compound as simple as hydrogen cyanide is only known as a nonexocrine defensive compound from some species in the lepidopterous family Zygaenidae.

Many aposematic species producing nonexocrine defensive compounds also possess distinct odors that are not identified with the allomones that have been detected in their blood and organs. This is particularly true of coccinellids and lampyrids, species of which are often highly odoriferous. It is not unlikely that these odoriferous compounds constitute another system evolved to signal to predators that their producers are distasteful or emetic prey. Since many aposematic arthropods are known to possess characteristic odors of nonexocrine origin, it may well be that a host of additional compounds remains to be identified which will ultimately be shown to constitute part of the chemical defenses utilized to deliver an emphatic message of unpalatibility.

Chapter 14

Proteinaceous Venoms

While poisonous species are widely distributed throughout the Arthropoda, not all of these invertebrates are venomous. It is important to distinguish cryptotoxic species from phanerotoxic animals which contain a venom apparatus (Russell, 1965). Cryptotoxic animals, which do not possess an external secretory apparatus, are poisonous only after being ingested, whereas the toxins of phanerotoxic invertebrates are generally delivered into the host by injection as part of an offensive or defensive reaction. Venomous (phanerotoxic) species are characterized by a venom apparatus consisting of a poison gland, reservoir, a venom duct, and generally a device for injecting the venom into another animal. These terminological criteria are used to separate the species with nonvenomous defensive secretions from those with true venoms. Thus, the quinone-rich exudates of tenebrionids may be poisonous, but they are not classified as venomous in the sense that the constituents of bee venom are.

Some of the pharmacologically active compounds present in the venomous secretions of arthropods are identical to those already present in the animals which are stung or bitten, and the validity of regarding these compounds as toxins may be questioned. However, when these compounds (e.g., histamine) are applied in toxic doses under nonphysiological conditions, they constitute true toxins (Vogt, 1970). As another example, the salivary secretions of many predatory Hemiptera are generally enriched with enzymes which appear to correspond to enzymes which are normally utilized in internal digestion. Therefore, although the salivary apparatus is not a venom gland in the sense of those of scorpions or spiders, it nevertheless has been biochemically adapted to function as an envenomating tool which converts normal digestive constituents into dangerous physiological agents. On the other hand, while the salivary products of trombiculid mites (Acarina) contain potent necrotizing substances (Vitzthum, 1929) and therefore constitute venomous secretions, these diminutive arachnids appear to utilize these salivary components only for feeding and not for defense. While venomous species are present in the arthropod classes Chilopoda, Arachnida, and Insecta, their venom-injecting apparatuses are highly variable. Spiders (Araneida) are the only arachnids with chelicerae which function as poison fangs. The paired chelicerae, located on the cephalothorax, inject venom into captured prey. The venom originates in glands located in the basal segment of these structures in true spiders and in the cephalothorax in the "tarantulas." In the centipedes (Chilopoda), the venom glands are situated in the telopodite of the paired poison jaws, which are the modified first pair of legs of the trunk, the maxillipeds. The stings of scorpions are present on the last segment of the postabdomen (metasoma), the telson, which contains paired poison glands. In the Insecta, the venom of hymenopterous species is injected through a sting that is connected to a poison gland located in the distal area of the abdomen. Salivary secretions of predatory insects are discharged through piercing structures which are highly variable in structure. The morphology of the venom apparatuses of arthropods has been most recently reviewed by Bücherl and Buckley (1971).

In general a venom gland that is located at the oral end of the animal is utilized solely for the acquisition of food, whereas an apparatus at the aboral end frequently indicates a purely defensive function. Thus, the venom jaws of centipedes, the chelicerae of spiders, and the salivary secretions of many hemipterous insects are primarily employed to immobilize prey prior to feeding. On the other hand, the stinging apparatuses of scorpicns and bees appear to be used strictly in defensive contexts. Those of ants and some wasps may be employed either for offense or defense. Indeed, it has not been unequivocally established whether or not social wasps normally envenom their prey during the process of capture.

Almost all arthropod venoms contain proteinaceous constituents, and this common denominator is used in this chapter as a unifying criterion for these diverse exocrine products. However, in addition to these macromolecular components of the venoms, the small biologically active constituents will also be comparatively treated so as to identify the compounds which are common to arthropod venoms. Since it appears that species in each order synthesize characteristic proteinaceous constituents, the venomous macromolecular composition of each arthropod order will be treated separately. Venoms will only be discussed in terms of the chemistry of their constituents, and no real attempt will be made to treat the multitude of papers which deal with the physiological or pharmacological effects of these secretions on animals. Excellent discussions of the pharmacological activities of these venomous products can be found in articles included in several comprehensive review volumes (Buckley and Porges, 1956; Welsh, 1964; Russell and Saunders, 1967; Bücherl and Buckley, 1971; Pavan and Dazzini, 1971; Bettini, 1978).

The proteinaceous venoms of arthropods are all delivered into their prey by injection and probably these secretions would generally not pack a deterrent "punch" if they were applied to the skin of potential predators. However, the urticating setae of a lepidopterous larva appear to be devoid of proteins (Picarelli and Valle, 1971), and there may be other arthropod venoms which lack proteinaceous constituents. Obviously, the formic acid-rich venoms of formicine ants are excluded from this chapter because these secretions, which are sprayed at their assailants, are devoid of proteins. However, although formicine venoms do not contain proteins, they may be fortified with other compounds such as small peptides or free amino acids (Osman and Brander, 1961; Hermann and Blum, 1968), and the venom is frequently sprayed into a cuticular site which has been disrupted by mandibular abrasion (Ghent, 1961). Thus, even the venoms of the Formicinae contain proteinaceous precursors and, since they can be delivered into a wound at such close range, in a sense these secretions are sometimes virtually injected, much like the venoms of their stinging hymenopterous counterparts.

The proteinaceous venoms of arthropods are frequently fortified with a variety of low molecular-weight constituents which often possess potent pharmacological activities. These compounds, which are listed in Table 14.1, are discussed under the taxa in whose venomous secretions they occur.

I. SCORPION VENOMS

Although histamine and a few indole compounds have been detected in the venoms of a few species of scorpions, low molecular-weight constituents have not been commonly encountered in these secretions. Histamine has been identified in the venom of the scorpion *Palamneus gravimanus* (Ismail *et al.*, 1975); 5–7.2 μg are present in each mg of dry

Table **14.1** Nonproteinaceous constituents in arthropod venoms

Name and Formula	Occurrence	Glandular Source	Comments
Trimethylenediamine H_2N–$CH_2CH_2CH_2$–NH_2 $C_3H_{10}N_2$ MW 74	Araneida Theraphosidae (Aviculariidae): *Grammostola mollicoma*, *Pamphobeteus tetracanthus*, *P. roseus*, *P. soracabae*, *Eurypelma vellutinum*, *Acanthoscurria atrox*, and *Lasiodura klugii* (Fischer and Bohn, 1957a).	A venom gland in the basal segment of each chelicera.	It is bound to several *p*-hydroxyaliphatic acids in at least two fractions and is a minor base accompanying bound spermine in the venoms.
γ-Aminobutyric acid H_2N–$CH_2CH_2CH_2$–COOH $C_4H_9NO_2$ MW 103	Araneida Theraphosidae (Aviculariidae): *Grammostola mollicoma*, *G. actaeon*, *G. pulchripes*, *Acanthoscurria atrox*, *Pamphobeteus soracabae*, *P. tetracanthus*, *P. roseus*, *Lasiodura klugii*, *Eurypelma vellutinum* (Fischer and Bohn, 1957a), and *Dugesiella hentzi* (Schanbacher *et al*., 1973). Dipluridae: *Atrax robustus* (Gilbo and Coles, 1964). Theridiidae: *Latrodectus mactans* (Bettini and Toschi-Frontali, 1962).	Theraphosidae: A venom gland in the basal segment of each chelicera. Dipluridae: A venom gland in the basal segment of each chelicera. Theridiidae: A venom gland in the cephalothorax adjacent to the basal segment of each chelicera.	In the venoms of the theraphosid species it is present at a concentration of 1.5-7.1 gm/100 gm venom. It is believed to act on the peripheral nervous system of mammals and may play a role in increasing the toxicity of the proteinaceous toxins in the venoms.

(*continued*)

Table 14.1 (Continued)

Histamine (Also see Miscellaneous compounds in arthropod defensive secretions and Nonexocrine defensive compounds of arthropod origin) 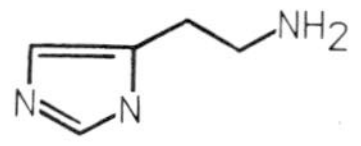 $C_5H_9N_3$ MW 111	Scorpinonida Scorpionidae: *Palamneus gravimanus* (Ismail *et al.*, 1975).	Scorpionidae: Paired poison glands in the distal segment of the postabdomen.	It is not a characteristic constituent of scorpion venoms. The venom of *P. gravimanus* contains 5-7.2 μg/mg of dried venom. Although it is not a common constituent in spider venoms, it is highly concentrated in the venoms of these two species (1-3.6%). It appears to be a characteristic component of hymenopterous venoms and it present at an average concentration of about 2%.
	Araneida Lycosidae: *Lycosa erythrognatha* (McCrone, 1969).	Lycosidae: A venom gland in the cephalothorax adjacent to the basal segment of each chelicera.	
	Ctenidae: *Phoneutria fera* (Fischer and Bohn, 1957b) and *P. nigriventer* (Diniz, 1963).	Ctenidae: A venom gland in the cephalothorax adjacent to the basal segment of each chelicera.	
	Lepidoptera Saturniidae: *Dirphia* sp. (Valle *et al.*, 1954).	Saturniidae: Larval setae on dorsal and lateral tubercles.	
	Hymenoptera Vespidae: *Polistes exclamans*, *P. annularis*, *P. fuscatus* (Prado *et al.*, 1966), *Vespula vulgaris* (Jaques and Schachter, 1954), *Vespa crabro* (Bhoola *et al.*, 1961), and *V. orientalis* (Edery *et al.*, 1972).	Vespidae: Poison gland of workers.	
	Apidae: *Apis mellifera* (Tetsch and Wolff, 1936; Schachter and Thain, 1954; Owen and Braidwood, 1974).	Apidae: Poison gland of workers.	
	Formicidae: *Myrmecia pyriformis* (de la Lande *et al.*, 1963) and *M. gulosa* (Cavill *et al.*, 1964).	Formicidae: Poison gland of workers.	

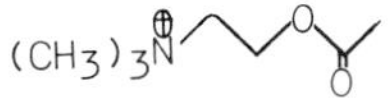 Acetylcholine $(CH_3)_3\overset{\oplus}{N}CH_2CH_2OCOCH_3$ $C_7H_{16}NO_2$ MW 146	Hymenoptera Vespidae: *Vespa crabro* (Bhoola *et al*., 1961) and *V. orientalis* (Edery *et al*., 1972).	Poison gland of worker.	These venoms are the richest natural source of this ester in the animal kingdom, containing up to 5% on a dry weight basis.
Dopamine 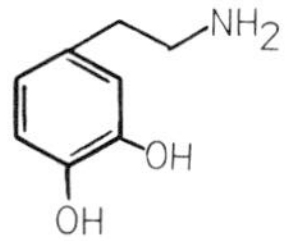 $C_8H_{11}NO_2$ MW 153	Hymenoptera Vespidae: *Vespula arenaria*, *V. maculata* (Owen, 1971), *Vespa orientalis*, *V. crabro*, *Paravespula germanica*, *P. vulgaris*, and *Dolichovespula saxonia* (Ishay *et al*., 1974). Apidae: *Apis mellifera* (Owen, 1971).	Vespidae: Poison gland of worker. Apidae: Poison gland of worker.	It was detected in extracts of the poison gland contents and is highly concentrated in honey bee venom (1 mg/gm) and some of the wasp venoms.
Noradrenaline 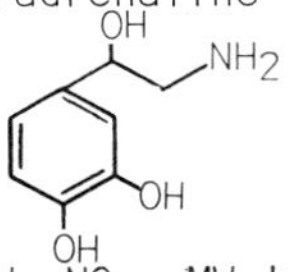 $C_8H_{11}NO_3$ MW 169	Hymenoptera Vespidae: *Vespula arenaria*, *V. maculata* (Owen, 1971), *Vespa orientalis*, *V. crabro*, *Paravespula germanica*, *P. vulgaris*, and *Dolichovespula saxonia* (Ishay *et al*., 1974). Apidae: *Apis mellifera* (Owen, 1971).	Vespidae: Poison gland of worker. Apidae: Poison gland of worker.	Honey bee venoms contain about 0.1 mg/gm whereas it is a minor constituent in the wasp venoms.

(continued)

Table 14.1 (Continued)

Adrenaline (Epinephrine) $C_9H_{13}NO_3$ MW 183	Hymenoptera Vespidae: *Vespa orientalis* (Edery *et al.*, 1972), *V. orientalis*, *V. crabro*, *Paravespula germanica*, *P. vulgaris*, and *Dolichovespula saxonia* (Ishay *et al.*, 1974).	Poison gland of worker.	It is the least concentrated of the five biogenic amines in the venoms.
Tryptamine $C_{10}H_{12}N_2$ MW 160	Scorpionida Scorpionidae: *Heterometrus scaber* (Nair and Kurup, 1975).	Paired poison glands in the distal segment of the postabdomen.	It is one of six indole compounds in the venom and is present at a concentration of 3.85 mg/gm dried venom.
Serotonin (5-Hydroxyxtryptamine) $C_{10}H_{12}N_2O$ MW 176	Scorpionida Buthidae: *Leiurus quinquestriatus* (Adam and Weiss, 1956) and *Buthotus minax* (Adam and Weiss, 1959). Scorpionidae: *Heterometrus scaber* (Nair and Kurup, 1975). Vejovidae: *Vejovis* sp. (Welsh and Moorhead, 1960) and *V. spinigerus* (Russell, 1967). Araneida Theraphosidae (Aviculariidae): *Acanthoscurria atrox*, *A. sternalis*, and *Pterinopelma vellutinum* (Welsh and Batty, 1963).	Buthidae: Paired poison glands in the distal segment of the postabdomen. Scorpionidae: Paired poison glands in the distal segment of the postabdomen. Vejovidae: Paired poison glands in the distal segment of the postabdomen. Theraphosidae: A venom gland in the basal segment of each chelicera.	Although it is not a common constituent of scorpion venoms, the secretions of *L. quinquestriatus* and *H. scaber* contain a high concentration (2-4.5 mg/gm) of this compound. It appears to be normally present in spider venoms, some of which are richly endowed with this amine. Venoms of ants and solitary wasps appear to lack it but those of social wasps are fortified with large

Lycosidae: *Lycosa erythrognatha* (Welsh and Batty, 1963). Ctenidae: *Phoneutria nigriventer* (= *fera*) (Welsh and Batty, 1963).	Lycosidae: A venom gland in the cephalothorax adjacent to the basal segment of each chelicera. Ctenidae: A venom gland in the cephalothorax adjacent to the basal segment of each chelicera.	amounts, often containing more than 1 μg/insect. The dried venom of *V. orientalis* contains ca. 8.2 μg/mg.
Chilopoda		
Scolopendridae: *Scolopendra viridicornis* (Welsh and Batty, 1963).	Scolopendridae: A venom gland in the basal articles of each telopodite of the postcephalic segment.	
Lepidoptera		
Saturniidae: *Automeris illustris* (Welsh and Batty, 1963).	Lepidoptera: Larval setae on dorsal and lateral tubercles.	
Hymenoptera		
Vespidae: *Polistes gallicus* (Erspamer, 1954), *P. versicolor* (Welsh and Batty, 1963), *P. exclamans*, *P. annularis*, *P. fuscatus* (Prado *et al*., 1966), *Vespula vulgaris* Jaques and Schachter, 1954), *Vespa crabro* (Bhoola *et al*., 1961; Ishay *et al*., 1974 *V. orientalis*, *Paravespula germanica*, *P. vulgaris*, *Dolichovespula saxonia* (Ishay *et al*., 1974), and *Synoeca surinama* (Welsh and Batty, 1963).	Vespidae: Poison gland of worker.	
Apidae: *Apis mellifera* (Welsh and Moorhead, 1960).	Apidae: Poison gland of worker.	

(*continued*)

Table 14.1 (Continued)

Adenosine 5'-phosphate (AMP) 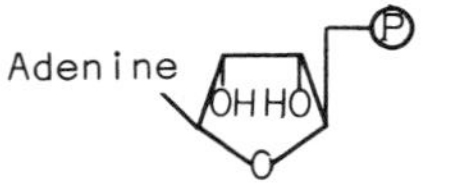 Adenine (OH HO) —Ⓟ $C_{10}H_{14}N_5O_7P$ MW 347	Araneida Theraphosidae (Aviculariidae): *Dugesiella hentzi* and *Aphonopelma* sp. (Chan *et al*., 1975).	A venom gland in the basal segment of each chelicera.	It is the least concentrated nucleotide in both venoms.
Adenosine 5'-diphosphate (ADP) 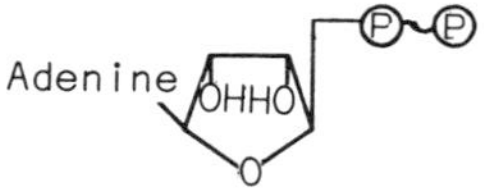 Adenine (OH HO) —Ⓟ~Ⓟ $C_{10}H_{15}N_5O_{10}P_2$ MW 427	Araneida Theraphosidae (Aviculariidae): *Dugesiella hentzi* and *Aphonopelma* sp. (Chan *et al*., 1975).	A venom gland in the basal segment of each chelicera.	It is one of three adenosine nucleotides in the venoms and is the second most concentrated nucleotide in both venoms.
Adenosine 5'-triphosphate (ATP) 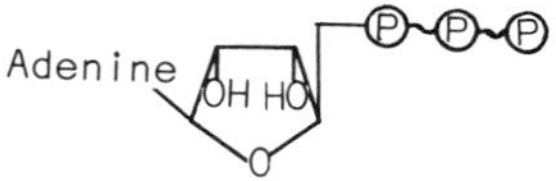 Adenine (OH HO) —Ⓟ~Ⓟ~Ⓟ $C_{10}H_{16}N_5O_{13}P_3$ MW 507	Araneida Theraphosidae (Aviculariidae): *Dugesiella hentzi* and *Aphonopelma* sp. (Chan *et al*., 1975).	A venom gland in the basal segment of each chelicera.	It is present at a concentration of 2.8% and 5.7% in the venoms of *D. hentzi* and *Aphonopelma* sp., respectively. It synergizes the toxicity of the *D. hentzi* venom necrotoxin.
Spermine  H_2N–N–N–NH_2 $C_{10}H_{26}N_4$ MW 202	Araneida Theraphosidae (Aviculariidae): *Pamphobeteus tetracanthus*, *P. roseus*, *P. soracabae*, *Grammostola pulchripes*, *Eurypelma vellutinum*, *Acanthoscurria atrox*, and *Lasiodura klugii*	Theraphosidae: A venom gland in the basal segment of each chelicera. Dipluridae: A venom gland in the basal seg-	It is in a bound state in all venoms and in those of the theraphosid species it is combined with phenolic acids. It is believed to be primarily responsible for

Spermine (con't)	(Fischer and Bohn, 1957a). Dipluridae: Atrax robustus (Gilbo and Coles, 1964).	ment of each chelicera.	the toxicity of the venom of A. robustus.
5-Hydroxytryptophan 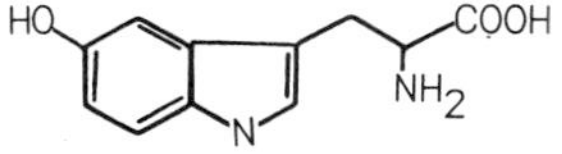 $C_{11}H_{12}N_2O_3$ MW 220	Scorpionida Scorpionidae: Heterometrus scaber (Nair and Kurup, 1975).	Paired poison glands in the distal segment of the postabdomen.	It is accompanied in the venom by tryptamine and 5-hydroxytryptamine and is present at a concentration of 2.45 mg/gm dried venom.
2-Ethyl-5-n-pentyl-Δ1,2-pyrroline 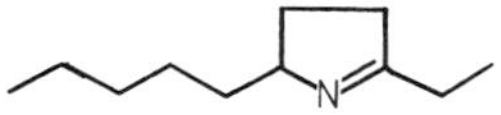 $C_{11}H_{21}N$ MW 167	Hymenoptera Formicidae: Solenopsis punctaticeps (Pedder et al., 1976).	Poison gland of worker.	It is one of six pyrrolines in the venom.
2-Ethyl-5-n-pentyl-Δ1,5-pyrroline 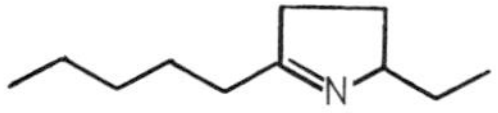 $C_{11}H_{21}N$ MW 167	Hymenoptera Formicidae: Solenopsis punctaticeps (Pedder et al., 1976).	Poison gland of worker.	It is a minor constituent which is accompanied by the Δ1,2-pyrroline.

(continued)

Table 14.1 (Continued)

2-Ethyl-5-*n*-pentyl-pyrrolidine 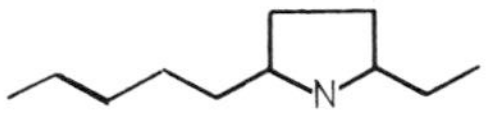 $C_{11}H_{23}N$ MW 169	Hymenoptera Formicidae: *Solenopsis punctaticeps* (Pedder *et al*., 1976).	Poison gland of worker.	It is a trace component and constitutes the least concentrated pyrrolidine in the venom.
2-Ethyl-5-*n*-heptyl-Δ1,2-pyrroline  $C_{13}H_{25}N$ MW 195	Hymenoptera Formicidae: *Solenopsis punctaticeps* (Pedder *et al*., 1976).	Poison gland of worker.	It is accompanied by the Δ1,5-pyrroline.
2-Ethyl-5-*n*-heptyl-Δ1,5-pyrroline  $C_{13}H_{25}N$ MW 195	Hymenoptera Formicidae: *Solenopsis punctaticeps* (Pedder *et al*., 1976).	Poison gland of worker.	It is a quantitatively important constituent in the venom.
2-*n*-Butyl-5-*n*-pentyl-Δ1,2-pyrroline 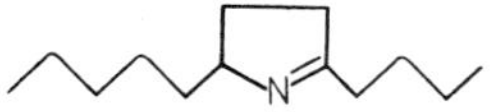 $C_{13}H_{25}N$ MW 195	Hymenoptera Formicidae: *Solenopsis punctaticeps* (Pedder *et al*., 1976).	Poison gland of worker.	It is accompanied by the Δ1,5-pyrroline.

2-*n*-Butyl-5-*n*-pentyl-Δ1,5-pyrroline $C_{13}H_{25}N$ MW 195	Hymenoptera Formicidae: *Solenopsis punctaticeps* (Pedder *et al.*, 1976).	Poison gland of worker.	It is one of three Δ1,5-pyrrolines in the venom.
5-Methyl-3-*n*-butyloctahydroindolizine $C_{13}H_{25}N$ MW 195	Hymenoptera Formicidae: *Monomorium pharaonis* (Ritter *et al.*, 1973).	Poison gland of worker.	It is the first indolizine isolated from insects and also appears to function as an attractant.
2-Ethyl-5-*n*-heptylpyrrolidine $C_{13}H_{27}N$ MW 197	Hymenoptera Formicidae: *Solenopsis punctaticeps* (Pedder *et al.*, 1976).	Poison gland of worker.	It is one of three pyrrolidines which dominate the venom.
2-*n*-Butyl-5-*n*-pentylpyrrolidine $C_{13}H_{27}N$ MW 197	Hymenoptera Formicidae: *Monomorium pharaonis* (Talman *et al.*, 1974) and *Solenopsis punctaticeps* (Pedder *et al.*, 1976).	Poison gland of worker.	It appears to be biogenically related to 5-methyl-3-*n*-butyloctahydroindolizine, another constituent in the venom of *M. pharaonis*.

(continued)

Table 14.1 (Continued)

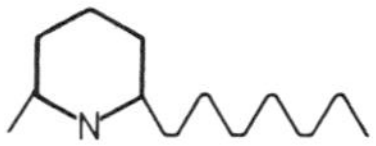 *cis*- and *trans*-2-Methyl-6-*n*-heptyl-piperidine $C_{13}H_{27}N$ MW 197	Hymenoptera Formicidae: *Solenopsis richteri* (MacConnell *et al*., 1974).	Poison gland of female.	It is a trace constituent in a venom dominated by 2-methyl-6-*n*-undecyl-piperidine.
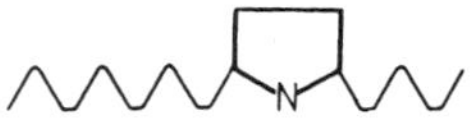 2-*n*-Butyl-5-*n*-heptyl-pyrrolidine $C_{15}H_{31}N$ MW 225	Hymenoptera Formicidae: *Solenopsis punctaticeps* (Pedder *et al*., 1976) and *S. fugax* (Blum *et al*., 1980).	Poison gland of worker.	It is the major constituent in the venom of *S. punctaticeps* and the only compound detected in the venom of *S. fugax*.
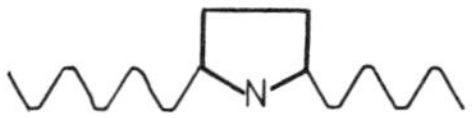 2-*n*-Pentyl-5-*n*-hexyl-pyrrolidine $C_{15}H_{31}N$ MW 225	Hymenoptera Formicidae: *Solenopsis texanas* and *S. molesta* (Jones *et al*., 1979).	Poison gland of worker.	It is the only nitrogen heterocycle detected in the venoms.

cis- and-*trans*-2-Methyl-6-*n*-nonyl-piperidine  $C_{15}H_{31}N$ MW 225	Hymenoptera Formicidae: *Solenopsis* sp. and *S. richteri* (MacConnell *et al*., 1974).	Poison gland of worker of *S*. sp. and poison gland of female of *S. richteri*.	Alkaloids containing the *n*-nonyl side chain appear to be very uncommon in *Solenopsis* venoms. Although workers of *S. richteri* do not produce these compounds, they synthesize alkaloids which are not present in the venom of the female.
2-Methyl-6-*n*-undecyl-Δ1,2-piperideine  $C_{17}H_{33}N$ MW 251	Hymenoptera Formicidae: *Solenopsis xyloni* (Brand *et al*., 1972).	Poison gland of worker.	It may represent a precursor of the 2-methyl-6-*n*-undecylpiperidines. It could not be detected in the venom of the female.
cis- and-*trans*-2-Methyl-6-*n*-undecyl-piperidine  $C_{17}H_{35}N$ MW 253	Hymenoptera Formicidae: *Solenopsis invicta* (=*saevissima* red form) (MacConnell *et al*., 1970, 1971), *S. richteri* (=*saevissima* black form), *S. xyloni*, *S. geminata* (Brand *et al*., 1972), and *S. aurea* (Blum *et al*., 1973c).	Poison gland of worker and female.	The venoms of *S. geminata* and *S. xyloni* consist essentially of these two isomers with the cis isomer predominating. These isomers are minor components in the venoms of *S. invicta* and *S. richteri* workers and the trans isomer is the major form. In the venom of all females examined, the cis isomer predominates and no other alkaloids are detectable

(*continued*)

Table 14.1 (Continued)

cis- and-*trans*-2-Methyl-6-*n*-undecyl-piperidine (con't)			although additional piperidines dominate the venoms of workers of the two species. The compounds have been assigned the trivial epithet of Solenopsin A.
cis- and-*trans*-2-Methyl-6-(*cis*-4-tridecenyl)piperidine  $C_{19}H_{37}N$ MW 279	Hymenoptera Formicidae: *Solenopsis invicta* (=*saevissima* red form) (MacConnell *et al*., 1971), *S. richteri* (=*saevissima* black form), *S. xyloni*, and *S. geminata* (Brand *et al*., 1972).	Poison gland of worker.	The trans isomer is the major constituent in the venom of *S. richteri* and a major constituent in the venom of *S. invicta*. Only the cis isomer is detectable in the venoms of *S. geminata* and *S. xyloni* as a trace component. Neither isomer is detectable in the venoms of females of the 4 species. The common name Dehydrosolenopsin B has been assigned to the isomers.
 *cis*- and-*trans*-2-Methyl-6-*n*-tridecyl-piperidine $C_{19}H_{39}N$ MW 281	Hymenoptera Formicidae: *Solenopsis invicta* (=*saevissima* red form) (MacConnell *et al*., 1971), *S. richteri* (=*saevissima* black form), *S. xyloni*, and *S. geminata* (Brand *et al*., 1972).	Poison gland of worker.	The cis isomer is a trace component in the venoms of *S. geminata* and *S. xyloni* in the absence of the trans isomer. The venom of *S. invicta* contains about 20% of the trans isomer whereas the cis isomer

			is a trace component, as it is in the venom of *S. richteri*. Solenopsin B has been used as a trivial name for the isomers.
cis- and-*trans*-2-Methyl-6-(*cis*-6-pentadecenyl)piperidine  $C_{21}H_{41}N$ MW 307	Hymenoptera Formicidae: *Solenopsis invicta* (=*saevissima* red form) (MacConnell *et al.*, 1971; Brand *et al.*, 1972).	Poison gland of worker.	The isomers are absent from the venoms of the four other species of *Solenopsis* (*Solenopsis*) in North America. The trans isomer is the major constituent produced whereas the cis isomer is present in trace quantities. The venom of the female contains neither of the isomers, which have been named Dehydrosolenopsin C.
cis- and-*trans*-2-Methyl-6-*n*-pentadecylpiperidine  $C_{21}H_{43}N$ MW 309	Hymenoptera Formicidae: *Solenopsis invicta* (=*saevissima* red form) (MacConnell *et al.*, 1971; Brand *et al.*, 1972).	Poison gland of worker.	The cis isomer is a trace constituent whereas the trans isomer comprises more than 20% of the alkaloidal mixture. The isomers are absent from the venom of the female. The highest molecular weight piperidines in fire ant venoms have been designated Solenopsin C.

venom. The venom of another scorpionid, *Heterometrus scaber*, contains several indole compounds (Nair and Kurup, 1975). 5-Hydroxytryptophan, tryptophan, serotonin, and tryptamine were identified in the venom and were accompanied by two unidentified indole compounds. 5-Hydroxytryptamine was a minor constituent relative to the other indole compounds (Table 14.1).

Lipids constitute nearly 2% of the dry venom of the buthid *Leiurus quinquestriatus* (Marie and Ibrahim, 1976), and this class of compounds may be characteristic of scorpion venoms. Phospholipids, which make up about a third of the available lipids, are accompanied by free cholesterol and its esters, free fatty acids and their methyl esters, and triglycerides. However, since the toxicity of the venom of *L. quinquestriatus* is not affected by removal of the lipids, the pharmacological significance of these compounds is unknown.

In general, scorpion venoms contain mixtures of proteinaceous compounds with molecular weights of about 8000 or less, and these substances appear to function primarily as neurotoxins. Scorpion toxins, which are generally basic compounds, are sometimes referred to as scorpamines, and they constitute some of the most pharmacologically potent compounds which have been isolated from animal venoms (Miranda *et al.*, 1964). Hemolysins appear to have a rather sporadic distribution in scorpion venoms and some of these venomous secretions contain few or no enzymatic constituents.

Glycosaminoglycans represent major constituents in the venoms of the scorpionid *H. scaber* (Nair and Kurup, 1973a). The dry venom contains nearly 2.5 mg of these carbohydrate macromolecules which are present in five different fractions. After papain digestion, fractions containing chondroitin sulfate-A, -B, and -C were detected, along with heparin sulfate and hyaluronic acid. The major glycosaminoglycans present were chondroitin sulfate-A and -B and heparin sulfate. Appreciable amounts of bound hexosamine and small quantities of free hexosamine are also present.

Subsequent investigations by Nair and Kurup (1975) demonstrated that the toxic protein in the venom of *H. scaber* is a glycoprotein with a molecular weight of 15,000. This protein contains glucosamine, galactosamine, sialic acid, fucose, and an unidentified sugar. The toxicity of the glycoprotein is low compared to these of the neurotoxic proteins isolated from the venoms of other scorpion species.

The toxins in the venoms of species in three genera in the family

Buthidae appear to constitute a family of homologous proteins. Miranda *et al.* (1970) purified 11 neurotoxins derived from the venoms of *Androctonus australis*, *Buthus occitanus*, and *Leiurus quinquestriatus*, and demonstrated that all proteins had a molecular weight of about 7000 and lacked methionine. The amino acid compositions and the chain lengths of these toxins were all very similar. Each protein consisted of a single polypeptide chain of 57–66 amino acid residues cross-linked by four disulfide bridges. Rochat *et al.* (1967) characterized two basic neurotoxins in the venom of *A. australis* and reported that they contained 63 and 64 amino acids, both ending in lysine at the N-terminal and both being highly toxic to mice. Subsequent research by Rochat *et al.* (1970) on the sequence of the first 22–26 amino acid residues from the N-terminal end of six additional neurotoxins has demonstrated that these compounds form a distinct set of proteins which can be divided into subgroups based on amino acid sequences. Thus, eight neurotoxins were divisible into three subgroups based on the presence of quasi-invariant amino acid residues in members of each subgroup. Small changes in amino acid residues were correlated with significant differences in the toxicities of the proteins (Rochat *et al.*, 1970). More recently, Sampieri and Habersetzer-Rochat (1975) have demonstrated that one or several amino acids are part of the active site of these neurotoxins.

Toledo and Neves (1976) characterized two toxins present in the venom of the buthid *Tityus serrulatus*. As in the case of other scorpion toxins, both proteins (= tityustoxins) are low molecular-weight compounds, i.e., 6400 and 7700. It appears that both tityustoxins are different aggregates of basic toxins with, possibly, nucleic acids. However, unlike most other scorpion toxins, the tityustoxins contain methionine and one of these proteins probably contains six disulfide bridges instead of the usual four.

Three neurotoxins in the venom of *Centruroides sculpturatus* have been recently isolated and characterized (Babin *et al.*, 1974). These toxins are homologous proteins that contain 65–66 residues, as is the case for the neurotoxins isolated from other scorpion venoms. Nearly all of the structural differences in the *Centruroides* proteins were found in the amino terminal one-third of the molecules. Few similarities were found between the amino acid sequences of the neurotoxins in the venoms of *C. sculpturatus* and the North African scorpion, *Androctonus australis*. Toxin I from the venom of *C. sculpturatus* has been shown to be cross-linked by four disulfide bridges (Babin *et al.*, 1975). This neurotoxin has nearly the same terminal amino acid sequence as toxin I in the venom of

A. australis and these two toxins are identical at 22 of their 64 residues. Thus, although the structures of toxins from these scorpion venoms have many similarities, they obviously are not closely related proteins.

Although scorpion venoms display a great proteinaceous heterogeneity, many of the compounds appear to be nontoxic to specific animals (Zlotkin *et al.*, 1978). El-Asmar *et al.* (1972) reported that the venom of the buthid *L. quinquestriatus* contained nine components, but only three of these possessed great toxicity when injected into mice. Similarly, seven protein fractions are present in the venoms of the Neotropical buthids *Tityus serrulatus* and *T. bahiensis*, but only two of the fractions are toxic to mice (Diniz and Gonçalves, 1960; Miranda *et al.*, 1966; Diniz and Gomez, 1968). Recently, Possani *et al.* (1977) fractionated the venom of *T. serrulatus* into 16 polypeptides and characterized five toxic proteins. One of these toxins, toxin λ, contains methionine and represents the first example of a fifth structural type of mammalian toxin from scorpion venom. Two potent mammalian neurotoxins have been isolated from the venoms of both *A. australis* and *B. occitanus* (Miranda *et al.*, 1966), and the venoms of both of these species contain several additional protein fractions which possess no demonstrable toxicity to mammals (Zlotkin *et al.*, 1972b). On the other hand, the venom of *A. australis* contains a proteinaceous constituent which is highly toxic to insects but lacks any demonstrable mammalian toxicity (Zlotkin *et al.*, 1971b). This protein, which produce rapid paralysis in insects, constitutes only 0.35% of the venom, whereas the three mammalian toxins, all of which are devoid of insect toxicity (Zlotkin *et al.*, 1971a), comprise 1.39% of the venom. The insect toxin blocks synaptic conduction in the central nerve cord of the American cockroach but mammalian toxins are without effect (D'Ajello *et al.*, 1972). The insect neurotoxin is not strongly basic like the mammalian toxins, and it may actually consist of two isotoxins, as was the case for a toxic mammalian fraction (Miranda *et al.*, 1970). Although the structure of the insect toxin somewhat resembles those of the mammalian toxins, it possesses different amino acid residues in several positions and may belong to a different family of homologous proteins. A similar toxin is present in the venom of A. *mauretanicus* (Zlotkin *et al.*, 1979b).

Zlotkin *et al.* (1979a) recently compared the insect toxins in the venoms of two *Androctonus* species and a *Buthotus* species (Buthidae) with those produced by the scorpionid *Scorpio maurus*. The toxins in the venoms produced by all three genera differed from each other in both chemistry and activity. Therefore, it appears that the insect toxins in scorpion venoms comprise a heterogenic group of polypeptides.

In addition to this insect toxin, the venoms of three species of scorpions contain additional insect toxins which do not possess the contraction–paralysis activity of the original insect neurotoxin (Zlotkin *et al.*, 1972b). The remarkable venom of *A. australis* also contains a protein which appears to be specifically toxic to crustaceans but lacks toxicity to mice, insects, and a scorpion (Zlotkin *et al.*, 1972a). It has been suggested that the crustacean neurotoxin has a specific affinity for the crustacean neural system (Pansa *et al.*, 1973). None of these protein fractions is toxic to arachnids, a fact that is consistent with the observation that the venom of the scorpion *Centruroides limpidus* is nontoxic to the spider *Aphonopelma smithi* (Wheeling and Keegan, 1972).

Zlotkin *et al.* (1975) purified this crustacean toxin and reported that it appeared to contain five disulfide bridges in contrast to the mammalian and insect toxins in the venom, which contain four bridges. This toxin possesses a molecular weight of nearly 8200 and is thus similar to the insect and three mammalian toxins in being a low molecular-weight protein.

Enzymes have been detected in the venoms of a few species of scorpions, but they appear to be absent from the secretions of the majority. Proteases are present in the venoms of the scorpionid *P. gravimanus* and the buthid *Buthus tamulus* (Master *et al.*, 1963). Phosphodiesterase was detected in the venom of *B. tamulus*, whereas that of *P. gravimanus* was rich in 5′-nucleotidase. The richest known enzyme source of scorpion venoms is that of *H. scaber*. The secretion of this scorpionid contains 5′-nucleotidase, acid phosphatase, ribonuclease, phospholipase A, and acetylcholinesterase (Nair and Kurup, 1973b). The venom of the scorpionid *Scorpio maurus* is reported to be a rich source of phospholipase A_2 (Lazarovici *et al.*, 1979). Weak acetylcholinesterase activity is also reported to be present in the venom of *Vejovis spinigerus* (Vejovidae) (Russell, 1967).

Hyaluronidase activity is reported to be present in the venoms of the scorpions in several families. Jaques (1956) reported high hyaluronidase activity in the venoms of *Scorpio maurus* and *B. occitanus* (Buthidae), and Tarabini-Castellani (1938) detected spreading factors in the venoms of *A. australis* (Buthidae) and *Euscorpius italicus* (Chactidae). The venoms of the buthids *T. serrulatus* and *T. bahiensis* also possess hyaluronidase activity (Diniz and Gonçalves, 1960; Possani *et al.*, 1977). Weak hyaluronidase activity is reported to be present in the venom of the scorpionid *H. scaber* (Nair and Kurup, 1975).

In general, scorpion venoms do not possess hemolytic activity, although the venoms of several species have been demonstrated to lyse red blood cells. Balozet (1962) reported that the venom of *S. maurus* was strongly hemolytic; the venoms of *Heterometrus indus* (Caius and Mhaskar, 1932) (Scorpionidae), *H. scaber* (Nair and Kurup, 1973b), and *Buthus martensis* (Mori, 1919) also contain hemolysins. The hemolysin in the venom of *H. scaber* exhibits particularly strong hemolytic activity in the presence of lecithin (Nair and Kurup, 1973b). More than one type of hemolytic agent appears to be present in different scorpion venoms; Rosin (1969) has demonstrated that the venom of *Nebo hierichonticus* (Diplocentridae) produces an α-hemolytic effect whereas the venom of the scorpionid *S. maurus* causes typical lysing of erythrocytes. Recently, Lazarovici *et al.* (1979) reported that the venom of *S. maurus* contained a direct lytic factor.

The highly toxic venom of *Buthus tamulus* is reported to contain a specific trypsin inhibitor (Chhatwal and Habermann, 1980).

II. SPIDER VENOMS

Biogenic amines are present in the venoms of several spiders (Table 14.1), but the toxic components of these secretions have not been chemically characterized to any great extent. Like the venoms of scorpions, those of spiders are characterized by multicomponent systems consisting of proteinaceous constituents.

The venom of the ctenid *Phoneutria nigriventer* contains 13 constituents, several of which appear to be associated with specific pharmacological properties of this venom (Pereira Lima and Schenberg, 1964). The venomous constituents, which appear to be polypeptides, have molecular weights between 5000–10,000 (Schenberg and Pereira Lima, 1971), the same approximate range as scorpion toxins. Similarly, McCrone and Hatala (1967) reported that the venom of the theridiid *Latrodectus mactans* contains seven proteins among which was a mammalian toxin with a molecular weight of about 5000. On the other hand, the venom of another population of *L. mactans* is enriched with five proteins, and their molecular weights apparently are in the range of 60,000 (Frontali and Grasso, 1964).

The venom of the tarantula *Aphonopelma rusticum* contains 10 protein fractions and, curiously, appears to be antigenically very similar to the venom of the scorpion *Centruroides sculpturatus* (Stahnke and Johnson, 1967). Lee *et al.* (1974) purified the necrotoxin in the venom of the tarantula *Dugesiella hentzi* and reported that it was a polypeptide with a molecular weight of 6700. This toxin, which is unusual in containing a high content of lysine, is toxic to both insects and white mice.

Proteases are commonly present in spider venoms and, although these enzymes are often highly concentrated in the exudates, they do not appear to be responsible for the toxic effects of these secretions. In addition, the recent demonstration that the proteins present in the venoms of three tarantula species constituted salivary contaminants (Perret, 1977) raises the question of the real significance of the reported presence of these proteins in spider venoms. Strong protease activity was detected in the venom of the tarantula *A. rusticum* (Stahnke and Johnson, 1967), the lycosid *Lycosa erythrognatha* (=*raptona*), the ctenid *Phoneutria* (=*Ctenus*) *nigriventer* (Kaiser, 1956), and the brown recluse spider *Loxosceles reclusa* (Jong *et al.*, 1979).

Hyaluronidase has been detected in the venoms of *L. erythrognatha*, *P. nigriventer* (Kaiser, 1953), *L. reclusa* (Wright *et al.*, 1973), *L. mactans* (Cantore and Bettini, 1958), and is a major constituent in the venom of the tarantula *D. hentzi* (Schanbacher *et al.*, 1973). Other enzymes which have been detected in spider venoms include weak RNase activity in *A. rusticum* (Stahnke and Johnson, 1967), strong esterase activity in *L. reclusa* (Wright *et al.*, 1973; Norment *et al.*, 1979), and negligible phosphodiesterase activity in the venoms of *L. mactans* and the diplurid *Atrax robustus* (Russell, 1966). Extracts of the venom apparatus of *L. reclusa* also contain alkaline phosphatase (Heitz and Norment, 1974; Norment *et al.*, 1979), 5′-ribonucleotide phosphohydrolase (Green *et al.*, 1976), and lipase (Norment *et al.*, 1979).

Green *et al.* (1975) reported that the necrotic activity associated with the venom of *L. reclusa* is not associated with the low molecular-weight fractions which include inosine and nucleotides. Subsequently, Green *et al.* (1976) isolated two proteins (molecular weight 34,000) that appear to be responsible for the toxicity of *L. reclusa* venom. Toxin I causes the characteristic lesions associated with this venom and is lethal to rabbits and mice. On the other hand, toxin II, which is not toxic to mice, is lethal to rabbits but does not induce the formation of necrotic lesions.

The venom of *L. mactans* appears to be functional as a mammalian neurotoxin, but only one of the five protein fractions possesses the strong toxicity associated with this venom when evaluated on guinea pigs (Frontali and Grasso, 1964). On the other hand, the mouse is susceptible to three protein fractions present in this venom, including the one which is toxic to guinea pigs (Granata *et al.*, 1972). However, two other fractions in this secretion are highly toxic to insects, the normal prey of this spider. Recent studies by Ornberg *et al.* (1976) demonstrated that homogenates of *L. mactans* venom glands contain 11 fractions. Two insect neurotoxins, with different modes of action were isolated, and one of these proteins (molecular weight 125,000) accounted for the major activity of the venom. Grasso (1976) also isolated a neurotoxin from this venom with a molecular weight of 130,000. This toxin was demonstrated to release norepinephrine from rat brain.

The eggs of another theridiid, *L. hesperus*, contain a mammalian toxin which is distinct from those in the venom of this species (Buffkin and Russell, 1971). The significance of this toxin is difficult to ascertain since it has no toxicity by oral administration, the presumed means by which it would gain access to the body of a predator.

ATP, ADP, and AMP have been detected in the venoms of the tarantulas *D. hentzi* and *Aphonopelma* sp. (Chan *et al.*, 1975). ATP, which is the major adenosene nucleotide in the venoms of both species, synergizes the toxic effects of the major toxin in the venom of *D. hentzi*.

III. CENTIPEDE VENOMS

A paucity of information is available on the chemistry of chilopod venoms and except for the identification of serotonin in *Scolopendra viridicornis* (Welsh and Batty, 1963) (Table 14.1), no venomous constituents have been characterized in detail. Bücherl (1946) studied the toxicities of the venoms of five scolopendrid and cryptopid species and concluded that these secretions contain neurotoxins, of probable proteinaceous nature.

IV. INSECT VENOMS

Hymenopterous species constitute the largest group of venomous arthropods but, in addition, insects in several other orders produce venoms fortified with potent pharmacological agents. However, virtually all investigations of insect venoms have been confined to a few species of bees, ants, and wasps, particularly those species which are of medical importance because of their propensity for stinging human beings. Nevertheless, although insect venoms mostly constitute *terra incognita*, these arthropods are known produce a multitude of highly distinctive compounds with significant biological activities. Results of studies on the chemistry of the venoms synthesized by species of Lepidoptera, Hemiptera, and Hymenoptera demonstrate clearly that these defensive products are characterized by diverse pharmacological substances of great chemical heterogeneity (Table 14.1).

A. Lepidoptera

Larvae in at least 13 families contain setae fortified with venom originating in an attached poison gland. Phanerotoxic species have been identified in the Arctiidae, Bombycidae, Eucleidae, Lasiocampidae, Lithosiidae, Lymantridae, Megalopygidae, Morphidae, Noctuidae, Notodontidae, Nymphalidae, Saturniidae, and Sphingidae (Pese and Delgado, 1971), and it appears that toxic setae occur much more frequently in moth larvae than those of butterflies. A number of qualitative investigations on the toxins in the setae of diverse species have been undertaken, and these have been succinctly reviewed by Picarelli and Valle (1971) and Pese and Delgado (1971). Detailed investigations on these caterpillar toxins have been very limited, and the identification of histamine in a *Dirphia* sp. (Saturniidae) (Valle *et al.*, 1954) represents the only study in which the structure of a toxic component has been determined. However, biogenic amines do not appear to be characteristic of these lepidopterous toxins, and the results of several investigations indicate that the setal agents are probably proteinaceous constituents.

Picarelli and Valle (1971) reported that the toxic setae of the saturniid *Megalopyge* sp. contained a very potent hemolysin which appeared to be proteinaceous in nature. The venomous constituents in the setae of *M. urens* have been demonstrated to consist of four protein fractions among which was a strong hemolytic component (Ardao *et al.*, 1966). In addition, other fractions possessed hyaluronidase activity and strong proteolytic and weak hemolytic activity. Proteolytic activity has also been demonstrated for the setal toxin of the saturniid *Automeris io* (Goldman *et al.*, 1960).

More recently, de Jong and Bleumink (1977a,b) demonstrated the presence of protease, esterase, and phospholipase A_2 activity in the venomous spicules of the brown tail moth, *Euproctis chrysorrhea*. Kawamoto and Kumada (1979) detected protease, esterase, and kininogenase activity in the spicules of another *Euproctis* species, *E. subflava*. No enzymatic activities were present in extracts of the spicules of the slug moths *Parasa consonia* and *Cnidocampa flavescens*.

B. Hemiptera

At least 2500 species of Heteroptera-Hemiptera are predatory (Edwards, 1962), and the salivary secretions of these insects have assumed the function of venoms which rapidly immobilize invertebrate prey. These salivary venoms may also cause very severe reactions when injected into vertebrate species and necrotic lesions of long duration may persist at the site of the bite (Edwards, 1961). Our knowledge of these salivary toxins is, however, essentially limited to the study of one species, the predatory reduviid *Platymeris rhadamanthus*, a strongly aposematic African species.

Both adults and larvae of *P. rhadamanthus* possess the ability to spit their venomous saliva (Vanderplank, 1958), sometimes for a distance up to 30 cm (Edwards, 1962). The saliva of this species, which is able to paralyze an adult of *Periplaneta americana* in 3–5 seconds, has a wide range of activities against insects in seven orders. The venom has no activity when administered either topically or orally, but when injected, causes violent contractions, cessation in systole, and a general contraction of tergal musculature. When applied to the central nervous system, the saliva produces intense discharges of giant fibers, leading to a cessation of conduction (Edwards, 1961); postsynaptic conduction is also terminated after salivary treatment. Potent lytic action on nerve, muscle, and fat body preparations is rapidly evident after exposure to the saliva.

The saliva of *P. rhadamanthus* is a mixture of 6–8 proteins, the main component of which is strongly proteolytic. Two weaker proteases are also present and the three proteolytic enzymes possess a wide range of substrate specificities (Edwards, 1961), similar to a combination of trypsin and chymotrypsin. Strong hyaluronidase and weak phospholipase activities (Table 14.2) were present in the salivary secretions, but no lipase or esterase activity was demonstrable.

Edwards (1961) concluded that the venom lacks a specific site of action but, rather, attacks many organs simultaneously. Proteolytic activity is primarily an external digestive process which occurs rapidly as the lytic action of the venom proceeds, but lipids are metabolized by enzymes in the midgut. Edwards believes that the potent protease and hyaluronidase activities of the saliva cannot explain its great toxicity, and he concludes

Table 14.2

Typical Constituents Identified in the Venoms of Some Groups of Arthropods[a,b]

Compounds	Scorpions	Spiders	True bugs	Insects (Hymenoptera)		
				Bees	Wasps	Ants
Histamine	−	+	−	+	+	+
Serotonin	+	+	−	−	+	−
γ-Aminobutyric acid	−	+	−	−	−	−
Spermine	−	+	−	−	−	−
Dopamine	−	−	−	+	+	+
Adrenaline	−	−	−	−	+	−
Noradrenaline	−	−	−	+	+	−
Acetylcholine	−	−	−	−	+	−
Alkaloids	−	−	−	−	−	+
Polypeptides	+	−	−	+	−	−
Acid phosphatase	−	−	−	+	+	−
Esterase	−	+	+	−	−	+
Proteases	−	+	+	−	−	−
Hyaluronidase	−	+	+	+	+	+
Phospholipase A	−	−	+	+	+	+
Mast cell-degranulating peptide	−	−	−	+	+	−
Kinins	−	−	−	−	+	−
Histidine decarboxylase	−	−	−	−	+	−
Lipase	−	−	−	−	−	+

[a] In cases where these constituents appear to rarely occur in the venoms of a group, they are not listed.

[b] Toxic proteins are characteristic of almost all venoms.

that, notwithstanding its weak activity, the salivary phospholipase may be critical to the rapid envenomation process.

Zerachia *et al.* (1972a) studied the salivary constituents of another predatory reduviid, *Holotrichius innesi*, and demonstrated that its saliva could kill a mouse within 30 seconds. The saliva contained 13 proteins and protease, esterase, and hyaluronidase activities could be easily detected. The venomous saliva of *H. innesi* also is fortified with a hemolytic constituent which possesses a direct lytic activity and is thus not identified with phospholipase A (Zerachia *et al.*, 1972b).

Baptist (1941) reported that the salivary glands of several predatory hemipterans contained both proteases and lipases, and he stressed the identity of these salivary enzymes with those of the midgut.

C. Hymenoptera

Detailed chemical and pharmacological investigations have been undertaken on the venoms of hymenopterous species in the families Apidae, Vespidae, and Formicidae. However, most of these studies have treated only a few species in each family and, in terms of the large number of venomous taxa in this order, it is impossible to generalize about the chemical composition of hymenopterous venoms. Indeed, it seems most appropriate to treat the venomous secretions of the species in the three families separately, especially since characteristic toxins are produced by members of each family.

1. Bee Venom

The venom of the honey bee (*Apis mellifera*) constitutes the most completely characterized insect venom. Numerous studies by different investigators have structurally delineated the major chemical constituents of this venom, and the pharmacological activities of several of the major components present are reasonably well established (Habermann, 1972). Hartter and Weber (1975) have developed a simple method for the isolation of three of the peptides in bee venom and demonstrated that after reduction, two of the peptides could be regenerated. Recently, Gauldie *et al.* (1976) fractionated a large quantity of honey bee venom and demonstrated that eight peptides were present, three of which had not been previously detected.

Aside from three biogenic amines (Habermann, 1972; Owen, 1971; Banks *et al.*, 1976) (Tables 14.1, 14.2), honey bee venom contains several major proteinaceous constituents (O'Connor *et al.*, 1967; Munjal and Elliot, 1971), most of which are small basic polypeptides. Fifty percent of the dry venom is identified with the polypeptide melittin, sometimes referred to as the "direct" hemolysin. It is present in a 20- to 50-fold molar excess over other venom polypeptides. This compound has a molecular weight of 2840 and its structure has been completely established (Habermann and Jentsch, 1967). This strongly basic polypeptide has recently been synthesized (Schröder *et al.*, 1971) and, in addition, a compound called melittin II has been synthesized and tested for hemolytic activity. The presence of melittin II, which contains an extra lysine residue, had been suggested by Habermann and Jentsch (1967). The structure of melittin in the venom of *Apis indica*, a species closely related to *A. mellifera*, is identical to that in the venom of the latter. On the other hand, melittin in the venom of the primitive species *A. florea* differs in five positions from that of *A. mellifera* (Kreil, 1973a) whereas that of *A. dorsata* differs in only three positions (Kreil, 1975).

Melittin possesses strong surface activity and is believed to increase the permeability of erythrocytes and other cells because of its physicochemical action on these cellular bodies. The direct cytolytic activity of melittin is distinct from those of other direct hemolysins such as digitalis and lysolecithin (Habermann, 1971).

Habermann and Jentsch (1967) observed that part of natural melittin is substituted at its NH_2 terminus. Kreil and Kreil-Kiss (1967) enzymatically degraded crude melittin and concluded that N-α-formylmelittin constituted the substituted polypeptide. Subsequently, Lübke *et al.* (1971) isolated this compound from bee venom and reported that it was about 80% as active as melittin as a hemolysin. It has been synthesized by Lübke *et al.* (1971).

Apamin, the smallest neurotoxic polypeptide known, constitutes only 2% of honey bee venom. This polypeptide contains two disulfide bridges and is comprised of 18 amino acids with an amide group at the carboxyl terminus (Shipolini *et al.*, 1967). Whereas other neurotoxins act at the neuromuscular junction, apamin affects synaptic pathways (Wellhöner, 1969). Some apamin molecules appear to be substituted on the amino terminus, as in the case of melittin (Habermann, 1971). Van Rietschoten *et al.* (1975) synthesized apamin by the solid-phase procedure. The synthetic peptide represents the first neurotoxin that has been synthesized with full chemical and pharmacological identity with the natural peptide.

Another minor component (1–2%) present in honeybee venom is mast cell-degranulating (MCD) peptide. Like melittin, this compound is capable of releasing histamine and is a powerful mastolytic agent (Fredholm and Haegermark, 1967). MCD peptide contains 22 amino acid residues as well as two disulfide bridges, and like melittin and apamin, the carboxyl terminus is in the amide form (Haux, 1969; Gauldie *et al.*, 1978). Billingham *et al.* (1973) reported that this compound, which they call peptide 401, possesses the powerful anti-inflammatory properties associated with honey bee venom. Using the rat paw edema test, it was demonstrated that peptide 401 was 100 times more active than hydrocortisone as an anti-inflammatory agent.

Gauldie *et al.* (1976) fractionated a large amount of crude venom and isolated, in addition to the known peptides, three new ones. Two of these peptides, melittin F and tertiapin, are trace constituents that probably would not have been detected if small amounts of venom had been similarly analyzed. The third peptide, secapin, constitutes 1% of the lyophilized venom.

Melittin F contains 19 amino acid residues and differs from melittin in having the first seven residues from the N terminus of the latter removed. It thus appears to be a fragment of melittin which contains residues 8–26 (Gauldie *et al.*, 1976, 1978).

Tertiapin contains 20 amino acid residues with alanine at the N terminus. This trace constituent, like MCD peptide, contains four cysteine residues.

Secapin is a highly basic peptide that contains 24 amino acid residues. It is characterized by a large proportion of proline and one disulfide bridge; its structure has recently been determined (Gauldie *et al.*, 1978). Secapin is essentially nontoxic to mice and does not possess any anti-inflammatory activity (Gauldie *et al.*, 1976). Recently, Jentsch and Mücke (1977) isolated secapin from bee venom and assigned it the name peptide M.

Minimine, originally characterized as another minor polypeptide, was isolated from honey bee venom by Lowy *et al.* (1971) and was reported to act as a growth inhibitor of *Drosophila* larvae without interfering with development. However, Habermann (1971) suggested that the activity of the compound may be attributable to phospholipase A, a compound which is isolated in the same fraction as minimine and Gauldie *et al.* (1976) reached the same conclusion. Lowy *et al.* (1976) confirmed that minimine was indeed phospholipase A_2.

Honey bee venom has also been reported to contain two peptides with histamine-terminal residues (Nelson and O'Connor, 1968; Peck and O'Connor, 1974). Both of these peptides were reported to contain only four amino acid residues bound to a terminal histamine group. One of these compounds, procamine, has been synthesized and shown to be identical to the natural peptide (Peck and O'Connor, 1974). However, Gauldie *et al.* (1976) were unable to detect either of these peptides in fractionated bee venom.

Phospholipase A_2 (lecithinase A), consisting of two enzymes (Jentsch and Dielenberg, 1972), is the major enzymatic constituent in honey bee venom and comprises about 12% of the dry weight of this secretion. It is reported to be the major allergen in this venom (Sobotka *et al.*, 1976). These enzymes attack structural phospholipids by hydrolyzing the 2-acyl bonds, thus generating structurally active lysophospholipids which appear to be crucial to the indirect lytic action of phospholipase A. Serious biochemical lesions can result from the ability of these enzymes to disrupt structure-bound enzymes or enzyme chains, and the free fatty acids released from phospholipids may exert pharmacological effects of their own. Although phospholipase A_2 cannot hemolyze erythrocytes directly, in combination with melittin, an indirect cytolysin, it is able to become an effective hemolytic agent (Habermann, 1958).

More recently, Lawrence and Moores (1975) reported that honey bee venom phospholipase A_2 was activated by fatty acids, aliphatic anhydrides, and glutaraldehyde. In addition, Drainas *et al.* (1978) have demonstrated that acylating agents activate phospholipase A_2 by modifying the enzyme and not by releasing fatty acids that interact with the substrate. These results strongly suggest that activators modify the enzyme, and not the substrate, as previously believed. The activity of this enzyme has also been demonstrated to be enhanced by a wide variety of peptides including melittin, direct lytic factors from cobra venom, and polymixin B (Mollay and Kreil, 1974).

Bee venom phospholipase A_2 has been characterized by Shipolini *et al.* (1971a) who identified isoleucine as the amino-terminal amino acid and demonstrated the presence of 12 half-cystines, probably interconnected by disulfide bridges. Each enzymatic molecule contained four hexosamine and eight mannose residues and a molecular weight of about 19,000 was indicated. However, amino acid analysis and sequence determination established a molecular weight of about 14,500, and this discrepancy was probably due to the carbohydrate content which increased the molecular-weight determination established by gel filtration. The sequence of amino acid residues of phospholipase A_2 has recently been established (Shipolini *et al.*, 1971b) and a simplified method for its purification has been described (Munjal and Elliot, 1971).

Hyaluronidase comprises about 2% of dried honey bee venom and crude venom containing this spreading factor is more active by weight than commercial preparations from bull testes (Habermann, 1971). The enzyme may also attack chondroitin sulfates A and C, heparin, and blood group-specific substances A and B (Barker *et al.*, 1963). The venom of another apid, *Bombus pratorum*, possesses a low hyaluronidase activity (Jaques, 1956).

Benton (1967) reported that honey bee venom contained an α- and β-esterase, as well as two alkaline and three acid phosphatases. Since increased levels of acid phosphatase precede the morphological signs of cell death, it has been suggested that the presence of these enzymes in bee venom may constitute an accidental inclusion (Owen and Bridges,

1976). However, Karpas *et al.* (1977) detected acid phosphatase in an allergenic fraction of bee venom but were unable to demonstrate alkaline phosphatase activity.

A protease inhibitor was demonstrated to be present in honey bee venom by Shkenderov (1973). This compound, which is a potent inhibitor of trypsin, was characterized as a basic protein with a molecular weight of about 9000. It was subsequently reported that this protein was capable of inhibiting the proteolytic and esterolytic activities of a wide range of compounds (Shkenderov, 1975). Recently, Shkenderov (1976) purified the inhibitor and studied its structure and inhibitory properties in great detail. This peptide which inhibited the amidase, proteolytic, and esterolytic activities of trypsin, was determined to be the fast acting-type of inhibitor.

2. *Wasp Venoms*

Virtually all studies on the chemistry of wasp venoms have been confined to species in the family Vespidae. Species in the genera *Vespula* and *Vespa* are referred to as yellow jackets (in the New World) and hornets, respectively, although several of the former species, e.g., *Vespula maculifrons*, are specifically called hornets. However, these two wasp genera are both taxonomically compact and closely related, and a name such as hornet, while it is descriptively useful for the larger species in these two genera, cannot be considered to possess any real taxonomic exactitude. Although there has been a widespread practice among venom researchers, when describing the compositions of wasp venoms, to rigorously segregate wasp and hornet venoms into separate categories (Habermann, 1971, 1972), in the subsequent discussion of these venoms no distinctions will be made between those derived from vespine hornets and those of other wasps, unless so indicated.

Five biogenic amines, histamine, serotonin, dopamine, noradrenaline, and adrenaline, have been identified in wasp venoms (Ishay *et al.*, 1974; Geller *et al.*, 1976; Owen, 1979b) (Tables 14.1, 14.2), and those of *Vespa* species appear to be the richest sources of acetylcholine in the animal kingdom. However, from a comparative biochemical standpoint, it is the presence of a group of pharmacologically active peptides, the kinins, which distinguishes wasp venoms from those of other hymenopterans. Kinins are hypotensive agents capable of increasing capillary permeability and relaxing most isolated smooth muscle preparations. Some kinins can be generated from mammalian α_2-globulin by the action of trypsin on snake venom (bradykinin) or kallikrein (kallidin), and they also occur in the skin of amphibians. Jaques and Schachter (1954) demonstrated that the venom of *Vespula vulgaris* contained a potent smooth muscle stimulant which produced a delayed, slow, contraction of the guinea pig ileum. Its similarity to bradykinin was noted, but it was subsequently established that, unlike bradykinin and kallidin, the activity of wasp kinin was reduced by trypsin (Holdstock *et al.*, 1957). Ultimately, Mathias and Schachter (1958) demonstrated that the venom of *V. vulgaris* contained three kinins with the same relative activities, but one compound accounted for 90% of the recovered activity. A single kinin is present in the venom of *Vespa crabro* (Bhoola *et al.*, 1961), but unlike the kinins of *V. vulgaris*, it is insensitive to the action of trypsin. Kinin(s) is also present in the venom of *Vespa orientalis* (Edery *et al.*, 1972).

The structure of one of the three kinins in the venoms of *Polistes annularis*, *P. fuscatus*, and *P. exclamans* has been determined (Prado *et al.*, 1966; Pisano, 1970). The genus *Polistes*, like *Vespula* and *Vespa*, is a taxon in the family Vespidae but is a member of the subfamily Polistinae whereas the other two genera belong to the Vespinae. *Polistes* kinin contains 18 amino acid residues including the same 9 amino acid residues that constitute the peptide bradykinin; each wasp contains about 0.43 μg. Pyroglutamic acid is at the N terminus and, like the kinin of *V. crabro*, activity is not diminished after incubation with trypsin (Prado *et al.*, 1966). However, uniformly high activity as a contractor of the rat uterus only occurred after tryptic digestion, and it has been established that the active kinin is a peptide containing 10 amino acid residues, glycylbradykinin. Thus, the native peptide, which is 2–20 times more active than bradykinin in various tests (Stewart, 1968), may actually constitute a kininogen. The kinin of *Polistes* is easily distinguished from that of *V. crabro* by pharmacological tests.

Two additional kinins have been identified in the venom of another *Polistes* species, *P. rothneyi iwatai* (Watanabe *et al.*, 1975, 1976). One of these kinins was identical to bradykinin except that one mole of serine was replaced by threonine. The other kinin was an undecapeptide that contained an alanylarginyl residue attached to the N terminus of a bradykinin residue that, like the other kinin, contained a threonine residue in place of serine. Recently, Hirai *et al.* (1980) identified a novel mast-cell degranulating tetradecapeptideamide, "polistes mastoparan," from the venom of *P. jadwigae*.

Two vespulakinins have been identified in poison gland extracts of the wasp *Vespula maculifrons* (Yoshida *et al.*, 1976). Both of these kinins are glycopeptide derivatives of bradykinin, one being a heptadecapeptide and the other a pentadecapeptide. These vespulakinins, which are the first reported vasoactive glycopeptides, both contain *N*-acetylgalactosamine and galactose attached to two threonine residues.

Kishimura *et al.* (1976) characterized the kinin in the venom of *Vespa mandarinia* as a novel hydroxyproline-containing compound. This kinin, vespakinin M, is a dodecapeptide containing bradykinin and is the first invertebrate peptide that has been demonstrated to contain a hydroxyprolyl residue. Yasuhara *et al.* (1977) identified another dodecapeptide, vespakinin X, in the venom of *V. xanthoptera.* This kinin contains the sequence alanylbradykinin-isoleucylvaline.

Two new mast cell-degranulating peptides have been identified recently in the venoms of vespid wasps. The venom of *Vespula lewisii* contains a tetradecapeptide amide, mastoparan, that mainly contains hydrophobic amino acids (Hirai *et al.*, 1979a). A similar peptide, mastoparan X, containing the same N- and C-terminal amino acids, has been characterized in the venom of *Vespa xanthoptera* (Hirai *et al.*, 1979b). Neither of these peptides are kinin-like in their pharmacological activities.

The venoms of wasps contain several enzymes (Table 14.2) but their roles in the intoxication process have not been clearly established. The venoms of *Vespula germanica* and *Vespa crabro* contain phospholipase A (Contardi and Latzer, 1928), and phospholipase B activity has also been demonstrated in these venoms (Ercoli, 1940). Recently, Rosenberg *et al.* (1977) detected very high levels of both phospholipase A (PhA) and phospholipase B (PhB) in the venom of *V. orientalis,* and Owen (1979b) reported that the venoms of the polistines *Polistes humilis* and *Ropalidia revolutionalis* are rich in phospholipase A_2. However, there is no proof that the phospholipase A and B activities of wasp venoms are not due to a single enzyme, and the hemolytic effects observed with wasp venoms may possibly reflect the presence of a single protein. Edery *et al.* (1972) reported that the venom of *V. orientalis* possessed hemolytic activity.

Hyaluronidase activity has been observed with the venoms of *Polistes omissa* (Said, 1960), *P. humilis* (Owen, 1979b), *V. vulgaris* (Jaques, 1956), *V. orientalis* (Edery *et al.*, 1972), and *Ropalidia revolutionalis* (Owen, 1979b) and appears to be greater than that in bee venoms. Potent hyaluronidase activity is also reported for the venoms of *V. crabro, V. germanica, Vespula saxonica, V. media, Polistes gallicus,* and *P. vulgaris* (Table 14.2). The enzymes resemble mammalian testicular hyaluronidase in their ability to depolymerize hyaluronic acid more effectively than chondroitin sulfate (Allalouf *et al.*, 1972). Slight acetylcholinesterase activity is present in the venom of *V. vulgaris,* but the venoms of other arthropods do not hydrolyze this choline ester (Jaques, 1955). The protease activity demonstrable in the venom of *V. orientalis* (Edery *et al.*, 1972) and *Vespula* sp. (Hoffman, 1978) appears to be unusual, since proteolytic enzymes have been rarely detected in other hymenopterous venoms. Esterase activity has been detected in the venom of *P. omissa* (Said, 1960) and acid phosphatase is present in the venom of *Vespula* spp. (Hoffman, 1978).

Geller *et al.* (1976) recently identified histidine decarboxylase in the venoms of *V. maculifrons* and *V. maculata* (Table 14.2). This enzyme, which resembled mammalian histidine carboxylase, was present in various levels in poison gland extracts of both species. It could function either by generating histamine in the venom or by producing this biogenic amine from the host's own histidine.

Slor *et al.* (1976) and Ring *et al.* (1978b) demonstrated that venom sac extracts of queen and workers of *V. orientalis* contained acid, neutral, and alkaline deoxyribonucleases (DNases). DNases have also been detected in venom sac extracts of *P. gallicus* (Ring *et al.*, 1978a). It has been suggested that these nucleases may be responsible for the increase in nucleic acid degradation products when cat blood is treated with venom sac extracts.

3. *Ant Venoms*

Detailed chemical investigations of the proteinaceous venoms of ants have been essentially limited to species in the myrmeciine genus *Myrmecia* and the myrmicine genera *Pogonomyrmex, Solenopsis,* and *Monomorium.* However, the compositions of the venoms produced by species in these genera are so extraordinarily different that any generalizations about formicid venoms are effectively negated. The poison gland secretions of two *Myrmecia* species and one *Pogonomyrmex* species are dominated by complex proteinaceous mixtures, whereas the venoms of *Solenopsis* and *Monomorium* species are comprised primarily of alkaloids. The minor aqueous phase in the venom of a *Solenopsis* species contains only a small amount of proteinaceous material consisting of three low molecular-weight allergenic constituents (Baer *et al.*, 1977, 1979).

The alkaloids present in *Solenopsis* venoms are reported to possess a wide range of physiological activities that include blocking neuromuscular transmission (Yeh *et al.*, 1975), releasing histamine from mast cells (Read *et al.*, 1978), inhibiting ATPase (Koch and Dessiah, 1975; Koch *et al.*, 1977), and uncoupling oxidative phosphorylation of mitochondria (Cheng *et al.*, 1977). Indeed, it seems certain that myrmicine venoms are generally proteinaceous (Blum, 1966) and as a consequence, the toxic dialkylpiperidine alkaloids in *Solenopsis* (*Solenopsis*) venoms (Tables 14.1, 14.2), which also possess potent antimicrobial (Blum *et al.*, 1958; Jouvenaz *et al.*, 1972), hemolytic (Adrouny *et al.*, 1959), and necrotoxic (Buffkin and Russell, 1972) activities, must, at this juncture, be regarded as constituting somewhat atypical toxins of ant venoms. However, the presence of dialkylpyrrolidines and -pyrrolines in the venoms of *Monomorium* (Talman *et al.*, 1974; Ritter *et al.*, 1975) and *Solenopsis* (*Diplorhoptrum*) species (Pedder *et al.*, 1976) demonstrates that alkaloids are characteristic of the poison gland secretions of species in at least two myrmicine genera (Tables 14.1, 14.2).

Histamine is the only low molecular weight compound detected in the venoms of *Myrmecia* species (Table 14.1). The poison gland secretions of both *Myrmecia pyriformis* and *M. gulosa* contain nonhistamine smooth muscle stimulants (de la Lande *et al.*, 1963; Cavill *et al.*, 1964) which are easily distinguished from the kinins present in wasp venoms. Phospholipase A activity is well developed in the venom of *M. pyriformis* (Lewis *et al.*, 1968) and in addition, the venoms of both *Myrmecia* species contain direct hemolytic factors which are heat labile. Two fractions possessing strong hyaluronidase activity are present in the venom of *M. gulosa* (Cavill *et al.*, 1964), and similar activity is detectable in the secretion of *M. pyriformis* (de la Lande *et al.*, 1963; Lewis and de la Lande, 1967).

The venom of the myrmicine *Myrmica ruginodis* contains, in addition to free amino acids and histamine, macromolecular constitutents including hyaluronidase, peptides, and a convulsive protein (Jentsch, 1969). No hemolytic constituents were detectable in this poison gland secretion.

The potently algogenic venom of the myrmicine *P. badius* contains histamine, a large number of free amino acids, and a host of enzymes (Schmidt and Blum, 1978a,b). Six classes of enzymes are detectable in the venom, and this poison gland secretion must be regarded as the richest enzymatic source among the characterized arthropod venoms. Very high levels of both phospholipase A (PhL-A_2) and phospholipase B (PhL-B) are present in this venom, the concentration of the former enzyme being twice that found in honey bee venom. The level of hyaluronidase is the highest recorded for any arthropod venom. Acid phosphatase, which has been rarely encountered in arthropod venoms, is also present at a very high level. Lipase, which had not been previously reported to be a constituent of any arthropod venoms, is highly concentrated in the venom of *P. badius*. Three esterases, which do not exhibit trypsin-like activity, are readily detected in this myrmicine poison gland secretion (Schmidt and Blum, 1978a,b).

The venom of *Pogonomyrmex barbatus* contains a hemolysin, barbatolysin, that bears a functional resemblance to melittin, the direct-acting hemolysin present in the venom of the honey bee (Bernheimer *et al.*, 1980). Barbatolysin is a basic polypeptide with a molecular weight of 3500 and contains 34 amino acid residues.

V. AN OVERVIEW

Some of the major constituents present in arthropod venoms are present in Table 14.2. From Table 14.2 it is obvious that these venoms contain a wide variety of compounds some of which constitute unique natural products. However, in view of the small number of venoms which have been critically examined, it seems premature to generalize about the toxic secretions possessed by thousands of venomous arthropods in many families. Investigations of the comparative toxinology of the Arthropoda are still at a very early developmental stage, but the results which have been obtained so far should spur considerable additional research in this field.

Among arthropods, the venoms of scorpions constitute the best-characterized toxins produced by species in one order, thanks to the research of Lissitzky and colleagues. Eight neurotoxins in the venoms of three species have been subjected to detailed structural analyses and they have been demonstrated to consist of three lines of homologous proteins (Rochat *et al.*, 1970). However, whereas these compounds constitute mammalian toxins of great potency, they are nontoxic to arthropods, and it has been subsequently demonstrated that additional proteins are specifically toxic to insects and crustaceans (Zlotkin *et al.*, 1971b, 1972a). Thus, these investigations emphasize the necessity of interpreting the toxicity of arthropod venoms in terms of their fractionated constituents rather than only as crude secretions. The ecological and pharmacological implications of the selective neurotoxicity of the different toxins in scorpion venoms are of great fundamental significance, but they never would have been apparent unless animals in different classes had been challenged with the isolated fractions.

The venom of *Androctonus australis* appears to be typical of scorpions in having mammalian neurotoxins of great potency (10–20 μg/kg) (Rochat *et al.*, 1967), which certainly must represent defensive compounds *par excellence.* The mammalian toxins, which may also be active against other classes of vertebrates (e.g., reptiles), have been presumably evolved to function against potential predators, but these basic proteins appear to be inactive against the invertebrates on which *A. australis* preys. On the other hand, although the venom contains about five times more vertebrate than insect neurotoxin, the greater potency of the latter more than compensates for its low concentration (Zlotkin *et al.*, 1971a). The presence of a specific crustacean neurotoxin in the venom of *A. australis* demonstrates that this scorpion has developed venomous constituents with specific activities for two groups of its most normally encountered prey, species in the Crustacea and Insecta. Thus, this venom, through the agency of multiple neurotoxic proteins, is able to discriminate between species in three classes. That scorpion venoms do not appear to be especially active against other scorpions (Zlotkin *et al.*, 1972a) or spiders (Wheeling and Keegan, 1972) may indicate great specificity of the neurotoxins present in these remarkable exocrine secretions.

The toxic proteins in scorpion venoms appear to constitute ideal compounds for studying the neurophysiology of vertebrates vis-à-vis that of invertebrates. The scorpamines have been subjected to detailed structural analyses, and, as a consequence, these rapidly acting toxins represent ideal vehicles for probing the comparative biochemistry and physiology of nervous and muscular receptors. The presence of different families of homologous proteins functioning as either mammalian or insect neurotoxins should permit the results of comparative neurophysiological investigations to be correlated with the amino acid sequences of the selectively toxic proteins. In addition, the question of the specific neurotoxicities of these proteins can be examined in terms of possible differences in either the basic neurophysiology of vertebrate and invertebrate nervous tissues or in the structures of the central nervous systems themselves. In short, the selective toxicities of the scorpion neurotoxins qualify these basic proteins as pharmacological gold mines rarely matched by any other group of natural products.

Scorpion venoms are further distinguished by their lack of the small biogenic amines which are virtually ubiquitous in other arthropod venoms. In addition, the toxic secretions of scorpions contrast to those of other arthropods in the highly selective distribution of enzymes and the sporadic occurrence of hemolysins. It thus appears that, among arthropod venoms, those of scorpions represent an unique evolutionary development which is predicated on the availability of a group of basic, low molecular-weight proteins, possessing potent activities against both vertebrates and invertebrates as a consequence of the presence of highly specific neurotoxins.

The venoms of another major group of arachnids, the spiders, are similar to those of scorpions in that they too contain complex mixtures of proteins or polypeptides. On the other hand, spider venoms are frequently enriched with biogenic amines (Table 14.1), which, while they are not considered to be identified with the toxic fractions, are capable of producing intense pain when injected into mammals. Spermine has been isolated only from spider venoms and may represent the toxic end product of an evolutionary specialization that has occurred in certain lines of spiders. Protease activity appears to be rather widespread in spider venoms, although this enzyme(s) is certainly not identified with the great toxicity of certain spider venoms.

Although the proteinaceous toxins in the poison gland secretions of spiders have generally not been characterized in any great detail, some of these toxins are in the same approximate molecular weight range as those of scorpions. In addition, some of the proteins in spider venoms share with those of scorpions the property of selective or specific toxicity to animals in different classes. For example, only one of the five protein fractions in the venom of *L. mactans* is toxic to guinea pigs, whereas two fractions are toxic to insects (Frontali and Grasso, 1964). The discovery that the venom of *L. mactans* contains three fractions toxic to mice and only one toxic to guinea pigs (Granata *et al.*, 1972) constitutes one of the few examples of toxic proteins with selective activities against different mammals in a single arthropod venom. The adaptive significance of this selective toxicity is not readily apparent, although it may reflect the development of a venom with a great defensive "punch" against the main potential vertebrate predators, in this case rodents. Whether selective toxicity is a general property of the toxic proteins in spider venoms remains to be determined, but it will be especially significant if structural studies demonstrate that these toxins belong to families of homologous proteins, as do those of scorpions.

The salivary venoms of predatory hemipterans appear to be typical, in their overall action, of the toxic salivas which are injected into invertebrate prey by insects in several orders. The bite of tabanid (Diptera) larvae is reported to produce an instantaneous paralysis in the victim (Beard, 1962), and potent salivary venoms are also characteristic of asilid (Diptera) adults (Whitfield, 1925), lampyrid and neuropterous larvae, as well as species in other taxa. Virtually nothing is known about the chemistry of these salivary venoms, and except for the excellent study of Edwards (1961) on the secretion of the reduviid *Platymerus rhadamanthus*, the nature of the toxins produced by predatory Hemiptera is virtual *terra incognita*. Edwards demonstrated that the saliva of *P. rhadamanthus* is qualitatively similar to snake venoms, sharing most of the enzymes (Table 14.2) that are present in the salivary secretions of these ophidians. However, the basis for the rapid paralytic activity of the venomous salivas of insects has not been established, and the investigation of this phenomenon constitutes an important avenue for examining the comparative neurophysiology of insects.

The few hymenopterous venoms that have been extensively studied provide cautious grounds for concluding that these toxic secretions share some common constituents as well as some rather unique peptides of great physiological activity. The venom of the honey bee is probably the most carefully studied insect poison gland secretion, but there is no reason to assume that the venom of this highly evolved apid is typical of that of bees. Honey bee venom contains both direct (melittin) and indirect (phospholipase A) lytic factors, and is thus comparable to cobra venom which contains an indirect and a direct cytolysin, cardiotoxin. Along with MCD peptide, melittin represents an excellent pharmacological agent for exploring membrane functions, considering its cytolytic potency. Apamin is the smallest neurotoxin known and, if the venoms of other bees also contain specific neurotoxic polypeptides, these compounds may play an important role in elucidating basic mechanisms of central nervous system function. The recent isolation (Gauldie *et al.*, 1976) of three additional peptides from honey bee venom further demonstrates the great complexity that may characterize these poison gland secretions.

The venoms of wasps are of particular pharmacological interest in that they contain concentrated solutions of many compounds normally found in mammalian tissues. Thus, it appears that mammals may have constituted one of the main selection pressures for the evolution of social wasp venoms which are rich in biogenic amines, acetylcholine, and kinins (Tables 14.1, 14.2). All these compounds are potent algogens and it seems significant that the great algogenic action of *Vespa* venoms, resulting from the presence of acetylcholine and histamine, is duplicated in the stinging nettles of urticaceous plants (Emmelin and Feldberg, 1947). Biogenic amines also serve as potent vasoconstrictive agents and, by maintaining a high concentration of venom at the sting site, prolong the local pharmacological effects of the secretion. The ultimate spread of the venom could be facilitated admirably by the hyaluronidase present in these venoms.

At this juncture, the kinins appear to be mainly restricted in their hymenopterous distribution to the venoms of wasps. It seems almost certain that the presence of these compounds in wasp venoms represents an adaptation for deterring vertebrate predators, particularly mammals. The kinins of *Polistes* spp. are chemically related to bradykinin and, like this mammalian polypeptide, possess potent pharmacological actions. In common with the catecholamines and serotonin, *Polistes* kinins are powerful algogens but, in addition, they increase capillary permeability and result in the contraction of smooth muscle organs. Among hymenopterous poison gland secretions, serotonin is only known from wasp venoms (Tables 14.1, 14.2), and it seems especially significant that this amine is reported to potentiate the pain-producing action of bradykinin (Sicuteri *et al.*, 1966). Thus, the wasp kinins may act in a combinative way with serotonin or other biogenic amines, or for that matter, with other venomous constituents. The demonstration that a native wasp kinin is probably a kininogen (Prado *et al.*, 1966) which must be hydrolyzed to an active kinin after injection may provide an explanation for the extended algogenic effects of wasp venoms at the sting site. Such a system might provide a prolonged kinin action at the site of injection and provide a further basis for regarding wasp venoms as some of the most elegant toxic secretions which have been evolved.

The presence of distinctive kinins in the venoms of *Polistes*, *Vespa*, and *Vespula* species emphasizes the widespread distribution of these compounds in wasp venoms (Table 14.2). Furthermore, since a different kinin has been identified in the venom of each species studied, it is not unlikely that a plethora of unique kinins remain to be characterized in wasp venoms. If these peptides are subsequently demonstrated to be ubiquitous poison gland products of social wasps, it will seem eminently reasonable to conclude that a host of selective pressures (predators?) were responsible for the evolution of these diverse mammalian vasoactive peptides.

The venom compositions of social wasps may represent an ideal compromise between toxic secretions which are admirably suited to both immobilize prey and to deter mammalian predators as a consequence of their potent algogenicity. High concentrations of biogenic amines, acetylcholine, and kinins, would appear to constitute an ideal series of algogens with which to discourage vertebrate predators, and the levels of antigenic proteins undoubtedly play a role in prolonging the effectiveness of the venoms. On the other hand, it is possible that the venoms of solitary wasps, which are utilized exclusively for immobilizing prey, generally lack the mammalian algogens which appear to be trademarks of social wasp venoms.

The paralyzing toxins in the venoms of the solitary wasps *Microbracon gelechiae* and *M. hebetor* have been characterized as very labile proteins with a molecular weight of about 62,000 (Piek *et al.*, 1974; Visser *et al.*, 1976). Rosenbrook and O'Connor (1964) analyzed the venom of the solitary wasp *Scelophron caementarium* and reported that it contained no low molecular-weight compounds, a very low level of protein, and a complete absence of the antigens found in social wasp and bee venoms. The stings of most solitary wasps are not painful to human beings, and it is unlikely that there has been any selection pressure on these insects to produce venoms which are painfully deterrent to vertebrates. If comparative analyses of the venoms of solitary and social wasps demonstrate that those of the former consistently lack vertebrate algogens, it will provide persuasive circumstantial evidence for concluding that the poison gland secretions of solitary wasps are selective invertebrate toxins.

The venoms of myrmeciine ants were initially believed to correspond most closely to those of some wasps (Cavill *et al.*, 1964) and, in particular, to that of the honey bee (Lewis and de la Lande, 1967). However, wasp venoms, which are now known to contain a variety of biogenic amines (Geller *et al.*, 1976) and unique kinins (Prado *et al.*, 1966; Yoshida *et al.*, 1976; Kishimura *et al.*, 1976), obviously differ considerably from *Myrmecia* venoms in terms of biochemical constituents. On the other hand, honey bee venom does share several major biochemical features with *Myrmecia* venoms. Wanstall and de la Lande (1974) have isolated a single fraction from the venom of *M. pyriformis* that exhibits a wide range of pharmacological activities. This fraction, which was enzyme-free, possesses histamine-releasing, red cell-lysing, and smooth muscle stimulant activities. This trio of activities also characterizes melittin, the major peptide present in honey bee venom. The *M. pyriformis* toxin possesses an apparent molecular weight of 11,000, after elution from a Sephadex gel column. Under similar conditions melittin elutes as four associated molecules giving an apparent molecular weight of 11,000. Thus, the major toxins in both honey bee and *M. pyriformis* venoms could be structurally related (Wanstall and de la Lande, 1974). On the other hand, it is simply impossible to attempt to correlate the compositions of the venoms of a few ant species belonging to the most primitive of the formicid subfamilies with those of ant species in the most highly evolved taxa. The chemistry of the venoms of stinging species in the Ponerinae, Dorylinae, Cerapachyinae, Pseudomyrmecinae, and Myrmicinae are essentially unknown and, until the comparative toxinology of the Formicidae is illuminated in some detail, the proteinaceous venoms of ants must be regarded as insufficiently characterized.

Only two protein-rich myrmicine venoms have been analyzed in any detail, and the results obtained in these investigations demonstrate that these secretions may be more complex than those of myrmeciine venoms. The venom of *Myrmica ruginodis* contains, in addition to histamine, hyaluronidase and peptides, a protein (convulsive factor) that constitutes the major component in the poison gland secretion (Jentsch, 1969).

The neurotoxic venom of *Pogonomyrmex badius* exhibits the highest toxicity to mice of any insect venom that has been similarly studied (Schmidt and Blum, 1978c). The venom contains both a direct and indirect hemolysin and heparin does not reduce the hemolytic properties of this secretion. Very high levels of phospholipase A_2, phospholipase B, hyaluronidase, acid phosphatase, and lipase are present (Table 14.2). In addition, three esterases are also readily detectable. In particular, the occurrence of both lipase and acid phosphatase in this venom further emphasizes that there is no basis for generalizing about the composition of myrmicine venoms. Indeed, whereas the venom of *M. ruginodis* is more toxic to insects than mammals (Jentsch, 1969), that of *P. badius* is far more toxic to mammals than insects (Schmidt and Blum, 1978c).

The chemistry of the myrmicine genus *Solenopsis* is so aberrant that it reduces generalizations about ant venoms to the absurd. Species in this myrmicine genus synthesize venoms rich in alkaloids (MacConnell *et al.*, 1971, 1976; Brand *et al.*, 1972; Pedder *et al.*, 1976) and proteinaceous constituents are present in relatively minor amounts. Furthermore, there are often major qualitative differences between the venoms of species in the *Solenopsis* subgenera *Solenopsis* and *Diplorhoptrum.* Whereas the secretions of *Solenopsis* (*Solenopsis*) species are dominated by a wide variety of 2,6-dialkylpiperidines (MacConnell *et al.*, 1976), those of *Solenopsis* (*Diplorhoptrum*) species are generally characterized by the presence of 2,5-dialkylpyrrolidines and -pyrrolines (Pedder *et al.*, 1976) (Table 14.2). Although the piperidine alkaloids have not been detected in the venoms of ant species in any other genera, some of the pyrrolidines have been identified in the venoms of *Monomorium* species (Talman *et al.*, 1974; Ritter *et al.*, 1975). The presence of indolizines in the venom of *Monomorium pharaonis* (Ritter *et al.*, 1973, 1975) emphasizes the highly distinctive character of this myrmicine poison gland secretion and suggests that species in different genera may synthesize idiosyncratic alkaloids.

Alkaloidal venoms do not seem to be typical of the subfamily Myrmicinae, and there are no indications of what factors may have acted to select for venoms rich in these nitrogen heterocycles and deficient in the proteinaceous components typical of hymenopterous venoms. It may well be that alkaloids are normally present in trace amounts in myrmicine venoms, and it is worth noting that a pyrrole has been identified as a trace component in the venom of another myrmicine species (Tumlinson *et al.*, 1971). Conceivably, concomitant selection for increased synthesis of toxic alkaloids and suppression of protein synthesis may have occurred in the evolution of *Solenopsis* and *Monomorium* venoms, and the analyses of the venoms of species in additional myrmicine genera may provide some supportive data for this suggestion.

Analyses of the venoms of a wide range of species within a genus may provide a basis for comprehending the biochemical evolution of these poison gland secretions. It has been suggested that the evolution of venoms within the genus *Solenopsis* proceeded from those in which *cis*-alkaloids were dominant to those in which the trans isomers predominated (Brand *et al.*, 1973a). This hypothetical construct in biochemical evolution was based on thermodynamic considerations of the cis and trans isomers and comparisons of the proportions of the different alkaloidal isomers in the venoms of queens and workers. Obviously, detailed comparative analyses of ant venoms (MacConnell *et al.*, 1976) can provide important insights into the evolutionary position of these hymenopterous secretions while at the same time providing a means of comprehending their *raison d'être* as defensive secretions.

PART II

Chemical Defenses of Arthropods in Perspective

Chapter **15**

Defensive Compounds and Arthropod Taxonomy

Botanical chemists have been evaluating the taxonomic significance of plant natural products for nearly a hundred years (Alston and Turner, 1963), and it seems certain that zoologists will encounter the same types of apparent contradictions and objections that botanists have met in attempting to utilize natural products as indicators of possible phylogenetic relationships. However, even at this early stage in arthropod chemical taxonomy, it appears that the distributional patterns of specific compounds may possess real value as an adjunct to taxonomy based on morphological characters.

At this juncture, too broad an application of the distribution of arthropod defensive compounds as chemical characters for already proposed phylogenetic treatments of specific arthropod taxa is impractical because (1) the species in too few major taxonomic groups have been chemically analyzed, and (2) very few in-depth investigations of species in minor taxa within major taxa have been carried out. When in the few instances in which investigations have been undertaken, the value of selected natural products as useful indicators for taxa in proposed tribal or subfamilial classifications has been considerable (Moore and Wallbank, 1968; Schildknecht *et al.*, 1968a,b; Tschinkel, 1975b).

In terms of biochemical systematics, the distribution of natural products possesses relatively limited value for assessing phylogenetic relationships. In contrast, amino acid sequences of various proteins can provide highly specific information about the relations of members of the lowest taxonomic categories, especially if genetic studies pertaining to the syntheses of these macromolecules have also been undertaken. In view of the fact that relatively few investigations have been made on the amino acid sequence of proteins derived from arthropods, the distributional patterns of exocrine products presently constitute the primary chemical data available for chemical taxonomy.

In itself, the presence of a single compound in species in different taxa may be of limited taxonomic value, a conclusion that has been emphasized by Brown (1967). Although parallel evolution of the same defensive compounds may occur as frequently in arthropods as it does in plants, this should not necessarily nullify the possible systematic value of the occurrence of the same compound in arthropod species belonging to related lower taxonomic categories. For example, the presence of formic acid in the defensive exudates of notodontid larvae, ants, and carabid beetles exemplifies the parallel evolution of the same allomone in taxa in three widely separated orders of endopterygote insects. Similarly, (*E*)-2-hexenal, which is utilized in the deterrent secretions of both exopterygotes (cockroaches, true bugs) and endopterygotes (ants, beetles), is another example of an evolutionary parallel in defensive glandular chemistry. Furthermore, the independent evolution of (*E*)-2-hexenal in widely separated phyletic groups does not necessarily imply that it was biosynthesized by the same metabolic pathway but, rather, that different arthropod lines were able to produce this aldehyde by extending some previously existing metabolic pathways. On the other hand, the independent evolution of the same exocrine product or associated morphological structures by unrelated arthropod lines is analogous, for example, to the evolution within unrelated angiosperm taxa of similar morphological and chemical characters. Thus, in itself, such biochemical congruency, as exemplified by the distribution of (*E*)-2-hexenal in insects, is not necessarily surprising.

In contrast to certain low molecular-weight compounds (e.g., formic acid), the biosynthesis of allomones that are more structurally complex probably requires the evolution of considerably more enzymes, and, as a consequence, it can be predicted, a priori, that such compounds will not be independently evolved as frequently. For example, pederin, one of the most complex defensive compounds so far identified in arthropods, is only known to be produced by beetles in the genus *Paederus* (Cardani *et al.*, 1965b). This amide appears to constitute an excellent chemical character for the species in this staphylinid taxon. Furthermore, even if pederin were subsequently identified in grasshoppers, this discovery would in no way diminish the value of this compound as a taxonomic indicator for species of *Paederus*. Similarly, vertebrate biochemists have recently encountered surprising natural products congruencies in the distribution of a complex compound in a variety of unrelated animals. Tetrodotoxin ($C_{11}H_{17}N_3O_8$), the polycyclic guanidinium-containing compound isolated from certain fish (Noguchi and Hashimoto, 1973), has proved to be identical to tarichatoxin, isolated from salamanders (Mosher *et al.*, 1964) and more recently from frogs (Kim *et al.*, 1975). Although this potent neurotoxin is known from fish, salamanders, and frogs, its presence constitutes an important chemical character for members of each of these taxa.

The possible diagnostic significance of a related series of compounds can be explored by examining the distribution of these compounds in the taxa under investigation. Analysis of the distribution of 1,4-quinones in the Arthropoda (Table 15.1) documents both the parallel and convergent evolution of this class of compounds. Quinones have been evolved independently in three arthropod classes, but members of each class produce characteristic compounds which may possess value as diagnostic indicators for species in certain taxa. In the Arachnida, for example, two dimethyl- and a trimethyl-1,4-benzoquinone fortify the defensive secretions, and none of these compounds has been frequently encountered as exocrine products of species in other arthropod classes. Furthermore, in the arachnid order Opiliones, benzoquinone-producing taxa appear to be limited to the suborder Laniatores, whereas in the suborder Palpatores, aliphatic ketones appear to be the dominant compounds in the defensive secretions (Blum and Edgar, 1971; Meinwald *et al.*, 1971; Jones *et al.*, 1976b, 1977). Thus, in the Arthropoda, the 1,4-benzoquinones of the Arachnida appear to be diagnostic for selected taxa both because they appear to be limited in occurrence to species in one taxon in the Opiliones and because some of these compounds (e.g., 2,3,5-trimethyl-1,4-benzoquinone) are chemically unique compared with the quinones produced by species in other classes.

In terms of suprafamilial taxonomic categories, the distribution of 1,4-quinones in the Diplopoda appears to be of considerable diagnostic relevance. Quinones are known to occur in three clearly related orders—Spirobolida, Julida, and Spirostreptida—and the combination of 2-methyl-1,4-benzoquinone and 2-methoxy-3-methyl-1,4-benzoquinone constitutes a characteristic quinonoid pair of diplopod defensive compounds (Table 15.1). In addition, the known distribution of 2,3-dimethoxy-1,4-benzoquinone is limited to the Parajulidae and Nemasomidae, two closely related families in the order Julida (Table

Table 15.1

Distribution of 1,4-Quinones in the Arthropoda

Class	Order and Suborder	Families and Genera	1,4-Quinone[a]									
			A	B	C	D	E	F	G	H	I	J
Arachnida	Opiliones (Phalangida)	Gonyleptidae										
		Heteropachyloidellus	–	–	–	–	+	+	–	–	+	–
		Cosmetidae										
		Paecilaemella	–	–	–	–	+	+	–	–	+	–
		Vonones	–	–	–	–	+	–	–	–	+	–
		Phalangiidae										
		Phalangium	–	–	–	–	–	–	–	–	–	+
Diplopoda	Julida	Julidae										
		Julus	+	–	–	–	–	–	–	–	–	–
		Archiulus, Brachyiulus, Cylindroilus	–	+	–	–	–	–	+	–	–	–
		Ophyiulus	+	+	–	–	–	–	+	–	–	–
		Nemasomidae										
		Blaniulus	–	+	–	–	–	–	+	–	–	–
		Nopoilus	–	+	–	–	–	–	–	+	–	–
		Parajulidae										
		Uroblaniulus	–	+	–	–	–	–	–	+	–	–
		Bollmaniulus, Ptyoilus Saiulus, Tuniulus, Oriulus	–	+	–	–	–	–	+	–	–	–
		Paeromopdidae										
		Paeromopus	+	+	–	–	–	–	+	–	–	–

Spirobolida	Atopetholidae										
	Atopetholus	+	+	−	−	−	−	+	−	−	−
	Floridobolidae										
	Floridobolus	−	+	−	−	−	−	+	−	−	−
	Pachybolidae										
	Pachybolus	−	+	−	−	−	−	−	−	−	−
	Leptogoniulus, *Trigoniulus*	+	+	−	−	−	−	+	−	−	−
	Rhinocricidae										
	Rhinocricus	+	+	−	−	−	−	+	−	−	−
	Eurhinocricus	+	+	+[b]	−	−	−	+	−	−	−
	Spirobolidae										
	Narceus	+	+	−	−	−	−	+	−	−	−
	Chicobolus	−	+	−	−	−	−	+	−	−	−
	Allopocockidae										
	Arolus	+	+	−	−	−	−	+	−	−	−
Spirostreptida	Cambalidae										
	Cambala	+	+	−	−	−	−	+	−	−	−
	Choctellidae										
	Choctella	+	+	−	−	−	−	+	−	−	−
	Harpagophoridae										
	Harpagophorus	−	+	−	−	−	−	+	−	−	−
	Odontopygidae										
	Prionopetalum	−	+	−	−	−	−	+	−	−	−
	Spirostreptidae										
	Spirostreptus	−	+	−	−	−	−	−	−	−	−
	Aulonopygus, *Rhapidostreptus*, *Peridontopyge*, *Doratogonus*, *Collostreptus*	−	+	−	−	−	−	+	−	−	−
	Orthopterus, *Archispirostreptus*	+	+	−	−	−	−	+	−	−	−

(continued)

Table 15.1 (Continued)

Class	Order and Suborder	Families and Genera	1,4-Quinone[a]									
			A	B	C	D	E	F	G	H	I	J
Insecta	Dictyoptera											
	Isoptera	Mastotermitidae										
		Mastotermes	+	+	–	–	–	–	–	–	–	–
		Termitidae										
		Hypotermes	+	–	–	–	–	–	–	–	–	–
		Odontotermes, *Macrotermes*	+	+	–	–	–	–	–	–	–	–
		Microtermes	–	+	–	–	–	–	–	–	–	–
	Blattaria	Blaberidae										
		Diploptera, *Blaberus*	+	+	+	–	–	–	–	–	–	–
	Orthoptera	Acrididae										
		Romalea	+	–	–	–	–	–	–	–	–	–
	Coleoptera	Tenebrionidae										
		Adelium	–	–	–	–	+	–	–	–	–	–
		Tribolium, *Diaperis*, etc.	+	+	+	–	–	–	–	–	–	–
		Eleodes	+	+	+	–	–	–	+	–	–	–
		Prionychus[c]	–	+	+	–	–	–	–	–	–	–
		Blapstinus, *Notibius*, *Conibius*	+	+	+	+	–	–	–	–	–	–
		Argoporis	+	+	+	–	–	–	–	–	–	+
		Staphylinidae										
		Lomechusa	+	+	+	–	–	–	–	–	–	–
		Drusilla	+	+	–	–	–	–	+	–	–	–
		Bledius	–	+	–	–	–	–	–	–	–	–

Carabidae										
Pherosophus, Stenaptinus	+	+	–	–	–	–	–	–	–	–
Clivina, Platycerozaena	+	+	+	–	–	–	+	–	–	–
Chlaenius, Mystropomus	+	+	+	–	–	–	–	–	–	–
Arthropterus	–	+	+	–	–	–	–	–	–	–
Goniotropus, Ozaena, Brachinus, Pachyteles	+	+	–	–	+	–	–	–	–	–
Homopterus	–	+	–	–	–	–	+	–	–	–
Metrius	+	–	–	–	–	–	–	–	–	–

[a] A, 1,4-benzoquinone; B, 2-methyl-1,4-benzoquinone; C, 2-ethyl-1,4-benzoquinone; D, 2-propyl-1,4-benzoquinone; E, 2,3-dimethyl-1,4-benzoquinone; F, 2,5-dimethyl-1,4-benzoquinone; G, 2-methoxy-3-methyl-1,4-benzoquinone; H, 2,3-dimethoxy-1,4-benzoquinone; I, 2,3,5-trimethyl-1,4-benzoquinone; J, naphthoquinone and 6-alkylnaphthoquinones.

[b] Identification based only on paper chromatographic R_f value of presumed quinol.

[c] Formerly considered to be a taxon in the Alleculidae, a family that has been combined with the Tenebrionidae.

15.1). Like the trimethylquinone of opilionids, dimethoxyquinone may constitute a distinctive chemical character for members of specific diplopod taxa.

The capacity to biosynthesize the same 1,4-benzoquinones has arisen independently in the Insecta several times, being present in species in at least two exopterygote and one endopterygote order (Table 15.1). The combination of 2-methyl-1,4-benzoquinone and 2-ethyl-(1,4)-benzoquinone in defensive secretions represents a characteristic insect quinonoid mixture; the ethyl homolog appears to be a particularly distinctive insect product. Since cockroaches are believed to have evolved from the termites, the occurrence of quinones in these two suborders of the Dictyoptera might be suggestive of a phylogenetic significance for these compounds. However, as is often the case with morphological characters, there is no way of determining whether the presence of quinones in the Isoptera and Blattaria represents convergent biochemical evolution or a real phylogenetic affinity between the members of the two quinone-producing suborders. Within the suborder Isoptera, quinone-producing species occur in the most primitive family (Mastotermitidae) and the most highly evolved (Termitidae) (Table 15.1). It is almost certain that the biosynthesis of quinones evolved separately in these two taxa, especially since quinones are not known to be produced either by members of the other subfamilies or most of the other genera in the family Termitidae.

1,4-Benzoquinones may have considerable importance as taxonomic characters in the order Coleoptera. 2-Methoxy-3-methyl-1,4-benzoquinone, a typical allomone of millipede species in three orders, has been identified in a few species in each of the three coleopterous families in which 1,4-benzoquinones have been detected (Table 15.1), and this compound cannot be considered a typical quinonoid constituent in insect defensive secretions. In the Staphylinidae, the synthesis of quinones has been independently evolved in both the tergal (*Lomechusa* and *Drusilla*) and pygidial (*Bledius*) glands, structures that are clearly not homologous. In the Carabidae, a crepitation mechanism for explosively discharging quinones has arisen several times. However, Moore and Wallbank (1968) have demonstrated that the uniting of the tribes Ozaeninae and Paussinae, suggested on morphological grounds, is amply supported by both chemical and biosynthetic similarities pertaining to the presence of the same crepitation mechanisms in members of both taxa.

In the Tenebrionidae, both the quinones and their glandular compartments have been demonstrated to be of considerable phylogenetic significance. As far as is known, there is an almost perfect correlation between the presence of external abdominal membranes and quinone-producing glands and reservoirs in the abdomen of members of the subfamily Tenebrioninae (Doyen, 1972). Furthermore, members of the closely related families Alleculidae, Lagriidae, and Nilionidae possess homologous glands (Kendall, 1968; Doyen, 1972), and it had been demonstrated that the alleculid *Prionychus ater* produces 2-methyl-1,4-benzoquinone in its abdominal glands (Schildknecht *et al.*, 1964). On the other hand, the other two tribes in the Tenebrionidae, the Tentyriinae and Asidinae, do not possess abdominal defensive glands. Based on these data, Doyen (1972) united the Tenebrioninae, Alleculidae, Lagriidae, and Nilionidae as subfamilies into a single taxon, the Tenebrionidae. The Tentyriinae and Asidinae, which lack quinone-producing defensive glands, were combined into the family Tentyriidae, coordinate with the newly constituted family Tenebrionidae. Thus, the combination of external and internal (defensive glands and reservoirs) morphology as well as the production of 1,4-benzoquinones was instrumental in establishing the taxonomic composition of the new family Tenebrionidae of Doyen.

An exhaustive comparative investigation of the chemistry of the Tenebrionidae (Tschinkel, 1975b) has established that at least a few quinones may be diagnostic for particular taxa. Whereas 2-methyl-1,4-benzoquinone and 2-ethyl-1,4-benzoquinone are common denominators of virtually all tenebrionid secretions, 2-propyl-1,4-benzoquinone is only known to be present in the secretions of the closely related genera *Blapstinus*, *Conibius*, and *Notibius* of the tribe Pederini (Tschinkel, 1975b). This alkylquinone, which constitutes about 85% of the secretion of *Notibius* spp., can be regarded as an important chemical character for these three tenebrionid taxa. It may also be significant that in those secretions in which 2-propyl-1,4-benzoquinone is predominant, 1,4-benzoquinone is present only as a trace component at best. Similarly, 2-methoxy-3-methyl-1,4-benzoquinone also appears to be diagnostically important, its known distribution in the Tenebrionidae being limited to the secretions of species in the subgenus *Blapylis* of *Eleodes* (Tschinkel, 1975b).

Tschinkel (1975b) has stressed the predominant quinonoid nature of tenebrionid secretions and has noted the virtual ubiquity of 2-methyl-1,4-benzoquinone and 2-ethyl-1,4-benzoquinone in these exudates. Many secretions contain additional compounds, particularly alkenes, and in many cases these admixtures of compounds may be sufficiently distinctive to be diagnostically significant. For example, 2-ethyl-1,4-

benzoquinone is absent from the secretions of *Eleodes beameri*, but this species in the subgenus *Holeleodes* produces a unique tenebrionid exudate which is enriched with saturated and unsaturated aldehydes (Tschinkel, 1975a). Tschinkel (1975b) has also demonstrated that in the defensive secretions of tenebrionids which contain only 1,4-benzoquinones (e.g., *Polopinus* spp.), 1,4-benzoquinone is always present in substantial amounts, whereas this compound is frequently absent from secretions that are also enriched with nonquinonoid constituents. Not surprisingly, the nonquinonoid compounds present in these secretions also appear to be taxonomically significant. For example, although 1-tridecene is often the only alkene present in these exudates, the presence of large amounts of 1-heptadecene and 1-nonadecene in *Cratidus* spp. appears to be characteristic of this and a few other genera. Similarly, *n*-octanoic acid, which is restricted to the secretions of some genera of Eleodini, is only a major constitutent (up to 20%) in species in the genera *Embaphion* and *Neobaphion* (Tschinkel, 1975b).

The quinonoidally rich secretions of tenebrionine species in the genus *Argoporis* appear to constitute some of the most diagnostically significant exudates of this family. In addition to the typical tenebrionid 1,4-benzoquinones, the abdominal secretions are dominated by 1,4-naphthoquinones (Tschinkel, 1972). The major constituent present in these secretions is 6-methyl-1,4-naphthoquinone, which is accompanied by the ethyl, propyl, and butyl homologs. The naphthoquinones appear to be characteristic products of the genus *Argoporis*, having been identified in the four species analyzed. Significantly, *Argoporis* is a taxon in the tribe Scaurini, the abdominal glandular products of which may possess considerable value as taxonomic indicators. Tschinkel (1972) has analyzed the secretions of four other scaurine genera and has demonstrated that the glandular products of each genus are distinctive because of the presence of compounds not detected in other tenebrionid secretions.

Examination of the arthropod 1,4-quinones by the "percentage of frequency rule" commonly utilized by botanical chemists (Hansel, 1956) demonstrates the value of correlating the chemical with the distributional incidence of these compounds (Fig. 15.1). It is immediately apparent that compounds such as 1,4-benzoquinone and 2-methyl-1,4-benzoquinone are of limited taxonomic value because of their widespread distribution in members of two higher taxonomic categories. 2-Ethyl-1,4-benzoquinone is virtually limited in its distribution to the Insecta, but its near ubiquity in the members of this class limits its further taxonomic value. Both 2-propyl-1,4-benzoquinone and 6-ethyl-1,4-naphthoquinone appear to

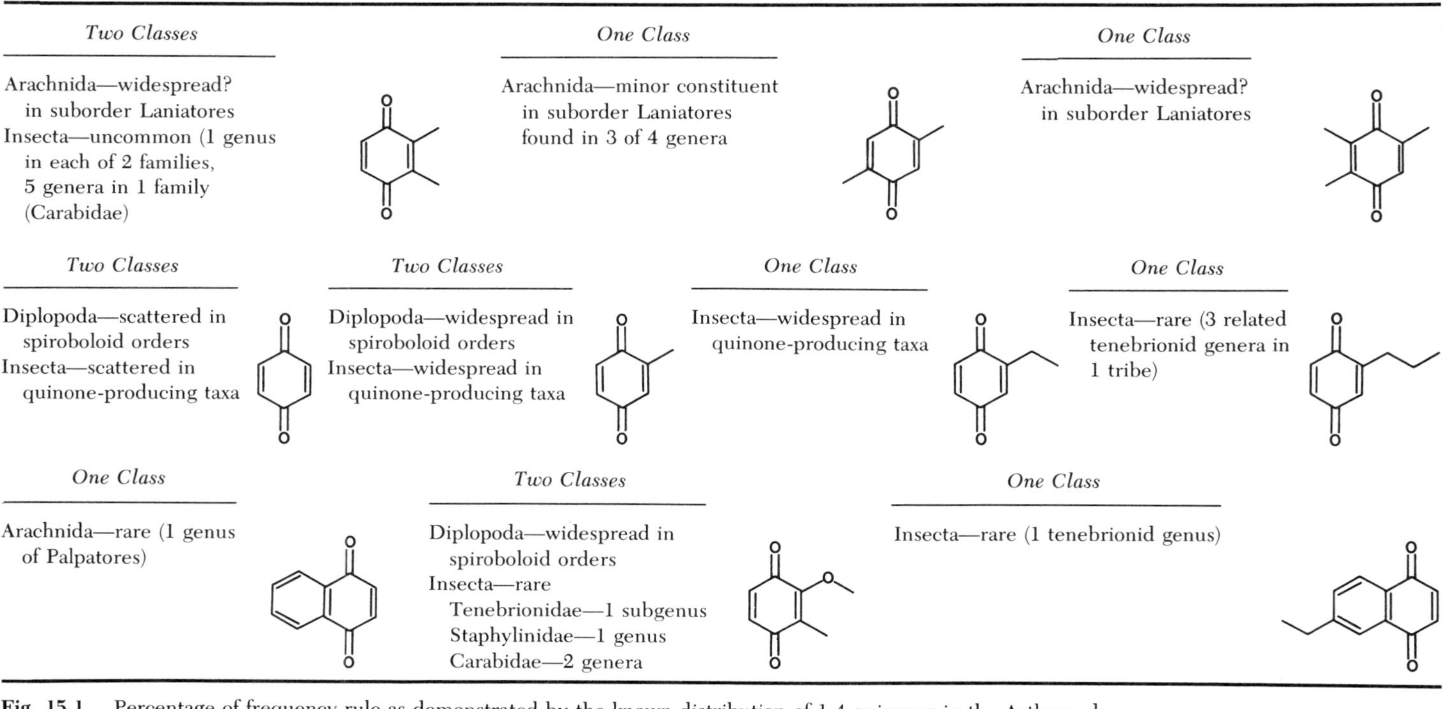

Fig. 15.1. Percentage of frequency rule as demonstrated by the known distribution of 1,4-quinones in the Arthropoda.

possess value as taxonomic indicators because of their circumscribed distribution in insects. In millipedes, 2-methoxy-3-methyl-1,4-benzoquinone appears to be almost an invariant component of the quinonoid defensive secretions and thus possesses little taxonomic significance among the orders in which it is produced. The low frequency of occurrence of 2,3-dimethoxy-1,4-benzoquinone may indicate that this compound is a valuable taxonomic indicator for species in selected diplopod families. Finally, one of the alkylquinones of opilionids, 1,4-naphthoquinone, which may ultimately be demonstrated to have a more widespread occurrence in the secretions of the species in a few quinone-producing families, nevertheless appears to be of some taxonomic significance as a characteristic natural product of selected taxa in this arthropod class (Fig. 15.1).

Analysis of the distribution of 1,4-quinones or other groups of chemically similar compounds must be considered as only a fundamental step in utilizing arthropod natural products as reliable biochemical adjuncts for systematic investigations. Ultimately, the use of these compounds as indicators of probable phylogenetic relationships must be predicated on biosynthetic analyses which can then provide the required framework for genetic studies pertaining to the expression of natural product phenotypes in species of arthropods. Redfield (1936) summed this up succinctly when he stated: "We must know not only what substances occur here and there, but also how they come to be where they are, from what they are made, and how their occurrence is determined." A defensive compound can have value as an indicator of phylogenetic affinity only if it can be demonstrated that its synthesis reflects parallel biogenetic evolution in two species being compared. If the same compound is biosynthesized by different metabolic pathways by two organisms, then its presence in these two species actually indicates that they are not closely related. Meaningful biochemical systematics require analyses of the biosynthetic pathways in terms of both their sequence and precursors in order to establish enzymatic homologies (Alston and Turner, 1963). Furthermore, true biosynthetic homologies would also require that the enzymes utilized by two species be identical or virtually so. If a defensive compound is synthesized in two species by the same enzymatic pathway, close systematic affinities could be inferred based on the compound being a phenotypic expression of clearly homologous genes.

The biosynthetic pathways for quinones would appear to constitute especially useful chemical characters with which to analyze the suitability of natural products as biochemical indicators. J. Meinwald *et al.* (1966a) have demonstrated that in the tenebrionid beetle, *Eleodes longicollis*, 1,4-benzoquinone is synthesized by a pathway completely different from that utilized to produce the concomitant alkylquinones. 1,4-Benzoquinone, which has a scattered distribution in the Arthropoda (Table 15.1) and is usually a minor constituent in quinone-rich secretions, is synthesized from precursors containing a preformed aromatic ring. Alkylquinones, on the other hand, arise by the acetate–malonate pathway in *E. longicollis*. If the substituted quinones present in other arthropod defensive secretions were also derived by acetate–malonate biogenesis, then a determination of the degree of parallel evolution of common quinonoid constituents should be amenable to biosynthetic studies. However, that *E. longicollis* may not necessarily represent a general case is indicated by the demonstration that both 1,4-benzoquinone and two alkylquinones are formed from aromatic amino acids by the millipede *Rhinocricus holomelanus* (S. S. Duffey, personal communication, 1974).

The presence or absence of 1,4-benzoquinone in arthropods may be of some diagnostic value because this compound is (1) apparently absent from arachnid secretions, (2) at best a trace constituent in most millipede secretions, and (3) of scattered occurrence in the defensive exudates of insects. If in selected taxa 1,4-benzoquinone arises in arthropod defensive secretions by a pathway distinct from that for producing substituted quinones, then its absence from specific taxa may indicate the lack of the necessary metabolic pathway. Investigations of the metabolic origins of 1,4-benzoquinones in arthropods promise to yield considerable information which may possess useful taxonomic correlates.

Biosynthetic studies of defensive compounds can provide information of considerable phylogenetic import especially since they can illuminate alterations in metabolic pathways that have been retained in more advanced groups. Based on a comparative investigation of lactate dehydrogenase in animals, Kaplan *et al.* (1960) have suggested that changes in enzyme structure may have been as significant in the establishment of new species as morphological of physiological changes. Furthermore, Blagoveshchenskii (1955) has emphasized that in more highly evolved animals the activation energy thresholds appear to be lower than they are for the same enzymes from more primitive species. More recently, Waley (1969) has suggested that, in the evolution of new metabolic pathways, extreme conservatism is characteristic of the conformations of sequential enzymes, which may have evolved from each other. Thus, more advanced species may have evolved by gaining and elaborating control mechanisms rather than generating new pathways. Therefore, detailed investigations of the structures, kinetics, and alteration of substrate affinity of enzymes involved in the synthesis of defensive compounds by different species should provide strong support for conclusions based on morphological grounds.

Comparative biochemistry, if used judiciously, can be regarded as a potentially powerful supplement to the more conventional systems employed for systematic investigations. Apparent systematic contradictions based on biochemical data can provide useful grounds for reexamining the affinities of the groups under consideration. On the other hand, apparent biochemical correlations, when none are actually present, may reflect the presence of cryptic selection pressures (Alston and Turner, 1963). Completely unrelated groups of organisms may utilize the same pathway to produce identical compounds, as for example the production of HCN, benzaldehyde, and β-cyanoalanine by both plants and millipedes (Duffey *et al.*, 1974, 1977; Duffey and Blum, 1977). The possible pitfalls of biochemical systematics based on nothing more than the synthesis of a common end product can be flagrant, as evidenced by the conclusion that geophilid centipedes and millipedes are related because they both produce HCN (Schildknecht *et al.*, 1968d)!

The last two decades have been characterized by a large number of investigations of arthropod natural products, especially those produced by species in the Insecta and Diplopoda. Although certain compounds have a particularly widespread distribution in defensive exudates, others appear to be so limited as to be diagnostic for species in particular taxa. However, before adumbrating the occurrence of these distinctive natural products, it seems advisable to indicate some of the factors which should be considered in evaluating the significance of the exocrine compounds in a particular defensive secretion.

I. FACTORS AFFECTING COMPOSITION OF DEFENSIVE SECRETIONS

A. Quantitative and Qualitative Variations

The quantitative compositions of the glandular secretions of arthropods appear to be so variable between individuals of the same species as to offer little or no value as taxonomic indicators. Regnier and Wilson (1969) reported that the minor defensive compounds produced by the ant *Lasius alienus* varied as much as 50% between individuals, and Tricot *et al.* (1972) found a similarly great individual variation in the two main mandibular gland products of the ant *Myrmica rubra.* Large quantitative variations in the alkaloids produced by individual fire ant workers from the same colony are also commonly encountered (Brand *et al.*, 1973c). Since determinations of the proportions of compounds found in defensive secretions are almost invariably based on pooled samples, the amount of individual variation is generally unknown.

When defensive secretions are primarily based on sequestered plant natural products, the compositions of exudates may be so variable as not to be recognized as being products of individuals of the same species. Jones and Blum (1979) demonstrated that the secretions derived from the mesothoracic spiracular defensive glands of the grasshopper *Romalea microptera* exhibit such compositional idiosyncrasies as to be essentially distinctive for each individual. Since individuals of this species, which appear to sequester most of their defensive compounds from ingested plant constituents, feed unpredictably on a wide range of host plants, it appears that the compositions of the glandular exudates have eclectic plant origins. In short, *there is no such thing as the defensive secretion for individuals of Romalea microptera but rather, idiosyncratic exudates.* This possibility should be considered whenever secretious containing plant natural products are analyzed.

B. Age

The age of an individual arthropod may be closely related to both the quantitative and qualitative chemistry of a defensive secretion. Recent investigations on the chemistry of honey bee natural products have clearly demonstrated this to be the case. Bachmayer *et al.* (1972) have demonstrated that maximum synthesis of melittin, the main toxic peptide in honey bee venom, does not occur until a worker is about 10 days of age. Indeed, until about 2 days of age, only the precursor of melittin, promelittin, is detectable in the venom, thus demonstrating a qualitative variation with age. More recently, Owen (1978) has demonstrated that maximum synthesis of histamine in the poison gland of a worker bee occurs up to 15 days of age; synthesis falls off after this until it is no longer detectable in 40-day-old bees. Boch and Shearer (1967) have shown that 2-heptanone, a mandibular gland defensive product of honey bees, is not synthesized until worker bees begin to forage. On the other hand, hyaluronidase levels in honey bee venom reach a maximum two days after eclosion and remain essentially at this level throughout the life of the bee (Owen, 1979a).

In male bumblebees synthesis of the labial gland products begins after eclosion and appears to be maximal after the fourth day of adult life (Agrén *et al.*, 1979). Active synthesis of the five main classes of exocrine compounds is evident during early adult life followed by a gradual decline in concentration of all classes of labial glandular constituents.

In general, the defensive glands of both immature and adult arthropods are not "turned on" maximally until an individual instar is at least several days old. Calam and Youdeowei (1968) found that fourth-instar larvae of the bug *Dysdercus intermedius* produced no secretion in their dorsal abdominal glands during the first two days after the molt. Linalool, a constituent in the lateral reservoir of the adult metathoracic scent gland of *D. intermedius*, reaches a maximum when the bugs are 14 days old (Everton *et al.*, 1979). 2-Hexenal and 2-octenal, present in the median reservoir and accessory gland, also reach a maximum at this time whereas alkenyl acetates are minimal after 2 weeks. Conversely, Bachmayer *et al.* (1972) have demonstrated that older honey bee workers may have reduced quantities of melittin in their poison gland secretions. Therefore, qualitative as well as quantitative surprises and apparent contradictions may be encountered if secretions from individuals of unknown age are analyzed.

C. Sex

Defensive compounds may be unexpectedly sex limited. Meyer *et al.* (1968) have demonstrated that whereas both male and female larvae of the meloid *Lytta vesicatoria* synthesize cantharidin, adult females are incapable of producing this compound. The ability of males to convert sodium [1-^{14}C]acetate to cantharidin contrasts to the failure of females to incorporate this precursor into this terpenoid anhydride. Whereas the presence of cantharidin in the blood of adult females had been attributed to carry over from larval synthesis, it has now been established that females derive this compound from the seminal ejaculate introduced during copulation.

Among social insects, defensive compounds, many of which simultaneously function as pheromones, are often limited to one sex. In the honey bee, 2-heptanone is produced in the mandibular glands of workers but not males (Shearer and Boch, 1965), whereas male bumblebees synthesize a variety of compounds in their labial glands which are absent from the secretions of females (Kullenberg *et al.*, 1970). Similarly, Brand *et al.* (1973a,b) have shown that male carpenter ants (*Camponotus* spp.) produce a variety of aromatic esters and aliphatic acids in their mandibular glands which are completely lacking in workers and females. These compounds possess both defensive and pheromonal functions.

D. Seasonal Variation

The proportion of compounds produced by a defensive gland may exhibit a considerable and unpredictable seasonal variation. Schreuder and Brand (1972) reported major quantitative change in some of the hydrocarbons synthesized in the Dufour's gland of the ant *Anoplolepis custodiens* when summer and winter ants were compared. In the winter, *n*-pentadecane increased by about 40-fold, whereas the concentrations of *n*-tridecane and tridecene decreased considerably.

Great seasonal variations in the concentration of norsesquiterpenes in the pygidial glands of the gyrinids *Dineutes assimilis* and *Gyrinus frosti* have been demonstrated (Newhart and Mumma, 1979a). The titers of these compounds are maximal in the spring and fall; minimal titers are encountered in the summer.

Steroid titers in the prothoracic glands of the dytiscid *Agabus seriatus* are maximum in midsummer and minimum in the fall (Miller and Mumma, 1973), whereas the converse occurs for the steroids in *Acilius semisulcatus* (Newhart and Mumma, 1979b). The concentrations of the pygidial gland products of *A. semisulcatus*—benzoic acid, *p*-hydroxybenzaldehyde, and methyl *p*-hydroxybenzoate—do not parallel those of the steroids and reach a maximum in the fall.

Seasonal variations in the concentrations of exocrine compounds, probably reflecting major physiological changes, may be of profound ecological significance.

E. Physiological State

Boch and Shearer (1967) have demonstrated that 2-heptanone, which is normally produced by honey bee workers when they begin to forage (ca. 14 days of age), is not synthesized in the mandibular glands if the bees are caged and not allowed free flight. This suppression of 2-heptanone production thus appears to be correlated with the physiological state of the workers rather than their chronological age. Similarly, Owen *et al.* (1977) have demonstrated that histamine concentration in the venom of worker bees maintained in a flight room is reduced compared to that of worker bees in the field. Whether the compositions of the defensive secretions of laboratory-reared or maintained arthropods relative to feral populations may frequently differ remains to be determined.

F. Instar

Papilionid larvae in the tribe Papilionini synthesize isobutyric and 2-methylbutyric acids in their osmeterial defensive glands. However, larvae of *Papilio aegus* produce these compounds only during the ultimate instar, and neither of these acids is detectable in the secretions of younger larvae (Seligman and Doy, 1972). On the other hand, β-hydroxybutyric acid is present in the osmeterial secretions of penultimate instar larvae, a finding whose significance is particularly enigmatic. Similarly, Burger *et al.* (1978) and Honda (1980) reported that whereas the osmeterial secretions of final-instar larvae of *P. demodocus* and *P. protenor*, respectively, were dominated by two fatty acids and esters, those of earlier larval instars contained a diversity of terpenes. These findings indicate that pooling the secretions obtained from different larval instars may produce analytical results which are both misleading and not necessarily representative of the secretions of particular instars. In addition, the probable ecological significance of qualitative variations in the exocrine secretions of different instars will be completely overlooked.

G. Caste

Surprising chemical differences may be encountered in analyzing the glandular constituents in the defensive secretions of individuals of different castes. Characteristically, hymenopterous species are represented by two kinds of females—reproductives and sterile workers. Although the chemistry of the defensive glands of females and workers of relatively few species have been compared, results indicate that major quantitative and qualitative differences may be encountered. Shearer and Boch (1965) reported that, whereas the mandibular glands of honey bee workers were a particularly rich source of 2-heptanone. this ketone could not be detected in queens. More recently, Brand *et al.* (1973c) compared the compositions of the venoms of queens and workers of four species of fire ants and found that the alkylpiperidines produced in the poison gland are characteristic of both the caste and the species. In some species major qualitative differences are present as well. For example, in one species the queen's venom lacked the four major constitutents found in her worker's venom, and the minor alkaloidal component in worker venom was essentially the only compound present in secretion of the queen. From an exocrine standpoint, it is obvious that queens and workers cannot necessarily be considered as equivalent.

Beyond caste idiosyncrasies, surprising differences may characterize the *secretions of subcastes* of both ants and termites. Minor workers of the ant *Oecophylla longinoda* produce nerol and geraniol in their mandibular glands, whereas these compounds are lacking in glandular exudates of major workers (Bradshaw *et al.*, 1975, 1979). Similarly, minor soldiers of the termite *Ancistrotermes cavithorax* synthesize ancistrodial in their frontal glands, in contrast to major soldiers which generate other sesquiterpenes (e.g., ancistrofuran) in this gland (Baker *et al.*, 1978). Prestwich (1978) also reported great variation in the terpenoid products produced by minor and major soldiers of a termite. Analyses of the two types of soldiers of *Trinervitermes gratiosus* demonstrated that both specific mono- and diterpenes were characteristic products of members of each of the two subcastes.

H. Series of Glands and Glandular Subdivisions

Many arthropods possess series of either single or paired defensive glands the members of which often do not discharge simultaneously. In some cases at least, the chemical compositions of the exudates from different glands may differ from each other to a strikingly qualitative degree. Larvae of the pyrrhocorid *Dysdercus intermedius* possess three dorsal abdominal defensive glands, the first two of which differ from the third both on morphology and chemistry (Youdeowei and Calam, 1969). The major component present in the secretion of the first two glands is *n*-tetradecane, whereas the exudate of the posterior gland is dominated by *n*-tridecane and aliphatic aldehydes. That these glands may possess different functions is indicated by the fact that, when disturbed, the larvae generally discharge secretion from only the posterior gland. More recently, Everton *et al.* (1979) reported that linalool was produced in the lateral reservoir of the metathoracic scent apparatus of *D. intermedius* but was virtually absent from the median reservoir and accessory glands.

Similarly, Games and Staddon (1973b) demonstrated that the secretions from the two dorsal abdominal glands of larvae of the bug *Oncopeltus fasciatus* were not usually discharged simultaneously and differed in their chemical composition. The exudate of the anterior gland is dominated by (*E*)-2-octenal whereas 4-oxo-(*E*)-2-octenal is the major constituent is the discharge of the posterior gland. Major differences also characterize the chemistry of the lateral and median reservoirs of the adult metathoracic scent gland of *O. fasciatus.* Whereas the secretion in the lateral reservoir chiefly contains alkenyl acetates, that of the median reservoir is dominated by alkenals in admixture with the alkenyl acetates (Everton and Staddon, 1979).

The defensive gland of a notodontid larva (*Schizura concinna*) consists of two sacs whose chemical compositions differ considerably. Whereas both the anterior and posterior sacs contain formic acid, the former contains decyl acetate, dodecyl acetate, and 2-tridecanone as well (Weatherston *et al.*, 1979).

The results of these investigations indicate that mixing the secretions from series of glands may obscure both the functions and chemistry of the individual glands.

I. Mutants

The exocrine chemistry of mutant strains of a species may differ drastically from that of a normal wild type. Engelhardt *et al.* (1965) have demonstrated that quinonoid production in young individuals of a mutant of the flour beetle *Tribolium confusum* is reduced to 1/20 of that found normally in the wild type. Furthermore, older mutant individuals do not even produce detectable quantities of quinones in their defensive glands. Although the probability of encountering a mutant population may not be ordinarily great, this possibility cannot be excluded when undertaking analyses of the defensive secretions of arthropods, especially if they are inbred lines and have had prolonged maintenance in the laboratory.

J. Concentrational Changes during the Regeneration of Secretion

Quantitative and possibly qualitative changes may result in the constituents in a defensive secretion during the regeneration of an exudate previously depleted by milking. Wallbank and Waterhouse (1970) have demonstrated that the concentration of 2-hexen-1-ol changed considerably one week after scent has been collected from the blattid *Polyzosteria limbata.* At the time of the initial milking, the cockroach secretion contained about 1.4% hexenol, but the concentration of this compound rose about threefold one week later. Indeed the defensive exudates of two individuals of *P. limbata* contained 5 and 12% hexenol one week after capture. These results indicate that major quantitative changes in defensive constituents may result when individual arthropods are milked over extended periods of time. Pooling samples collected in this way may produce a quantitative, and possibly qualitative, picture of the secretion which differs considerably from that present in newly collected individuals.

Recently, Owen (1978) reported that honey bee workers, after venom depletion, synthesize histamine actively starting at an age of five days. Maximum synthesis occurs in 15-day-old bees, falling off rapidly thereafter. Therefore, as in the case of the defensive compounds in the cockroach secretion, histamine replenishment would be suppressed in older bees that had been previously depleted of venom.

K. Populational Variations in Secretions

Defensive compounds may exhibit extraordinary qualitative and quantitative variations when different populations of the same arthropod species are analyzed. For example, Tschinkel (1975b) has demonstrated that 1,4-benzoquinone is frequently lacking from the secretions of certain populations of the tenebrionids *Eleodes extricata* and *Alobates pennsylvanicus*. Similarly, Daloze and Pasteels (1979) have shown that great quantitative and qualitative variations characterize the cardenolide-rich secretions of chrysomelid beetles in the genus *Chrysolina* when populations of two species in southern Europe are compared to those from western Europe.

Social insects may characteristically exhibit extreme variations in their defensive allomones when different populations are compared. Bradshaw *et al.* (1975) reported that, whereas three simple 1-alkanols were present in the mandibular gland secretions of major workers of one population of the ant *Oecophylla longinoda*, only one of these alcohols (1-hexanol) was detectable in the secretions of workers from two other populations. Similarly, Prestwich (1978) noted great quantitative and qualitative variation in the mono- and diterpenes produced in the frontal glands of termite soldiers (*Trinervitermes gratiosus*) in three different populations.

II. TAXONOMIC CORRELATIONS UTILIZING DEFENSIVE COMPOUNDS

Chemical characters have now been applied to several equivocal systematic problems and in a few cases analyzed in terms of proposed phylogenies or utilized to help construct a tentative evolutionary sequence. These preliminary analyses probably constitute the forerunners of more frequent studies utilizing defensive compounds as diagnostic indicators that will become possible as more comparative exocrinological data become available. Although many obvious examples of the independent evolution of the same defensive compounds are already apparent, it is also evident that arthropods in certain taxa are identified with characteristic defensive compounds whose value as chemical characters may be considerable. Thus, whereas biochemical convergence for specific natural products may have occurred frequently among members of diverse arthropod lines, these compounds may still be useful as distinctive characters that are restricted to specific taxa. The distribution of (*E*)-2-hexenal within the Insecta is indicative of the systematic utility of a compound that is produced by several unrelated groups of arthropods.

(*E*)-2-Hexenal has been identified in insects belonging to the orders Hemiptera, Dictyoptera, Coleoptera, and Hymenoptera. This aldehyde appears to be a rather characteristic defensive compound produced by species belonging to the two main branches of the suborder Hemiptera. Thus, its value as a chemical character in the Hemiptera is limited to the subordinal level. In the Dictyoptera, this compound has only been identified in species in the Blattaria, one of three suborders in this order. However, within the Blattaria, its known distribution is limited to the subfamily Polyzosteriinae of the family Blattidae. In the Coleoptera, 2-hexenal has been identified only in the defensive secretion of one species of *Eleodes*, *E. beameri*, a taxon in the family Tenebrionidae. However, it may also prove to be significant that *E. beameri* is the only species in the subgenus *Holeleodes* whose defensive secretion has been analyzed. Finally, in the Hymenoptera, the known distribution of this carbonyl compound is restricted to the species in the subgenus *Crematogaster* of the genus *Crematogaster* of the Formicidae. Therefore, among the four insect orders in which this compound is found, 2-hexenal presently is a diagnostic indicator at four different taxonomic levels: suborder (Hemiptera), subfamily (Dictyoptera), genus (Coleoptera), and subgenus (Hymenoptera).

In general, the idiosyncratic occurrence of a considerable number of compounds in selected species appears to be diagnostically useful in categories at all taxonomic levels. At the ordinal level, for example, investigations of a large number of arthropod secretions produced by species in diverse taxa indicate that 4-oxo-(*E*)-2-hexenal (Hemiptera), 16-hexadecanolide (Hymenoptera), geranylgeraniol (Hymenoptera), 1,2-

dimethyl-4(3*H*)-quinazilinone (Glomerida), and polyzonimine (Polyzoniida), may represent good chemical characters. Indeed, these compounds may be somewhat representative of biosynthetic emphases which characterize the species in these orders. Hemipterans, for example, produce a multitude of conjugated aldehydes, some of which are shared with other orders, but 4-oxo-(*E*)-2-hexenal as well as 4-oxo-(*E*)-2-octenal appear to be distinctive of species in a major taxon in this order. Geranylgeraniol and 16-hexadecanolide may constitute good diagnostic chemical characters of selected hymenopterans, particularly since they exemplify the ability of ants and bees, respectively, to synthesize diterpenes and macrocyclic lactones. Both the quinazilinones synthesized by glomerid millipedes and the azaspirane synthesized by polyzoniid millipedes contrast markedly with the phenols, quinones, and benzaldehyde produced by species in other diplopod orders.

Several compounds produced by members of the Formicidae appear to possess considerable taxonomic value at the subfamily level. The cyclopentanoid monoterpenes, characteristic natural products of the anal glands of the Dolichoderinae (Cavill, 1970), appear to be limited in their formicid distribution to species in this subfamily and as such constitute a useful chemical character. Similarly, formic acid is restricted to species in the subfamily Formicinae in contrast to the proteinaceous themes that generally characterize the poison gland products of species in other formicid subfamilies (Blum, 1966). Besides constituting an excellent formicine character, the ubiquity of both formic acid and the same type of poison gland among the genera in this subfamily indicate that little biochemical or morphological change has occurred in the evolution of the formicine venom gland.

Beetles are particularly distinctive in their ability to produce characteristic compounds which are limited in their known distribution to species in single families. Some of these apparent chemical characters of coleopterous families are listed in Table 15.2. The characteristic compounds produced by species in most of these families are rather typical of the natural products synthesized by members of each taxon. For example, testosterone is one of 19 steroids identified in the prothoracic gland secretions of dytiscids, and it appears that these beetles have a monopoly on this class of steroidal exocrine products in the Arthropoda. In the Carabidae isocrotonic acid exemplifies the diversity of conjugated carboxylic acids identified as pygidial gland products of these beetles. On the other hand, cantharidin is the sole defensive compound detected in the blood of meloids in all characterized genera; similarly, gyrinidal and/or

Table 15.2

Defensive Compounds Limited in Their Distribution to Beetle Species in Single Families

Family	Compound	Related compounds in family
Dytiscidae	Testosterone[a]	Many other steroids
Meloidae	Cantharidin[b]	None known
Cantharidae	Dihydromatricaria acid[c]	None known
Gyrinidae	Gyrinidal[d]	Isogyrinidal, gyrinidone, gyrinidione
Carabidae	Isocrotonic acid[e]	Angelic acid, crotonic acid
Tenebrionidae	2-*n*-Propyl-1,4-benzoquinone[f]	2-Methyl- and 2-ethyl-1,4-benzoquinone
Staphylinidae	5,8-Tetradecadienal[g]	5-Tetradecenal
Staphylinidae	γ-Dodecalactone[h]	None known

[a] Schildknecht *et al.*, 1967a.
[b] Kobert, 1906.
[c] Meinwald *et al.*, 1968b.
[d] Meinwald *et al.*, 1972.
[e] Moore and Wallbank, 1968.
[f] Tschinkel, 1975b.
[g] Brand *et al.*, 1973b.
[h] Wheeler *et al.*, 1972a.

gyrinidone fortify the defensive exudates of all gyrinid beetles analyzed. Finally, the cardenolides and lucibufagins of chrysomelids and lampyrids, respectively, appear to be especially diagnostic of some of the taxa in these two families.

A diversity of species in other families produce highly distinctive compounds which appear to have a very limited arthropod distribution. The unsaturated C_{14} alcohols dominating the larval mandibular gland secretion of the cossid *Cossus cossus* (Trave *et al.*, 1966) may represent good family characters, or even generic indicators, especially since they are absent from the exudate of *Zeuzera pyrina* (Marchesini *et al.*, 1969), a species in another taxon in this lepidopterous family. Detailed analyses of the defensive secretions of large numbers of species in various arthropod families indicate that certain compounds are characteristically synthesized by the members of selected genera. Some of these genus-specific compounds are presented in Table 15.3.

Several of these idiosyncratic generic compounds may be of considerable value as taxonomic indicators. 1-Tetradecen-3-one, a frontal gland product of the rhinotermitid *Schedorhinotermes putorius* (Quennedey *et al.*, 1973), is one of several ketones in this secretion, which contrasts considerably with the terpene-rich exudates of soldiers in termitid gen-

Table 15.3

Genus-Specific Defensive Compounds of Arthropods

Compound	Family	Genus
2-Methyl-4-heptanone[a]	Formicidae	*Tapinoma*
Manicone[b]	Formicidae	*Manica*
1-Tetradecen-3-one[c]	Rhinotermitidae	*Schedorhinotermes*
Pederin[d]	Staphylinidae	*Paederus*
6-Ethyl-1,4-naphthoquinone[e]	Tenebrionidae	*Argoporus*
β-Selinene[f]	Papilionidae	*Battus*
Leiobunone[g]	Phalangiidae	*Leiobunum*
1-Nitro-1-pentadecene[h]	Rhinotermitidae	*Prorhinotermes*
Nitropolyzonamine[i]	Polyzoniidae	*Polyzonium*
2-Methyl-6-*n*-pentadecylpiperidine[j]	Formicidae	*Solenopsis*

[a] Trave and Pavan, 1956.
[b] Fales *et al.*, 1972.
[c] Quennedey *et al.*, 1973.
[d] Cardini *et al.*, 1965a.
[e] Tschinkel, 1969.
[f] Eisner *et al.*, 1971c.
[g] Meinwald *et al.*, 1971.
[h] Vrkoč and Ubik, 1974.
[i] Meinwald *et al.*, 1975.
[j] MacConnell *et al.*, 1971.

era. 1-Nitro-1-pentadecene, the major constituent detected in the secretion of the termite *Prorhinotermes simplex* (Vrkoč and Ubik, 1974), also appears to be a highly specific chemical character of species in this rhinotermitid genus. Similarly, the terpene 2-methyl-4-heptanone is an anal gland product of dolichoderines in the genus *Tapinoma*, whereas species in other genera analyzed produce different methyl ketones. Species in other formicid genera also appear to synthesize highly distinctive exocrine compounds. For example, manicone appears to be a particularly characteristic compound, since it constitutes one of several ketones limited in their known distribution to the secretions of species in the genus *Manica*, the members of which were recently removed from the closely related genus *Myrmica* (Creighton, 1950). Significantly, the mandibular gland secretions of *Myrmica* spp. lack manicone and two other ketones characteristic of species of *Manica* (Crewe and Blum, 1970a; Fales *et al.*, 1972).

In several instances, novel compounds are only known to fortify the defensive exudates of one species in a genus. Provided that subsequent investigations support the species specificity of these defensive compounds, their uniqueness will constitute a chemical character *par excellence*. A selection of the compounds whose known distribution is confined to the secretions of single species is listed in Table 15.4 along with examples of the compounds found in the secretions of other species in the same genera.

Several of these arthropod allomones are especially indicative of the biosynthetic peculiarities of the species producing them vis-à-vis other generic members. Dendrolasin, a unique furanoterpene, is accompanied by another furan, perillene, in the mandibular gland secretion of *Lasius fuliginosus*, a species in the subgenus *Dendrolasius* (Bernardi *et al.*, 1967). On the other hand, another member of this subgenus, *L. spathepus*, produces citronellal (Kistner and Blum, 1971) in its mandibular glands, a compound with a widespread distribution in the genus. Similarly, the piperideine identified in the venom of the fire ant *Solenopsis xyloni* (Table 15.4) is absent from that of other species in this genus including *S. aurea* (Blum *et al.*, 1973c), a species that was separated from *S. xyloni* fairly recently (Creighton, 1950). 2-Pentanol, one of two saturated alcohols in the chemical bullet "fired" by the cockroach *Platyzosteria armata*, is an especially distinctive constituent of a secretion which differs from the enal-rich exudates of other generic members (Wallbank and Waterhouse, 1970). The presence of 2-phenylethyl 2-methyl butyrate, along with the isobutyrate ester, in the secretion of *C. interrupta* (Blum *et al.*, 1972) seems equally diagnostic for this species in light of the identification of salicylaldehyde as the only detectable component in the exudates of other *Chrysomela* species.

III. THE EXPANDING UTILIZATION OF DEFENSIVE COMPOUNDS AS DIAGNOSTIC INDICATORS

Increasingly, arthropod deterrent allomones have been utilized as taxonomic adjuncts for species at a variety of classificatory levels. For example, Waterhouse and Gilby (1964) studied the morphology and chemistry of the defensive glands of eight species of Coreidae and identified four especially characteristic compounds. In addition, the defensive secretion of a hycocephalid species, a taxon with either coreoid or lygaeoid affinities, was examined. Since both the chemistry and morphology of its defensive glands were congruent with those of the

Table 15.4

Species-Specific Compounds in the Defensive Secretions of Arthropods

Compound	Family	Species	Typical compounds of genus
Dendrolasin[a]	Formicidae	*Lasius fuliginosus*	Citronellal
1-Nonen-3-one[b]	Tenebrionidae	*Eleodes beameri*	2-Ethyl-1,4-benzoquinone
2-Methyl-6-*n*-undecyl-$\Delta^{1,2}$-piperideine[c]	Formicidae	*Solenopsis xyloni*	2-Methyl-6-*n*-undecylpiperidine
8-Hydroxyisocoumarin[d]	Tenebrionidae	*Apsena pubescens*	2-Methyl-1,4-benzoquinone
4,6-Pregnadiene-12β,20α-diol-3-one[e]	Dytiscidae	*Cybister lateralimarginalis*	Cortexone
Isodihydronepetalactone[f]	Formicidae	*Iridomyrmex nitidus*	Iridodial
2-Phenylethyl 2-methyl butyrate[g]	Chrysomelidae	*Chrysomela interrupta*	Salicylaldehyde

[a] Quilico *et al.*, 1957b.
[b] Tschinkel, 1975a.
[c] Brand *et al.*, 1972.
[d] Lloyd *et al.*, 1978a.
[e] Schildknecht, 1971.
[f] Cavill and Clark, 1967.
[g] Blum *et al.*, 1972.

coreids, the placement of the Hyocephalidae in the Coreoidea was strongly supported. Numerous other examples of arthropod natural products providing useful taxonomic information are now available.

A. Honey Bees

A defensive product of honey bees has been especially useful as a taxonomic indicator. The domesticated Asiatic honey bee, *Apis indica*, had been regarded either as a valid species or a subspecies of *Apis mellifera*, the common European honey bee. Morse *et al.* (1967) demonstrated that, whereas 2-heptanone is a major product of the mandibular glands of *A. mellifera* workers, it is not produced by those of *A. indica*, a fact that clearly favors the recognition of these two forms as separate species.

More recently, Kreil (1973b) compared the amino acid sequences in the peptide melittin, the main venomous toxin, in a primitive species of *Apis*, *A. florea*, with those produced by *A. mellifera* and *A. indica*. The sequence was identical in the melittins synthesized by *A. mellifera* and *A. indica* but that produced by *A. florea* was considerably different. These results support the recognized phylogeny of the genus *Apis*, since *A. florea* is regarded as a very primitive species, whereas *A. mellifera* and *A. indica* are considered to be two very closely related species which have separated fairly recently. Kreil (1975) has also studied the structure of the melittin produced by the fourth species of *Apis*, *A. dorsata*. This peptide differs from that produced by *A. mellifera* and *A. indica* by three residues and from that produced by *A. florea* by five residues. Thus, these sequence data support the previous phylogenetic treatment of this genus based on morphological characters in which *A. dorsata* was considered to be more closely related to *A. mellifera* and *A. indica* than to *A. florea*. The structures of the melittins produced by these four species of *Apis* support the intermediary placement of *A. dorsata* between *A. florea* and the two more recently evolved species, *A. mellifera* and *A. indica*.

B. Beetles

Moore and Wallbank (1968) examined the pygidial gland chemistry of a large number of carabid species and concluded that chemical characters were especially useful from the tribal to the generic level. For example, conjugated carboxylic acids such as angelic and isocrotonic are characteristic of species in the tribe Scaritinae. In addition, the presence of the same crepitation mechanism for generating 1,4-benzoquinones in the

Ozaeninae and Paussinae provided valuable chemical and morphological support for the placement of these two taxa in a single subfamily.

C. Cockroaches

There is a tendency for certain species of nonhemipterous insects to produce secretions that are qualitatively comparable to those generated by true bugs. Hemipterous exudates are generally characterized by the presence of α,β-unsaturated and saturated aldehydes including 2-hexenal, 2-octenal, 2-decenal, *n*-hexanal, and *n*-octanal (Waterhouse *et al.*, 1961; Waterhouse and Gilby, 1964; Tsuyuki *et al.*, 1965). In many pentatomids (*E*)-2-hexenal is the major constituent in the secretion and is usually accompanied by the C_8 and C_{10} enals (Schildknecht *et al.*, 1962). Similarly, the defensive secretions of cockroaches in the genus *Polyzosteria* are dominated by (*E*)-2-hexenal, and those of two species, *P. cuprea* and *P. pulchra,* also contain minor quantities of (*E*)-2-octenal and *n*-octanal (Wallbank and Waterhouse, 1970). Indeed, the exudate of an unidentified cockroach species in the genus *Platyzosteria* contained the (*E*)-isomers of 2-hexenal, 2-octenal, and 2-decenal, and is thus a carbon copy of some hemipterous exudates (Wallbank and Waterhouse, 1970). However, the abdominal glandular secretion of the tenebrionid *Eleodes beameri* represents the ultimate example of a buglike secretion produced by a nonhemipterous species. In addition to the characteristic C_6–C_{10} enals, the glandular exudate of this beetle is enriched with (*E*)-2-heptenal and (*E*)-2-nonenal, as well as the C_6–C_8 alkanals (Tschinkel, 1975a). Only one of the typical 1,4-benzoquinones of tenebrionids is present in the "hemipterous" exudate of *E. beameri*.

Detailed investigations of many species in lower taxa may expose the presence of anomalous secretions whose significance can only be guessed at. For example, 2-heptanone, a common glandular product of dolichoderine and myrmicine ants (Blum *et al.*, 1963; Moser *et al.*, 1968), is a major constituent produced in the ventral abdominal gland of the cockroach *Platyzosteria armata* (Wallbank and Waterhouse, 1970). Since the glandular exudates of the other species of *Platyzosteria* (*Platyzosteria*) spp. contain at least 90% (*E*)-2-hexenal, the ketonic secretion of *P. armata* is particularly anomolous. Indeed, the glandular products of virtually all polyzosteriine cockroaches examined are identified with unsaturated aliphatic aldehydes. Similarly, the occurrence of 2-pentanone and 2-heptanone in the pygidial gland exudate of the carabid *Dychirius wilsoni* demonstrates that the defensive chemistry of this genus differs considerably from that of other genera in the subfamily Scaritinae (Moore and Brown, 1979). These findings, along with others, indicate that the systematic position of *Dyschirius wilsoni* needs to be reevaluated, especially since all other investigated scaritines produce aliphatic acids or benzoquinones in their pygidial glands.

D. Hemipterans and Aromatic Compounds

Other exceptions to the aliphatic aldehydic theme characterizing the metasternal scent gland products of hemipterans have been exposed. For example, the defensive exudates of the naucorid *Ilyocoris cimicoides* and the notonectid *Notonecta glauca* are dominated by two aromatic compounds, methyl *p*-hydroxybenzoate and *p*-hydroxybenzaldehyde (Staddon and Weatherston, 1967; Pattenden and Staddon, 1968). These compounds also constitute the aromatic duet which enriches the pygidial gland secretions of dytiscid beetles in at least nine genera (Schildknecht, 1970). Since the naucorids, notonectids, and dytiscids are all aquatic groups, it could be argued that the presence of both the aromatic aldehyde and ester in the secretions of species in all three taxa is correlated with an adaptation for living in water. Indeed, Schildknecht (1970) believes that these compounds function as antimicrobial agents for the dytiscids in contradistinction to the steroids in the prothoracic glands, which are regarded as vertebrate deterrents. While this explanation might also satisfy the hemipterous requirement for protection against aquatic microorganisms, it would not seem to endow these insects with any protection against vertebrate predators. As a matter of fact, both *p*-hydroxybenzaldehyde and methyl *p*-hydroxybenzoate are effective repellents for predatory fish (Hepburn *et al.*, 1973). Furthermore, Maschwitz (1971) has reported that the pleid *Plea leachi* produces only hydrogen peroxide in its metasternal scent gland, whereas the belastomatid *Lethocerus indecus* synthesizes (*E*)-2-hexenyl acetate in the same gland (Butenandt and Tam, 1957). The presence of only aliphatic compounds in the secretions of some hemipterans demonstrates clearly that the adaptive defensive allomones produced by aquatic bugs are frequently nonaromatic.

E. Diverse Arthropods and Benzaldehyde

The scattered occurrence in arthropod secretions of another aromatic aldehyde, benzaldehyde, persuasively documents the distributional surprises that may be encountered with arthropod natural products. This compound is a characteristic product in the defensive secretions of polydesmid millipedes (Guldensteeden-Egeling, 1882; Blum and Woodring, 1962; H. E. Eisner *et al.*, 1963), being generated from a cyanogenic precursor at the instant of secretion (T. Eisner *et al.*, 1963a; Duffey *et al.*, 1974). Benzaldehyde and HCN have been detected in the exudates of species in at least seven polydesmoid families (Table 12.1), and this allomonal duet has also been identified in the secretions of larval chrysomelid beetles (Moore, 1967). The beetle secretory products are almost certainly derived from a cyanogenic precursor similar or identical to that utilized by polydesmoid millipedes, indicating that the two groups may have evolved the same types of pathway for generating HCN and the aldehyde. Benzaldehyde also occurs in the defensive exudates of a carabid beetle (Moore and Brown, 1971a) but has not been demonstrated to be accompanied by HCN. Similarly, benzaldehyde occurs in the apparent absence of HCN in the mandibular gland secretions of stingless bees in the subgenus *Scaptotrigona* (Blum *et al.*, 1973b; Luby *et al.*, 1973), in the secretions of a species of myrmicine ant (*Veromessor;* Blum *et al.*, 1969b), and in two species of dolichoderine ants (*Azteca;* Blum and Wheeler, 1974) (Table 4.2). The absence of this aromatic aldehyde from the secretions of large number of other species in these two ant genera further emphasizes the unpredictable distribution of this compound in the glandular discharges of arthropods.

F. Social Insects and Other Arthropods as Terpenoid Specialists

Distributional anomalies of arthropod natural products have been encountered in particular with a variety of terpenoid constituents. Whereas monoterpene hydrocarbons are relatively common frontal gland products of termitid soldiers in at least five genera (Moore, 1964, 1968; Nutting *et al.*, 1974; Prestwich, 1978) and thus can be regarded as characteristic exocrine constituents of nasute termites, these compounds have been detected in only a few nontermitid arthropods. The poison gland secretion of workers of the myrmicine ant *Myrmicaria natalensis* is the qualitatively richest source of monoterpene hydrocarbons currently known among arthropods (Quilico *et al.*, 1962; Brand *et al.*, 1974), and constitutes the only terpene-rich venom known to be produced by insects. The venoms of ant species in the subfamily Myrmicinae are characteristically proteinaceous or, in the case of species in a few genera, dominated by alkaloids (MacConnell *et al.*, 1971; Pedder *et al.*, 1976; Jones *et al.*, 1979). The monoterpenoid theme stressed by the poison gland of *M. natalensis* constitutes an extraordinary digression from the mainstream of myrmicine venom chemistry. Similarly, the monoterpene hydrocarbon-rich defensive secretions of rhopalid bugs (Aldrich *et al.*, 1979) contrast markedly with those produced by hemipterans in other families.

A large variety of oxygenated terpenes are produced by insects and, with few exceptions, these compounds constitute major elements in the chemical defenses of hymenopterans. The terpene alcohols identified in the defensive exudates of arthropods are currently restricted in their distribution to species in a few bee and ant genera. Whereas 2,6-dimethyl-5-hepten-2-ol has only been identified as a mandibular gland constituent of species in three genera of formicine ants (Law *et al.*, 1965; Lloyd *et al.*, 1975; Duffield *et al.*, 1977), the remaining mono-, sesqui-, and diterpene alcohols are primarily restricted to the exudates of bees. The sesquiterpenes 2,3-dihydro-6-(*E*)-farnesol and all (*E*)-farnesol are limited in their distribution to *Bombus, Psithyrus* (Kullenberg *et al.*, 1970), and *Andrena* (Bergström and Tengö, 1974) spp., and the diterpenoids geranylgeraniol and geranylcitronellol are restricted to species of *Bombus, Psithyrus* (Kullenberg *et al.*, 1970), and *Formica* (Bergström and Löfqvist, 1973) species (Table 3.2).

Although the terpene aldehydes, like the alcohols, have a widespread distribution in hymenopterous defensive secretions, they also occur unpredictably in the exudates of beetles. Isovaleraldehyde has been identified in the pygidial gland secretions of gyrinids (Schildknecht *et al.*, 1972b), staphylinids (Bellas *et al.*, 1974), and carabids (Scott *et al.*, 1975). The norsesquiterpenes gyrinidal and isogyrinidal are similarly limited in their known arthropod distribution to the exudates of the pygidial glands of gyrinid beetles (Meinwald *et al.*, 1972; Miller *et al.*, 1975; Newhart and Mumma, 1978). On the other hand, some of the cyclopentanoid monoterpene aldehydes characteristic of dolichoderine ants have a sporadic occurrence in beetle defensive secretions and, in two cases, those of phasmids. An isomer of dolichodial, a major constituent of the anal gland secretions of some *Dolichoderus* and *Iridomyrmex* spp. (Cavill and Hinterberger, 1960), constitutes the main chemical bullet "fired" from the prothoracic glands of the phasmid *Anisomorpha buprestoides* (Meinwald *et al.*, 1962). Iridodial, another cyclopentanoid monoterpene that has been detected in the exudates of four genera of dolichoderine ants (Cavill *et al.*, 1956a,b; Trave and Pavan, 1956; McGurk *et al.*, 1968; Wheeler *et al.*, 1975), is also synthesized by staphylinid (Abou-Donia *et al.*, 1971; Bellas *et al.*, 1974), carabid (Moore and Brown, 1979), and cerambycid (Vidari *et al.*, 1973) beetles. Citral, which is widely distributed in hymenopterous secretions, also is present in the diversified pygidial gland secretion of staphylinid species in the genus *Bledius* (Wheeler *et al.*, 1972a). Similarly, citronellal, a common mandibular gland product of formicine ants (Chadha *et al.*, 1962; Blum *et al.*, 1968b), has also been encountered in the pygidial gland exudates of staphylinid beetles (Bellas *et al.*, 1974). By contrast, the sesqui- and diterpene aldehydes identified in arthropod secretions are virtually restricted to the mandibular gland secretions of species of formicine ants in the genus *Lasius*. Farnesal, geranylgeranial, and geranylcitronellal have been identified solely as *Lasius* natural products (Bernardi *et al.*, 1967; Bergström and Löfqvist, 1970) whereas 2,3-dihydro-6-(*E*)-farnesal is also generated in the labial glands of males of the bumblebee *Bombus jonellus* (Bergström and Svensson, 1973). As is the case for the diterpene alcohols, the arthropod distribution of the diterpene aldehydes is limited to the secretions of a few species of bees and ants. Similarly, the novel diterpenes identified as termitid frontal gland products (Prestwich, 1979) are not known to be produced by insects or any other taxa.

G. Insects as Idiosyncratic Acid Producers

Isobutyric acid exemplifies the uneven distribution of several fatty acids in the glandular outpourings of a wide variety of arthropods (Table 6.2). This compound may represent a chemical character for hemipterous species in the reduviid subfamily Triatominae, having been detected in the Brindley's glandular exudates of all species examined (Pattenden and Staddon, 1972; Games *et al.*, 1974). Outside of these reduviid secretions, isobutyric acid has been detected in only one other hemipterous exudate, that of the alydid *Megalotomus quinquespinosus* (Aldrich and Yonke, 1975). The distribution of this acid in the defensive secretions of carabid beetles in six subfamilies (Schildknecht *et al.*, 1968a,b) is so spotty as to nullify any diagnostic significance for this compound in these taxa. On the other hand, isobutyric acid, along with 2-methylbutyric acid, appear to be valid chemical characters, at least for the last larval instar, for lepidopterous species in the genera *Papilio*, *Baronia*, *Eurytides*, and *Graphium* (Eisner *et al.*, 1970; Crossley and Waterhouse, 1969; Burger *et al.*, 1978), being present in the osmeterial exudates of all species that have been analyzed in these genera. However, the occurrence of this acid in the secretion of the myrmicine ant *Myrmicaria natalensis* (Quilico *et al.*,

1962) again demonstrates that, as is the case for a variety of other compounds, the arthropod distribution of isobutyric acid is remarkably unpredictable.

H. Arthropods and 4-Methyl-3-heptanone

The identification of 4-methyl-3-heptanone in the defensive exudates of opilionids (Blum and Edgar, 1971; Meinwald *et al.*, 1971; Jones *et al.*, 1976a,b, 1977), mutillids (Schmidt and Blum, 1977; Fales *et al.*, 1980), and ant species in three subfamilies (McGurk *et al.*, 1968; Duffield and Blum, 1973; Duffield and Blum, 1975c) serves to emphasize the ability of unrelated arthropods to biosynthesize the same compound. While this ethyl ketone may be diagnostically significant for opilionids in the suborder Palpatores, at this juncture it is impossible to evaluate its utility as a chemical character for the hymenopterans. Its occurrence in the mandibular gland exudates of *Dasymutilla* species (Schmidt and Blum, 1977; Fales *et al.*, 1980) may indicate that it is a characteristic natural product of mutillids in selected genera. On the other hand, its hymenopterous distribution clearly indicates that its biosynthesis has been evolved independently in at least two families. Although it is a rather characteristic natural product of myrmicine ants (McGurk *et al.*, 1968; Blum *et al.*, 1968a; Fales *et al.*, 1972), it appears to be a rather exceptional compound in the defensive exudate of ponerine (Duffield and Blum, 1973) and formicine (Duffield and Blum, 1975c) ants. Although it may be eventually established that 4-methyl-3-heptanone possesses diagnostic value for species in selected genera (*Neoponera*) or subgenera (*Myrmothrix* of *Camponotus*), at this juncture the formicid distribution of this compound must be regarded as too spotty to be useful.

I. Arthropods as Aliphatic Hydrocarbon Chemists

Several hydrocarbons possess such a widespread, but uneven distribution, as to indicate strongly that although their production is obviously highly adaptive, it is of limited chemotaxonomic value. *n*-Tridecane is the most widespread hydrocarbon in arthropod defensive exudates, having been identified as a glandular product of many species in two classes. It is a major constituent in the alkane-rich secretion of a population of the opilionid *Phalangium opilio* (Blum *et al.*, 1973d) but it is in the defensive exudates of the Insecta that the widespread but uneven distribution of this hydrocarbon becomes particularly evident. This alkane is a major constituent in both the metasternal and larval dorsal abdominal glandular secretions of bugs in four families (Table 2.2). Indeed, it is the only compound reported to be present in the defensive exudate of the plataspidid *Ceratocoris cephalicus* (Baggini *et al.*, 1966). *n*-Tridecane has been identified in the pygidial gland secretions of a variety of carabids (Moore and Wallbank, 1968; Schildknecht *et al.*, 1968a; Eisner *et al.*, 1977), usually as a concomitant of a variety of aliphatic acids or quinones. Both the tergal and pygidial gland secretions of staphylinids are enriched with this alkane (Brand *et al.*, 1973b; Bellas *et al.*, 1974) which is in admixture with a variety of aliphatic and alicyclic aldehydes. Curiously, this C_{13} hydrocarbon has been detected only in the defensive exudates of species in one hymenopterous family—the Formicidae. *n*-Tridecane appears to be a normal constituent in the Dufour's gland secretions of formicine ants (Bergström and Löfqvist, 1968, 1970, 1973) but it has not been commonly encountered in the exudates of species in other subfamilies. It has not been detected in the Dufour's gland secretions of numerous species of bees and if this alkane is not detected in any nonformicid hymenopterans, it will constitute a good chemical character for the ants.

Eisner *et al.* (1971c) analyzed the constituents present in the osmeterial secretion of the papilionid *Battus polydamas* and identified two sesquiterpenes, β-selinene and selin-11-en-4-ol. These defensive compounds are radically different from isobutyric and 2-methylbutyric acids, the constituents identified in the secretions of last-instar larvae of all other papilionid species. Significantly, *Battus* is a member of the tribe Troidini, whereas the other acid-producing species are members of other tribes. These results raise the possibility that the sesquiterpenes synthesized by *Battus* larvae may constitute a useful diagnostic character for this genus.

J. Ants and Subgenerically Specific Compounds

Recent investigations on the mandibular gland chemistry of male carpenter ants (*Camponotus* spp.) have established that certain compounds may be of considerable taxonomic significance at the subgeneric level. The secretions of species in the subgenus *Camponotus* are dominated by the lactone mellein, methyl 6-methylsalicylate, and 10-methyldodecanoic acid (Brand *et al.*, 1973e), whereas methyl anthranilate and 2,4-dimethyl-2-hexenoic acid are major constituents of the exudates of species in the subgenus *Myrmentoma* (Brand *et al.*, 1973d). Although in general the species in these two subgenera are rather easily assigned to one or the other of these two taxa, a few related forms do not fit readily into either of them. Creighton (1950) suggested that additional morphological investigations might establish that these aberrant forms should be placed in a separate subgenus. An investigation of the mandibular gland chemistry of one of these equivocal forms, *Camponotus schaefferi*, has demonstrated that Creighton's reservations about its subgeneric placement were well founded.

Males of *C. schaefferi* produce none of the aromatic compounds or carboxylic acids identified in the secretions of species in the subgenera *Camponotus* and *Myrmentoma*. Indeed, males, females, and workers of *C. schaefferi* synthesize the same compounds in their mandibular glands, 3-octanone and 3-octanol (Duffield and Blum, 1975b). Thus, on natural products grounds, this species is not closely related to species in either subgenus in which it has been placed. Furthermore, the lack of both caste and sex specificity in the mandibular gland products of *C. schaefferi* contrast to the general restriction of detectable glandular products to males of *Camponotus* and *Myrmentoma* species. Based on both mandibular gland chemistry and the distribution of the same compounds in individuals of both sexes and castes, *C. schaefferi* cannot be readily assigned to either the subgenera *Myrmentoma* or *Camponotus*.

K. Beetles and Subfamilial Specific Compounds

Pasteels *et al.* (1973) concluded that the distribution of alkaloids in the Coccinellidae accorded well with the accepted phylogeny of the subfamilies in this taxon. Alkaloids have been detected in members of widely separated tribes which belong to the more highly evolved subfamilies (Pasteels, 1977). Furthermore, the same or closely related alkaloids are synthesized by genera in the same subfamily, a further indication that these taxa may share common biosynthetic pathways.

L. Insect Defensive Allomones as Phylogenetic Characters

Crewe and Blum (1972) examined the chemistry of the mandibular gland products of a series of genera in the myrmicine tribe Attini, a taxon for which the phylogeny is rather clearly defined. *Cyphomyrmex*, the most primitive genus in the tribe, grades into *Trachymrymex*, a transitional genus which grades into *Acromyrmex* and ultimately *Atta*, the most highly evolved taxon in this tribe (Creighton, 1950). The only detectable mandibular gland product in *Cyphomyrmex* is 3-octanol, an alcohol that is present in species in all four genera. However, analyses of the glandular products of *Trachymyrmex septentrionalis* and a putative subspecies, *T. septentrionalis seminole*, demonstrated that these two forms produced qualitatively different blends of compounds in their mandibular glands. Whereas *T. septentrionalis* synthesizes 3-octanol and 3-octanone, *T. septentrionalis seminole* produces, in addition to these two compounds, 4-methyl-3-heptanol and 4-methyl-3-heptanone as major constituents. These results, combined with major morphological differences, clearly indicate that these two forms represent two distinct species, *T. septentrionalis* and *T. seminole* (Crewe and Blum, 1972). A species in the more highly evolved genus *Acromyrmex* produces 3-octanol and 3-octanone, whereas the *Atta* species generate these two compounds in addition to 4-methyl-3-heptanol, 4-methyl-3-heptanone, as well as 2-heptanol, 2-heptanone, and 3-heptanol (Riley *et al.*, 1974). An attine phylogeny based on these chemical characters is presented in Figure 15.2. In addition, the mandibular gland chemistry of a further attine genus, *Mycocepurus*, is included.

Qualitatively, the simplest mandibular gland secretion is produced by *Cyphomyrmex*, presumably the most primitive of the attine genera. The two *Trachymyrmex* species would give rise to both *Acromyrmex* from an ancestor like *T. septentrionalis* or from *T. seminole* after evolutionary reduction of mandibular gland products (Fig. 15.2). On the other hand, from elaboration of the mandibular gland products of a form like *T. seminole*, the genus *Atta*, which synthesizes C_7 compounds, could be evolved. This phylogeny (Fig. 15.2), which is based on biochemical characters consistent with those developed on morphological grounds, favors the elaboration of new mandibular gland compounds as a major development in the evolution of the tribe Attini. By contrast, *Mycocepurus*, which is considered to have diverged from the main line of attine evolution (Weber, 1958), has clearly done so in terms of synthesizing a mandibular gland product, *o*-aminoacetophenone (Blum *et al.*, 1974a), which greatly contrasts with the aliphatic compounds synthesized by species in other attine genera.

M. Ant Alkaloids and Phylogenetic Phenoclines

Defensive compounds have proven to be especially useful for proposing a phylogenetic phenocline for *Solenopsis* (*Solenopsis*) species. The venoms of these fire ants are thoroughly dominated by 2-methyl-6- alkylpiperidines and thus differ from the protein-rich venoms produced by species of stinging ants in other genera (Cavill *et al.*, 1964; Blum, 1966;

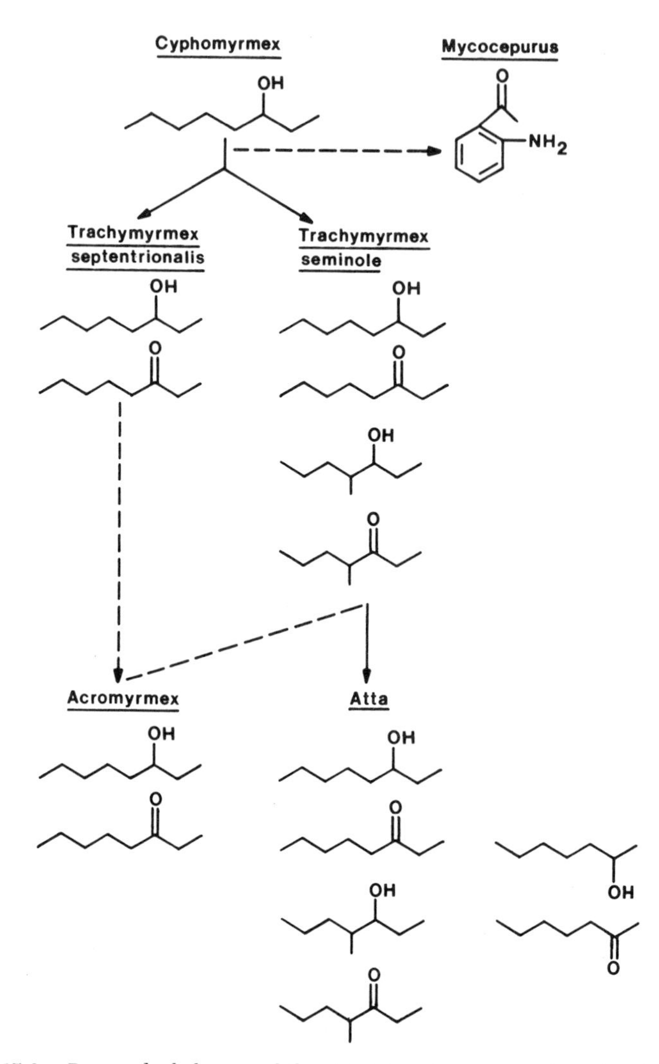

Fig. 15.2. Proposed phylogeny of the Attini based on morphological and chemical characters. Modified from Crewe and Blum (1972) with additions from Blum *et al.* (1974) and Riley *et al.* (1974).

Schmidt and Blum, 1978a,b). Comparisons of the alkaloidal compositions of the venoms produced by workers of different species have demonstrated that both major qualitative and quantitative differences may be encountered (Brand *et al.*, 1972). The alkaloids are often present as cis–trans mixtures with one or the other isomers invariably predominating. The 6-alkyl group on the piperidine ring may be C_9, C_{11}, C_{13}, $C_{13:1}$, C_{15}, or $C_{15:1}$ (MacConnell *et al.*, 1971, 1974, 1976), depending on the species and, surprisingly, the caste. Brand *et al.* (1973c) have demonstrated that, whereas workers of different species may produce alkaloids containing C_{13} or C_{15} side chains, fire ant queens do not synthesize appreciable amounts of any piperidines with side chains greater than C_{11}.

Considering proteinaceous venoms as primitive, the alkaloid-rich venoms of fire ants could be regarded as a more advanced character (Brand *et al.*, 1973a). Therefore, these alkaloids could constitute newly evolved components, which, when arranged in the order of their increasing complexity, reflect the pattern of evolution of the species in this genus. Since the venoms of some species contain exclusively C_{11} alkaloids which are present predominantly in the thermodynamically favorable cis form, these species could constitute the most primitive members in the genus. A switch to the synthesis of mainly *trans*-C_{11} alkaloids would reflect a derived condition favoring the biosynthesis of energetically unfavorable compounds. Further evolution would first result in the elaboration of alkaloids with C_{13} and $C_{13:1}$ side chains and, ultimately, forms with C_{15} and $C_{15:1}$ side chains would be added.

Brand *et al.* (1973a) presented a hypothetical construct in biochemical evolution which is based on both the alkaloidal compositions of the venoms and morphological studies. In terms of the alkaloids, both the qualitative and steroisomeric requirements for this evolutionary scheme were met by the presence of *Solenopsis* species which satisfied the steps in this proposed model of evolution. Although further investigations and testing will be required to test the validity of this construct, in itself it serves as an example of the potential of biochemical systematics. A detailed analysis of this model has been recently made (Brand, 1978).

MacConnell *et al.* (1976) analyzed the venoms of about 15 species of *Solenopsis* in the subgenus *Solenopsis.* Based on morphological considerations, the species were divided into three species groups and the compositions of the venoms of the taxa in each group were compared. This investigation demonstrated that, in general, the quantitative and qualitative nature of the dialkylpiperidines possessed significant taxonomic value. The venoms of all *Solenopsis* (*Solenopsis*) spp. were rich in 2-methyl-6-alkylpiperidines and, for the most part, closely related species contained similar alkaloidal compositions. Furthermore, the venoms from different populations of single species were quantitatively and qualitatively comparable over wide geographical areas. It was also noted that the venom composition of a *Solenopsis* sp. from Mato Grosso, Brazil was identical to that of *S. invicta*, the red imported fire ant (MacConnell *et al.*, 1971), as were the morphologies of the ants from the two populations. These chemical data provide support for the belief that *S. invicta* originated in Brazil (Buren *et al.*, 1974).

MacConnell *et al.* (1976) also demonstrated that both the qualitative and quantitative compositions of the venoms of several forms of *Solenopsis* differed considerably from those of species previously characterized (Brand *et al.*, 1972). Again, the chemical data were consistent with morphological considerations which indicated these were probably cryptic species which had not been previously described. It was also noted that the chemical data supported the elevation of a subspecies of *S. xyloni* to specific status. The venom of *S. xyloni* is distinctive among *Solenopsis* venoms in containing 2-methyl-6-*n*-undecylpiperideine as a minor constituent (Brand *et al.*, 1972). On the other hand, the venom of *S. aurea* lacks this piperideine (Blum *et al.*, 1973c), a fact that strongly supports the designation of *S. aurea* as a valid species and not a subspecies of *S. xyloni* (Creighton, 1950).

The taxonomic value of the 2,6-dialkylpiperidines in *Solenopsis* (*Solenopsis*) venoms appears to be of further significance in view of the recent demonstration that the venoms of at least some *Solenopsis* species in another subgenus contain different nitrogen heterocycles. Pedder *et al.* (1976) have shown that the poison gland secretion of *Solenopsis punctaticeps* is fortified with a large number of 2-ethyl-5-alkylpyrrolidines and -pyrrolines. No dialkylpiperidines were detected in the venom of this species. However, *S. punctaticeps* is not included in the subgenus *Solenopsis*, the true fire ants, but is considered to be more closely related to the species in the subgenus *Diplorhoptrum*, a taxon that primarily includes small ants not noted for their stinging propensities. Analyses of the venoms of two other *Diplorhoptrum* species showed that they also contained pyrrolidines (Jones *et al.*, 1979), thus indicating that five-membered nitrogen heterocycles may be characteristic venom components of many species in this subgenus.

Certainly the results of some of these studies demonstrate that unexpected biochemical congruencies may characterize the constituents in the defensive secretions of unrelated arthropods. Notwithstanding this possibility, the usefulness of arthropod natural products as adjuncts for assessing the relationships of various taxa is now amply supported by the results of diverse investigations that demonstrate that these compounds can play a meaningful role as indicators of possible phylogenetic relationships.

Chapter **16**

Biosynthesis of Defensive Compounds

The metabolic origins of most defensive compounds constitute *terra incognita.* Indeed, most of the biosynthetic investigations that have been undertaken with arthropods have simply determined whether the candidate compound is labeled after administration of a few common building block radiolabeled precursors. Generally, neither the enzymes responsible for the syntheses of defensive compounds nor their metabolic intermediates have been elucidated. In many cases, putative precursors, of unknown metabolic origin have been suggested, based both on their apparent structural relationships to the candidate defensive compounds and their occurrence in the same glandular exudates. Not surprisingly, the majority of biogenetic studies of these exocrine products have been modeled after comparable investigations on the synthesis of the same compounds in nonarthropod animals. Unfortunately, a paucity of data exists on the biosynthesis of the host of novel compounds identified in the defensive secretions of arthropods.

Although herbivorous arthropods may sometimes fortify their exocrine exudates with plant-derived compounds (von Euw *et al.*, 1967), there are no strong grounds for concluding that arthropods generally sequester the compounds in their defensive secretions from food sources. This fact is emphasized by the widespread distribution of the same exocrine compound in arthropods with completely different diets as well as the relative quantitative and qualitative consistencies of secretions of the same species feeding on a wide variety of diets (Waterhouse *et al.*, 1961). It seems certain that the diverse chemical defenses of these invertebrates reflect a biosynthetic versatility which has been barely illuminated heretofore.

A variety of arthropods are capable of biosynthesizing exocrine compounds with great chiral specificity (Brand *et al.*, 1979). Therefore, the possibility that an arthropod may produce only one enantiomer of a compound emphasizes the need to employ appropriate radioprecursors when investigating the biogenesis of potentially chiral molecules. A wide variety of hydrocarbons, ketones, and aldehydes that fortify the defensive secretions of arthropods contain chiral centers and the strong possibility that these products are synthesized with chiral exactitude cannot be excluded.

I. BIOGENESIS

A. Hydrocarbons

In one of the earliest investigations of the biosynthesis of compounds in defensive secretions, Gordon *et al.* (1963) demonstrated that the pentatomid *Nezara viridula* produced highly labeled *n*-tridecane after injection with sodium [1-^{14}C]acetate. *n*-Dodecane was also labeled, but less strongly. After injection of adult bugs with either sodium [1-^{14}C]decanoate, sodium [1-^{14}C]caproate, or sodium [1-^{14}C]propionate, *n*-tridecane was also found to contain significant radioactivity.

Waldner *et al.* (1969) fed sodium mevalonate[2-^{14}C] to the ant *Lasius fuliginosus* and isolated *n*-undecane which was highly radioactive. In *L. fuliginosus*, this compound is a major exocrine product of the Dufour's gland.

B. Aliphatic Aldehydes and Alcohols

The green vegetable bug, *Nezara viridula*, incorporates injected sodium [1-^{14}C]acetate into both (*E*)-2-hexenal and (*E*)-2-decenal in significant amounts (Gordon *et al.*, 1963). Injection of adult bugs with sodium [1-^{14}C]decanoate, sodium [1-^{14}C]caproate, or sodium [1-^{14}C]propionate resulted in significant labeling of hexenal but decenal was barely radioactive. Since there is about 25 times more decenal than hexenal in the secretion (Gilby and Waterhouse, 1965), these results suggest that these acids are not important precursors of the C_{10} enal.

Recently, Aldrich *et al.* (1978) demonstrated that hexanal, the main— and most toxic—constituent in the secretion of the coreid *Leptoglossus phyllopus*, was produced in a storage reservoir of the metathoracic scent gland. The secretion also contains *n*-hexyl acetate but this relatively non-toxic ester is synthesized in the primary accessory glands only to be secreted into the storage reservoir of the metathoracic gland. The acetate ester is then hydrolyzed by an esterase that is generated in the secondary accessory glands before being discharged into the storage reservoir. Hydrolysis of *n*-hexyl acetate produces another component of the secretion, 1-hexanol, which is converted to hexanal by a pair of dehydrogenases that also arise in the secondary accessory glands. These results support the suggestion by Gilby and Waterhouse (1967) that the synthesis of the most reactive defensive constituents produced by adult true bugs, i.e., aldehydes, occurs in the storage reservoir and not in the cells of the defensive gland. In addition, the results of Aldrich *et al.* (1978) provide an explanation for the *raison d'être* of proteins in such nonvenomous secretions. It is likely that the proteinaceous constituents found in many other nonvenomous secretions are also identified with anabolic enzymes that produce the ultimate defensive allomones.

C. Aromatic Compounds

The glucoside of *p*-isopropyl mandelonitrile was identified more than 30 years ago by Pallares (1946) from the millipede *Rhysodesmus* (=*Polydesmus*) *vicinus*. This compound is degraded to cuminaldehyde which appears to constitute an important defensive allomone of this millipede. The glucoside is quite toxic to rabbits when orally administered but possesses no demonstrable toxicity when introduced intravenously unless emulsin is also injected. Since the glucoside and its degradation products could only be introduced into the body of a mammal by ingestion, it is of obvious selective advantage to the millipede for its toxic secretory products to exhibit maximum pharmacological activity after oral administration. Significantly, Mexican Indians use these millipedes in order to prepare an arrow poison, but always in admixture with plant parts (Pallares, 1946). Presumably, the plant tissues contribute an emulsin-like enzyme which serves to hydrolyze the glucoside to its toxic moieties *in vivo*.

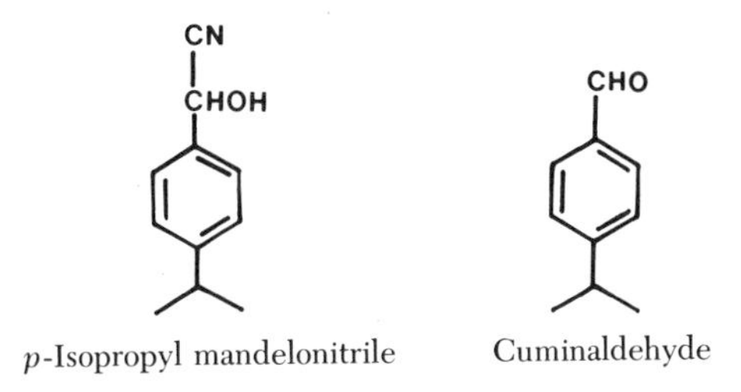

p-Isopropyl mandelonitrile Cuminaldehyde

Cuminaldehyde is an atypical defensive product since other species of polydesmid millipedes have been demonstrated to secrete benzaldehyde along with HCN from their defensive glands when irritated (Guldensteeden-Egeling, 1882; Blum and Woodring, 1962; H. E. Eisner *et al.*, 1963). As is the case for *R. vicinus*, the precursor of the defensive compounds in *Pachydesmus crassicutis* may be a glucoside (Blum and Woodring, 1962), whereas in *Polydesmus collaris*, HCN and benzaldehyde are reported to be primarily generated from mandelonitrile benzoate and apparently mandelonitrile as well (Casnati *et al.*, 1963).

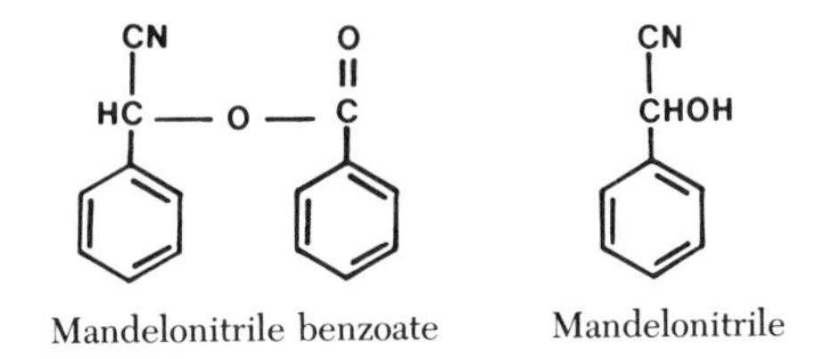

Mandelonitrile benzoate Mandelonitrile

The cyanogenic precursor of the polydesmid *Apheloria corrugata* is mandelonitrile, which is stored in the reservoirs of bicompartmentalized defensive glands (T. Eisner *et al.*, 1963a). Breakdown of the cyanogen is effected by a lyase which is present in the other glandular compartment, a vestibule into which the reservoir evacuates its contents. Thus, both benzaldehyde and HCN are generated only after the millipede has been tactually stimulated and the contents of the two compartments are mixed prior to discharge to the exterior.

In one of the most thorough investigations of the biosynthesis of arthropod defensive compounds to date, Duffey and his colleagues have illuminated the metabolic origins of benzaldehyde and HCN in millipedes. Towers *et al.* (1972) had demonstrated that the polydesmid *Oxidus gracilis*, which secretes HCN and benzaldehyde when irritated, does not generate these compounds from prunasin, the glucoside of mandelonitrile. Rather, these defensive products are presumably derived from mandelonitrile, a cyanogen that is synthesized from ingested phenylalanine. Administration of DL-[2-^{14}C]phenylalanine resulted in the subsequent production of radioactive HCN, whereas benzaldehyde was labeled if millipedes were fed DL-[3-^{14}C]phenylalanine or DL-[1-ring-^{14}C]phenylalanine. These results are consistent with the conversion of phenylalanine to mandelonitrile and subsequent breakdown to yield benzaldehyde and HCN. Administration of DL-[1-^{14}C]phenylalanine re-

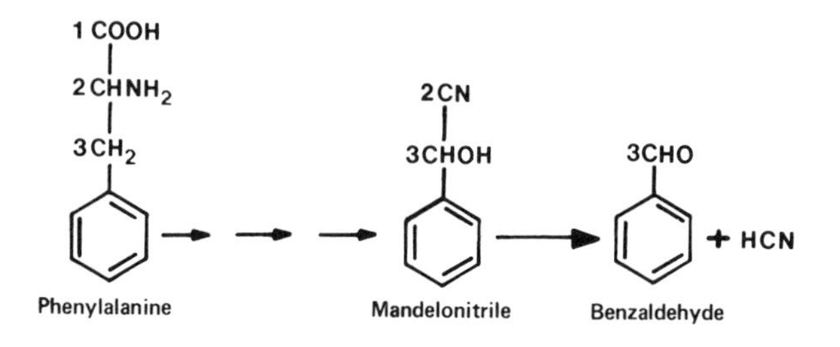

sulted in the formation of nonradioactive cyanogen, a result that would be expected as a consequence of the decarboxylative loss of 1-^{14}C. DL-[2-^{14}C]Tyrosine was not utilized as a cyanogen precursor as indicated by the lack of radioactivity in the HCN liberated by millipedes administered this amino acid (Towers *et al.*, 1972).

The metabolic pathway from phenylalanine to mandelonitrile has been explored in great depth by Duffey *et al.* (1974). A large series of candidate radioactive precursors were injected into adults of the polydesmid *Harpaphe haydeniana* and the labeling in both benzaldehyde and HCN determined. It was first established that as in *O. gracilis*, phenylalanine is ultimately converted to mandelonitrile which serves as the source of both benzaldehyde and HCN (Table 16.1). HCN arises from carbon 2 of the amino acid side chain and benzaldehyde is derived from carbon 3 and the ring carbons. However, based on both the low specific activities of these two end products and the great dilution in their radioactivities relative to the labeled amino acid precursor, phenylalanine is not effectively utilized to produce mandelonitrile. However, it is likely that metabolic diversions of this amino acid are probably responsible for its poor utilization as a cyanogen precursor.

Duffey *et al.* (1974) have established unequivocally that mandelonitrile

Table 16.1

Incorporation of Radioactive Precursors into Benzaldehyde and HCN by the Millipede *Harpaphe haydeniana*

Radioactive precursor	Specific activity (μCi/μmole)	Amount injected/animal (μmoles)	(μmoles)		Specific activity (μCi/μmole)		% Conversion of precursor	Dilution of radioactivity
			φCHO	HCN	φCHO	HCN		
DL-[2-^{14}C]Phenylalanine	89	0.15	3.2	2.7	0	0.010	0.002	8.900
DL-[3-^{14}C]Phenylalanine	82	0.31	4.0	4.6	0.009	0	0.002	9.000
DL-*N*-[2-^{14}C]Hydroxyphenylalanine	92	0.14	3.4	2.3	0	0.77	2.8	60
Phenylpyruvic acid [2-^{14}C]oxime	100	0.19	5.5	5.0	0	0.031	0.08	3230
[1-^{14}C]Phenylacetaldoxime	443	0.026	4.3	4.6	0	1.5	6.3	295
[1-^{14}C]Phenylacetonitrile	212	0.057	4.3	5.8	0	4.4	22.0	48
DL-2-[2-^{3}H]Hydroxyphenyl-acetaldoxime	34	0.24	7.5	3.8	1.0	0	9.7	34
DL-*N*-[2-^{14}C]Hydroxyphenylalanine + phenylpyruvic acid oxime	92	0.16 +0.22	5.8	7.6	0	0.011	0.65	8350
Phenylpyruvic acid [2-^{14}C]oxime + phenylacetaldoxime	100	0.19 +0.22	7.4	9.3	0	0.032	0.16	3130
[1-^{14}C]Phenylacetaldoxime + phenylacetonitrile	62	0.32 +0.22	7.2	7.6	0	0.014	0.80	4030

rather than mandelonitrile glucoside constitutes the main storage form of benzaldehyde and HCN. The stored product, which can be easily hydrolyzed to mandelic acid, is present as the D-(*R*)-enantiomer, as determined by circular dichroism (S. S. Duffey, personal communication, 1974). Indeed, S. S. Duffey (personal communication, 1974) has been able to isolate droplets of pure mandelonitrile from the storage vestibule of the two-chambered gland of *H. haydeniana*. Furthermore, crude enzyme extracts prepared by ammonium sulfate precipitation of glandular extracts contain a potent mandelonitrile lyase that would be capable of generating the quick bursts of HCN that are characteristic of polydesmid millipedes.

Duffey *et al.* (1974) studied the biosynthesis of mandelonitrile in *H. haydeniana* by administering several compounds that are known to be precursors of this cyanogen in plants. *N*-Hydroxyphenylalanine is an efficient precursor of benzaldehyde and HCN (Table 16.1). On the other hand, phenylpyruvic acid oxime is poorly utilized, and in view of the similarly poor utilization of phenylalanine, it is unlikely that the hydroxy acid is incorporated into mandelonitrile via its disproportionation into the free amino acid and the oxime (Kindl and Undergill, 1968). Both the efficient conversion of *N*-hydroxyphenylalanine to mandelonitrile and its low dilution value (Table 16.1) favor the N-hydroxylation of phenylalanine as an enzymic rather than chemical reaction in the biosynthesis of cyanogen by *H. haydeniana*.

On the other hand, it was not possible to determine if phenylpyruvic acid oxime is an intermediate in the biosynthetic pathway for mandelonitrile. The oxime is poorly utilized as indicated both by the specific activity of the benzaldehyde generated from mandelonitrile and the high dilution of radioactivity (Table 16.1). However, since this α-oximino acid is better utilized than phenylalanine and, in addition, it reduces the utilization of *N*-hydroxyphenylalanine when simultaneously fed (Table 16.1), it is possible that phenylpyruvic acid oxime is a functional intermediate in the biogenesis of mandelonitrile. Conceivably, the oxime enters this pathway by its direct conversion to phenylacetonitrile, a conclusion that is consistent with the inability of phenylacetaldoxime to suppress the incorporation of this oximino acid into mandelonitrile. It is nevertheless possible that phenylpyruvic acid oxime enters the biosynthetic pathway by being converted to an intermediate, rather than being a part of the metabolic sequence.

Phenylacetaldoxime, phenylacetonitrile, and 2-hydroxyphenylacetaldoxime are all very effectively utilized for the biosynthesis of cyanogen (Table 16.1), and these compounds would appear to be outstanding candidates as metabolic intermediates in the production of mandelonitrile (Fig. 16.1). Duffey *et al.* (1974) were not able to establish whether 2-hydroxyphenylacetaldoxime and/or phenylacetonitrile are biological precursors of mandelonitrile. Both radioactive phenylacetaldoxime and phenylacetonitrile have been isolated from *H. haydeniana*, thus demonstrating that mandelonitrile is generated by hydroxylation of the nitrile (Fig. 16.1). Significantly, since there is a direct relationship between the degree of incorporation of a precursor and its metabolic proximity to the cyanogenetic end product, (Table 16.1, Fig. 16.1), the pathway proposed by Duffey *et al.* (1974) appears to be biogenetically very sound.

Harpaphe haydeniana poorly utilized phenylacetamide, phenylethylamine-HCl, and phenylacethydroxamic acid when these compounds were administered at low concentrations (Duffey *et al.*, 1974). However, the inability of the millipede to incorporate these compounds when they were administered at higher doses argues against regarding them as likely intermediates in the biosynthesis of mandelonitrile.

The requirement of an exogenous source of aromatic compounds is indicated by the failure of *H. haydeniana* to utilize shikimic acid as a precursor of the cyanogenetic end product (Duffey *et al.*, 1974).

S. S. Duffey (personal communication, 1974) has purified both a β-glycosidase and a mandelonitrile lyase from extracts of the defensive

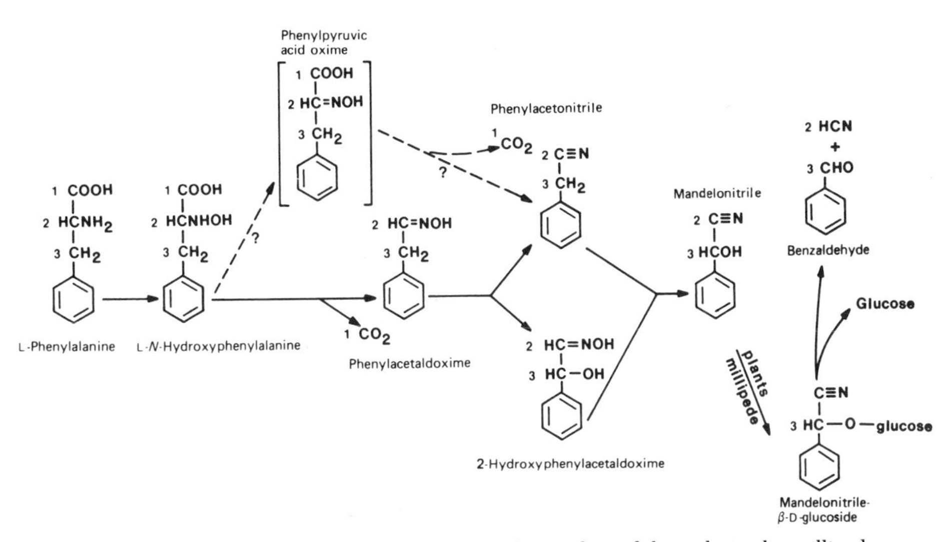

Fig. 16.1. Proposed pathway for the biosynthesis of mandelonitrile in the millipede *Harpaphe haydeniana* leading to the production of HCN and benzaldehyde. (After Duffey *et al.*, 1974.)

glands of *H. haydeniana*. The lyase is highly active and is capable of generating HCN from either mandelonitrile or *p*-hydroxymandelonitrile. On the other hand, the β-glycosidase, which liberates HCN from cyanogenetic heterosides with (taxiphyllin) or without (sambunigrin) hydroxylated aglycones, is not very active. This result would indicate that glycosidase activity is not important in the rapid generation of HCN, but rather in the slow resynthesis of glycosidic cyanogen. However, the presence of even weak glycosidase activity in the defensive glands demonstrates that the millipede biosynthetic pathway is essentially identical to that of plants (Tapper *et al.*, 1972), an interesting example of parallel metabolic evolution between plants and animals.

Jones *et al.* (1976a) identified mandelonitrile, benzaldehyde, benzoic acid, benzoyl cyanide, and HCN in the defensive secretions from the ventral segmental glands of several species of geophilomorphous centipedes. All of these compounds are probably derived from aromatic cyanogenetic precursors, as they are in millipedes.

Benzoyl cyanide and benzoic acid have also been detected in the defensive exudates of several species of polydesmoid millipedes (Connor *et al.*, 1977; Duffey *et al.*, 1977) and in some cases they are accompanied by mandelonitrile benzoate. Benzoyl cyanide may constitute a pivotal compound in the biosynthesis of most of the aromatic compounds detected in these centipede and millipede secretions. This cyanogen, which can arise from the dehydrogenation of mandelonitrile, can be converted to benzoic acid and HCN by hydrolysis. Acylation of benzoic acid by benzoyl cyanide would produce mandelonitrile benzoate, a major constituent in the secretions of several xystodesmid millipedes (Duffey *et al.*, 1977).

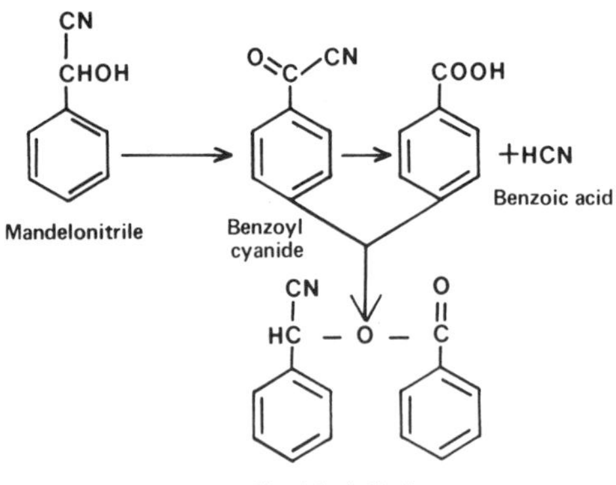

Phenolics have also been identified as minor constituents of some species of polydesmoid millipedes (Monteiro, 1961; Blum *et al.*, 1973a; Duffield *et al.*, 1974; Duffey *et al.*, 1977). In the polydesmoid millipedes *Oxidus gracilis*, *Euryurus maculatus*, and *Pseudopolydesmus erasus* both phenol and guaiacol are produced from tyrosine by the enzyme tyrosine phenol lyase (Duffey and Blum, 1977). Guaiacol is presumably produced from a methoxy derivative of dihydroxyphenylalanine (DOPA); DOPA is a well-known intermediate in the metabolism of tyrosine to quinones by arthropods.

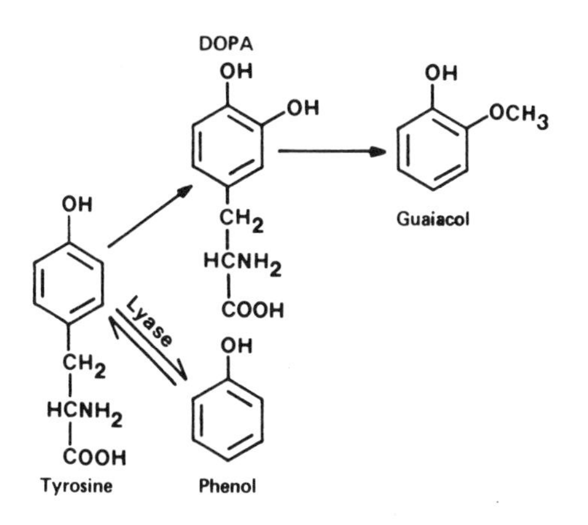

Larval chrysomelids in the genera *Paropsis* and *Chrysophtharta* secrete benzaldehyde, HCN, and glucose when tactually stimulated (Moore, 1967), and it is not unlikely that these compounds are generated from an aromatic cyanogenetic glucoside.

D. 1,4-Quinones

In arthropods 1,4-quinones appear to be generated on demand from stored phenolic glucosides followed by oxidation of the free quinols. Roth and Stay (1958) have demonstrated that the second tracheal abdominal glands of the cockroach *Diploptera punctata*, which are a rich source of *p*-quinones, contain both a glucoside and a polyphenol oxidase. Although no β-glucosidase was identified in tracheal gland extracts, the high concentration of glucoside combined with the well-developed polyphenol oxidase activity are consistent with the genesis of the *p*-quinones from bound glucosidic precursors.

Happ (1968) studied the histochemistry of the quinone-producing glands of the tenebrionids *Tribolium castaneum* and *Eleodes longicollis* and established that the cytoplasm of the defensive gland is isolated from the reaction chambers in which the quinones are actually generated. The biosynthesis of *p*-quinones occurs in a two-chambered secretory unit that contains, in addition to secretory cells and their associated vesicles, cuticular organelles and efferent tubules in which the quinones are sequentially generated. Stepwise segregation of reactors requires: (1) the presence of reaction compartments, (2) addition of appropriate enzymes to these compartments, and (3) a unidirectional flow of the reaction products from one compartment to another. All these conditions are met in the secretory apparatuses of the tenebrionids (Happ, 1968).

The quinonoidal secretory unit and the proposed scheme for the partial biosynthesis of *p*-quinones are illustrated in Fig. 16.2 (Happ, 1968). In the more apical cell a phenolic β-glucoside contained in the secretory cells is secreted into the lumen where it is hydrolyzed by a β-glucosidase. The newly formed quinol is transferred to the isolated vesicular organelles where a portion of it is oxidized to 1,4-quinones by polyphenol oxidase. The phenolic–quinonoidal mixture is passed along the efferent duct to the more distant secretory unit and finally to the free efferent duct where additional quinol is oxidized by a hemoprotein peroxidase. All enzymatic steps culminating in the production of highly reactive *p*-quinones are carried out in the vesicular organelle which is essentially isolated from the

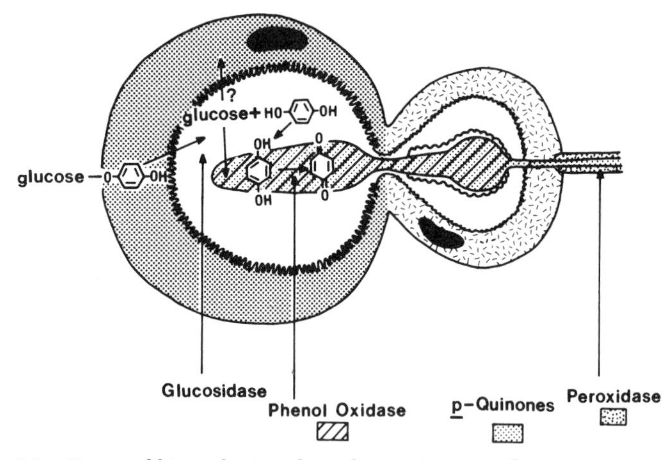

Fig. 16.2. Proposed biosynthetic pathway for *p*-quinones in the secretory apparatus of a tenebrionid beetle. (After Happ, 1968.)

sensitive cytoplasm of the secretory unit. Thus, these morphological compartments appear to constitute "reaction" chambers in which the final enzymatic steps leading to the generation of toxicants take place (Eisner *et al.*, 1964).

Schildknecht (1970) has elucidated the biosynthesis of 1,4-quinones in *Brachinus creptians* and *B. explodens*, two species of carabid beetles that expel their defensive products with an audibly explosive force. Like the tenebrionids, *Brachinus* species produce the quinones in a bicompartmentalized secretory apparatus. The reservoir of the apical cells contains a 25% solution of hydrogen peroxide and a 10% solution of hydroquinones. The mixture is transferred to the explosion chamber, the lumen of which is fortified with both catalases and peroxidases. Four catalases (H_2O_2:oxidoreductase) decompose the peroxide into water and oxygen whereas the three polyphenol peroxidases (polyphenol:H_2O_2:oxidoreductase) oxidize the hydroquinones to their corresponding 1,4-quinones (Schildknecht, 1970). Both groups of enzymes are more stable against H_2O_2 than most previously characterized catalases and peroxidases and in addition, are remarkably heat stable. The temperature optima of the catalases lie between 70°–80°C and those of catalases are about 70°–77°C. Significantly, with an increase in substrate concentration, the catalases produce an unlimited increase in activity and the reaction rate is further accelerated by the exceptionally high reaction temperature. Ultimately, the explosive discharge of quinones results from the production of high concentrations of free oxygen which are generated in the final reaction chamber.

The biosynthetic pathways for the six-membered aromatic rings of 1,4-benzoquinones were studied by J. Meinwald *et al.* (1966a). Adult tenebrionid beetles (*Eleodes longicollis*) were injected with or fed DL-[2-^{14}C]tyrosine, DL-[ring-^{14}C]phenylalanine, L-[1-^{14}C]tyrosine, sodium [2-^{14}C]acetate, sodium [2-^{14}C]propionate, and sodium [2-^{14}C]malonate. The labeling in the synthesized quinones—1,4-benzoquinone, methyl-1,4-benzoquinone and ethyl-1,4-benzoquinone—demonstrated that the alkylated compounds were synthesized from the aliphatic acids whereas 1,4-benzoquinone was generated from the preformed aromatic rings of the amino acids. The extensive incorporation of sodium malonate supports the formation of alkylated benzoquinones by acetate condensation, specifically the condensation of malonate units with acylcoenzyme A.

The patterns of isotopic incorporation into methyl-1,4-benzoquinone and ethyl-1,4-benzoquinone were defined by J. Meinwald *et al.* (1966a) by degrading the compounds after collection from beetles previously injected with sodium [1-^{14}C]acetate and sodium propionate-1-^{14}C (Fig. 16.3).

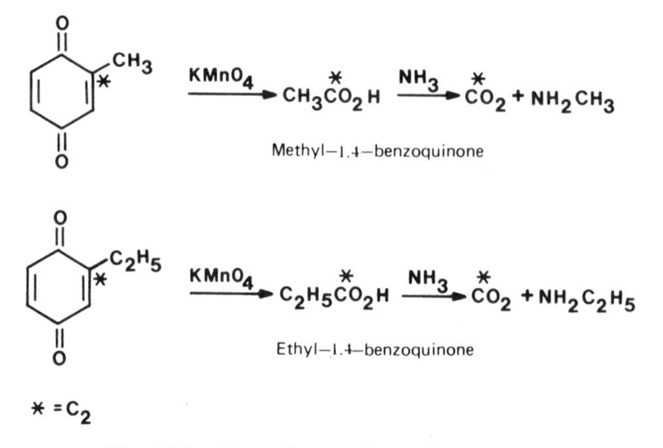

Fig. 16.3. Degradation scheme for quinones.

After administration of sodium acetate almost 30% of the total activity of methyl-1,4-benzoquinone appeared at C-2 and less than 0.1% of the activity was present in the methyl group. Therefore the other five ring carbon atoms must contain the remaining 40% of the activity. In the case of ethyl-1,4-benzoquinone, about 10% of the activity was found in the C-2 ring carbon and virtually all the remaining activity was present in the other ring carbons. On the other hand, sodium propionate was not efficiently utilized as a precursor of methyl-1,4-benzoquinone but was significantly incorporated into the ethyl homolog. About 95% of the total activity of ethyl-1,4-benzoquinone was localized at C-2 of the ring. Since sodium propionate is poorly incorporated into the methyl homolog of 1,4-benzoquinone and five of the remaining ring carbons of the ethyl homolog, it appears that this precursor is only slightly converted into C-2 units. Furthermore, condensation of propionyl-CoA with malonate units and subsequent cyclization would be readily compatible with the observed distribution of labeling (J. Meinwald *et al.*, 1966a). The origins of the ring-substituted alkyl groups remain to be determined. The significance of this tenebrionid maintaining distinct metabolic pathways for the biosynthesis of 1,4-benzoquinone and its alkylated homologs has yet to be established.

Weatherston and Meinwald (in Weatherston, 1971) have demonstrated that the methoxy group present in 2-methoxy-3-methyl-1,4-benzoquinone, a defensive compound produced by the millipede *Narceus gordanus*, is derived from methionine. After injection of labeled methionine, about 95% of the label was found in the methyl group in the alkoxy group. In addition, it was established that 6-methylsalicylic acid was the best precursor administered for both 2-methoxy-3-methyl-1,4-benzoquinone and its concomitant, toluquinone. This latter result would seem to indicate that these two quinones probably arise via the acetate pathway. However, Duffey and Blum (1976) have demonstrated that the millipedes *Rhinocricus holomelanus*, *Narceus annularis*, and *Cambala annulata* required the preformed aromatic ring of tyrosine in order to synthesize benzoquinone, toluquinone, and 2-methoxy-3-methyl-1,4-benzoquinone. On the other hand, no labeled quinones were produced if the millipedes were treated with either radiolabeled acetate or malonate. In contrast, the tenebrionid beetle *Zophobas rugipes* produces benzoquinone from tyrosine whereas toluquinone is derived from acetate (Duffey and Blum, 1976), results that are in agreement for the biosyntheses of these two quinones by the tenebrionid *E. longicollis* (J. Meinwald *et al.*, 1966a).

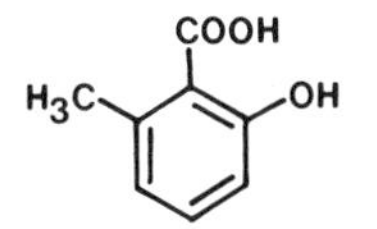

6-Methylsalicyclic acid

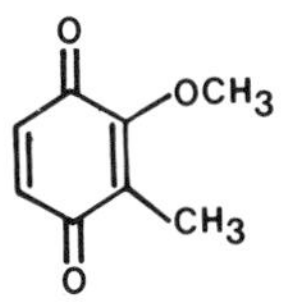

2-Methoxy-3-methyl-1,4-benzoquinone

Duffey and Blum (1976) suggest that the greater quinonoid diversity of insect exocrine secretions vis-à-vis those of millipedes may be explained by the biosynthetic pathways that appear to be available to these two groups of arthropods. Whereas the quinonoid diversity of millipedes would be limited because these compounds would have to be biosynthesized from the preformed aromatic rings of amino acids (e.g., tyrosine), insects would share this metabolic restriction only for the synthesis of the nonalkylated quinone, 1,4-benzoquinone. Since insects appear to derive alkylated quinones from the acetate–malonate pathway, the biogenesis of compounds such as 2-propyl-1,4-benzoquinone (Tschinkel, 1975b) could be easily realized. In essence, the availability of two distinct metabolic pathways in insects for the generation of 1,4-quinones would enable these arthropods to exploit quinonoid diversity with a virtuosity denied the millipedes whose quinonoid repertoire appears to be limited by their contrasting biosynthetic conservatism.

E. Mono- and Sesquiterpenes

The metabolic derivations of a small potpourri of terpenoids in arthropods have been studied in the last decade. Although arthropods have been demonstrated to produce a variety of polyisoprenoids including diterpenes (Goodfellow *et al.*, 1972), they are not able to cyclize squalene oxide (Goodfellow *et al.*, 1973) and thus are unable to synthesize cholesterol. However, the presence of enzymes between mevalonate kinase and squalene cyclase (Goodfellow *et al.*, 1973) is reflected in the ability of diverse arthropods to biosynthesize a multitude of often distinctive mono-, sesqui-, and diterpenes as components of their defensive secretions.

1. *Monoterpene Aldehydes and Lactones*

Happ and Meinwald (1965) fed both sodium [1-^{14}C]acetate and [2-^{14}C]acetate as well as mevalonic acid [1-^{14}C]lactone and [2-^{14}C]-lactone to workers of the ant *Acanthomyops claviger* and determined the amount of labeling in the mandibular gland compounds citral and citronellal (Chadha *et al.*, 1962). With the exception of [1-^{14}C]-mevalonic acid, the precursors were extensively incorporated into the monoterpene aldehydes. The inability of the monoterpenes to be radiolabeled after the administration of [1-^{14}C]mevalonic acid is readily reconciled with the decarboxylative loss of the radioactive carbon atom during its conversion to Δ^3-isopentylpyrophosphate in the well-established mevalonic acid pathway. These results clearly establish the

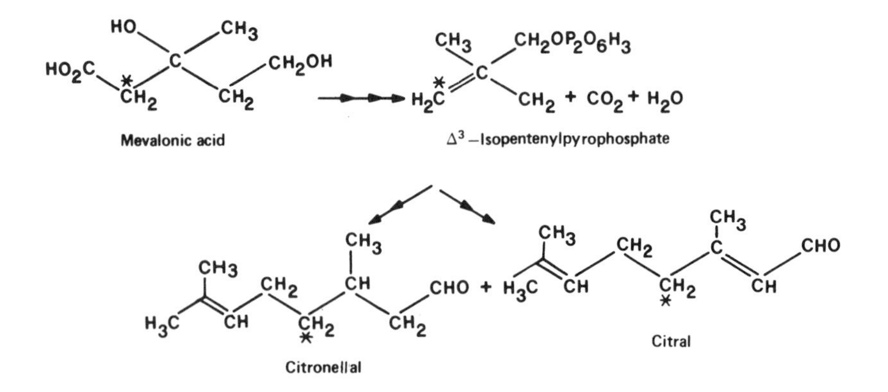

ability of this species to biosynthesize monoterpenes by acetate condensation.

Subsequently, J. Meinwald *et al.* (1966b) demonstrated that adults of the phasmid *Anisomorpha buprestoides* synthesized labeled anisomorphal, a cyclopentanoid monoterpene, after injection with sodium [1-^{14}C]acetate or [2-^{14}C]acetate, sodium [2-^{14}C]malonate, and mevalonic acid [2-^{14}C]lactone. Thus, this species shares the ability with ants to utilize

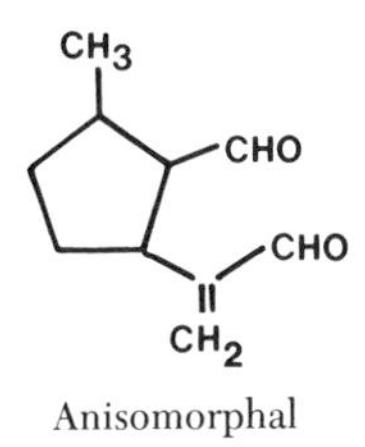

Anisomorphal

normal isoprenoid precursors for the biosynthesis of terpenes.

Anisomorphal appears to be an epimer of dolichodial, a cyclopentanoid monoterpene previously identified as an anal gland product of dolichoderine ants (Cavill and Hinterberger, 1961). A variety of monoterpene ketones accompany the cyclopentanoid monoterpenes in dolichoderine ants, and Cavill and Hinterberger (1960, 1962) have proposed a biogenetic scheme for the production of all these ant-derived compounds (Fig. 16.4). Citral occupies a key position in this terpenoid pathway, being converted to either acyclic ketones, or after stereospecific reduction, to L-citronellal. An allylic oxidation and Michael addition would yield the key cyclopentanoid monoterpene iridodial which, by the equivalent of a Cannizzaro reaction, would be converted to the iridolactones—iridomyrmecin and isoiridomyrmecin. On the other hand, β-hydroxylation of the propional side chain followed by dehydration would yield dolichodial. Nepetalactone, a product of the catnip plant, could also be produced by oxidation of the enol-lactol tautomer of iridodial whereas ammoniation of iridodial would yield actinidine.

The biosynthetic pathway proposed by Cavill and Hinterberger (1960, 1962) is an extension of a biogenetic scheme suggested by Clark *et al.* (1959) who converted citral to iridodial and the iridolactones. The pathway delineated by Cavill and Hinterberger is particularly attractive since it can be rationalized with the presence of the major dolichoderine anal gland products—three ketones, iridodial, the two iridolactones, dolichodial (Cavill, 1970), and actinidine (Wheeler *et al.*, 1977b) in the anal

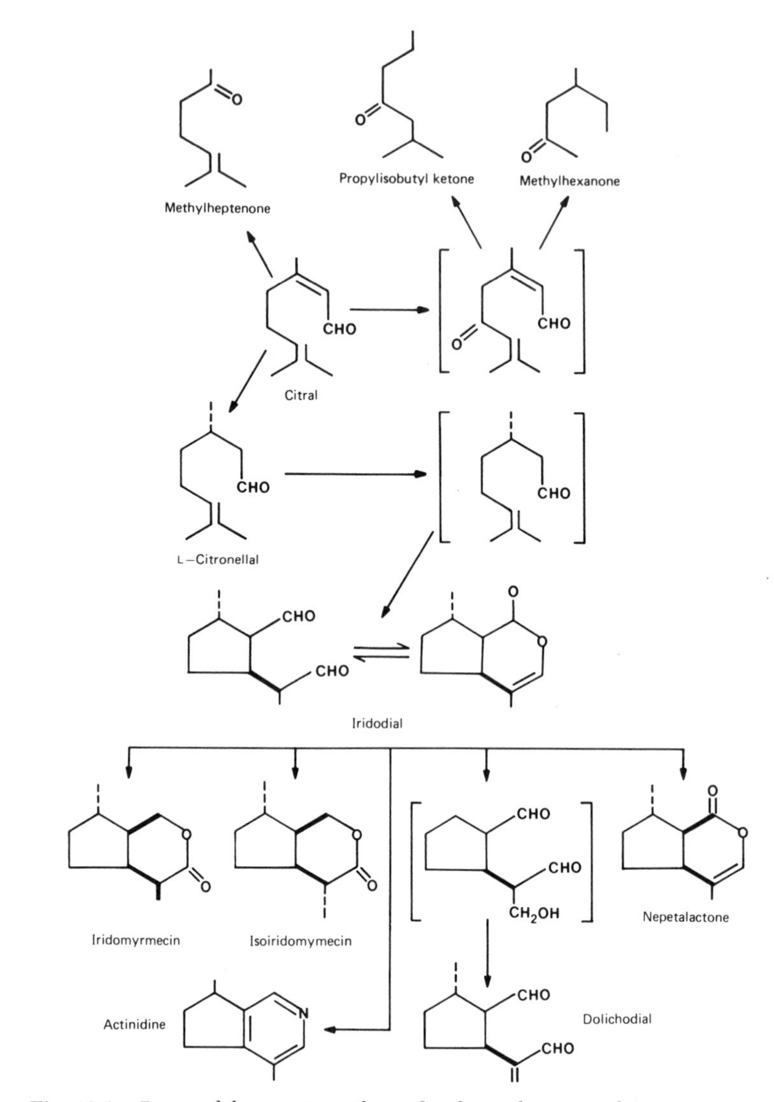

Fig. 16.4. Proposed biogenetic pathway for the cyclopentanoid monoterpenes from acyclic precursors. (After Cavill and Hinterberger, 1960, 1962.)

glands of dolichoderine ants. On the other hand, neither citral nor citronellal have been identified in any dolichoderine species, and the apparent absence of these key intermediates may indicate that these ants biosynthesize their cyclopentanoid monoterpenes from some uncharacterized precursors. However, since Bellas *et al.* (1974) have identified geranial, citronellal, iridodial, and significantly, actinidine in the pygidial gland secretions of rove beetles and Smith *et al.* (1979) characterized nepetalactone in a stick insect, it is not unlikely that a biosynthetic pathway similar to the one proposed by Cavill and Hinterberger (1962) is operational in dolichoderine ants.

2. *Cantharidin*

The monoterpene anhydride cantharidin, a powerful vesicant and inducer of priapism, is only known to be produced by meloid beetles. The presence of such an unusual cyclohexanoid skeleton in this compound indicates that its biogenesis may not follow the classical terpene pathway, a conclusion that has been supported by the results of biosynthetic investigations.

Meyer *et al.* (1968) demonstrated that adult males, but not females, of the meloid *Lytta vesicatoria* synthesized labeled cantharidin after injection with sodium [1-^{14}C]acetate. Since larvae were also capable of producing tagged cantharidin after administration of labeled acetate, it was initially concluded that the presence of this compound in adult females was at least partially due to holdover from larval synthesis.

Degradation studies of radiolabeled cantharidin synthesized by males of *L. vesicatoria* established the presence of an unexpected labeling pattern in this monoterpene (Schlatter *et al.*, 1968). After injection with sodium [1-^{14}C]acetate, two-thirds of the labeling occurred at C-2 and C-3 whereas one-third of the activity was equally divided between the pairs of carbon atoms at C-8 to C-9 and C-10 to C-11 following sodium mevalonate-2-^{14}C treatment. These results do not support the biogenesis of cantharidin by the usual head-to-tail linkage of two isoprenoid units but suggest head-to-head condensation.

Based on further biosynthetic studies, Peter *et al.* (1972) concluded that cantharidin is produced from three isoprenoid units rather than two. The apparent precursor of cantharidin in meloids is the C_{15} sesquiterpene alcohol farnesol, which must be extensively degraded in order to produce the monoterpene anhydride.

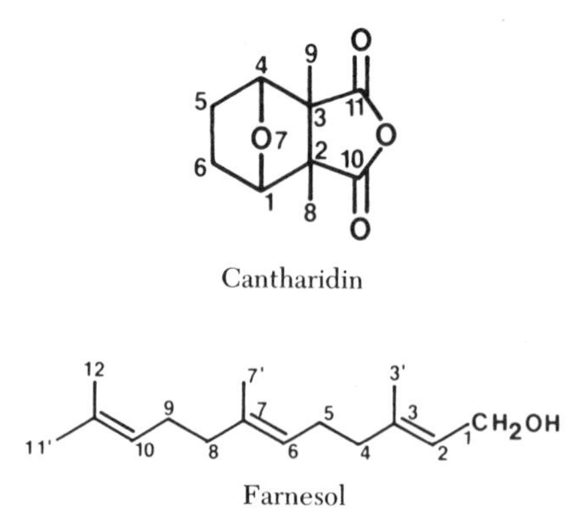

Cantharidin

Farnesol

Farnesol could only be converted to cantharidin after cleavage of carbon atoms 1, 5, 6, 7, and 7′, a conclusion that was consistent with the results of degradation studies. The administration of [2-^{14}C]farnesol and [11,12-^{14}C]farnesol resulted in the synthesis of specifically labeled cantharidin whereas it was not labeled if either [1-^{14}C]farnesol or [7-^{14}C]farnesol were administered. Degradation studies following the administration of sodium [2-^{14}C]mevalonate established that one-third of the activity resided in each of the carbon atoms 1 and 6 and one-sixth in each of the positions 9 and 11. No incorporation occurred if variously labeled geraniols were administered to the beetles.

Virtually all the cantharidin biosynthesized by males of *L. vesicatoria* is transferred to females during copulation (Sierra *et al.*, 1976). Males injected with either radiolabeled farnesol or methyl farnesoate 24–30 hr prior to copulation transferred about 95% of the synthesized cantharidin after prolonged copulation (ca. 20 hr). If males were injected with cantharidin while they were *in copula* about 10% of the label was subsequently detected in the female reproductive organs. Adult females are unable to biosynthesize cantharidin from appropriate terpenoid precursors. They appear to sequester this compound in their reproductive organs after it has been introduced into their reproductive tract as part of the seminal ejaculate. Thus, it appears that cantharidin is selectively biosynthesized in the male accessory glands (Sierra *et al.*, 1976). Utilization of the seminal ejaculate as a vehicle for transferring defensive compounds to females has not been previously reported in any arthropod species.

3. *Dendrolasin*

Dendrolasin is a furanoterpene which is synthesized in the mandibular glands of the ant *Lasius fuliginosus* (Quilico *et al.*, 1957b). Castellani

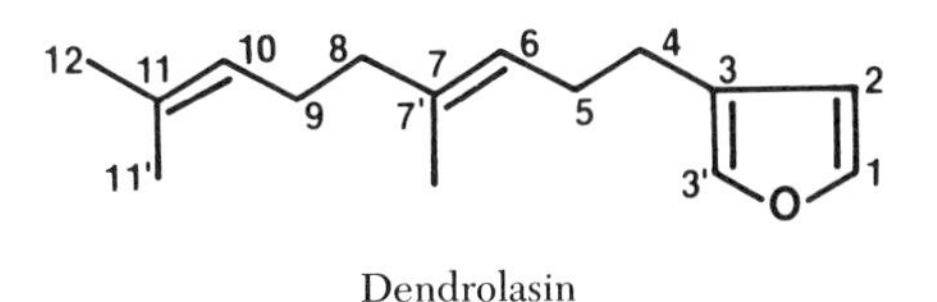

Dendrolasin

and Pavan (1966) isolated labeled dendrolasin after feeding [2-^{14}C]mevalonate to worker ants, thus demonstrating that this sesquiterpene could be biosynthesized from normal terpenoid precursors. Subsequently, Waldner *et al.* (1969) either fed or injected ants with several isoprenoid precursors and determined the labeling pattern in isolated dendrolasin by degradation. Administration of sodium [1-^{14}C]acetate, sodium [2-^{14}C]-mevalonate, [U-^{14}C]glucose, and [1-^{14}C]glucose resulted in a labeling pattern that demonstrated that insertions were spread all over the molecule in a manner consistent with the usual terpene biosynthetic pathway. Mevalonate was incorporated into carbon atoms 3, 4, 5, 6, 7, 7′, 8, 9, 10, 10′, 11, 11′, and 12; the carbon atoms of both acetate and glucose were similarly inserted. However, whereas these precursors were extensively incorporated into hydrocarbons, they were very poorly utilized for the biogenesis of dendrolasin. Thus, Waldner *et al.* (1969) suggest that these ants may normally utilize a precursor other than mevalonate to specifically produce dendrolasin.

F. Carboxylic Acids

The defensive secretions of arthropods are fortified with a wide variety of carboxylic acids that function as excellent agents of deterrence. Although the biosyntheses of only a few of these compounds have been investigated, it appears that they generally follow well-established metabolic pathways.

1. *Formic Acid*

Although formic acid was isolated from ants over 300 years ago (Wray, 1670), its biogenesis in these insects was investigated only recently. This compound appears to be an ubiquitous poison gland product of formicine ants and some species produce secretions containing up to 65% aqueous formic acid (Osman and Brander, 1961).

The biosynthesis of formic acid in the ant *Formica lugubris* was studied by feeding DL-[3-^{14}C]serine to workers and combusting the isolated acid to CO_2 which was assayed for radioactivity as a $BaCO_3$ derivative. Considerable incorporation of the label was present in the derivative, thus establishing that serine served as an excellent precursor for the synthesis of formic acid. Castellani *et al.* (1969) believe that formylglycine and formaldehyde are intermediates in the biogenesis of formic acid from serine.

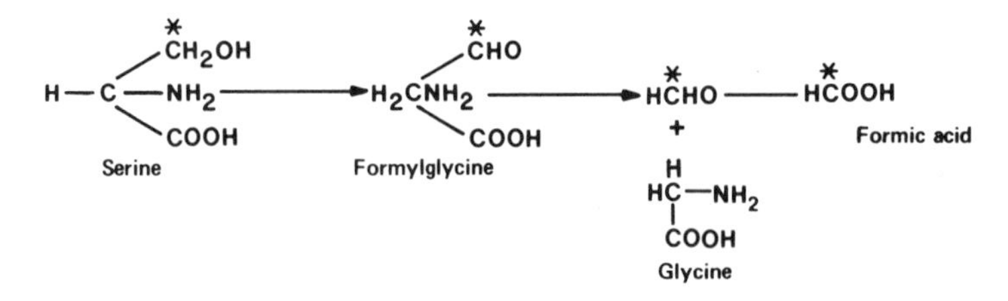

This conclusion is based on their reported isolation of labeled formylglycine from ants that had fed on a food source fortified with [3-^{14}C]serine. However, this proposed pathway differs considerably from the accepted metabolic scheme for the biosynthesis of formic acid in both plants and animals. Serine is known to generate one-carbon units through its initial conversion to methylene tetrahydrofolate which is ultimately converted to 10-formyltetrahydrofolate, the direct precursor of formic acid (Blakley, 1969).

More recently, Hefetz and Blum (1978a,b) studied the synthesis of formic acid in the poison glands of several species of formicine ants in the genera *Formica* and *Camponotus*. It was demonstrated that the carbon of formic acid could be derived from either the α or β carbons of serine or the α carbon of glycine. Thus serine can be converted to glycine, after a β-elimination reaction, and the latter can contribute its α carbon to formic acid after being totally degraded. In addition, it was established that the C-2 of the imidazole ring of histidine was also incorporated into formic acid, probably via *N*-formylglutamate and formiminotetrahydrofolate (Tabor *et al.*, 1951). Histidine was a less efficient precursor than serine, a fact that can be attributed both to the multiple steps (derivatives) involved in transferring this carbon and the relatively small pool of free histidine in the poison gland, It was concluded that any compound that contributes a C-1 fragment to tetrahydrofolate (H_4 folate) can serve as a potential precursor of formic acid.

Hefetz and Blum (1978a,b) further showed that for the synthesis of formic acid, there was an absolute requirement for H_4 folate in a cell-free system. Although the existence of the highly unstable intermediate 5,10-methylene H_4 folate could only be demonstrated indirectly, the synthesis of [5,10^{14}C]methenyl H_4 folate by poison glands incubated with [3-^{14}C]serine was unambiguously established. Since 5,10-methenyl H_4

folate can only arise from 5,10-methylene H_4 folate, the H_4 folate pathway is clearly operational in the formicine poison gland. Furthermore, since [^{14}C]formic acid is the final product when poison glands and [3-^{14}C]serine are incubated, it is obvious that 5,10-methenyl H_4 folate is converted to 10-formyl H_4 folate prior to the cleavage of the formic acid moiety (Fig. 16.5).

The proposed pathway for the biosynthesis of formic acid from serine was also supported by the results of enzymatic studies (Hefetz and Blum, 1978a,b). The four enzymes catalyzing the transfer of the β-carbon of serine to formic acid—serine hydroxymethyltransferase, 5,10-methylene H_4 folate dehydrogenase, 5,10-methylene H_4 folate cyclohydrolase, and 10-formyl H_4 folate synthetase—were all demonstrated to be present in high concentrations in the poison gland. With the exception of serine hydroxymethyltransferase, these enzymes were present in much higher concentrations in the poison gland than in any other tissue examined, and 10-formyl H_4 folate synthetase, which directly catalyzes the formation of formic acid, was the most active enzyme present.

The accumulation of a large amount of highly concentrated formic acid in the poison gland reservoir would seem to pose a particularly difficult regulatory problem because this acid is very reactive and is incorporated readily into more complex molecules. However, it is possible to readily accommodate this accumulation in terms of both the metabolic pathway

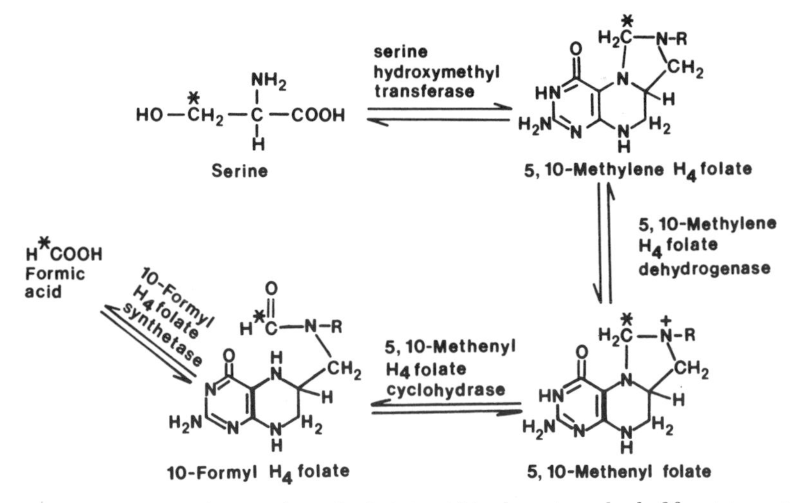

Fig. 16.5. Biosynthetic pathway for formic acid in the poison gland of formicine ants. (After Hefetz and Blum, 1978a,b.)

for formic acid and the morphological compartmentalization of the poison gland. Serine, after penetrating the glandular cells, is converted to formic acid via H_4 folate intermediates. The last step in this pathway generates H_4 folate and ATP, the latter compound probably being utilized to transport formic acid through the cell membrane to the gland lumen.

Accumulation of formic acid in the poison gland reservoir would create a back pressure which would ultimately fill the gland lumen with formic acid, thereby saturating the carrier system. Since the equilibrium between formic acid and 10-formyl H_4 folate is 1:20, favoring the latter, formic acid production would cease. As a consequence, the concentration of formic acid in the glandular cells will always be small and this acid can only be accumulated by being transferred to another compartment, i.e., the poison gland reservoir. Thus, the cytotoxic effects of formic acid would be avoided, since the concentrated acid would be stored in a reservoir containing an impermeable cuticular intima. Reactivation of this biosynthetic storage system would only occur if the ant ejected formic acid from the reservoir, being switched off again when the glandular reservoir was refilled.

2. *Methacrylic Acid*

Adults of the beetle *Carabus taedatus* are typical of many carabid species in producing a defensive secretion in the pygidial glands which is dominated by carboxylic acids. The secretion of *C. taedatus* consists primarily of methacrylic and ethacrylic acids with the former consisting of about 85% of the mixture (Benn *et al.*, 1973). Labeled methacrylic acid

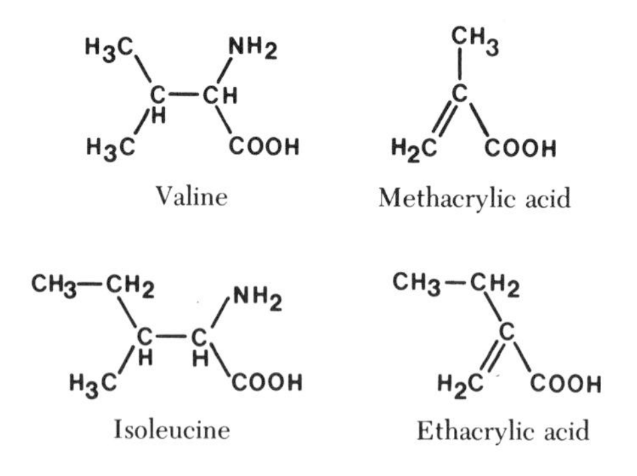

Valine Methacrylic acid
Isoleucine Ethacrylic acid

was isolated from the secretion of beetles previously injected with DL-[4-^{14}C]valine, thus establishing the role of this amino acid as a precursor.

Benn *et al.* (1973) propose that valine is converted to the acid in a manner analogous to that for the production of other short-chain unsaturated acids from isoleucine (Robinson *et al.*, 1956). This metabolic pathway was further rationalized with the demonstration that about half of the radioactivity present in the terminal methyl groups of valine was isolated in the methylene carbon of methacrylic acid. In all probability, ethacrylic acid is generated from isoleucine by utilizing a similar metabolic pathway.

3. *Isobutyric, α-Methylbutyric, and β-Hydroxybutyric Acids*

The osmeterial secretion of the ultimate larval instar of *Papilio aegus* contains isobutyric and α-methylbutyric acids (Seligman and Doy, 1972). On the other hand, the secretions of earlier instar larvae contains neither of these acids. Younger larvae produce β-hydroxybutyric acid in their osmeteria and only switch to the synthesis of the branched-chain acids subsequent to the final larval molt. Both [U-^{14}C]leucine and sodium [2-^{14}C]acetate were incorporated into β-hydroxybutyric acid but the corresponding β-keto acid could not be detected in the secretion (Seligman and Doy, 1972). Isobutyric acid was readily synthesized from either DL-[^{3}H]valine or [U-^{14}C]valine after injection into larvae or *in vitro* with isolated osmeteria (Seligman and Doy, 1973). [U-^{14}C]Isoleucine was converted into α-methylbutyric acid, thus demonstrating that both compounds in the ultimate larval secretion are derived from branched-chain amino acids, as they are in carabid beetles.

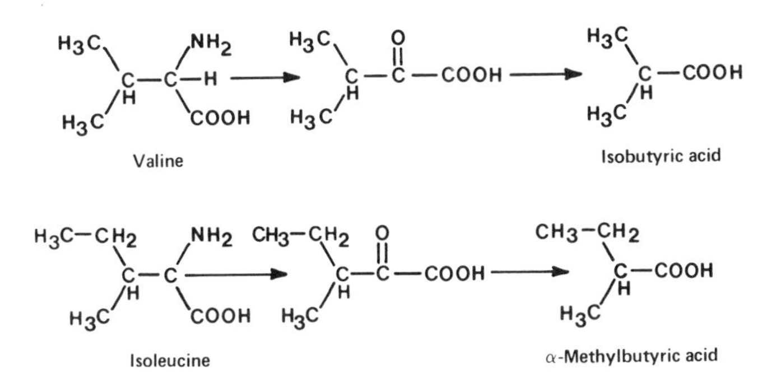

The osmeteria of last-instar larvae but not earlier ones, were rich in branched-chain amino acid transaminase activity (Seligman and Doy,

1973). α-Ketoglutarate was readily transaminated by osmeterial enzymes when either valine, isoleucine, or leucine were utilized as substrates. However, valine, the precursor of the major osmeterial compound—isobutyric acid—was transaminated less efficiently than leucine or isoleucine. On the other hand, valine was transaminated more efficiently than either leucine or isoleucine when pyruvic acid was present instead of α-ketoglutarate. Nevertheless, based on the great activity of the transaminating enzymes with leucine, the absence of its metabolite in the secretion of last-instar larvae is unexpected. The significance of different metabolic pathways in the osmeteria of various instars is equally enigmatic but it probably will be demonstrated to have an ecological correlate vis-à-vis predators when analyzed in detail. Recently, it has been reported that variation in osmeterial chemistry between younger and older larvae characterizes the defensive secretions of other *Papilio* species (Burger *et al.*, 1978; Honda, 1980).

G. Steroids

Arthropods lack a sterogenic pathway and thus are unable to biosynthesize cyclic triterpenes. Although squalene has been identified in insect extracts, it has been demonstrated that squalene synthetase is lacking in these insects and this acyclic triterpene must have originated from exogenous sources (Goodfellow *et al.*, 1972). Notwithstanding their inability to biosynthesize steroids, some species are able to produce a dazzling variety of these compounds, many of which are identical to well-known vertebrate hormones. Indeed, the prothoracic glands of dytiscid beetles constitute a virtual treasure trove of steroidal compounds that are generated from ingested cholesterol or related compounds.

The pygidial gland secretion of adults of the beetle *Acilius sulcatus* contains cortexone, 6,7-dehydrocortexone, cybisterone, 6,7-dihydrocybisterone, and 6,7-dehydroprogesterone (Schildknecht *et al.*, 1967b).

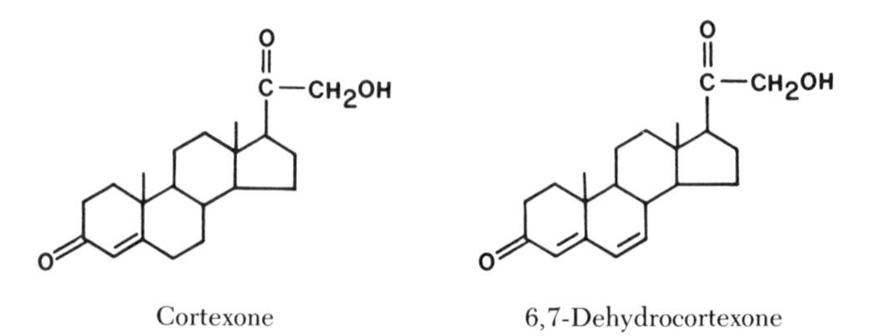

Cortexone 6,7-Dehydrocortexone

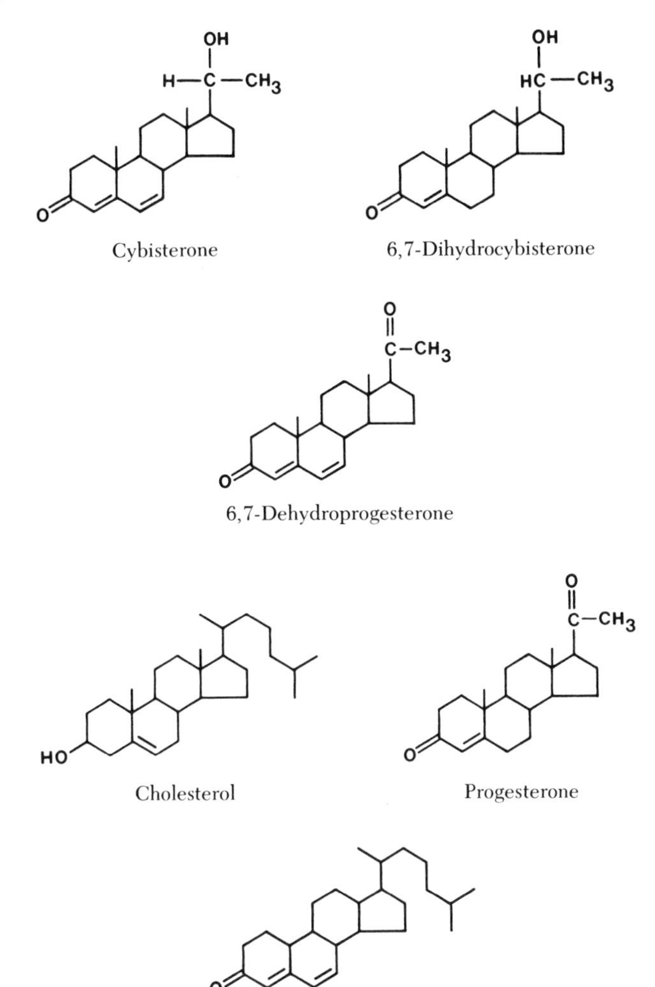

Cybisterone

6,7-Dihydrocybisterone

6,7-Dehydroprogesterone

Cholesterol

Progesterone

4,6-Cholastadien-3-one

As expected, injected [2-^{14}C]mevalonate was not incorporated into any of the *Acilius* steroids, whereas both [4-^{14}C]cholesterol and [4-^{14}C]progesterone served as precursors for these compounds (Schildknecht, 1970). [4-^{14}C]Cholesterol was utilized equally well for conversion into either enones or dienones, but [4-^{14}C]progesterone was very poorly incorporated into the dienones. Presumably, the biosynthetic pathways for these two groups of steroidal compounds branch prior to the biogenesis of progesterone, the apparent precursor of cortexone and dihydrocybisterone. The greater incorporation of 4,6-[^{14}C]cholestadien-3-one into the dienones than the enones would support the presence of separate pathways for the mono- and diunsaturated steroids (Schildknecht, 1970). Significantly, dehydroprogesterone was not labeled, and since this steroid appears to be biosynthesized by a different pathway from that of the other dienones, it is not converted to either dehydrocortexone or cybisterone.

H. Esters

Seligman (1972) has established that the alcoholic moiety of isopentyl acetate, an alarm pheromone and defensive product of the honey bee, is derived from a branched-chain amino acid. Injection of adult bee workers with [^{14}C]leucine resulted in the incorporation of labeling into this ester, thus demonstrating that amino acids are converted into isoalcohols by some insects.

Recently, Graham *et al.* (1979) demonstrated that the acetate moiety of decyl acetate, a Dufour's gland constituent of *Formica schaufussi*, is labeled when ant workers are fed [^{14}C]acetate.

I. Quinazolinones

The defensive exudate of the glomerid millipede *Glomeris marginata* is dominated by two quinazolinones, 1-methyl-2-ethyl-4(3*H*)-quinazolinone and 1,2-dimethyl-4(3*H*)-quanazolinone (Y. Meinwald *et al.*, 1966). These compounds have been assigned the trivial names homoglomerin and glomerin respectively (Schildknecht *et al.*, 1967c), and more recently, their biosyntheses have been investigated. [^{14}COOH]Anthranilic acid, which has been demonstrated to be a precursor of a plant-derived quinazolinone, was either injected or fed to millipedes and the specific activities of the two glomerins in the defensive secretion were subsequently determined (Schildknecht and Wenneis, 1967). Both quinazolinones were extensively labeled and the orally administered precursor was much more effectively incorporated than the injected acid. Alkaline hydrolysis of labeled homoglomerin yielded *N*-methyl anthranilic acid which possessed virtually the same level of specific activity as had been present in the quinazolinone. Thus, the carboxylic acid group in anthranilic acid provided the entire label for the biosynthesized homoglomerin.

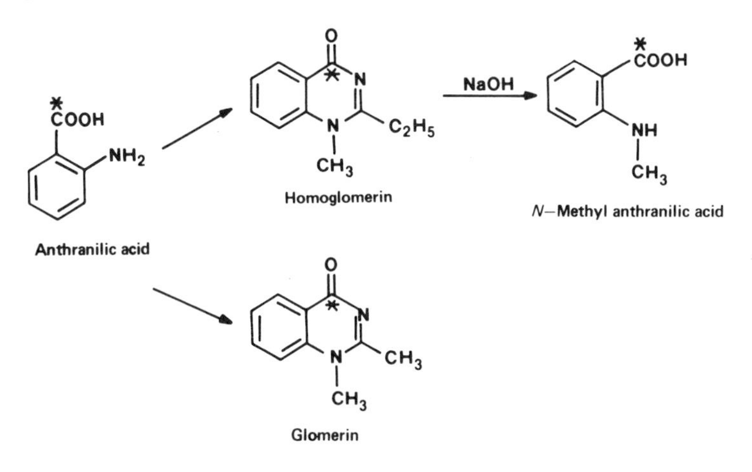

J. Pederin

The biosynthesis of pederin, a complex secondary amide produced by the beetle *Paederus fuscipes* (Cardani *et al.*, 1965a,b), has been investigated by feeding radioactive precursors and subsequently degrading the isolated toxin (Cardani *et al.*, 1973). Beetles were fed sodium [1-^{14}C]ace-

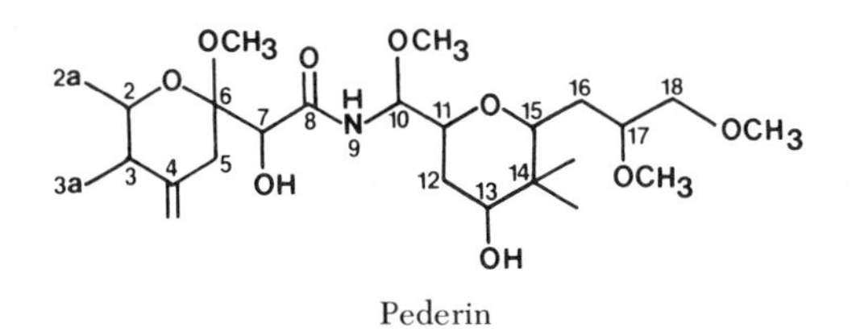

Pederin

tate, sodium [2-^{14}C]acetate, [1,2-^{14}C]glycine, and sodium [2-^{14}C]propionate and it was established that all precursors were incorporated into pederin. Specific degradation studies were undertaken with pederin derived from beetles which had ingested sodium [1-^{14}C]acetate or [2-^{14}C]acetate. The two acetates provided the carbon atoms at position 2 and 2a more efficiently than they did at carbon atoms 5, 6, 7, 8, 10, and 11. A large percentage of the incorporation also occurred in carbon atoms 10–18 after ingestion of the labeled acetates. Furthermore, carbon 2 of pederin was derived almost exclusively from the C-2 atom of acetate. The general labeling observed after administration of the acetates indicates that the pederin skeleton is derived primarily via polyketide biosynthesis (Cardani *et al.*, 1973). This conclusion was supported by the incorporation of sodium [2-^{14}C]propionate into pederin.

K. Peptides and Proteins

The venom of the honey bee (*Apis mellifera*) constitutes the only proteinaceous defensive secretion of arthropods for which the structures of most of the individual constituents have been determined in detail (Habermann, 1972; Gauldie *et al.*, 1976, 1978). A toxic peptide, melittin, comprises about 50% of the dried venom and the complete amino acid sequence of this venomous constituent has been determined (Habermann and Jentsch, 1967). Melittin, which is a highly hemolytic peptide containing 26 amino acid residues, is also accompanied by a peptide that is formylated at the amino end, and "formyl-melittin" was originally suspected of being the precursor of melittin (Kreil and Kreil-Kiss, 1967). However, recent *in vivo* studies on melittin biosynthesis have indicated that the partial formylation of melittin was a late step in the biogenesis of bee venom and as a consequence, "formyl-melittin" did not seem like a likely precursor of the toxic peptide (Kreil and Bachmayer, 1971). Indeed, Kreil and Bachmayer have detected a melittin-like peptide whose biosynthesis precedes that of melittin and this compound, designated as promelittin, is considered to be the precursor of the main toxin in the venom.

By feeding radioactive leucine, isoleucine, lysine, and valine—four of the main amino acids in melittin—to adult honey bees, it was established that significant incorporation into the peptide occurred (Kreil and Bachmayer, 1971). Furthermore, these amino acids were also extensively incorporated into promelittin, which reached its peak production before it was converted to melittin in significant amounts. Enzymatic degradation studies indicated that promelittin contained all the amino acids present in melittin, and based on analyses of the fragments produced by proteolytic enzymes a structural homology between promelittin and melittin was demonstrated. Kreil and Bachmayer (1971) established that promelittin differs from melittin at the amino end, the former peptide containing an acidic fragment which is lacking in melittin. Since the toxicity of melittin appears to be related to the asymmetric distribution of apolar and basic amino acids, the latter being concentrated mostly at the C-terminal end, the presence of acid residues at the amino end may result in a peptide of neglible toxicity. Thus, modifications at the N-terminus may ensure that the ultimately toxic peptide is never present on the ribosomes during the stepwise growth of the peptide chain (Kreil and Bachmayer, 1971).

Promelittin Glu-Pro-Glu-Pro-Asp-Pro-Glu-Ala-Gly-Ile-Gly-Ala-Val-Leu-Lys-Val-Leu-Thr-Thr-Gly-Leu-Pro-Ala-Leu-Ile-Ser-Trp-Ile-Lys-Arg-Lys-Arg-Gln-Gln-NH_2

Melittin Gly-Ile-Gly-Ala-Val-Leu-Lys-Val-Leu-Thr-Thr-Gly-Leu-Pro-Ala-Leu-Ile-Ser-Trp-Ile-Lys-Arg-Lys-Arg-Gln-Gln-NH_2

Administration of a variety of radioactive amino acids to bee workers resulted in the production of labeled promelittin whose structure was ultimately determined after analyses of the fragments produced by proteolytic enzymes and acidic hydrolyses (Kreil, 1973a). Promelittin contains the entire amino acid sequence of melittin, but is inhomogenous at the N-terminus. The main promelittin component contains eight additional acidic amino acids and proline residues, and additional components containing two less or one more amino acid have also been detected. In all probability, these different peptidic species represent intermediates in the activation of promelittin to melittin.

Significantly, pronase was the only protease that would attack the acidic amino acids and proline sequence at the N-terminus of promelittin (Kreil, 1973a). On the other hand, melittin and the corresponding part of the promelittin sequence were both highly susceptible to cleavage by a wide variety of common endopeptidases. Therefore, promelittin activation should only be catalyzed by a highly specific enzyme(s) which cannot hydrolyze any of the susceptible bonds in the melittin chain.

Bachmayer *et al.* (1972) have investigated the rates of synthesis of promelittin and melittin in both queen and worker bees by analyzing these peptides after feeding [4,5-^{3}H]leucine. Promelittin synthesis was detectable in newly emerged workers and reached its peak at about 10 days of age. Melittin, on the other hand, could not be detected in the venom of workers during the first two days after emergence but its rate of production from promelittin increased rapidly thereafter to reach a peak in 10-day-old bees. The rate of conversion of promelittin to melittin increased steadily up to 20 days after emergence; negligible synthesis of either peptide was detectable in aged winter bees.

In queen bees, on the other hand, the rate of conversion of promelittin to melittin was at a maximum at the time of emergence (Bachmayer *et al.*, 1972). Thus, queens, unlike workers, appear to possess a functional venom at the time of eclosion, an adaptation that may be correlated with the fatal encounters which are commonplace between newly emerged sister queens in the milieu of the hive. The presence of very different control systems for the rate of synthesis and conversion of the venom peptides in queens and workers is thus well established.

The likelihood that promelittin is a primary gene product with a small messenger RNA seems especially great since this peptide contains only 35 residues. Furthermore, based on the composition of the venom (Gauldie *et al.*, 1976) and the observed labeling patterns of melittin (Kreil, 1973a), it seems likely that melittin messenger RNA should be the predominant messenger RNA in the venom. Indeed, Kindås-Mügge *et al.* (1974, 1976) have demonstrated that melittin messenger RNA from the poison glands of queen bees is the major poly(A)-containing RNA present in venom glands. This RNA is a rather small but otherwise typical eukaryotic messenger that constitutes about 1–2% of 20 μg of total RNA isolated from a poison gland.

The translation of melittin messenger RNA has been analyzed employing three different systems. Injection of frog oocytes with poison gland RNA resulted in the production of a polypeptide which was very similar to promelittin (Kindås-Mügge *et al.*, 1974). Analysis of this oocytic product demonstrated that more than two-thirds of its sequence was identical to that of promelittin. The oocytic polypeptide contained a frayed amino terminus like promelittin, but differed at the carboxyl end from the poison gland product.

In order to eliminate or reduce posttranslational reactions, Suchanek *et al.* (1975) studied the translation of melittin messenger RNA employing a cell-free mammalian system. A polypeptide, larger than promelittin, was detected, which contained the sequences characteristic of melittin and promelittin. This compound, prepromelittin, contained the majority of the extra amino acids linked to the amino terminus of promelittin. Prepromelittin was also produced by a wheat germ system (Suchanek and Kreil, 1977), and was demonstrated to contain methionine as the amino terminal residue. If methionine is the initiating residue, then prepromelittin, containing more than 30 residues before the promelittin sequence starts, would constitute the primary product of translation.

Kreil *et al.* (1977) have analyzed the synthesis of melittin in considerable detail. The first step in the reaction sequence involves the cleavage of about 30 residues from the amino end to yield promelittin, with an extended carboxyl end. The amino-terminal region of prepromelittin may act as a "signal" that permits the vectorial growth of the polypeptide chain through the membrane of the endoplasmic reticulum, thus ensuring that the product is transported.

The formation of the carboxy terminal amide appears to constitute the second step leading to the formation of melittin (Kreil *et al.*, 1977). The conversion of promelittin to melittin probably represents the final step in the biosynthesis of melittin and must take place outside of the poison gland cells because of the pronounced lytic effects of the end product. More recently, Kaschnitz and Kreil (1978) demonstrated that prepromelittin, obtained by translation of melittin mRNA in a cell-free system from wheat germ, was converted to promelittin by a subcellular fraction of rat liver. Processing activity was present in the microsomal and mitochondrial/lysosome fractions and was membrane bound. Kreil and co-workers have thus provided persuasive evidence for regarding melittin, a simple polypeptide, as a paradigm for studying different posttranslational reactions of biologically important molecules.

Bee venom phospholipase A, in contrast to the vertebrate enzyme, is not reported to have an enzymatically inactive precursor (Jentsch and Dielenberg, 1972).

L. Alkyl Sulfides

Alkyl sulfides are only known as animal natural products because of the occurrence of compounds including dimethyldisulfide and dimethyltrisulfide in the mandibular gland secretions of a few ponerine ant species such as *Paltothyreus tarsatus* (Casnati *et al.*, 1967). The biosynthesis of these two sulfides was studied by Crewe and Ross (1975a, b) who demonstrated that methionine served as a precursor for both compounds. Dimethyldisulfide was demonstrated to be synthesized from the intact methanethiol group of methionine, probably in much the same way as it is in microorganisms (Crewe and Ross, 1975b). On the other hand, although the CH_3—S—moiety appears to be also incorporated intact into dimethyltrisulfide, the origin of the third sulfur atom is unknown.

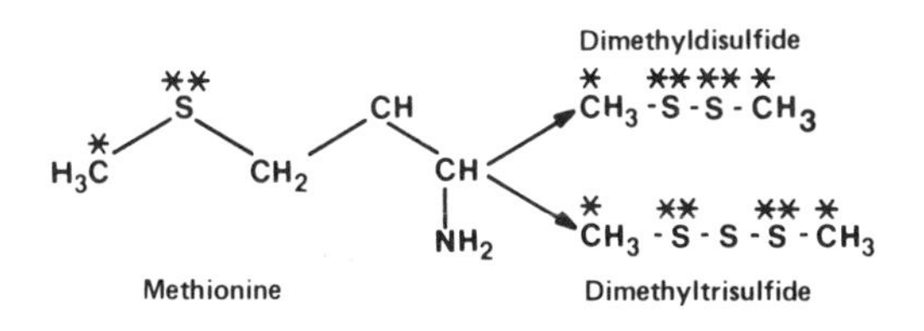

M. Coccinelline

Tursch *et al.* (1975) studied the biosynthesis of the alkaloid coccinelline after feeding sodium [1-^{14}C]acetate and sodium [2-^{14}C]acetate to adults of the coccinellid *Coccinella septempunctata*. Both precursors were incorporated into coccinelline, and after derivatization and oxidation to acetic acid (corresponding to carbon atoms C-10 and C-2), it was found that 16% of the original activity was present. These results can be readily reconciled with the formation of the alkaloid by linear combination of seven acetate units. The formation of a hypothetical intermediated by decarboxylation of the original condensation product would permit the facile formation of precoccinelline as a logical intermediate in the synthesis of coccinelline. This polyacetate pathway could also account for the biosynthesis of the other coccinellid alkaloids as well (Tursch *et al.*, 1975).

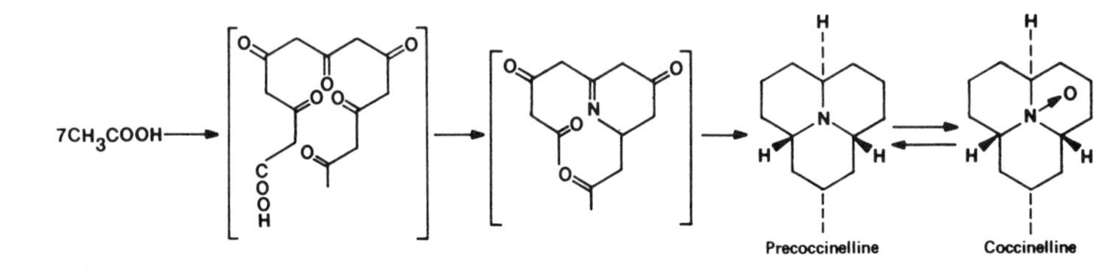

N. Hydrogen Cyanide*

Hydrogen cyanide, which is generated from aromatic glucosides by millipedes (see Section C), is produced from two aliphatic glucosides in females of the zygaenid moth *Zygaena filipendulae* (Davis and Nahrstedt, 1979). Both linamarin and lotaustralin have been identified in extracts of gravid females; hydrogen cyanide appears to be produced enzymatically from these two glucosides. Although these cyanogens are present in some species of plants fed upon by the larvae, it appears that the glucosides are produced *de novo* since hydrogen cyanide is generated from moths whose larvae had been reared on acyanogenic plants.

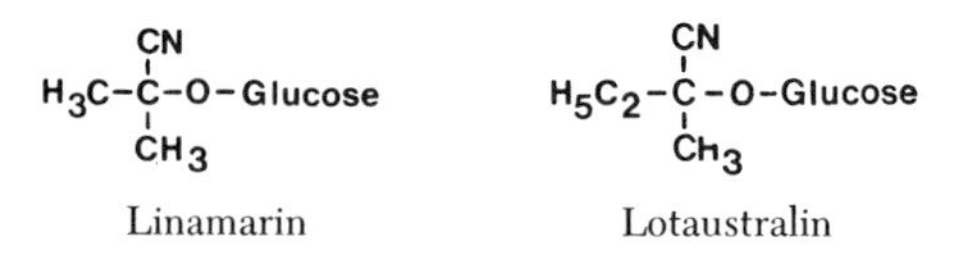

Linamarin Lotaustralin

II. SUGGESTIVE BIOSYNTHETIC INTERRELATIONSHIPS

Arthropod defensive secretions often contain constituents that appear to be related. Although biogenetic investigations have not been under-

*See also Section I,C, Aromatic Compounds.

taken in order to substantiate most of these implied pathways, the presence of these end products and their potential precursors should offer cogent grounds for undertaking future biosynthetic studies.

A. Hydrocarbons

Cavill and Williams (1967) reported that the Dufour's gland secretion of the ant *Myrmecia gulosa* was dominated by *cis*-8-heptadecene, pentadecane, and heptadecane. These compounds correspond both in structure and proportion to the main fatty acids identified in this ant—oleic, palmitic, and stearic (Cavill *et al.*, 1970)—and the possible role of these fatty acids as precursors of the Dufour's gland hydrocarbons has been suggested (Cavill, 1970). Similarly, Morgan and Wadhams (1972) also reported a congruency in the proportions of the same three hydrocarbons and fatty acids in the ant *Myrmica rubra*.

1-Undecene has been identified in the pygidial gland secretions of the staphylinids *Bledius mandibularis* and *B. spectabilis*, along with γ-dodecalactone (Wheeler *et al.*, 1972a). It has been suggested that both 1-undecene and the lactone could be formed from a β,γ-unsaturated C_{12} acid, by decarboxylation, or addition across the double bond, respectively.

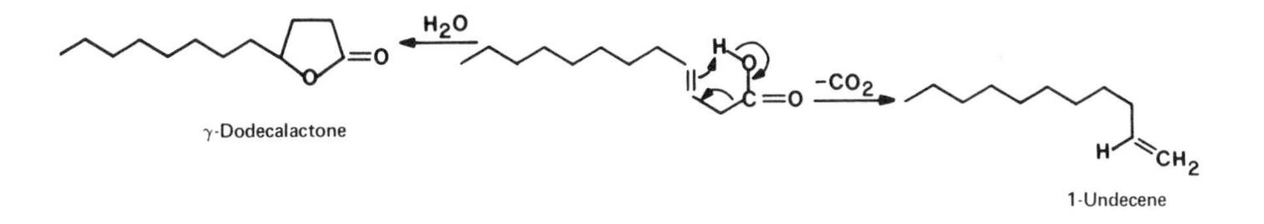

B. Aliphatic Aldehydes

Cockroaches in the genus *Platyzosteria* discharge defensive secretions which are dominated by branched aldehydes and alcohols of the same series (Waterhouse and Wallbank, 1967). The secretions of four species of *Platyzosteria* are dominated by 2-methylenebutanal and 2-methylenebutan-1-ol, and in addition, contain minor amounts of 2-methylbutanal and 2-methylbutan-1-ol. Since these compounds are analogous by reduction of the double bond or oxidation or reduction of the terminal group, Waterhouse and Wallbank suggest that their biosyntheses are clearly linked.

C. Alcohols and Ketones or Aldehydes

The exocrine products of many arthropods, particularly those of ants and bees, are fortified with both aliphatic carbonyl compounds and their corresponding alcohols. It is tempting to suggest that ketones, which are generally the dominant compounds in the secretions, are derived from the alcohols, although ketonic reduction to the carbinols is equally possible. At any rate, the following ketones or aldehydes, along with their corresponding alcohols, have been identified in the defensive secretions of ants and bees: 4-methyl-3-heptanone (*Pogonomyrmex* spp., McGurk *et al.*, 1966), 3-octanone (*Myrmica* spp., Crewe and Blum, 1970a), 6-methyl-3-octanone (*Crematogaster* spp., Crewe *et al.*, 1972), 2-hexenal (*Crematogaster* spp., Crewe *et al.*, 1972), and C_7–C_{17} methyl ketones (*Trigona* spp., Blum, 1970).

D. Lactones

4-Hydroxyoctadec-9-enolide, a major component in the Dufour's gland secretion of the ant *Lasius flavus*, is almost certainly derived from the concomitant 4-hydroxyoctadec-9-enoic acid (Bergström and Löfqvist, 1970).

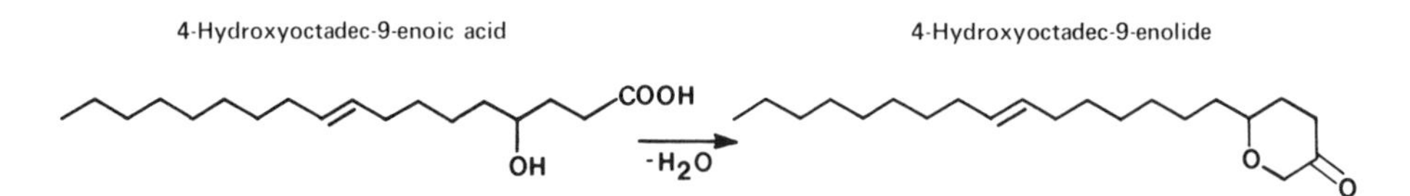

E. Octahydroindolizine

Ritter *et al.* (1973) have identified 5-methyl-3-butyloctahydroindolizine as a poison gland product of the ant *Monomorium pharaonis*. Subsequently, Talman *et al.* (1974) isolated 2-pentyl-5-butylpyrrolidine from the poison gland secretion. The pyrrolidine would appear to constitute a key precursor for the indolizine.

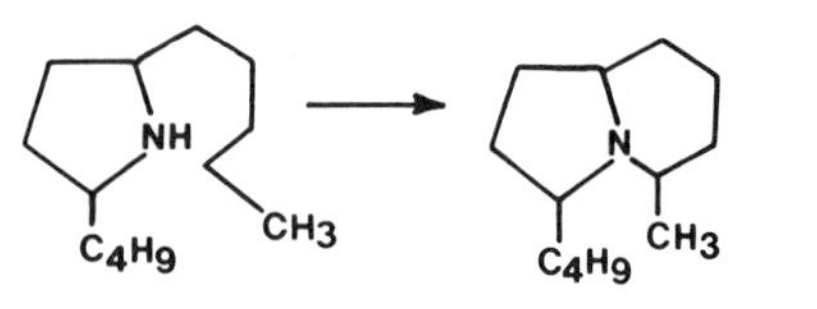

F. Romallenone

One of the major constituents in the spiracular defensive secretion of the grasshopper *Romalea microptera* is the allenic sesquiterpenoid romallenone (Meinwald *et al.*, 1968a). This compound can be produced by photosensitized oxygenation of an allenic ketodial produced from β-ional and it has been suggested that romallenone is derived from a comparable metabolic pathway of carotenoid metabolism (Isoe *et al.*, 1971).

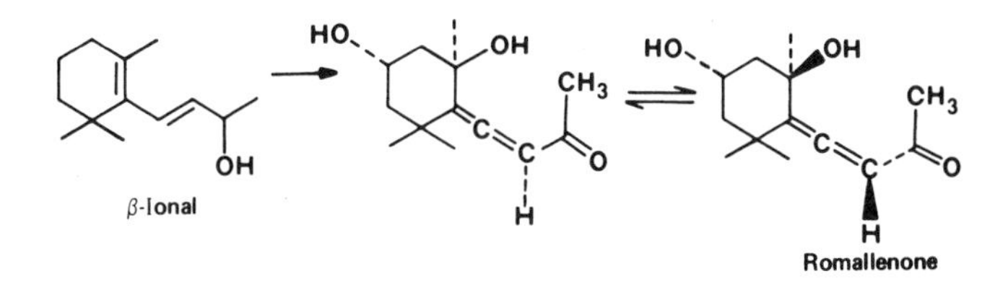

G. Norsesquiterpenes

The pygidial glands of gyrinid beetles synthesize an unsual aldehyde, gyrinidal, which is a potent repellent for fish (Schildknecht *et al.*, 1972b; Meinwald *et al.*, 1972). Subsequently, Wheeler *et al.* (1972b) identified gyrinidone, a cyclopentanoid norsesquiterpene, from the secretion of another gyrinid species and noted that gyrinidone could be the precursor of gyrinidal by addition across the conjugated system. Indeed, Miller *et al.* (1975) have identified both compounds in the secretion of two gyrinid species along with two additional norsesquiterpenes, gyrinidione and isogyrinidal. These authors suggest that all four compounds could be derived from farnesal.

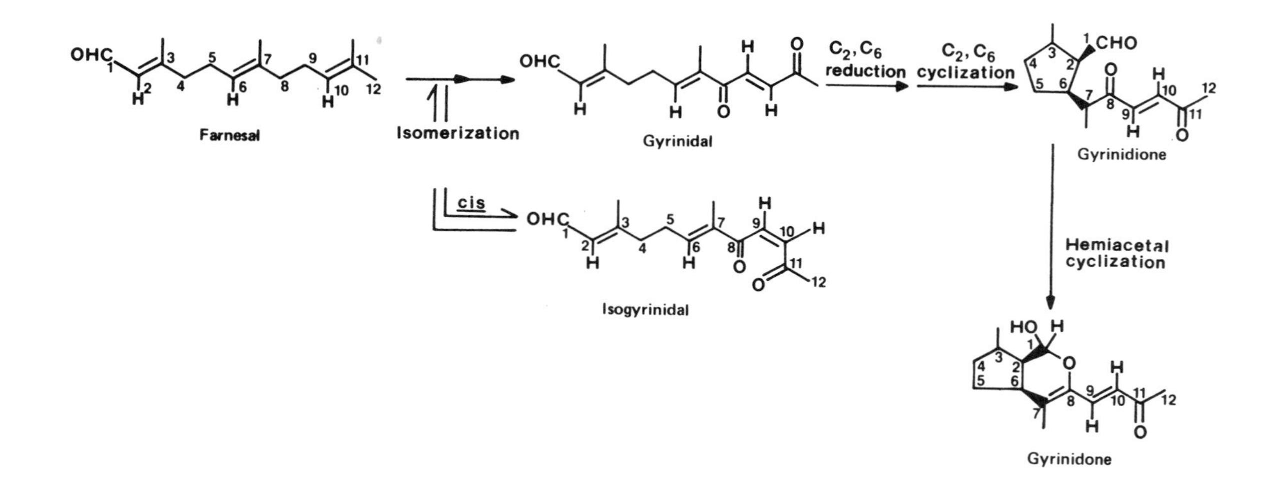
Farnesal
Isomerization
cis
Gyrinidal
Isogyrinidal
C_2, C_6 reduction
C_2, C_6 cyclization
Gyrinidione
Hemiacetal cyclization
Gyrinidone

Chapter 17

Insects and Toxic Plants

Plants are the natural products chemists *par excellence*, and an extraordinarily wide variety of so-called secondary plant substances are now known to be characteristic of members of many taxa. One of the multitudinous roles of these chemically diverse natural products is evidently to function as an antipredator device (Fraenkel, 1959; Ehrlich and Raven, 1965; Levin, 1976) by rendering the plant either unattractive, indigestible, or in some cases, toxic to herbivores. Since however, a single potent defense can be, and has been, overcome by the evolution of a better offense, many, if not all, of the so-called toxic plants are fed upon by specialized herbivores, whose diets are in many cases limited to plant species containing common or closely related natural products (Ehrlich and Raven, 1965). Obviously the term toxic is relative at best since selected insect species have evolved the physiological capability not only to cope with the presence of these natural products in their diets, but to sequester these compounds in their own bodies as well. In terms of natural products chemistry, these insects have become an extension of their host plants, utilizing their hosts' metabolic products as deterrents to their own potential predators.

In a penetrating review of insects which feed on toxic plants, Rothschild (1972) lists species in at least six orders that sequester ingested plant-derived compounds. However, while the sequestration of natural products (allomones) by insects is well established, the physiological bases for the animal's ability to persevere while ingesting the "good with the bad" are largely *terra incognita* (Duffey, 1980). The same can be said for the insect's ability to incorporate the ingested natural products into its own body, often retaining them as an adult after they have been sequestered as a larva. Thus, while intensive research in the past two decades has documented what many early naturalists have suggested—that many insects store plant products known to be toxic to animals—we know virtually nothing about the physiological mechanisms responsible for this sequestration.

Before describing the arthropods which have been proved to sequester plant natural products, it seems appropriate to examine some of the known mechanisms which have been implicated for enabling phytophagous insects to deal with toxic compounds that occur in their food. In so doing, a few evolutionary strategies utilized by these animals to overcome some of the plants' primary antiherbivorous mechanisms can at least be exposed.

I. DETOXICATION OR SELECTIVE ELIMINATION OF TOXIC NATURAL PRODUCTS

Insects have evolved a diversity of strategies for tolerating the presence of plant toxins in their diets. It is almost certain that many species utilize multiple mechanisms for preventing intoxication from the ingestion of potential toxins. Furthermore, these devices for avoiding the toxic effects of plant allomones are probably not mutually exclusive; more than one may be utilized simultaneously.

Sequestration obviously constitutes one effective mechanism for removing high concentrations of ingested plant toxins from circulation in the bodies of a wide range of insect species. However, there is simply no way of predicting whether a species will sequester ingested allomones in addition to excreting and/or detoxifying them. Rothschild (1972) lists nearly as many species in as many orders that do not sequester plant toxins as those that do incorporate these toxins into their body tissues. Often species in the same genus (e.g., *Papilio*) are either effective sequestrators of natural products or show no evidence of the capacity to store these compounds. Thus, if we are to ultimately comprehend how an insect copes with toxic plants, it appears that we will have to analyze each insect and its associated host plant as an individual evolutionary case.

A. Mixed-Function Oxidases

Microsomal mixed-function oxidases (MFO) promote the metabolic attack on a dazzling variety of organic compounds (Gilette *et al.*, 1969). A diversity of compounds is epoxidized, dealkylated, hydroxylated, and oxidized, and these metabolic alterations often result in the modified compounds being detoxified and rapidly excreted. These MFO systems may play a significant role in enabling insects to deal with a multitude of potentially toxic compounds that fortify the plant parts upon which they feed. However, it should be kept in mind that MFO can also convert compounds into more toxic forms (hypertoxication), and it is likely that their primary function is the conversion of lipophilic compounds into water-soluble metabolites that can be readily excreted, as first suggested by Brodie *et al.* (1958).

In a study of the epoxidase activities present in the midgut tissues of species of lepidopterous larvae from 11 families, it was demonstrated that MFO activity was correlated with the degree of herbivory of the species (Krieger *et al.*, 1971). Utilizing the efficacy of epoxidative conversion of aldrin to dieldrin as a criterion of enzymatic activity, it was established that monophagous species possessed about one-fifteenth the activity of polyphagous forms (20 vs. 294). Oligophagous larvae, which are intermediate in their herbivorous propensities between monophagous and polyphagous species, also possessed intermediate levels of MFO.

Recently, Brattsten *et al.* (1977) demonstrated that a wide variety of plant allomones are potent inducers of MFO in larvae of the armyworm *Spodoptera eridania*, a polyphagous insect. Monoterpene hydrocarbons such as β-pinene and myrcene were outstanding MFO inducers and the degree of induction was determined to be dose dependent. Induction was shown to be a graded response to increasing concentrations of secondary plant substances, and it occurred in significant amounts within 30 minutes after ingestion of compounds such as β-pinene. Furthermore, larvae with an induced MFO system were better protected from toxins such as nicotine than were uninduced control larvae. Isolation of microsomes from the midgut of induced larvae demonstrated that both MFO activity and cytochrome *P*-450 titer were considerably higher than those of uninduced control larvae (Brattsten *et al.*, 1977). However, the recent demonstration that MFO levels in an insect may vary considerably depending on what food plant had been fed upon (Brattsten, 1979) indicates that absolute correlations between the degree of feeding specialization and the activity levels of these enzymes may not be as firm as previously believed.

MFO may comprise one of the main classes of enzymes responsible for detoxifying many of the allomones that are ingested by herbivores. However, the fate of these ingested natural products may sometimes represent an evolutionary compromise between their toxicity to the herbivore and their value as deterrents to the predators of these phytophagous insects. An insect whose gut is fortified with plant toxins may be distasteful or even toxic to predators as a consequence, and the rapid conversion of these plant compounds to nontoxic metabolites could thus be counteradaptive. Furthermore, as Eisner (1970) has emphasized, these natural products can function as excellent deterrents to predators when they are enterically discharged from either the mouth or anus. Therefore, an insect, independent of its own repellents or sequestered plant toxins, may be rendered distasteful as long as it can maintain an efficacious level of plant-derived allomones in its alimentary canal. In addition, the possibility that gut enzymes can convert these natural products to more biologically active compounds should not be overlooked, especially in view of the wide range of chemical modifications manifested by the MFO system. The increase in biological activity resulting from the oxidation of thioethers and epoxidation of selected olefins by these enzymes (Krieger *et al.*, 1971) clearly demonstrates the hypertoxic potential of these metabolic alterations in insects.

Although the MFO systems of insects appear to be of importance in helping these animals to cope with biologically active compounds present in ingested plant parts, these enzymes should not be regarded as a panacea for insects feeding on toxic plants. Polyphagy has its limits and the presence of high titers of MFO enzymes in a species does not guarantee that it can feed with impunity on an infinite variety of plant toxins. Similarly, although these enzymes have been demonstrated to be important in the detoxication of a wide variety of insecticides, it is nevertheless true that this protection is far from complete, since insects continue to be killed by these toxicants.

B. Metabolism of Plant Allomones by Insects

Limited evidence indicates that insects metabolize a variety of plant natural products to compounds which are either excreted or ultimately sequestered. Unfortunately, the physiological significance of these metabolic alterations is obscure, except in cases where it has been established that the metabolites represent products of detoxication. However, it is probable that in ingesting a veritable potpourri of plant toxins, many insects will simultaneously detoxify and excrete some, while at the same time sequestering others which may or may not be metabolically altered in the process. Therefore, a diversity of physiological and metabolic strategies will probably be discovered from in-depth analyses of how an insect treats the allomonal concomitants that accompany the nutrients it processes in its digestive tract.

Nicotine, an alkaloid produced by *Nicotiana tabacum,* is rapidly converted to cotinine by some insects that normally feed on tobacco leaves and some that do not (Guthrie *et al.*, 1957; Self *et al.*, 1964a). Cotinine, which is virtually nontoxic to insects, is the major metabolite produced by larvae of the tobacco wireworm (*Conoderus vespertinus*), the cigarette beetle (*Lasioderma serricorne*), and the differential grasshopper (*Melanoplus differentialis*) (Self *et al.*, 1964a). In addition to cotinine, small amounts of unidentified metabolites are produced by larvae of *L. serricorne* and *M. differentialis.* Cotinine may arise from oxynicotine (Boit, 1961), a compound that is easily formed during the fermentation of nicotine alkaloids (Frankenburg and Gottscho, 1955). While it has not been established if any of these nicotine-feeding insects sequester nicotine, it is highly unlikely that this toxic alkaloid is readily stored by these invertebrates.

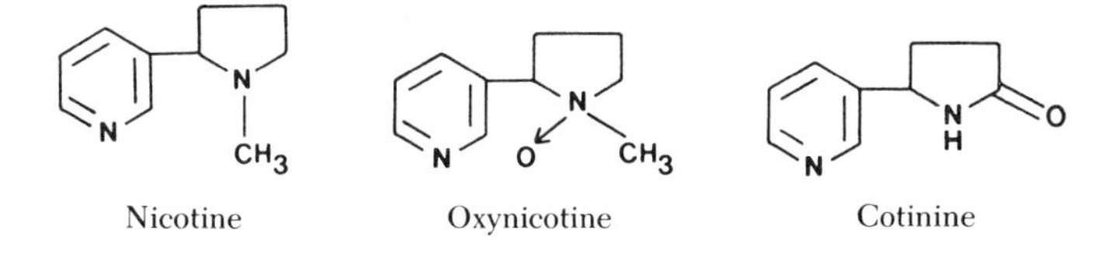

Nicotine Oxynicotine Cotinine

Cotinine is also produced by the American cockroach (*Periplaneta americana*), German cockroach (*Blatella germanica*), and the housefly (*Musca domestica*), three species that do not feed on tobacco (Guthrie *et al.*, 1957; Self *et al.*, 1964a). In contrast, the southern armyworm (*Spodoptera* (=*Prodenia*) *eridania*) converted ingested nicotine to a large series of unidentified metabolites (Guthrie *et al.*, 1957). On the other hand, no evidence was obtained for the metabolism of any of the tobacco alkaloids ingested by three tobacco pests—the tobacco budworm (*Heliothis virescens*), the cabbage looper (*Trichoplusia ni*), and the tobacco hornworm (*Manduca* (=*Protoparce*) *sexta*) (Self *et al.*, 1964a,b). These three species all rapidly excreted nicotine, although in the case of *P. sexta,* a small percentage of the ingested alkaloid was detected in the blood. While nicotine was essentially absent from the blood six hours after ingestion, the possibility that this compound was sequestered was not examined. Self *et al.* (1964b) suggest that intoxication from absorbed nicotine is avoided by preventing a toxic dose from accumulating at the unidentified site of action.

Teas *et al.* (1966) demonstrated that larvae of the tiger moth *Seirarctia echo* accumulated cycasin in the blood after ingestion of cycad leaves. Subsequently, Teas (1967) reported that when the aglycone of cycasin, methylazoxymethanol, was fed to larvae of *S. echo*, they converted it to the β-glycoside cycasin which was sequestered. Cycasin is relatively nontoxic compared to the aglycone. Teas (1967) noted that β-glucosidase,

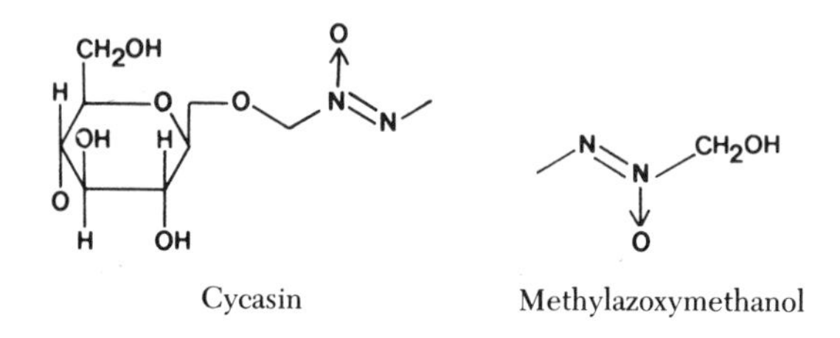

Cycasin Methylazoxymethanol

the enzyme responsible for cycasin synthesis and hydrolysis, was limited to the gut of *S. echo* larvae. On the other hand, cycasin was present primarily in the blood and Malpighian tubules, tissues that lacked the glucosidase capable of generating the toxic aglycone. That cycasin is the preferred storage form for azoxyglucosides is indicated by the fact that when larvae were fed on leaves of the cycad *Zamia floridana*, which contains an azoxyglucoside different from cycasin, they hydrolyzed the former and converted it to the latter which was then sequestered.

Another glycoside, sinigrin, may also manifest its toxicity after hydrolysis to its aglycone (Erickson and Feeny, 1974). Sinigrin or allylglucosinolate, which is found in many cruciferous plants, yields the mustard oil allylisothiocyanate upon hydrolysis, and the latter is known to be toxic to insects. Blau *et al.* (1978) demonstrated that larvae of the swallowtail *Papilio polyxenes* experienced acute toxicity when fed on their normal umbelliferous host plant to which sinigrin had been systemically introduced. The aglycone, allylisothiocyanate, was present in all larval body tissues that were analyzed.

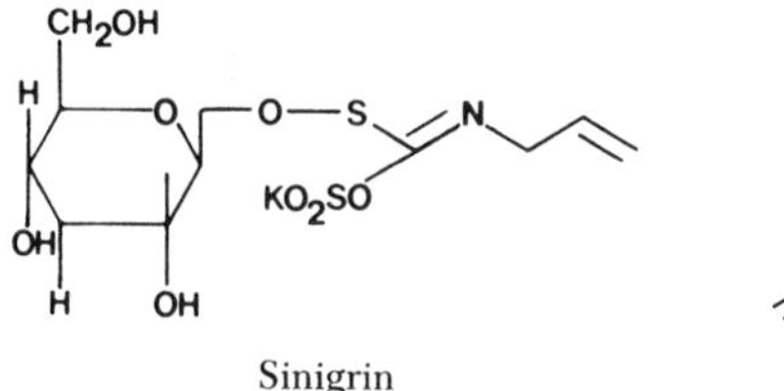

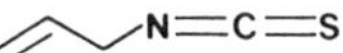

Sinigrin Allylisothiocyanate

Aplin and Rothschild (1972) reported that larvae of the arctiid moth *Tyria jacobaeae* sequestered some of the alkaloids present in their food plant, the ragwort *Senecio jacobaeae*. However, whereas seneciphylline is the major alkaloid sequestered, this compound is present as the *N*-oxide in the ragwort. Apparently, efficient sequestration by the larvae

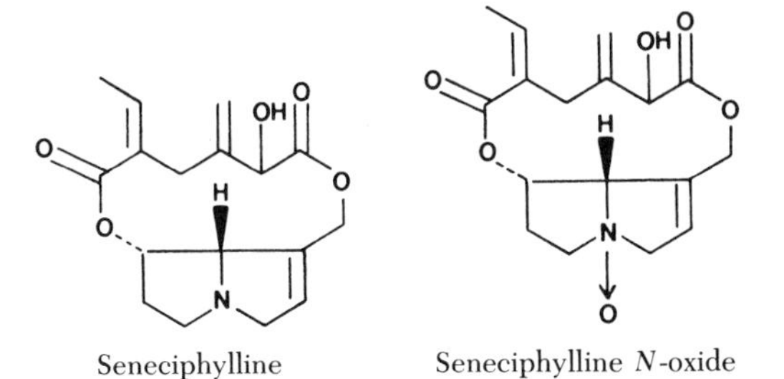

Seneciphylline   Seneciphylline *N*-oxide

requires that the water-soluble *N*-oxide of seneciphylline be converted to the lipophilic free alkaloid. When fed on groundsel, *Senecio vulgaris*, larvae, pupae, and adults of the arctiids *T. jacobaeae* and *Arctia caja* sequester a metabolite with the empirical formula $C_{15}H_{23}NO_5$ (Aplin *et al.*, 1968). This compound appears to contain a modified ester substituent attached to a retronecine nucleus and is not found in the host plant.

Von Euw *et al.* (1967) suggest that the grasshopper *Poekilocerus bufonius* metabolizes ingested cardenolides and Rothschild and Aplin (1971) similarly believe that larvae of *A. caja* convert cardenolides of *Digitalis* to aglycones in the gut. The propensity of the lygaeid *Oncopeltus fasciatus* for sequestering predominantly polar cardenolides in the dorsolateral spaces in the thorax and abdomen is reflected in its ability to metabolize nonpolar to polar cardenolides (Duffey and Scudder, 1974; Duffey and Blum, 1976; Duffey *et al.*, 1978). When milkweed seeds (*Asclepias syriaca*) were treated with two nonasclepiadaceous cardenolides, ouabain and digitoxin, the immediate fates of the two glycosides were quite different. Ouabain rapidly diffused into the blood and accumulated in the dorsolateral space fluid. On the other hand, digitoxin, although it rapidly entered the blood, was slowly sequestered in the dorsolateral spaces (Duffey and Scudder, 1972; Duffey and Blum, 1976; Duffey *et al.*, 1978). Although the specific details have not been worked out, the sequestered ouabain was either minimally or not metabolically altered, whereas only polar metabolites were sequestered in experiments using digitoxin (Duffey and Scudder, 1974; Duffey *et al.*, 1978). These results mirror the response of *O. fasciatus* to the natural cardenolides in its food and indicate that the process of sequestration of these steroids is associated with the polarity of these compounds as found in the food, as well as with the insect's ability to convert them to more polar molecules that are more amenable to storage. This subject will be developed in more detail in the section dealing with sequestration in the Hemiptera (Section II,F).

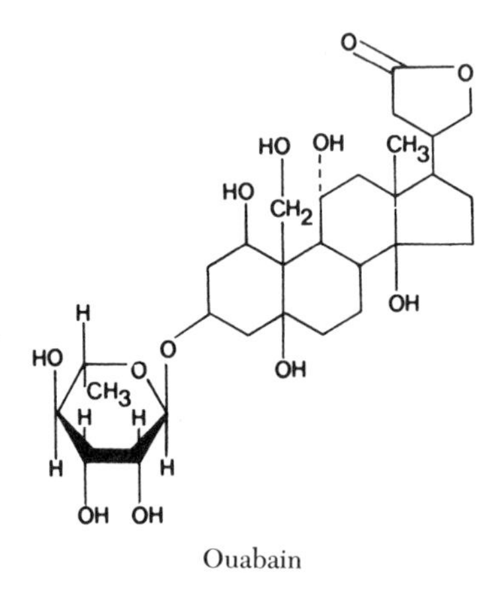

Ouabain

C. Enzymatic Specificity with Toxic Metabolites

Many species of plants, especially in the family Leguminosae, synthesize amino acids with structural similarities to amino acids that are important metabolic intermediates or protein constituents. Utilization or "counterfeit incorporation" of these toxic analogs as surrogates for the compounds that they mimic often results in serious biochemical lesions that may ultimately prove to be fatal. For example, introducing L-canavanine, a structural analog of L-arginine, into an artificial diet inhibited growth of tobacco hornworm larvae (*Manduca sexta*) by reducing ingestion. In addition, the efficiency of conversion of ingested and assimilated food into body mass (Dahlman and Rosenthal, 1975; Dahlman, 1977) was reduced. Furthermore, ovarian development of adult females derived from canavanine-fortified diets was considerably less than that of controls. It thus appears that this arginine analog can decrease fitness in sensitive insects by prolonging the developmental period and reducing fertility and/or fecundity.

However, some insect-feeding specialists have avoided the deleterious effects of these structural analogs by evolving the metabolic capacity to selectively discriminate between the normal biochemical and the potentially toxic surrogate. The bruchid beetle *Caryedes brasiliensis* is a monophagous species that subsists solely on the cotyledons produced from mature seeds of the legume *Dioclea megacarpa*. The seeds of this legume contain large amounts of canavanine, a compound that is limited in its distribution to species in a few subfamilies of the Leguminosae (Bell, 1971; Bell *et al.*, 1978). This amino acid analog is highly toxic when

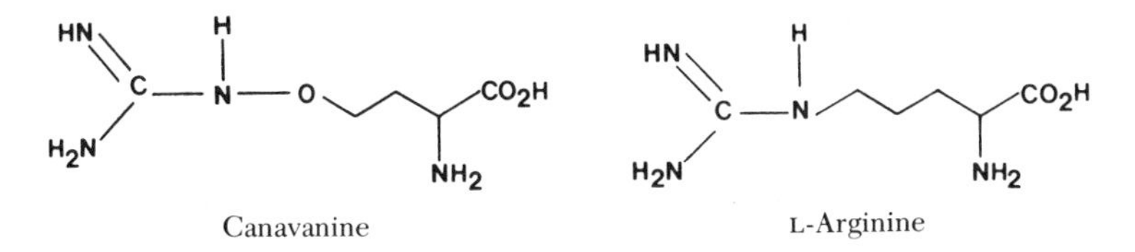

Canavanine L-Arginine

orally administered to a variety of phytophagous insects (Vanderzant and Chremos, 1971; Rehr *et al.*, 1973; Janzen, 1975) and thus probably constitutes an effective defense against most species of herbivorous insects.

The toxic effects of canavanine are identified with the synthesis of canavanyl proteins which disrupt DNA and RNA metabolism as well as protein synthesis (Schachtele and Rogers, 1965). In some species this guanidinooxy structural analog of arginine is esterified to arginine transfer RNA ($tRNA^{Arg}$) by arginyl-tRNA synthetase (Attias *et al.*, 1969), resulting in the incorporation of transmogrified proteins into the nascent polypeptide chain. Larvae of the tobacco hornworm, *Manduca sexta*, for which canavanine is highly toxic (Dahlman and Rosenthal, 1975), incorporate large amounts of this compound into proteins and at least 3.5% of the injected dosage can be recovered as CO_2 after acid and enzymatic hydrolysis of larval proteins (Rosenthal *et al.*, 1976). Furthermore, the ability of L-arginine kinase to mediate the phosphorylation of canavanine to form L-phosphocanavanine has been correlated with the known sensitivity of *M. sexta* larvae to this amino acid analog (Rosenthal *et al.*, 1977a). Canavanine toxicity appears to be well correlated with sensitive reactions related to protein synthesis or genome expression.

Larvae of *C. brasiliensis* do not synthesize canavanyl proteins even in the presence of large amounts of canavanine. The ability of these larvae to feed on canavanine-rich food appears to result from the presence of a ligase, an amino acid-activating enzyme, capable of discriminating between this amino acid analog and arginine (Rosenthal *et al.*, 1976). In addition, this bruchid larva metabolizes about 60% of the ingested canavanine (Rosenthal *et al.*, 1977b), and obviously utilizes this compound as an important, if not major, source of dietary nitrogen. The larvae contain appreciable levels of arginase, an enzyme that can convert arginine or canavanine to urea and ornithine, or L-canaline, respectively. In addition, these bruchids possess extraordinarily high levels of urease (Rosenthal *et al.*, 1977b), an enzyme rarely reported to be present in insects in significant amounts. It was established that 70% of the degraded canavanine was ultimately converted to CO_2 and ammonia, thus providing the larvae with an important supply of nitrogen for fixation into organic linkages.

The unusually high levels of urease in larvae of *C. brasiliensis* function to generate, from urea, utilizable ammonia, thus permitting these insects to mobilize the appreciable nitrogen stored in canavanine (Rosenthal *et al.*, 1977b). However, urea formation from canavanine also results in the biosynthesis of another toxic amino acid, canaline, with which these larvae must cope. Canaline has been determined to be a potent growth inhibitor for larvae of *Manduca sexta* and, in addition, increases mortality and causes adult malformation (Rosenthal and Dahlman, 1975). On the other hand, larvae of *C. brasiliensis* circumvent the potential toxicity of canaline by reductively deaminating this nonprotein amino acid to L-homoserine and ammonia (Rosenthal *et al.*, 1978). This reaction, which completes the detoxication of canavanine, is highly adaptive for the larvae, since it conserves the carbon skeleton of this amino acid while utilizing the generated urea to produce ammonia, thus increasing the mobilization of canavanine's stored nitrogen by 50%.

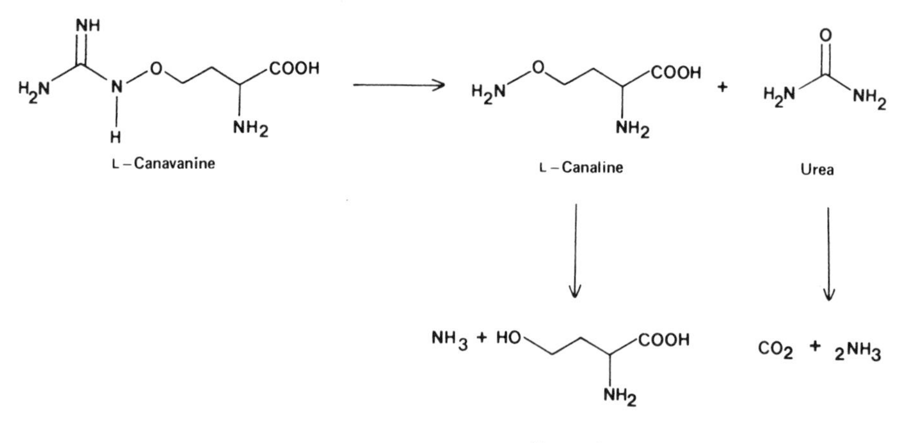

Larvae of *C. brasiliensis* have exploited the seeds of *D. megacarpa* as a food source and avoided the potentially toxic effects of canavanine by being able to recognize this compound as foreign and to rigorously exclude it from the anabolic pathways in which arginine participates. It will be interesting to determine if the capacity of other insects to selectively feed on plants similarly enriched with the many potentially toxic amino acid analogs identified in the Leguminosae (Bell, 1971) is also identified with the presence of highly specific amino acid-activating enzymes that fail to react with these compounds.

II. SEQUESTRATION OF PLANT NATURAL PRODUCTS IN NONSECRETORY STRUCTURES

That many species of insects eliminate or detoxify the plant allomones that they encounter in their diets is indicated by the large number of species that do not contain detectable amounts of these compounds (Rothschild, 1972; Rothschild and Reichstein, 1976). Even in some of the cases in which toxic plant natural products have been identified in extracts of the bodies of herbivores, it has not always been possible to rule out the possibility that these compounds are simply present in the animal's digestive tract rather than having been sequestered. This may be especially true of pupal extracts which may contain, in admixture with accumulated larval nitrogenous excretory products (meconium), plant toxins that will be excreted by the freshly emerged adult. Notwithstanding the impreciseness that sometimes characterizes the derivation of these ingested plant toxins, it has been possible to establish that insect herbivores are highly selective in terms of which of the plant's natural products are retained in their bodies. An examination of representatives of different insect orders that are sequestrators of a wide variety of plant compounds demonstrates that virtually each species of arthropod treats these compounds in a highly distinctive, if not unpredictable, manner.

A. Homoptera

The diaspidid *Aspidiotus nerii*, which feeds on oleander (*Nerium oleander*), a rich source of cardiac glycosides, contains low levels of adynerine, odoroside-H, and strospeside in its body (Rothschild *et al.*, 1973a). Adynerine, which is the dominant cardiac glycoside stored, had not been detected as a sequestration product in any other insect species that feeds on oleander. Another oleander-feeding insect, *Pseudococcus adonidum*,

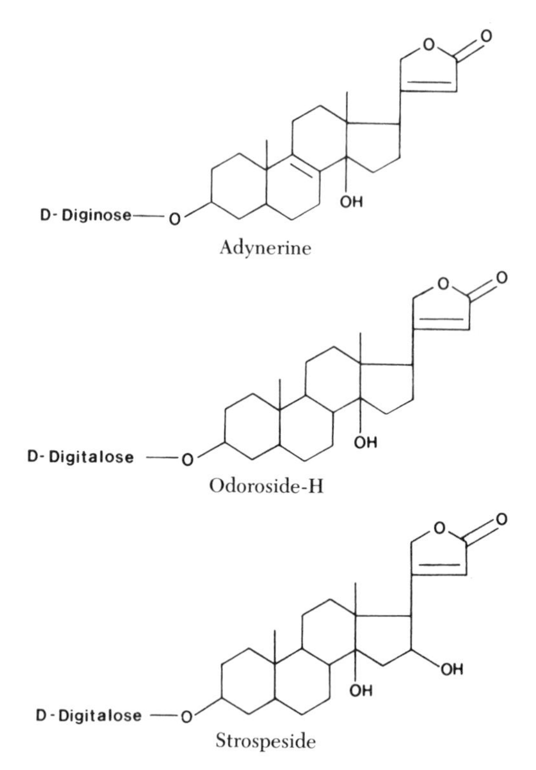

Adynerine

Odoroside-H

Strospeside

did not contain detectable quantities of cardenolides (Rothschild *et al.*, 1973a).

The same three cardiac glycosides have been identified in extracts of the aphid *Aphis nerii* that had been feeding on oleander (Rothschild *et al.*, 1976). On the other hand, when these aphids had fed on leaves of the milkweed *Asclepias curassavica*, two of its main cardenolides, calotropin and proceroside, were identified in extracts of this insect (Rothschild *et al.*, 1970a). It has not been established that the cardiac glycosides identified in extracts of *A. nerii* feeding on either plant source were sequestered rather than simply present in the gut.

B. Neuroptera

Pupae of the lacewing *Chrysopa carnea* contained cardiac glycosides which were obtained by the larvae after feeding on the oleander aphid *A. nerii* (Rothschild *et al.*, 1973a). In this case the lacewing is appropriating the plant cardenolide secondarily from a herbivore, assuming that these compounds are truly sequestered rather than representing waste products accumulated in the pupal meconium.

C. Diptera

Aqueous extracts of pupae of the tachinid fly *Zenilla adamsoni* possess strong digitalis-like activity when tested on a frog heart preparation (Reichstein *et al.*, 1968). The larvae of *Z. adamsoni* were reared from larvae of the monarch butterfly, *Danaus plexippus*, and the polar cardenolides in the parasite must have been appropriated from its lepidopterous host.

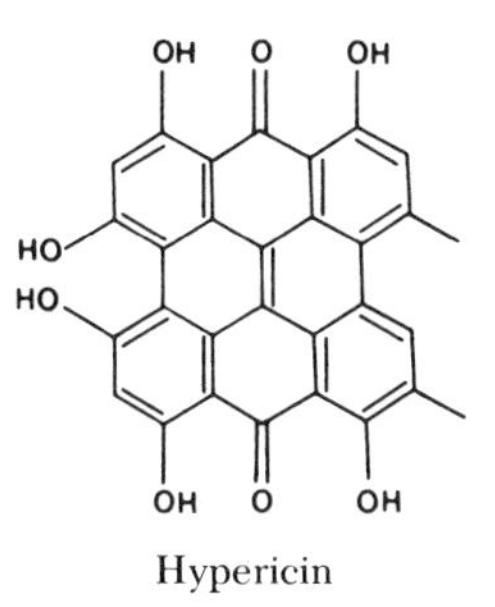

Hypericin

D. Coleoptera

Adults of *Tetraopes oregonensis* and *T. tetrophthalmus*, both of which feed on milkweeds, contain small amounts of cardiac glycosides (Duffey and Scudder, 1972; Isman *et al.*, 1977a). Since the larvae of these cerambycids feed on the roots of milkweeds and the adults feed on the leaves, it is conceivable that the latter may contain cardenolides derived from both portions of the plant. It seems probable that the cardenolides detected in *Tetraopes* extracts represent sequestered compounds rather than gut constituents, since Jones (1937) has demonstrated that, even after starvation for five days, adults of *T. tetrophthalmus* as well as two other chrysomelids are rejected by both birds and ants, assuming that these beetles do not contain other distasteful compounds as well.

Adults of the chrysomelid *Chrysolina brunsvicensis* sequester hypericin (mycoporphyrin), hexahydroxydimethylnaphthodianthrone, from the flower heads of their food plant *Hypericum hirsutum* (Rees, 1969). Each beetle contains about 13.3 μg of hypericin, a compound that is known to be toxic to mammalian herbivores (Brockmann *et al.*, 1942; Thomson, 1957). Interestingly, although adults of seven *Chrysolina* species synthesize cardenolides, those feeding on *Hypericum* spp. do not contain detectable amounts of these steroids (Pasteels and Daloze, 1977).

Adults of two aposematic species of chrysomelids, *Chrysochus cobaltinus* and *Labidomera clivicollis*, were reported to contain little or no cardenolides in their bodies after feeding on milkweeds (Isman *et al.*, 1977a).

The ladybird *Coccinella undecimpunctata*, which was predatory on the aphid *A. nerii*, contained cardiac glycosides (Rothschild *et al.*, 1973a). The aphids, which had been feeding on a rich source of these compounds, oleander, appear to sequester at least three of these steroids (Rothschild *et al.*, 1970a). Whether larvae and adults of *C. undecimpunctata* sequester all the cardenolides appropriated from *A. nerii* has not been determined. In contrast, adults of *C. septempunctata*, which also feed on the aphid *A. nerii*, did not contain detectable amounts of cardenolides (Rothschild *et al.*, 1970a).

E. Orthoptera

The hemolymph of the pyrgomorphid grasshopper *Poekilocerus bufonius* contains high concentrations of cardenolides which are derived from asclepiadaceous food plants (von Euw *et al.*, 1967). Two cardenolides were identified in the eggs, calotropin and calactin, and these two compounds are presumably present in the hemolymph. *Phymateus* spp. also sequester cardenolides (Rothschild, 1972).

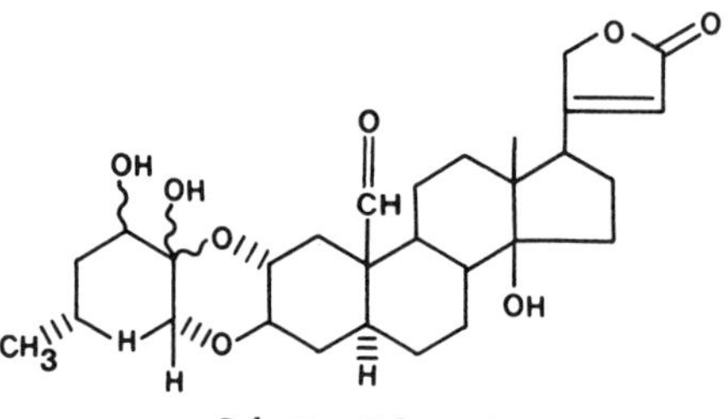

Calactin→Calotropin

Another pyrgomorphid grasshopper, *Zonocerus variegatus*, sequesters pyrrolizidine alkaloids from a *Crotalaria* species (Bernays *et al.*, 1977). The free alkaloid monocrotaline and its *N*-oxide were sequestered in large amounts and based on the percentages of these two compounds in the leaves, it appears that the grasshopper converts a considerable amount of the former to the latter. No free alkaloid and only traces of the *N*-oxide were detected in the frass.

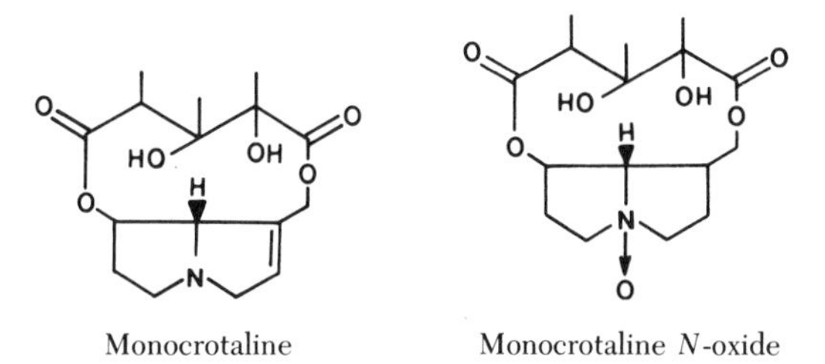

Monocrotaline   Monocrotaline *N*-oxide

Larvae of *Zonocerus elegans* store cannabinoids when fed on *Cannabis sativa* (Rothschild *et al.*, 1977). Although *Z. elegans* has not been recorded as feeding on *C. sativa* in nature, the grasshoppers fed sparingly on two strains of marijuana for six weeks. One strain contained Δ^1-tetrahydrocannabinol, a psychoactive compound, whereas the other contained the inactive cannabidiol. Small amounts of Δ^1-tetrahydrocannabinol were detected in both the insects and their exuviae, whereas no cannabidiol could be recovered from either source. Both compounds were demonstrated to be predominantly in the frass, and there was an indication that *Z. elegans* efficiently metabolized cannabidiol.

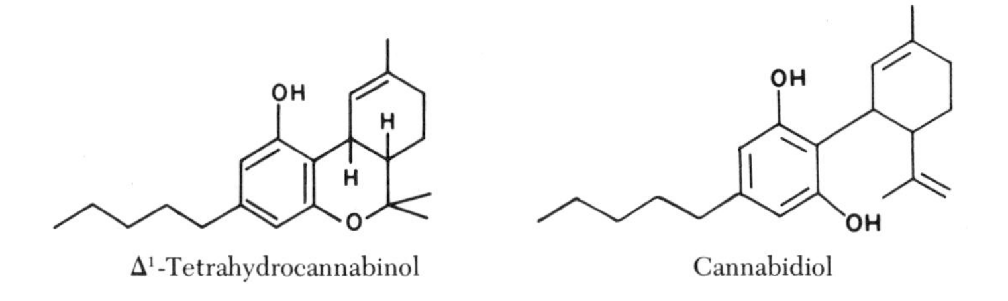

Δ^1-Tetrahydrocannabinol   Cannabidiol

F. Hemiptera

Hemipterous species, particularly members of the family Lygaeidae, feed on toxic plants belonging to a variety of families. Many species have evolved series of dorsolateral spaces on the abdomen and/or thorax which contain high concentrations of sequestered cardenolides (Scudder and Duffey, 1972). However, in addition, many lygaeids appear to sequester cardenolides in nonsecretory structures.

Duffey and Scudder (1972) demonstrated that the lygaeids *Oncopeltus fasciatus*, *O. sandarachatus*, *Lygaeus kalmii kalmii*, and *L. k. angustomarginatus* sequestered cardenolides when fed on another milkweed,

Calotropis procera, as well as leaves of *Nerium oleander*. It was estimated that each adult of *O. fasciatus* contained about 70–110 μg of cardenolides. Although a large percentage of these steroids are sequestered in the dorsolateral spaces, at least some of these compounds are present in nonsecretory structures.

Scudder and Duffey (1972) established that species in 27 genera in the Lygaeidae contained cardiac glycosides, and in those in 10 genera, very high concentrations of these compounds were present. *Oncopeltus fasciatus* was included among the cardenolide-rich bugs, but when this insect was reared on cardenolide-free sunflower seeds, the resulting adults did not contain detectable quantities of these steroids.

Although dorsolateral spaces are present in lygaeids in 12 genera, these are absent from those in 20 others (Scudder and Duffey, 1972). However, species in the latter genera probably sequester cardenolides in other structures. Blood contains substantial quantities of cardenolides when species of either *Oncopeltus* or *Lygaeus* are feeding on asclepiadaceous plants, and there is a predominance of polar steroids in the hemolymph of *O. fasciatus* adults. Cardenolides likely bind nonspecifically to some of the blood proteins (Duffey and Scudder, 1974). While these compounds are present in the hemolymph as an emulsion, it is not known whether this emulsion phase encompasses protein binding (Duffey *et al.*, 1978).

Highly nonpolar cardenolides are also present in adults of *O. fasciatus;* these compounds are absent from the dorsolateral fluid, blood, and metathoracic gland fluid (Duffey *et al.*, 1978). In all probability, some of these lipophilic compounds are dissolved in moderate quantities in the fat body (Duffey *et al.*, 1978). Similarly, fifth-instar larvae contain large quantities of cardenolides (59–153 μg), but the concentration of steroids cannot be accounted for in terms of cardenolide content of the middorsal gland fluid and hemolymph (Duffey and Scudder, 1974).

Von Euw *et al.* (1971) identified several cardenolides in the lygaeids *Caenocoris nerii* and *Spilostethus pandurus*, both of which were feeding on *Nerium oleander*. The major compounds present in *C. nerii* were adigoside and neritaloside, whereas odoroside-A, nerigoside, and strospeside were present as minor concomitants. Six unidentified cardenolides were also present in the bodies of *C. nerii* adults. Adults of *S. pandurus* contained detectable quantities of two cardenolides, odoroside-H and nerigoside. Neither species contained oleandrin, the principle cardenolide

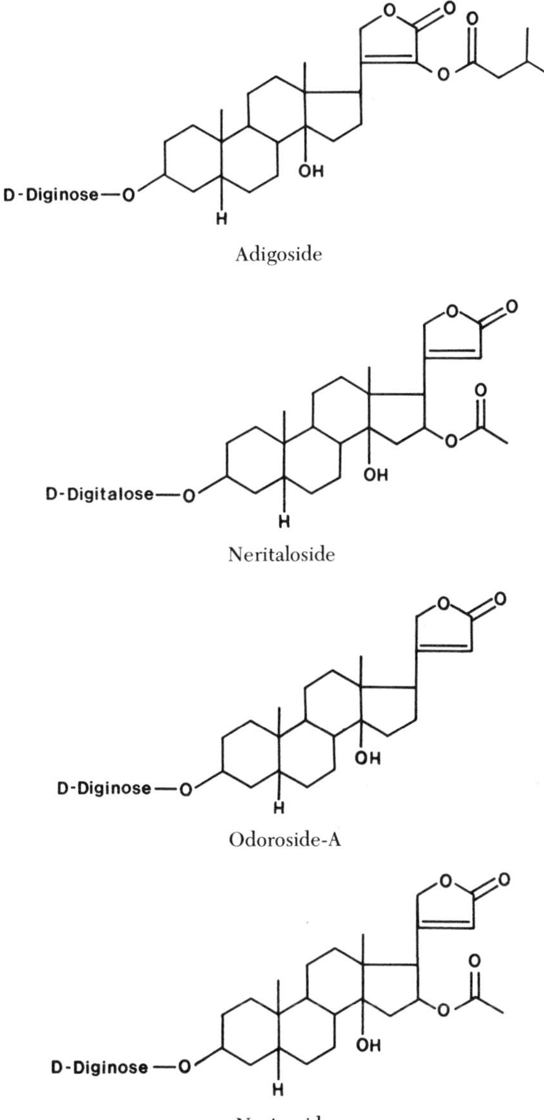

Adigoside

Neritaloside

Odoroside-A

Nerigoside

in *N. oleander*. Scudder and Duffey (1972) also detected cardenolides in extracts of both of these species.

At least some of the cardenolides identified in extracts of *S. pandurus* are probably present in the dorsolateral space fluid, especially since Scudder and Duffey (1972) identified these glands in this species, and Abushama and Ahmen (1976) demonstrated that *S. pandurus* possesses a pair of dorsolateral spaces whose fluid contains cardenolides. On the other hand, Scudder and Duffey (1972) reported that adults of *C. nerii* lacked these dorsolateral spaces.

Duffey *et al.* (1978) studied the uptake of [^{3}H]ouabain and [^{3}H]digitoxin in the milkweed bug *O. fasciatus* and demonstrated that the fates of two steroids are very different subsequent to ingestion. In the presence of natural cardenolides in the seeds, [^{3}H]ouabain is rapidly ingested, probably because of the great solubility of this polar compound in the saliva. Ouabain is very rapidly sequestered in the dorsolateral space fluid, accumulating in this medium as the cardenolide is preferentially taken up from the blood. Injection of ouabain, which results in high concentrations of this compound in the hemolymph, is rapidly followed by sequestration in the dorsolateral space fluid. [^{3}H]Digitoxin, on the other hand, is probably ingested more slowly probably because of the poorer solubility of this nonpolar cardenolide in the saliva. Injection of digitoxin results in increased levels of this compound in the blood from which some of it is taken up by the dorsolateral space fluid. While the digitoxin concentration in the space fluid mirrors that of the blood, substantial amounts of this steroid are also transferred from the hemolymph to lipophilic tissues such as the fat body where it is presumably sequestered. Digitoxin is also metabolized to more polar compounds which are then facilely sequestered in the dorsolateral space fluid. However, whereas *O. fasciatus* can sequester limited quantities of digitoxin in its lipidic tissues (e.g., fat body), this insect has the capacity to sequester inordinately high levels of polar cardenolides such as ouabain (Duffey *et al.*, 1978).

The sequestration of ouabain in the dorsolateral spaces appears to result from physical rather than enzymatic processes occurring in the aqueous phase of this emulsive system. It has been established that the initial accumulation of ouabain in the saliva is a double log relationship of seed dose versus amount in solution (Duffey *et al.*, 1978). Subsequent penetration of the gut results in some ouabain being sequestered in the hemolymph, whereas the majority of the ouabain is sequestered in the dorsolateral space fluid. This fluid contains a high concentration of emulsion phase droplets which constitute a dispersed phase. The majority of the sequestered polar cardenolide (ouabain) resides in this emulsion phase, and the amount sequestered is proportional to free hemolymph concentrations. In essence, this is a physical process, not an enzymatic one, so that sequestration of most of the polar cardenolides is identified with an emulsion in the aqueous phase of the dorsolateral space fluid that discriminates polarity of the cardenolides and appears to appropriate them independent of any energetic expenditures (Duffey *et al.*, 1978).

G. Lepidoptera

Adults of moths and butterflies are ideal subjects for sequestration studies since they generally feed on nectar, a plant secretion that usually lacks most of the compounds found in the leaves and other tissues. Therefore, the presence of these natural products in the bodies of adult lepidopterans must reflect the ingestion and sequestration of these compounds during the larval stages. Not surprisingly, the bulk of the data on the storage of plant compounds by insects has been obtained with lepidopterous species.

1. *Noctuidae*

Adults of *Xanthopastis timais* contain amaryllidaceous alkaloids in their bodies (Aplin and Rothschild, 1972). Larvae of this species feed on the leaves of *Ficus* species.

2. *Ctenuchidae*

The polka-dot moth, *Syntomeida epilais*, whose larvae probably had fed on *Nerium oleander*, contains several cardiac glycosides, one of which, oleandrin, had not been previously identified as a sequestration product of insects (Rothschild *et al.*, 1973a). This compound, the main steroid present in oleander leaves, had not been detected in extracts of coccids and aphids which had fed on oleander. Another major compound

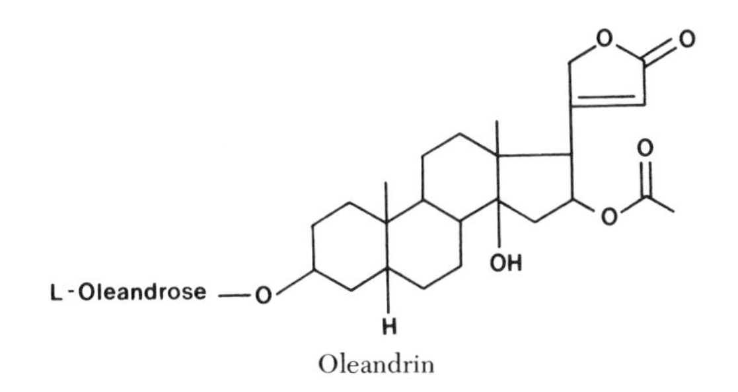

Oleandrin

in *S. epilais* is nerigoside. Adynerine, strospeside, and several unidentified cardenolides appear to be minor constituents.

3. *Arctiidae*

Larvae of tiger moths feed on a wide variety of toxic plants in unrelated families, and many species of these arctiids can feed on plants containing either alkaloids or cardenolides. Among the moths, arctiids appear to constitute one of the most efficient groups of natural products sequestrators that have been studied.

Larvae of *Seirarctia echo* sequester cycasin from *Cycas* leaves, and this compound can be subsequently detected in all life stages (Teas *et al.*, 1966). β-Glucosidase, which is capable of converting the glucoside to the potent alkylating agent methylazoxymethanol, is concentrated in the midgut, and the relatively nontoxic cycasin accumulates in the hemolymph (Teas, 1967). When fed methylazoxymethanol, larvae of *S. echo* convert it to cycasin and thus presumably avoid the potentially intoxicating effects of the former (see Section I,B). Larvae which had fed on a cycad (*Zamia* sp.) containing a glucoside other than cycasin appeared to hydrolyze it and then convert it to the latter which was then sequestered (Teas, 1967).

The African arctiid *Amphicallia bellatrix*, when reared on *Crotalaria semperflorans*, sequesters the *Crotalaria* alkaloids crispatine and trichodesmine (Rothschild and Aplin, 1971). Although the major alkaloid in *C. semperflorans* is an octonecine ester—crosemperine—adults of *A. bellatrix* contain primarily trichodesmine in their bodies.

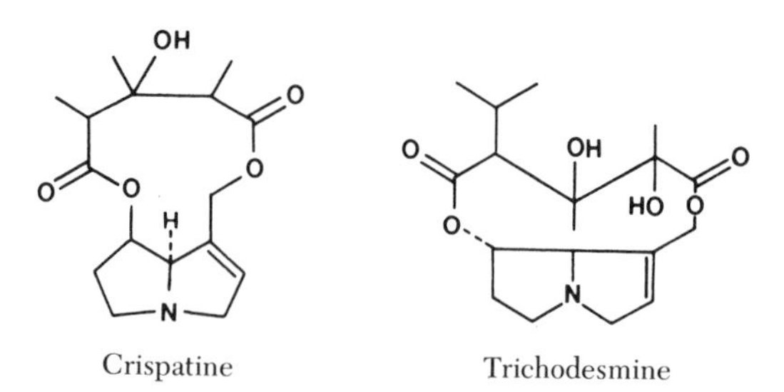

Crispatine Trichodesmine

The arctiid *Arctia caja* stores cardenolides when reared on *Digitalis* and pyrrolizidine alkaloids when fed on *Senecio* (Rothschild and Aplin, 1971). When reared first on *Digitalis* and then transferred to *Senecio* at the fourth instar, the larvae sequester both cardenolides and pyrrolizidine alkaloids (Rothschild, 1972). When larvae of *A. caja* develop on the groundsel *Senecio vulgaris*, the distribution of sequestered alkaloids mirrors that of the host plant (Aplin and Rothschild, 1972). Both larvae and pupae contain more than 50% seneciphylline along with integerrinine (ca. 25%), senecionine (0.5%), and a metabolite that appears to be an ester of retronecine. The metabolite, which is not found in the leaves of the plant, possesses a mass spectrum characteristic of the *Senecio* alkaloids. The percentage of alkaloids (dry weight) in both the adults and pupae is about 4 times greater than that found in the leaves of *S. vulgaris*. The great efficiency of *A. caja* larvae as sequestrators of these pyrrolizidine alkaloids is indicated by the fact that the frass contains very low amounts of these compounds (Aplin and Rothschild, 1972).

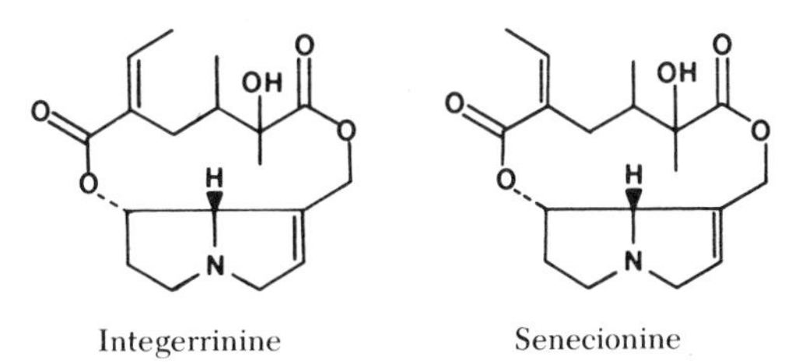

Integerrinine Senecionine

Like those of *A. caja*, larvae of the cinnabar moth *Tyria jacobaeae* are very efficient sequestrators of pyrrolizidine alkaloids. Aplin *et al.* (1968) reported that both pupae and adults of this moth store about 8–12 times (dry weight) more alkaloids than are present in the leaves of the ragwort *Senecio jacobaeae*. Larvae that had fed on groundsel, *S. vulgaris*, were only slightly less efficient in sequestering these compounds. The major alkaloid stored in the pupae was senecionine, with seneciphylline and integerrinine also constituting quantitatively important constituents. Jacobine, jacozine, and jacoline were minor concomitants, whereas the retronecine metabolite often constituted about 20% of the alkaloidal pool (Aplin and Rothschild, 1972). The eggs of the cinnabar moth lacked detectable pyrrolizidine alkaloids. The larval frass contained minor amounts of these compounds, a fact that further emphasizes the great efficiency of these insects as sequestrators of *Senecio* alkaloids. Significantly, the concentration of the different alkaloids stored by *T. jacobaeae* differs considerably from that present in its host plant, demonstrating this insect is highly selective as a sequestrator. For example, senecionine, which is a

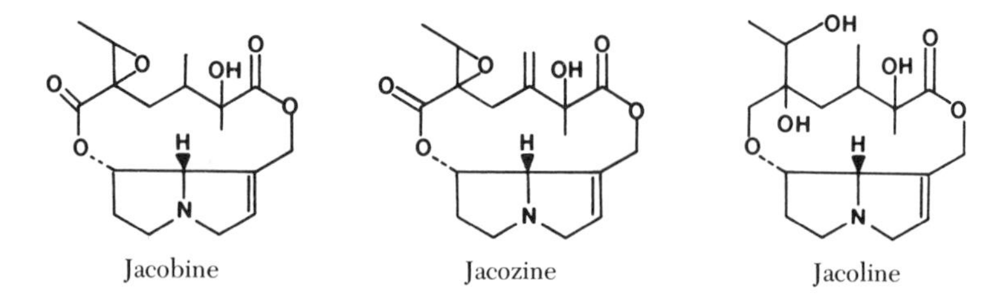

trace constituent in plant leaves, is the major compound stored in pupae of this species (Aplin *et al.*, 1968; Aplin and Rothschild, 1972).

Pyrrolizidine alkaloids are also sequestered by the arctiids *Utethesia bella*, *U. pulchelloides*, and *Argina cribraria* which had fed on *Crotalaria* or *Heliotropium* species (Rothschild, 1972). Larvae of *A. cribraria*, which were reared on a *Crotalaria* species, sequestered monocrotaline from this plant.

Larvae of *A. caja*, which are not reported to feed on *Cannabis sativa* in nature, are able to sequester cannabinoids when forced to feed on different strains of marijuana (Rothschild *et al.*, 1977). However, these polyphagous larvae, which normally feed on a variety of toxic plants, are unable to complete development when reared on a strain rich in Δ^1-tetrahydrocannabinol (THC). Larvae that fed on a strain containing this psychoactive cannabinoid were stunted in growth and failed to survive beyond the third instar. About 80% of the recovered THC was observed in the frass, demonstrating that most of the detectable cannabinoid was excreted. Since far less was found in the frass than in the plant, it appears that most of the ingested THC was metabolized. Large amounts of THC were also found in the exuviae (cast skins).

A cannabidiol (CBD)-rich strain of *C. sativa* proved to be more suitable for development of *A. caja* larvae (Rothschild *et al.*, 1977). Although larvae developed more slowly on this strain than they did on the controls, successful pupation was achieved in about 7–8 weeks. CBD, which is not reported to be psychoactive, was only detected in the larvae in trace amounts, along with an unidentified metabolite. Large amounts of CBD were present in the frass and traces were detected in the exuviae.

Larvae which were transferred from the CBD strain to the THC strain successfully developed and pupated (Rothschild *et al.*, 1977). These larvae stored more CBD than those reared exclusively on the CBD strain, and contained substantial quantities of THC as well, along with some decarboxylated THC. In addition, the exuviae of these larvae contained

2.5 times more CBD than those reared on the CBD strain. Evidence of extensive metabolism of the ingested cannabinoids was again obtained.

4. *Pieridae*

Butterfly larvae are also capable of sequestering a diversity of natural products after ingestion. Pupae of the large white butterfly (*Pieris brassicae*) and the small white (*P. rapae*) contain allylisothiocyanate, which is derived from the glucoside sinigrin that is present in the cruciferous food plants (Aplin *et al.*, 1975). While both the mustard oil and sinigrin were always detected in pupae of *P. brassicae*, the latter was not always readily evident in pupae of *P. rapae*. Since adults of the small white did not contain detectable quantities of allylisothiocyanate, it appears that this pierid is either only an emphemeral sequestrator of these compounds at best (see Section I,B), or that the plant-derived products are actually present in the pupal meconium.

Rothschild (1972) has reported that adults of the pierid *Catopsilia pomona* contain monoterpene alkaloids derived from *Cassia* species.

5. *Papilionidae*

Nitrophenanthrenes are sequestered by papilionid species in several genera, often in very high concentrations. These pharmacophagous swallowtails often feed on *Aristolochia* species which characteristically contain aristolochic acids, compounds that are noted for their bitter and toxic qualities. Von Euw *et al.* (1968) identified aristolochic acid-I from extracts of the papilionid *Pachlioptera aristolochiae* and reported that each adult contained about 100 mg of this compound. In addition to *P. aristolochiae*,

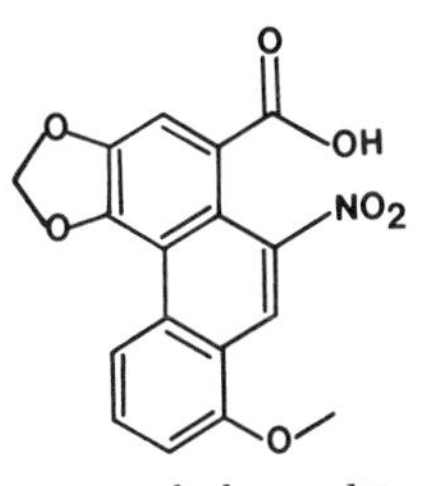

Aristolochic acid-I

this nitrophenanthrene has been detected in extracts of adults of *Battus philenor* and *B. polydamas* as well as a pupa of *Ornithoptera priamus* (Rothschild, 1970). On the other hand, several other *Papilio* species and *Troides* species lacked aristolochic acid-I. *Zerynthia polyxena* (Parnassiinae), a flamboyantly aposematic species, sequestered aristolochic acid-Ia, a previously undescribed nitrophenanthrene, as well as aristolochic acid-C. Each butterfly stored about 0.15 mg of these acids, but no traces of the acid characteristic of the genus *Aristolochia*—aristolochic Acid-I—were evident.

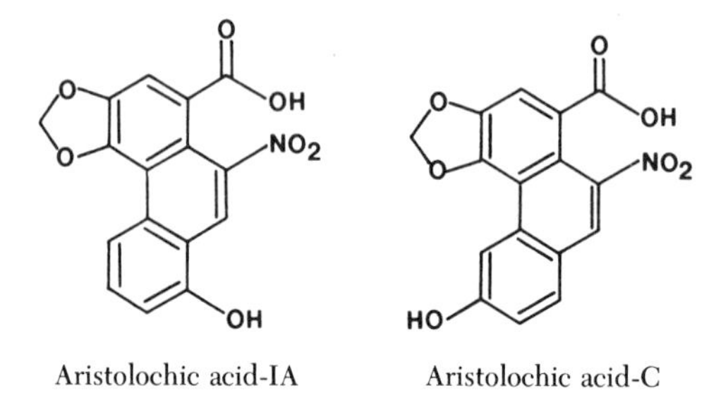

Aristolochic acid-IA Aristolochic acid-C

That papilionids are not limited to the sequestration of aristolochic acids is demonstrated by the fact that adults of *Papilio antimachus* (food plant unknown) contain large quantities of unidentified cardenolides (Rothschild, 1970).

6. *Danaidae*

The first demonstration of the sequestration of plant toxins by an insect was undertaken on the monarch butterfly *Danaus plexippus* after it had been established that extracts of adults contained a digitalis-like toxin (Parsons, 1965). Subsequently, Reichstein *et al.* (1968) identified several cardenolides in extracts of monarch adults and thus provided proof for the long-held belief that some insects sequester toxic plant products in their own bodies which render them unpalatable and/or poisonous to predators (Slater, 1877; Haase, 1896; Poulton, 1916). The demonstration that cabbage-reared monarchs were palatable to an avian predator, the blue jay (*Cyanocitta cristata*), whereas those that were reared on the milkweed *Asclepias curassavica* were not, provided additional circumstantial evidence for the probable role of sequestered plant natural products as predator deterrents (Brower *et al.*, 1967). Since monarchs were palatable when their larval stages fed on a milkweed lacking cardenolides (*Gonolobus laevis*), the possible importance of these compounds as defensive agents against vertebrate predators was further emphasized.

Reichstein *et al.* (1968) identified calotropogenin, calotoxin, calotropin, and its isomer, calactin, as the main cardenolides stored in the eggs and the bodies of the adults of *D. plexippus* that had been reared on *Asclepias curassavica*. In addition, uzarigenin and five other relatively nonpolar cardenolides were present in extracts of adults. With the exception of uzarigenin, the structures of all the asclepidaceous cardenolides are hypothetical (Crout *et al.*, 1963, 1964; Brüschweiler *et al.*, 1969).

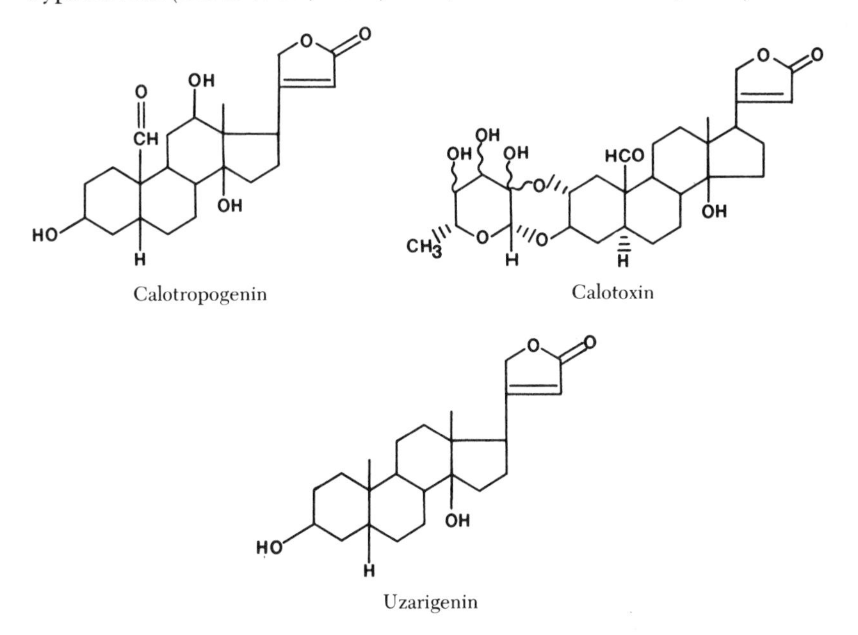

Calotropogenin

Calotoxin

Uzarigenin

Analyses of various morphs of *Danaus chrysippus* from Africa demonstrated that these butterflies are poor sequestrators of cardenolides at best (Rothschild *et al.*, 1975). Indeed, whereas a morph from one area sequestered cardiac glycosides, the same morph from another area did not, although larvae from both populations were reared on the same steroid-rich asclepiadaceous species. Furthermore, some morphs contained detectable levels of cardenolides in the pupal stage, but the emergent adults contained essentially none of the nonpolar cardenolides sequestered during the larval state. In general, for *D. chrysippus* adults, there was no correlation between the presence of nonpolar steroids in the food plant and the sequestration of these compounds by the adults. This species is a poor sequestrator vis-à-vis the monarch, *D. plexippus*, and for unknown reasons, eliminates most of the ingested cardenolides either as a larva or during pupal-adult transformation (Rothschild *et al.*, 1975).

Great variation also exists in the cardiac glycoside content of migrants of *D. plexippus* from different areas in North America (Brower *et al.*, 1972). Contrary to prediction, both the concentration mean and variance of cardenolides decrease southward, notwithstanding the fact that the number of milkweed species increases in the southern part of North America. Brower *et al.* (1972) suggest that monarchs selectively sequester more toxic cardiac glycosides in the southern areas of the continent and are highly emetic as a consequence. However, in the absence of qualitative analyses of the sequestered steroids, the adaptive significance of these unexpectedly low levels of cardiac glycosides is impossible to ascertain. Furthermore, the higher cardenolide content of monarchs reared in more northern latitudes may simply reflect the longer developmental periods of the insects as compared to their southern counterparts. Long developmental times may result in increased cardenolide sequestration because of the ingestion of larger quantities of cardenolides than would occur in larvae growing more rapidly at higher temperatures (Rothschild and Reichstein, 1976). Although this suggestion was not supported by the results of Dixon *et al.* (1978) with monarch larvae reared at two temperatures, it is in agreement with the findings of Isman *et al.* (1977a) who reported that milkweed bugs reared at lower temperatures sequestered more cardenolides than those reared at higher ones.

Brower and Glazier (1975) reported that, although monarchs contain higher concentrations of cardenolides in the wings than in the abdomen, those in the latter body region are considerably more emetic than those in the wings. They believe that cardenolide-rich wings would initially deter avian predators; highly emetic cardiac glycosides in the abdomen would effectively reinforce the unpalatable nature of the monarch if a bird tasted the wing of a pursued butterfly. However, since birds may reject monarchs solely on the basis of unpalatability, it is equally possible that cardenolides deposited (=excreted?) in the wings may be highly adaptive as a predator defense *in the absence of any emetic response.*

Cardenolide concentrations in adult monarchs may vary significantly depending on sex, as well as the geographical region from which the butterflies originated. For example, Brower and Moffitt (1974) reported that adult butterflies from California contained a much lower mean cardenolide content than monarchs from Massachusetts: in both populations females contained more cardenolides than males. However, whereas virtually all Massachusetts butterflies contained significant amounts of cardenolides, those from California were essentially dimorphic, with about 50% of the individuals containing essentially no detectable cardenolides. Significantly, the California butterflies were more emetic than their Massachusetts counterparts at all emetic concentrations. In the absence of data pertaining to the identity of the larval food plants, the qualitative and quantitative nature of the plants' cardenolides, and the fate of the sequestered steroids during the transformation of the larva into the adult, it is impossible to appreciate the many factors which undoubtedly have ultimately contributed to the cardenolide patterns analyzed by Brower and Moffitt.

Another danaid, *Euploea core*, sequesters cardenolides when reared on species of either Asclepiadaceae or Moraceae, but extracts of adults and pupae do not produce emetic effects when fed to pigeons by gavage (Marsh *et al.*, 1977). These results demonstrate that *E. core* is not as effective a sequestrator of these compounds as some other danaid species (e.g., *Danaus plexippus*). However, adults of *E. core*, in spite of their lack of emetic qualities, are rejected by birds, demonstrating that these butterflies possess compounds which render them eminently distasteful.

7. *Nymphalidae*

When larvae of *Hypolimnas bolina* were reared on *Ipomoea batatas* (Convolvulaceae), the resultant adults contained cardioactive substances (Marsh *et al.*, 1977). Extracts of dried specimens of adult *H. bolina*, which had developed on a plant lacking cardioactive compounds, lacked cardioactivity, whereas living specimens produced a slight response. It appears that intrinsic toxicants generally disappear after death in contrast to sequestered plant toxicants which often can be detected in extracts of dried specimens several years old.

In some geographic areas *H. bolina* is designated as a mimic of danaids in the genus *Euploea*, species of which sequester cardenolides after feeding on plants in the families Asclepiadaceae and Moraceae (Marsh *et al.*, 1977). Since *H. bolina* can also sequester cardioactive compounds, this supposed Batesian mimic of *Euploea* species may actually constitute a Müllerian mimic, depending on the access of the larvae to toxic food plants. Indeed, in areas where *Euploea species* and *H. bolina* resemble each other, it seems likely that species in either genus can constitute the model or mimic, depending on the food plants available (Marsh *et al.*, 1977). The existence of mimetic relationships predicated on the opportunistic sequestration of pharmacologically active plant compounds would be consistent with the observed unpredictability of several aposematic species as sequestrators of natural products (Rothschild *et al.*, 1975). The ability of an aposematic herbivore to develop on plant species which either contain or lack sequesterable compounds could result in individuals developing into either potential Batesian or Müllerian mimics as circumstances dictate.

III. SEQUESTRATION OF PLANT NATURAL PRODUCTS IN SECRETORY STRUCTURES

It will not prove surprising if it is ultimately determined that a variety of insects feeding on toxic plants commonly sequester many of the natural products in the secretions of glands that synthesize defensive compounds *de novo*. Although few examples of this phenomenon can yet be cited, analyses of defensive secretions have seldom included a search for the possible presence of plant-derived constituents. If the deterrency of a defensive exudate is augmented by the presence of plant natural products, it is very likely that arthropods have commonly evolved the ability to sequester these diverse plant compounds in the reservoirs of defensive glands isolated from the sensitive tissues of the body.

A. Hemiptera

Several species in the family Lygaeidae sequester cardenolides, derived from milkweed seeds, in their defensive glands, which are also a rich source of aliphatic carbonyl compounds synthesized by the insects. Cardenolides have been detected in the middorsal abdominal glands of larvae and ventral metathoracic glands of *Lygaeus kalmii angustomarginatus*, *L. k. kalmii*, *Oncopeltus fasciatus*, and *O. sandarachatus* (Scudder and Duffey, 1972). Subsequently, Duffey and Scudder (1974) reported that the levels of cardenolides in the middorsal and metathoracic glands were higher than those of the hemolymph. In addition, the cardenolides sequestered in the gland fluids consisted primarily of polar compounds, as they did in the case of the dorsolateral space fluid. The concentration of cardiac glycosides in the glandular secretions was at least 10 $\mu g/\mu l$, indicating relatively high levels of these steroids in the fluids of these defensive glands.

B. Orthoptera

The defensive secretion in the bilobed poison gland of the grasshopper *Poekilocerus bufonius* contains, in addition to acetylcholine, high concentrations of cardiac glycosides sequestered from asclepiadaceous food plants (von Euw *et al.*, 1967). The secretion contains at least 10× more calactin than its isomer calotropin, but if the insects were fed on plants that did not produce cardenolides, the concentration of these compounds in the secretion was very low. *Poekilocerus bufonius* is obviously a selective sequestrator since only two of the six cardenolides present in leaves of the host plant (*Calotropis procera*) were present in the defensive gland fluid. The grasshoppers may excrete or metabolize the nonsequestered steroids or possibly convert them to calotropin and calactin.

Plant natural products are almost certainly appropriated in the defensive secretion of the acridid *Romalea microptera*, an aposematically colored grasshopper. This insect audibly discharges an odoriferous froth from glands in its mesothoracic spiracles, and this exudate constitutes an excellent repellent for small predators such as ants (Eisner *et al.*, 1971a). Among the compounds identified in the secretion are such well-known plant products as verbenone, isophorone, and 2,6,6-trimethylcyclohex-2-ene-1,4-dione. However, the main constituent present in the secretion is

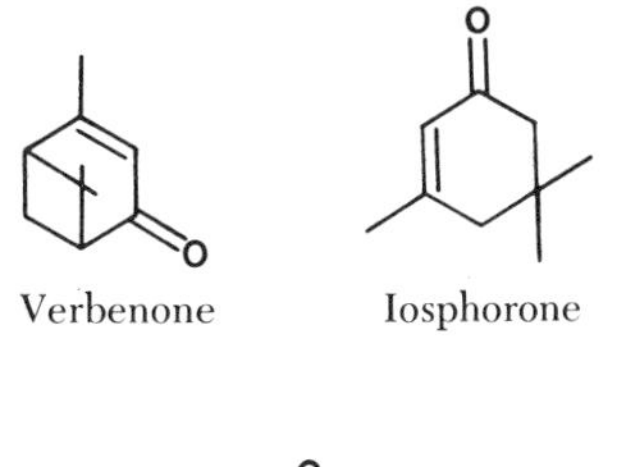

Verbenone Iosphorone

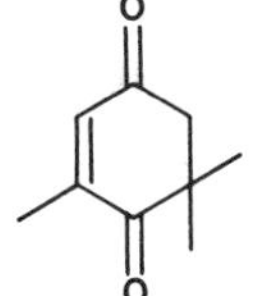

2,6,6-Trimethylcyclohex-2-ene-1,4-dione

romallenone, an allenic sesquiterpenoid that is probably derived from carotenoid pigments (Meinwald *et al.*, 1968a). Although the synthesis of this ketodiol from allenic precursors supports its probable biogenesis from carotenoids (Isoe *et al.*, 1971; Mori, 1973), it is impossible to ascertain whether romallenone is produced by the plant or the insect.

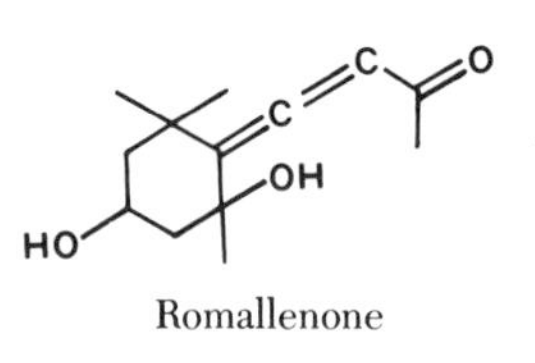

Romallenone

C. Hymenoptera

Larvae of the sawfly *Neodiprion sertifer* sequester both monoterpenes and sesquiterpenes from their pine host (*Pinus silvestris*) and utilize these compounds as effective deterrents for predators (Eisner *et al.*, 1974b). The compounds are stored in capacious diverticular pouches of the foregut and can be discharged even when the larva is enclosed in its cocoon prior to pupating. Young, first-instar larvae feeding on only pine needles, sequestered only α-pinene, β-pinene, and pinafolic acid. On the other hand, older larvae feed on both pine needles and their basal fascicled portions, which results in their ingesting resin acids produced in the resin canals of the branches. Thus, in addition to the three pine-needle compounds, older larvae sequester pimaric acid, levopimaric acid, palustric acid, dehydroabietic acid, abietic acid, and neoabietic acid. The two pinenes are excellent repellents and their evaporative loss is retarded by the resin acids, thus prolonging the deterrent action of the two monoterpene hydrocarbons. The resin acids probably also function to entangle would-be predators, therefore providing a dual protective function (Eisner *et al.*, 1974b). Since none of the compounds sequestered from the pine host are detectable in the feces, the larvae have obviously evolved an effective means of avoiding the potential toxic effects of these plant constituents.

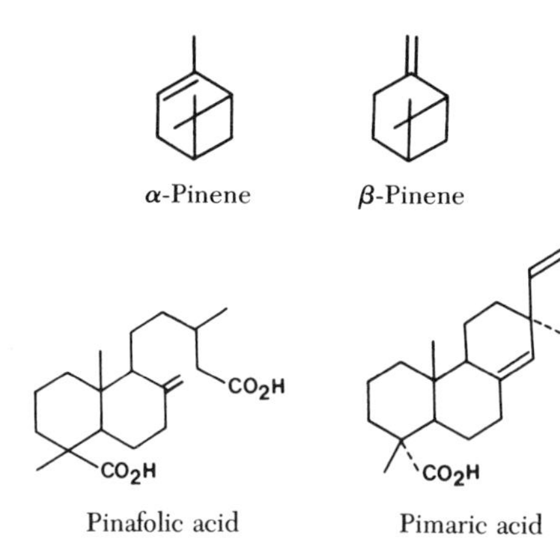

α-Pinene β-Pinene

Pinafolic acid Pimaric acid

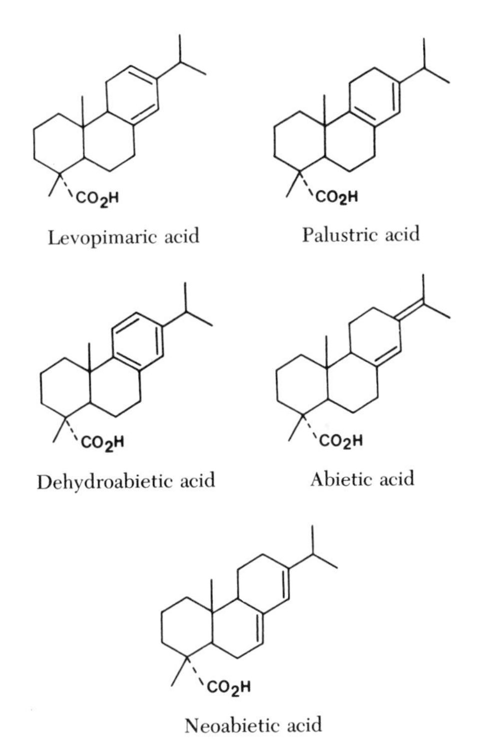

Levopimaric acid

Palustric acid

Dehydroabietic acid

Abietic acid

Neoabietic acid

Australian sawfly larvae in the family Pergidae similarly store oil from eucalyptus leaves in a diverticular pouch of the foregut (Morrow *et al.*, 1976). The oil, which consists of mono- and sesquiterpenes, is stored in a single pouch and is utilized as a deterrent to both vertebrate and invertebrate predators. However, unlike *N. sertifer*, the pouch is not retained in the cocoon of pergid pupae, but rather is incorporated into the wall of the cocoon. The presence of a similar defensive system in sawflies in unrelated families may indicate that the storage of plant resins in cuticular pouches is widespread in these insects.

Similarly, larvae of some predatory insects utilize enteric discharges, containing host-derived allomones, as animal deterrents, only in these cases the repellents originate from arthropod prey rather than plants. For example, larvae of the pyralid *Laetilia coccidovara*, feeding on coccids containing the anthraquinone carminic acid, repel ants with the quinone-rich oral discharges from their capacious crops (Eisner *et al.*, 1980) in much the same way as sawfly larvae deter predators with regurgitates consisting of ingested plant products.

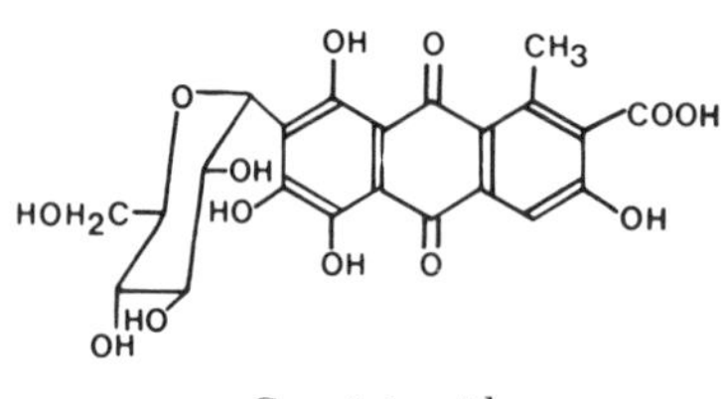

Carminic acid

IV. RELATIONSHIP OF INSECTS TO TOXIC PLANTS—AN OVERVIEW

Obviously the term "toxic plant" is relative at best, since plants serve as a food resource for a multitude of competitors which include, besides other plants and microorganisms, both invertebrate and vertebrate herbivores (Sondheimer and Simeone, 1970; Harborne, 1972; Wallace and Mansell, 1976). Indeed, it is not improbable that species of insects will be found to feed on all species of plants known to synthesize compounds that possess toxicity when administered to vertebrates. In short, notwithstanding the great toxicity of diverse plant natural products to mammals, selected groups of invertebrate species have evolved physiological mechanisms for coping with these compounds and, in so doing, have been able to exploit food resources that constitute a "forbidden fruit" for most herbivores. In general, when aposematic species develop on toxic plants, they selectively sequester some of the ingested compounds, whereas cryptic species, often feeding on the same species, do not.

The demonstration that some insects can sequester specific plant-derived compounds in their bodies (Reichstein *et al.*, 1968) appears to be illustrative of the capacities of many other species to similarly store potential plant toxins. Research in the last decade has focused on the abilities of diverse insects to sequester a variety of plant natural products (Rothschild, 1972) without providing much insight into the *modus operandi* of this storage process. However, whereas a considerable amount of research has now been undertaken on the sequestration of plant toxins by insects (Reichstein *et al.*, 1968; Rothschild *et al.*, 1973b; Aplin and Rothschild, 1972; von Euw *et al.*, 1971; Duffey and Scudder, 1974; Rothschild and Reichstein, 1976; Seiber *et al.*, 1980), the question of why a multitude of species do not store these compounds has been virtually ignored (Duffey, 1980). Rothschild (1972) lists many species that develop on toxic plants but do not contain detectable quantities of their hosts' natural products in their bodies. Similarly, Isman *et al.* (1977a) demonstrated that only a few of the temperate milkweed-feeding insects examined sequester cardiac glycosides. In many cases, species that do not sequester natural products belong to the same genera as species that do, and often both sequestrators and nonsequestrators have fed on the same species of plants. While the ability to sequester ingested natural products is obviously not adaptive for all herbivores, neither is the basis for this nonstorage nor its ecological implications vis-à-vis predators clearly evident. In short, the question of why many insects do not store plant compounds is as important as why some do, if the relationships of these animals to toxic plants is to be comprehended.

Although many species of insects are limited to feeding on toxic plants in a single family or feed on species in other families containing similar toxins [e.g., *Aphis nerii* can feed on *Nerium oleander* (Apocynaceae) and *Asclepias curassavica* (Asclepiadaceae)—both of which contain cardenolides], other groups of insects feed on unrelated plant species containing several classes of toxic compounds. For example, many lygaeid bugs and arctiid moth larvae can develop on plants in genera as diverse as *Digitalis* (Scrophulariaceae) and *Senecio* (Compositae).

Significantly, some herbivores, e.g., *Arctia caja*, can sequester the unrelated toxic compounds produced by *Digitalis* (cardenolides) and *Senecio* (pyrrolizidine alkaloids) (Rothschild and Aplin, 1971). Indeed if a larva is switched from *Digitalis* to *Senecio*, it sequesters both cardenolides and pyrrolizidine alkaloids. Therefore, at least in the case of some insect species, an understanding of the physiological bases for the storage of plant toxins must be predicated on an analysis of a herbivore's ability to sequester unrelated plant natural products. It is one thing to conclude that the arctiid *A. caja* can exploit a diversity of toxic plants as food sources but quite another to analyze this polyphagous propensity not only in terms of tolerating a variety of plant allomones in the diet but of absorbing these compounds and appropriating them in body tissues as well. It is possible that high levels of mixed-function oxidases are frequently correlated with the ability of insects to tolerate ingested plant toxins, as has been recently demonstrated for a polyphagous species, the southern armyworm, *Spodoptera eridania* (Brattsten *et al.*, 1977). However, since there is no evidence that *A. caja* appreciably metabolizes the sequestered alkaloids or cardenolides (Rothschild and Aplin, 1971; Aplin and Rothschild, 1972), there is no compelling reason to invoke the detoxicative abilities of the mixed-function oxidases to explain why this insect can feed on plants containing high levels of these known animal toxins.

The predilection of certain polyphagous insects for feeding on a wide variety of unrelated toxic plant species does not mean that these insects can ingest all poisonous plants with impunity. *Arctia caja,* an arctiid with catholic tastes for toxic plants, readily sequesters both cardenolides and pyrrolizidine alkaloids from plants containing these classes of compounds (Rothschild and Aplin, 1971). However, when larvae of *A. caja* are forced to feed on a strain of *Cannabis sativa* that is rich in the psychoactive compound Δ^1-tetrahydrocannabinol, their growth is stunted and they do not survive beyond the third instar (Rothschild *et al.*, 1977). Presumably, for *A. caja,* the presence of high concentrations of the cannabinoid in its diet constitutes a physiological liability that exceeds its well-developed capacity to tolerate plant toxins. Notwithstanding the capacity of these larvae to metabolically cope with and sequester dissimilar toxic natural products, there are obviously limits to both the variety and concentration of toxins that can be safely ingested.

Arctia caja may be characteristic of *generalist feeders* in being able to develop at an acceptable rate on plants containing a wide variety of natural products, provided that these compounds are not present in unusually high concentrations. Similarly, growth of another generalist feeder, *Spodoptera eridania,* is not inhibited by normal concentrations of sinigrin in crucifer leaves, whereas higher concentrations of this glycoside retard its growth (Blau *et al.*, 1978). In general, it appears that generalist feeders exhibit an intermediate tolerance to specific natural products in comparison to the specialist feeders that appear to be completely adapted for the characteristic natural products that occur consistently in their food plants. For example, growth of larvae of the cabbage white, *Pieris rapae,* which feed on cruciferous plants often containing sinigrin, is unaffected by even inordinately high concentrations of this compound in their food. The basis for the resistance of *P. rapae* and *S. eridania* larvae to sinigrin is unknown, although in the latter species, it has been demonstrated that this compound stimulates the induction of microsomal mixed-function oxidative enzymes (Brattsten *et al.*, 1977).

In general, it appears that chewing insects preferentially sequester nonpolar compounds from the plants on which they are feeding. Von Euw *et al.* (1968) and Reichstein *et al.* (1968) demonstrated that both the grasshopper *Poekilocerus bufonius* and the butterfly *Danaus plexippus* selectively stored two nonpolar cardenolides—calactin and calotropin—after ingesting a large series of these steroids from their asclepiadaceous host plants. The cinnibar moth, *Tyria jacobaeae,* converts the polar *N*-oxide of senecephylline to the free base which is sequestered along with several other nonpolar pyrrolizidine alkaloids (Aplin and Rothschild,

1972). These results indicate that insects feeding on toxic plants have a tendency to exhibit selective sequestration, favoring the retention of nonpolar compounds in various tissues or blood.

Since nonpolar cardenolides are considerably more emetic than polar compounds (Hoch, 1961; Okita, 1967; Wilbrandt, 1962), the adaptive significance of such selective sequestration is probably great. Indeed, until the specifics of an insect's storage of plant natural products vis-à-vis the available compounds ingested is determined, it will be impossible to really comprehend the ecological significance of the relationships of predators to insects feeding on toxic plants. Conclusions based on semiquantitative analyses of unidentified plant natural products, stored by insects feeding on unknown toxic plants, can hardly be expected to provide much insight into the question of how insects cope with either their toxic host plants or, for that matter, their predators.

How do insects avoid the deleterious effects of compounds that are reported to be highly toxic to animals? The physiological and/or biochemical bases for the tolerance of insects to such diverse classes of compounds as cardiac glycosides and pyrrolizidine alkaloids are unknown, notwithstanding the fact that these compounds are not only ingested by arthropods, but they are absorbed into the blood and subsequently sequestered (Duffey, 1980). Similarly, virtually nothing is known about the fate of sequestered compounds during the development of the insect. Do sequestered compounds undergo tissue reorganization during the major histolytic changes that characterize larval–pupal–adult transformations in insects? Is there constant leakage from storage sites which results in constant excretion and replacement? Rothschild *et al.* (1975) reported that some morphs of *Danaus chrysippus* eliminated all detectable cardenolides during pupal–adult transformation, demonstrating that the cardenolide content of the adult butterfly may bear no relationship to what compounds the larval stages ingested or sequestered. Thus, the conclusion by Brower *et al.* (1975) that the absence of detectable levels of cardiac glycosides in wild-caught adults of *D. chrysippus* indicates that the larvae fed on milkweeds lacking cardenolides is not necessarily valid.

It is possible that insects that normally ingest toxic plant natural products may excrete large amounts of these compounds, along with the exuviae, when they molt (Rothschild *et al.*, 1977). Large amounts of Δ^1-tetrahydrocannabinol are concentrated in the cast skins of larvae of *A. caja* that fed on *Cannabis sativa*. Similarly, monarch larvae and pupae eliminate large amounts of cardenolides with the cast skins (Nishio and Blum, 1980). It may be significant that the pupal molting fluid of the monarch is rich in these compounds. Accumulation of plant toxins in the cuticle prior to molting would present insects with an alternate means of eliminating these compounds while at the same time providing them with an antipredator device at a time when they are particularly vulnerable. If aposematic insects that feed on toxic plants as immatures routinely concentrate plant toxins in the cuticle, it would be especially adaptive for species that are cryptic as adults and do not store plant toxins (Marsh and Rothschild, 1974).

The ability of insects to feed with impunity on plants containing potentially toxic compounds reflects the presence of a potpourri of physiological mechanisms that include, among others, the excretion, metabolism, selective absorption, and sequestration of these plant products (Rothschild and Reichstein, 1976). Presumably, an analysis of the interplay of these factors, while it may not explain why these compounds do not induce toxicosis in the insect, will provide some insight into the evolutionary strategy utilized by a species to process the allomonal concomitants of the nutrients that have been consumed (Duffey, 1980). It is not unlikely that many insects avoid the toxic effects of a given plant toxin by excreting most of the ingested compound without absorbing it. However, when substantial amounts of a toxic compound are absorbed through the gut, sequestration may be an inevitable phenomenon as a prerequisite for coping with the potentially intoxicative effects of the allomone. Indeed, it would seem to be energetically maladaptive to absorb substantial quantities of plant toxins only to immediately excrete and metabolize them in the Malpighian tubules.

Phytophagous insects may not be metabolically limited by cardenolides, as are vertebrates, if these arthropods contain ionic pumps in selected tissues that are insensitive to these compounds. Alkali metal transport across the midgut epithelium of larvae and adults of three lepidopterous species, *Manduca sexta*, *Hyalophora cecropia*, and *Danaus plexippus*, occurs via energy-linked processes not requiring Na^+, K^+-ATPases (Jungreis and Vaughan, 1977). On the other hand, neuronal tissues of these insects contain these enzymes which are sensitive to the cardenolide ouabain (Vaughan and Jungreis, 1977). However, these ATPases of the milkweek-feeding monarch, *D. plexippus*, are 300 times less sensitive than those of the other two insects, neither of which feed on cardenolide-containing plants. Vaughan and Jungreis suggest that the *in vivo* insensitivity of lepidopterans to cardenolides is due to the high levels of potassium in the blood which prevent these steroids from binding to neuronal ATPases. Presumably, these compounds could be excreted or sequestered before they reach ATPase-sensitive concentrations in the blood.

Excretory processes may play a key role in allowing insects to exploit food plants that contain potentially toxic compounds. Isolated Malpighian tubules of *Zonocerus variegatus*, a species that feeds on toxic plants and sequesters toxins, excreted cardiac glycosides more efficiently than those of *Locusta migratoria*, an acridid that does not feed on toxic plants (Rafaeli-Bernstein and Mordue, 1978). Furthermore, it was possible to induce a higher rate of excretion of cardiac glycosides by isolated tubules by feeding cardiac glycosides to adults of *Z. variegatus*. These results indicate that the ability of insects to utilize toxic plants may be correlated, among other things, with a very efficient system for excreting the toxins, provided that they are not present in the blood in ordinately high levels.

An examination of the fates of ingested phytoecdysones in the silkworm *Bombyx mori* demonstrates that several strategies are utilized by insects to prevent the development of high levels of these potential growth regulators. Hikino *et al.* (1975a) showed that ingested ecdysterone is very slowly absorbed and rapidly excreted. Furthermore, rapid catabolism reduced the levels of absorbed ecdysterone. When another phytoecdysone—ponasterone A—was ingested, it was rapidly converted to either ecdysterone or inokosterone, which were then metabolized to poststerone, a compound with negligible molting-hormone activity (Hikino *et al.*, 1971, 1975b). Thus, the concurrence of slow absorption and rapid catabolism–excretion ensures that high internal levels of ecdysterones are not readily obtained. In this context, it may be significant that before high levels of ecdysone are secreted in tobacco hornworm larvae, prior to molting, the larvae cease feeding and purge their guts (Riddiford, 1976). If such behavior is common in herbivorous insects, it would constitute an effective means of removing exogenously derived molting hormones from the gut at a critical developmental period. Indeed, gut purging prior to molting would constitute a highly adaptive means of eliminating plant toxins at an especially crucial period in development.

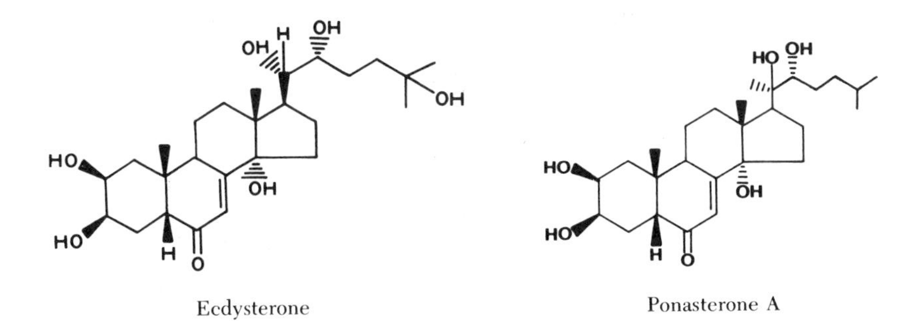

Ecdysterone

Ponasterone A

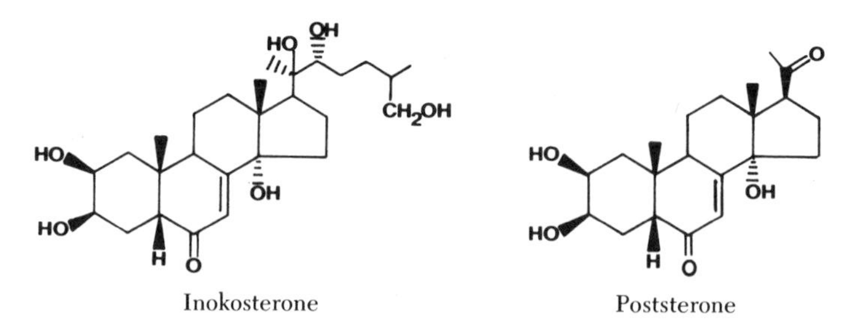

Inokosterone Poststerone

While few investigations have been undertaken on the fates of ingested compounds that are sequestered (von Euw *et al.*, 1967; Reichstein *et al.*, 1968; Duffey and Scudder, 1974; Duffey *et al.*, 1978), it has nevertheless been possible to conclude that the sequestration fingerprint exhibited by an insect reflects a combination of metabolism, excretion, and selective absorption–sequestration. While both the qualitative and quantitative natural products composition of the host plant will be critical determinants of what compounds are ultimately sequestered, the problem is further complicated by unknown genotypic factors which regulate the ability of the herbivores to either sequester or retain stored plant compounds (Rothschild *et al.*, 1975; Duffey, 1980).

Selective alkaloidal sequestration, as exemplified by the cinnibar moth *Tyria jacobaeae* feeding on ragwort *Senecio jacobaeae*, may be typical of a diversity of insects that feed on plants containing pyrrolizidine alkaloids. The alkaloids jacobine, jacozine, and jacoline, which are dominant in the plant, constitute minor products of sequestration in the insect (Aplin and Rothschild, 1972). On the other hand, senecionine and seneciphylline, which are present as either trace or minor constituents in the ragwort, represent the major alkaloids stored by the insect. This selective absorption and sequestration may be further reflected in the low alkaloidal content of the frass which probably is composed predominantly of the major alkaloids present in the plant. However, in addition to selective absorption–sequestration and excretion, the composition of the stored alkaloids is greatly influenced by the metabolic capacities of *T. jacobaeae*.

Seneciphylline, which is ingested as a water-soluble *N*-oxide, is converted to a lipophilic free base presumably more amenable to sequestration in selected body tissues (fat body?). Furthermore, when fed on groundsel, *Senecio vulgaris*, the larvae of *T. jacobaeae* metabolize the *Senecio* alkaloid(s) to a metabolite that constitutes about 20% of the stored compounds (Aplin and Rothschild, 1972). Although the significance of the production of this metabolite, which contains a modified ester substituent attached to a retronecine type nucleus is unknown, it may well represent a true detoxication process. Obviously, its role as a potential predator deterrent cannot be overlooked.

In contrast to *T. jacobaeae*, the pyrgomorphid grasshopper *Zonocerus variegatus* metabolizes ingested pyrrolizidine alkaloids to their *N*-oxides. Whereas the leaves of the food plant, *Crotolaria retusa*, contain about 45% of the free alkaloid monocrotaline and 55% of the *N*-oxide, the bodies of the pyrgomorphids contain nearly 75% of the *N*-oxide, the free alkaloid content being about 25% (Bernays *et al.*, 1977).

The lygaeid *Oncopeltus fasciatus* also metabolizes some ingested plant toxins and selectively sequesters others (Duffey and Scudder, 1974). The polar cardenolide ouabain can be absorbed unchanged and sequestered in the dorsolateral space fluid, whereas the nonpolar cardiac glycoside digitoxin is converted to at least two polar metabolites that are ultimately sequestered. *Oncopeltus fasciatus* shows a predilection for absorbing and sequestering polar cardenolides, a development partly reflected in the greater solubility of these compounds in the saliva utilized to dissolve them from the seeds (Duffey *et al.*, 1978). Whereas nonpolar steroids are also ingested, they may prove more toxic to bugs than their polar counterparts. Thus, conversion of nonpolar to polar cardenolides by *O. fasciatus* may reflect a detoxication process that also produces compounds facilely sequestered in the dorsolateral space fluid. However, in addition to the cardenolide-containing fluid of the adult dorsolateral glands, adults of *O. fasciatus* possess metathoracic defensive glands that synthesize aldehydic compounds of proven repellency to predators. Thus, these insects are capable of utilizing defensive secretions which contain compounds of both intrinsic (aldehydes) and extrinsic (cardenolides) origin. Furthermore, since the metathoracic gland fluid of *O. fasciatus* adults also sequesters cardenolides, these plant-derived toxins are present in all the defensive secretions utilized against predators by these insects. Therefore, the distinction between the deterrent effects of *de novo* synthesized allomones and those of plant origin is difficult to make when evaluating the defensive secretions of insects like *O. fasciatus* or other species with comparable defensive systems.

The lack of host–plant fidelity by some generalist herbivores may also contribute to unexpected variations in sequestered plant compounds. Recent studies by Jones and Blum (1979) have demonstrated that the defensive secretions generated by the polyphagous grasshopper *Romalea microptera* are highly idiosyncratic. It appears that the great variations in the chemistry of the defensive exudates of individual grasshoppers are correlated with the unpredictable polyphagy manifested by each grasshopper. Since the defensive secretions of this species appear to primarily contain sequestered plant compounds (Eisner *et al.*, 1971a), their compositions should be greatly influenced by the plant species fed upon. Thus, it seems probable that the qualitative nature of the diet of polyphagous insects will constitute a critical correlate of the chemical composition of defensive secretions containing sequestered plant natural products.

Individual plants or plant populations can exhibit great variation in their capacity to biosynthesize natural products and, as a consequence, there are simply no grounds for generalizing about what compounds will be sequestered by an insect after feeding on a given species of plant. Superimposed on this variation in plant chemistry are genetic differences in the insect's ability to sequester and retain plant toxins (Rothschild *et al.*, 1975; Duffey, 1980). Plant–herbivore "coevolution" can be represented as a series of adaptations–counteradaptions in which herbivore breakthroughs are countered by effective recombinations drawn from the diverse gene pool available to the plants (Atsatt and O'Dowd, 1976). Thus, herbivore susceptibility can be best ensured through host nonuniformity that disrupts evolutionary specialization or tracking by the herbivores.

Dolinger *et al.* (1973) have similarly concluded that individual variation in natural products of plants is a mechanism to avoid resistant strains of specialist herbivores. In an investigation of the predation of the lycaenid *Glaucopsyche lygdamus* on inflorescences of three species of lupines, it was concluded that the butterfly placed a frequency-dependent selection pressure on the plant populations. Lupines that grow in areas ecologically unfavorable to the insects produce only low amounts of single alkaloid, whereas plants that suffer only minor predation contain variable levels of three to four alkaloids. On the other hand, lupines that are heavily preyed upon produce high levels of nine alkaloids with little variation displayed between individual plants. These results strongly indicate that the quantitative and qualitative alkaloidal composition of the plants may reflect predatory pressure from the lycaenid larvae. The escape of the lupines from their herbivores, both in terms of time and space, would seem to reflect the recombinative potential of the diverse gene pool that controls the biogenesis of alkaloids.

Since the concentrations of plant natural products are also drastically affected by environmental stresses (del Moral, 1972; Moore *et al.*, 1967; Wender, 1970), it can be anticipated that the natural products chemistry of plants may be highly variable. Feir and Suen (1971) have also demonstrated that the concentrations of plant toxins may alter during development and aging of a plant. Overall, the combination of herbivore pressure, environmental stresses, and plant developmental stage appear to be able to drastically affect both the natural products fingerprint of a plant and that of its insect sequestrator. This certainly appears to be the case for several asclepiadaceous species. For example, Brower (1969) concluded that, since monarchs reared on *Asclepias syriaca*, *A. incarnata*, and *A. tuberosa* were palatable to predators, the milkweeds lacked cardiac glycosides. However, Duffey (1970) demonstrated that these northern species of North American milkweeds are rich in cardenolides and that monarchs reared on these plants sequester these steroids. Rothschild *et al.* (1975) also reported great variability in cardenolide patterns of *Asclepias* species, further noting that the cardenolide content of the plant can constitute a critical determinant of what the adult butterfly ultimately sequesters. The presence of known and unknown factors that affect the natural products chemistry of plants certainly militates against deriving conclusions about the specific identity of the food plant an insect developed upon based on quantitative approximations of the amount of plant-derived compounds sequestered by wild-caught butterflies (Brower *et al.*, 1975).

Recent investigations by Roeske *et al.* (1976) have demonstrated that larvae of *D. plexippus* exhibit distinctive cardenolide sequestration patterns depending on the species of milkweed fed upon. Although larvae may assimilate cardenolides from different asclepiadaceous species with about equal efficiency, they exhibit considerable differences in their capacity to store cardenolides. Furthermore, the efficiency of both storage and egestion of cardenolides varies with the plant species utilized as food, so that the ultimate sequestration fingerprint manifested by the adult reflects the idiosyncratic reaction of the insect to the particular blend of cardenolides ingested. Monarch larvae can sequester individual cardenolides more efficiently at lower than higher dosages, favoring the sequestration of glycosides over genins (Seiber *et al.*, 1980). Furthermore, some cardenolides (e.g., uscharidin) are converted to other compounds before storage. These results reemphasize that the cardenolide sequestration pattern exhibited by the adult may in no way mirror the cardenolide content of its host plant.

Insects may be very conservative sequestrators in terms of the natural products potential of their food plants. Isman *et al.* (1977b) have demonstrated that the field-collected lygaeids *Oncopeltus fasciatus* and *Lygaeus kalmi kalmi* vary widely in their cardenolide content, often containing next to no detectable cardenolides. This variation appears to be correlated with interspecific and intraspecific differences in the content of the host plant species as well as differences in the content of plant organs on which the insects were feeding. These insects often feed on nectar, which may lack cardenolides entirely, and these lygaeids would not have access to the cardenolide-rich portions of the plants. Furthermore, these bugs often colonize species of *Asclepias* that contain a low concentration of these compounds. It is difficult to interpret the significance of these results vis-à-vis predators, since the predators of these lygaeids are largely unknown. Nevertheless, it is obvious that low concentrations of cardenolides, or other natural products, in an insect may not simply indicate that it has fed on a plant with a low cardenolide content. Rather, it may also be a consequence of the insect selectively feeding on plant parts that are deficient in the steroids. In short, the sequestration fingerprint exhibited by an insect appears to be correlated with a variety of factors, both plant and insect derived.

Idiosyncratic feeding patterns may have geographical correlates as well, so that an insect population may have a propensity for feeding on different plant parts in different parts of its range. Ralph (1976) has demonstrated that in eastern North America, survival of larvae of *O. fasciatus* is promoted by their feeding aggregatively on milkweed seeds. Since asclepiadaceous seeds are often rich in cardenolides, this feeding behavior can promote the ability of larvae to ingest and possibly sequester these steroids. On the other hand, in California, the development of larvae of *O. fasciatus* on milkweeds lacking seeds will probably affect the amount of cardenolides that are ingested. Whether this shift from feeding on seeds to vegetative portions of the milkweed plant is correlated with high cardenolide concentrations in the latter has not been determined.

It seems remarkable that the specific storage sites in insects of sequestered plant compounds are largely unknown. This is especially true for lepidopterous species that have obtained these natural products as larvae. Brower and Glazier (1975) report that the monarch butterfly contains more cardenolides in the wings than either the thorax or abdomen, but the emetic potency of the abdominal cardenolides is much greater than that in the other body regions. However, no information is given as to specific tissues in which the cardenolides are stored.

Recently, Nishio and Blum (1980) provided the first detailed study on the anatomical distribution of sequestered plant natural products in an insect. Cardenolides are widely distributed in the body of the adult monarch with the wing scales being a particularly rich source of these compounds. Wing muscles contained only traces of these natural products whereas both the fat body and gut were moderately rich in cardenolides. Surprisingly, the hemolymph contained a higher concentration of these steroids than any other tissue except the wing scales. The significance of high levels of cardenolides in the hemolymph is unknown, although this finding demonstrates that the pool of these compounds in the body of the adult is not in a static state.

In general, except for the few cases in which it has been demonstrated that specific structures have been evolved as cardenolide sinks (Scudder and Duffey, 1972; Duffey and Scudder, 1974) or that glands evolved to synthesize allomones also sequester cardenolides as well (von Euw *et al.*, 1967; Scudder and Duffey, 1972; Duffey and Scudder, 1974), our knowledge of the specifics of sequestration by insects is woefully inadequate. Until both the fates and sites of stored plant natural products are illuminated, the physiological bases for sequestration will remain obscure at best.

Although it has been repeatedly stressed that compounds such as sequestered cardenolides constitute the *sine qua non* for the rejection of insects by predators (Brower *et al.*, 1967; Brower and Moffitt, 1974), there are really no grounds for eliminating the possibility that these sequestrators may also derive unpalatability from other classes of compounds that are simultaneously stored. Many plant species can be many splendored toxin producers, subjecting their herbivores to a potpourri of pharmacologically active natural products. For example, in addition to cardenolides, *Asclepias syriaca* synthesizes nicotine (Marion, 1939), thus possibly subjecting its herbivores to an alkaloidal toxin in addition to the cardiac glycosides. But even if the assumption that an insect species sequestered specific, identifiable compounds were correct, one must confront the real possibility that the insect has synthesized its own natural products which may function as highly effective deterrents to the specific predators with which it contends.

It has been emphasized that aposematic Lepidoptera that sequester plant natural products contain additional repellents or toxins that are not of plant origin (Rothschild *et al.*, 1975). Species of butterflies in the nymphalid genus *Amauris* feed on milkweeds lacking cardenolides but are rejected by all avian predators evaluated (Rothschild *et al.*, 1970b). These butterflies possess an extremely unpleasant odor that may be related to their unpalatability. Another danaid, *Euploea core*, which sequesters cardenolides—but not in emetic quantities—has been found to be highly distasteful to a wide variety of birds (Marsh *et al.*, 1977). Similarly, Rothschild *et al.* (1975) concluded that deterrent factors other than sequestered cardenolides were associated with the unpleasantness or toxicity of at least some of the morphs of *Danaus chrysippus*. High concentrations of choline esters in the tissues of arctiid moths (Morley and Schacter, 1963; Frazer and Rothschild, 1962; Bisset *et al.*, 1960) may play key roles in providing these insects with highly distasteful or toxic properties. In addition, the presence of powerful odors in insects such as danaids may provide predators with an "early warning" system for avoiding unpleasant gustatory or emetic experiences with these potentially toxic insects.

Ultimately, the adaptive significance of sequestering plant toxins has been interpreted in terms of the protection that the sequestrators derive from predators. In the case of the monarch, this has led to the development of a palatability spectrum that is reported to reflect the emetic potential of the insect to its laboratory predator, the blue jay, *Cyanocitta cristata bromia* (Brower *et al.*, 1972; Brower and Moffitt, 1974). However, what is actually being measured is a spectrum of putative emetic activity rather than a gustatory (palatable) reaction on the part of the blue jays. Swynnerton (1919) noted that the danaid *Danaus chrysippus* was rejected by birds on taste alone, but this is certainly not the case for either the blue jay (Brower and Brower, 1964) or the British jay *Garrulus glandarius* (Reichstein *et al.*, 1968). Thus, the term "palatability spectrum" is not terminologically accurate, since the reaction of predators to insects of differing toxicities is not predicated on gustatory sensitivity. Indeed, it is not certain that it is a true emetic reaction which is being observed (Brower, 1969), since emesis constitutes a regurgitative response based on stimulation of higher nerve centers rather than simple gastric irritation. It is not known whether the vomiting encountered with blue jays that have been fed cardenolide-rich monarchs (Brower and Moffitt, 1974) actually reflects true pharmacological emesis or simply stimulation of the highly enervated gastric mucosa.

Evaluations of the efficacies of sequestered plant compounds as predator deterrents will ultimately depend on the sensitivity of the candidate predator to the test insect. Rothschild (1970) reported considerable variation among individual birds in their reactions to danaid butterflies. Some birds rejected the insects based on taste alone, whereas others required an emetic experience before they learned to avoid the butterflies. Certain mammals ate all danaids with avidity without experiencing any adverse reactions. Since a wide spectrum of avian species is reported to feed on monarchs (cardenolide content unknown) (Petersen, 1964), it is obvious that conclusions about the evolutionary value of sequestration as a defensive mechanism will ultimately depend on the reaction of the predators specifically selected for analysis. As a corollary, it is conceivable that rejection at a gustatory or olfactory level, rather than at a gastric one, well may be of fundamental importance in comprehending the evolution of mimetic complexes. Conceivably, Batesian mimics could evolve odors similar to their protected models and thus derive protection from predators which reject the models at an olfactory level.

It has been suggested that the storage of cardenolides imposes a physiological penalty on the sequestrator in terms of reduced viability (Brower *et al.*, 1972). However, Erickson (1973) demonstrated that when monarch larvae were reared on *Asclepias curassavica*, the most cardenolide-rich species utilized, the larvae exhibited an accelerated growth rate and were the most efficient in converting digested matter into biomass compared to larvae feeding on other *Asclepias* species. The findings of Dixon *et al.* (1978) also demonstrate that there is no "physiological cost" to monarch larvae developing on cardenolide-rich plants. Similarly, Smith (1977) reported that when larvae of *Danaus chrysippus* were reared on five species of milkweeds, four rich in cardenolides and the other cardenolide-free, the former larvae grew much more rapidly and produced larger adults than those that developed on the milkweed lacking these steroids. Isman (1977) reached a similar conclusion after studying the dietary influence of cardenolides on the growth and development of *Oncopeltus fasciatus*. There was no indication that the cardenolide content of milkweed seeds in any way imposed a physiological cost in terms of larval growth or development. Beyond the fact that these results do not support the hypothesis of Brower *et al.* (1972), it is important to realize that since we have no information on the mechanism of sequestration of cardenolides by either larvae or adults of *D. plexippus*, it is simply impossible to relate storage of these steroids to any presumed physiological or biochemical lesions. Indeed, it may well be that cardenolide sequestration in this species reflects processes that do not result in either an appreciable expenditure of energy or a metabolic liability, as appears to be the case for the lygaeid *Oncopeltus fasciatus* (Duffey *et al.*, 1978; Duffey, 1980).

Indeed, if *O. fasciatus* is typical of herbivores of asclepiadaceous plants in sequestering cardenolides primarily by a physical process rather than an enzymatic one, then it will certainly not be necessary to regard the uptake of these compounds in terms of a metabolic liability. For example, if the cardenolides in the wings of the monarch butterfly are sequestered by a physical process, then their storage (=excretion?) can be affected with little metabolic expenditure. These polar cardenolides, which represent the major pool of sequestered steroids, would be expected to be of limited emetic potential, which is precisely what Brower and Glazier (1975) demonstrated in their study of the distribution of cardenolides in adult butterflies. On the other hand, the sequestration of highly emetic cardenolides in lipid pools would result in these compounds being selectively concentrated in the fat body, which is concentrated in the abdomen, a conclusion consistent with what Brower and Glazier (1975) reported. Again dissolution of these cardenolides would not represent an energetic liability and would not result in toxicosis since these steroids are effectively prevented from reaching sensitive tissues at concentrational levels that are metabolically harmful. Thus, the monarch may have successfully coevolved with its milkweed host by employing physical processes to capture high levels of potentially toxic compounds in tissues that appropriately reflect the polarity of these compounds.

If we are to comprehend the evolutionary bases for the coevolution of herbivores and their toxic host plants, it will ultimately be necessary to comprehend how the herbivore processes the different classes of compounds that are introduced into its gut. It may eventually develop that a lifestyle restricted to many toxic plants is neither perilous nor energetically prohibitive, but rather simply the evolutionary compromise of being able to channel a proportion of the absorbed potential toxins to benign storage sites, in the absence of any great metabolic expenditures, while continuously excreting the balance. However, even if this proves to be the case, it will tell us nothing about how the insects avoid the potentially toxic effects of many of these compounds, some of which may be produced as a result of insect damage to the plant (Woodhead and Cooper-Driver, 1979). If virtually each plant species can be visualized as producing a highly distinctive array of natural products (Levin, 1976), then it will be necessary to explore how the herbivore processes the individual compounds that it has ingested. The selective absorption, metabolism, and excretion by insects of plant allomones must be understood if the evolutionary triangle of the plant, insect, and insect predator is to be illuminated. The sequestrative unpredictability of some populations of plant herbivores (Rothschild *et al.*, 1975) emphasizes how poorly we comprehend the interrelationship of the elements in these biological triangles.

The mixed-function oxidases may well be of great importance in enabling some insects to cope with plant toxins (Brattsten *et al.*, 1977; Brattsten, 1979) but studies on the fates of specific natural products after ingestion have generally not been implemented. The lack of detailed investigations on the metabolic idiosyncrasies of monophagous species vis-à-vis the plant natural products in their diets renders it impossible to analyze the biochemical characteristics that are identified with each species' ability to process these compounds once they are in the gut. The ability of a species to sequester compounds as disparate as cardenolides and alkaloids (Rothschild, 1972) may indicate that this insect has evolved multiple metabolic devices for coping with these diverse compounds, but it is equally possible that these plant toxins are metabolized by a common system (e.g., mixed-function oxidases). Beyond that, we must recognize that we do not understand how a single species can sequester both steroids (cardenolides) and alkaloids, or for that matter, how this species can selectively store members of either class of compounds.

The last two decades have produced a host of significant papers which demonstrate clearly that insects are versatile sequestrators of a wide variety of plant allomones (Reichstein *et al.*, 1968; Rothschild *et al.*, 1970b; Rothschild, 1972; Duffey, 1980). The time has come to elucidate the physiological and biochemical adaptations that have characterized these evolutionary developments. It seems appropriate to now analyze the plant, insect, and insect predator relationship in detail so that its significance is comprehended as the exciting ecological phenomenon that it surely represents.

Chapter **18**

Recently Identified Defensive Compounds*

I. LACTONES

Meinwald *et al.* (1977) and Sugawara *et al.* (1979b) identified plagiolactone ($C_{10}H_{12}O_2$) in the thoracic and abdominal glandular exudates of larvae of the chrysomelid beetles *Plagiodera versicolora* and *Linaeidea aenea;* epiplagiolactone ($C_{10}H_{12}O_2$) is also produced by these species. These lactones are structurally related to chrysomelidial, the main component in the secretion. The absolute configuration of plagiolactone has been determined by synthesis (Meinwald and Jones, 1978).

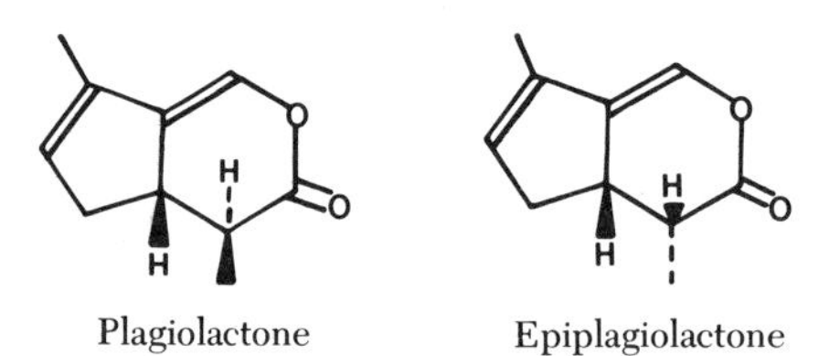

Plagiolactone   Epiplagiolactone

Octadecanolide ($C_{18}H_{34}O_2$), eicosanolide ($C_{20}H_{38}O_2$), and 22-docosanolide ($C_{22}H_{42}O_2$) were identified as Dufour's gland constituents of halictine bees in the genera *Agapostemon*, *Augochlorella*, *Dialictus*, *Evylaeus*, and *Lasioglossum* (Hefetz *et al.*, 1978).

*Structural formulas are only given for compounds not already presented in the tables.

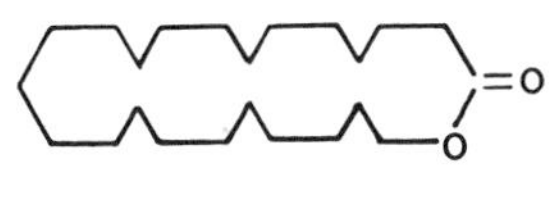

22-Docosanolide

In addition to the typical 1,4-benzoquinones, the sternal gland secretion of the tenebrionid beetle *Apsena pubescens* is fortified with 8-hydroxyisocoumarin ($C_9H_6O_3$) and 3,4-dihydro-8-hydroxyisocoumarin ($C_9H_8O_3$) (Lloyd *et al.*, 1978a).

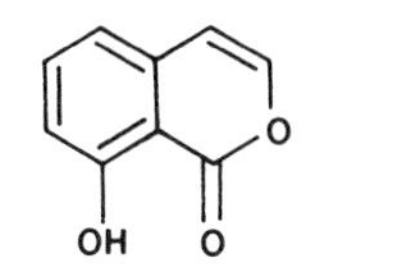

8-Hydroxyisocoumarin

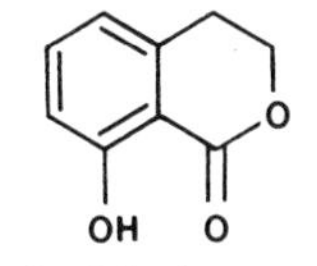

3,4-Dihydro-8-hydroxyisocoumarin

Smith *et al.* (1979) reported that nepetalactone ($C_{10}H_{14}O_2$) was a major constituent in the prothoracic glandular secretion of the stick insect *Graeffea crouani*.

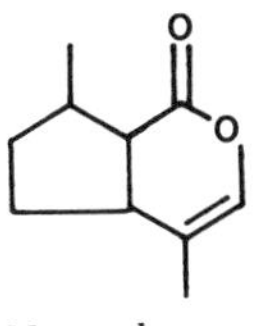

Nepetalactone

γ-Decalactone ($C_{10}H_{18}O_2$) has been identified as a minor constituent in the mandibular gland secretion of the ant *Atta sexdens* (Schildknecht, 1976).

The metasternal scent gland secretion of the beetle *Phoracantha synonyma* (Cerambycidae) contains a series of macrocyclic lactones which include decan-9-olide ($C_{10}H_{18}O_2$), (Z)-dec-4-en-9-olide ($C_{10}H_{16}O_2$), and 11-hydroxytetradec-5-en-13-olide ($C_{14}H_{24}O_3$) (Moore and Brown, 1976).

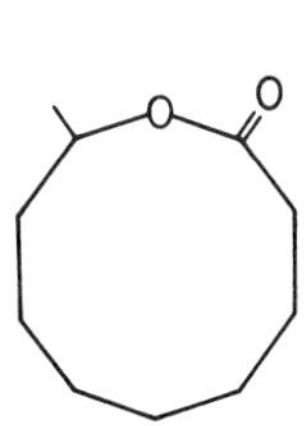

Decan-9-olide

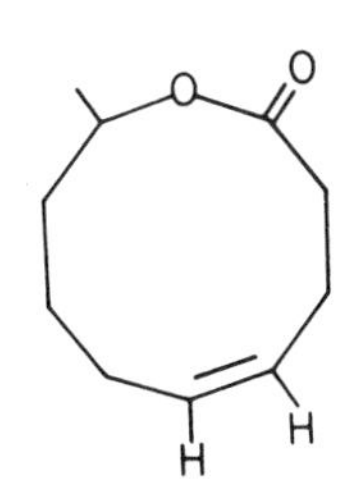

(Z)-Dec-4-en-9-olide

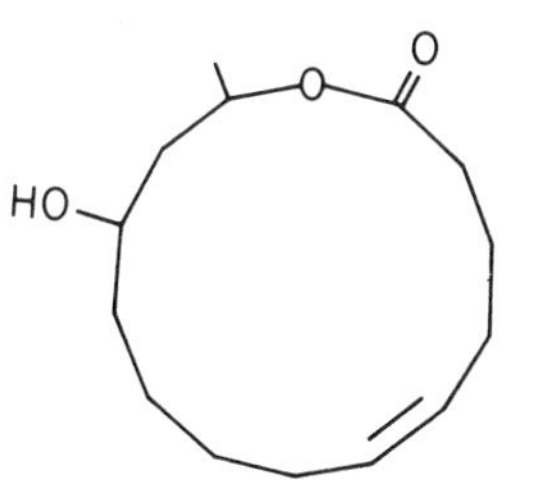

11-Hydroxytetradec-5-en-13-olide

II. QUINONES

2,5-Dimethyl-1,4-benzoquinone ($C_8H_8O_2$), 2,3-dimethyl-1,4-benzoquinone ($C_8H_8O_2$), and 2,3,5-trimethyl-1,4-benzoquinone ($C_9H_{10}O_2$) are present in the glandular secretion of the opilionid *Paecilaemella eutypta* (Opiliones:Cosmetidae); the exudate of *P. quadripunctata* contains only the dimethylquinones (Jones *et al.*, 1977). 2,3-Dimethyl-1,4-benzoquinone has also been identified in the cosmetids *Cynorta nannacornuta*, *Eucynortula albipunctata*, and the gonyleptids *Nesopachylus monocerus* and *Zygopachylus albimarginis* (Roach *et al.*, 1980). All species except *Z. albimarginis* also produce 2,3,5-trimethyl-1,4-benzoquinone.

Five different quinones have been detected in the explosive discharges of bombardier beetles (Eisner *et al.*, 1977; Roach *et al.*, 1979). *p*-Benzoquinone ($C_6H_4O_2$) is produced by *Brachinus quadripennis*, *B. sublaevis*, *Stenaptinus insignis*, *Metrius contractus*, *Goniotropis nicaraguensis*, *G.* sp., *Pachyteles* spp., *P. striola*, *P. longicornis*, *Ozaena magna*, *Physea hirta*, and *Platycerozaena panamensis*. All these species, with the exception of *M. contractus*, produce toluquinone ($C_7H_6O_2$) as does *Homopterus arrowi*. 2,3-Dimethyl-1,4-benzoquinone accompanies *p*-benzoquinone and toluquinone in the exudates of *B. quadripennis*, *B. sublaevis*, *G. nicaraguensis*, *G.* sp., *Pachyteles* spp., *P. hirta*, *P. panamensis*, and *Ozaena magna*. *Homopterus arrowi* was the only species of bombardier beetle that discharged a secretion containing 2-methoxy-3-methyl-1,4-benzoquinone ($C_8H_8O_3$) whereas *P. panamensis* was singular in producing 2-ethyl-1,4-benzoquinone ($C_8H_8O_2$).

Hydroquinone ($C_6H_6O_2$) has been identified as a pygidial gland product of a large number of dytiscid species especially in the subfamily Colymbetinae (Dettner, 1979).

Two naphthoquinones, 6-methylnaphthoquinone ($C_{11}H_8O_2$) and naphthoquinone ($C_{10}H_6O_2$), constitute the major defensive compounds produced by the opilionid *Phalangium opilio* (Wiemer *et al.*, 1978).

The secretion of *Gastrolina depressa* contains juglone ($C_{10}H_6O_3$), which is probably derived from the host plant (Matsuda and Sugawara, 1980).

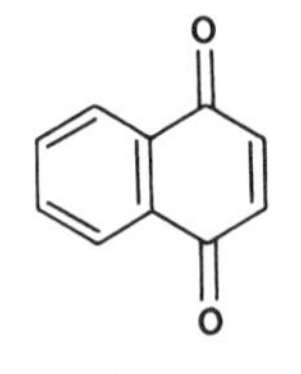

Naphthoquinone

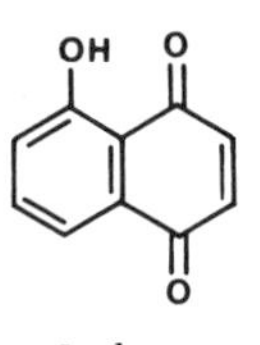

Juglone

III. PHENOLS

The opilionid *Cynorta astora* (Opiliones:Cosmetidae) produces 2,3-dimethylphenol ($C_8H_{10}O$) and 2-methyl-5-ethylphenol ($C_9H_{12}O$) in its paired defensive glands, whereas the secretion of *Zygopachylus albimarginis* (Gonyleptidae) contains only the dimethylphenol (Eisner *et al.*, 1977). The cosmetid *Eucynortula albipuncta* also produces 2,3-dimethylphenol and 2-methyl-5-ethylphenol whereas the gonyleptid *Zygopachylus albimarginis* is distinctive in synthesizing 2,3,4-trimethylphenol ($C_9H_{12}O$) (Roach *et al.*, 1980).

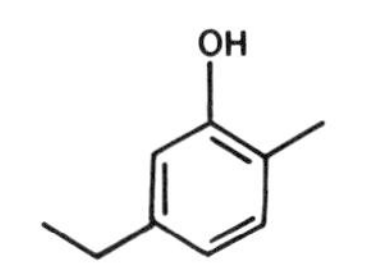

2-Methyl-5-ethylphenol

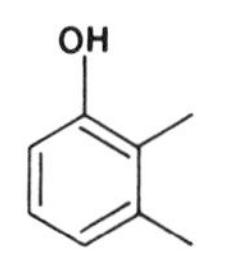

2,3-Dimethylphenol

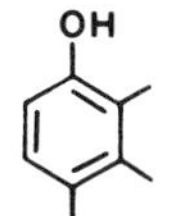

2,3,4-Trimethylphenol

Phenol (C_6H_6O) has been identified in the defensive exudates of the millipedes *Pseudopolydesmus erasus* and *Euryurus maculatus* (Duffey *et al.*, 1977). The secretions of these two species, plus that of *Oxidus gracilis*, also contain guaiacol ($C_7H_8O_2$) (Duffey and Blum, 1977).

p-Cresol (C_7H_8O) and phenol constitute the main defensive products in the secretion of the cockroach *Archiblatta hoeveni* (Maschwitz and Tho, 1978).

IV. KETONES

4-Methyl-3-hexanone ($C_7H_{14}O$) has been identified as a concomitant of iridodial in the pygidial gland secretion of the staphylinid beetle *Staphylinus olens* (Fish and Pattenden, 1975).

The mandibular gland secretion of workers of the ant *Atta sexdens* contains 6-methyl-5-hepten-2-one ($C_8H_{14}O$) as a trace constituent (Schildknecht, 1976). The same compound is present in the secretion of a *Calomyrmex* species (Brown and Moore, 1979).

6-Methyl-5-hepten-2-one ($C_8H_{14}O$) thoroughly dominates the mandibular gland secretions of the ants (workers) *Formica perpilosa*, *F. lasioides*, *F. neogagates*, *F. nitidiventris*, *F. schausfussi*, and three unidentified *Formica* species (Duffield *et al.*, 1977). 3-Octanone ($C_8H_{16}O$) is the major constituent in the mandibular gland secretion of a *Pheidole* sp. and *Tetramorium sericeiventre* and *T. guineense* (Longhurst *et al.*, 1979b).

The opilionid (Phalangiidae) *Leiobunum nigripalpi* discharges a defensive exudate containing (*E*)-4-methyl-4-hexen-3-one ($C_7H_{12}O$) and 4-methyl-3-hexanone ($C_7H_{14}O$) (Jones *et al.*, 1977). *Leiobunum leiopenis* produces (*E*)-4-methyl-4-hepten-3-one ($C_8H_{14}O$) and 4-methyl-3-heptanone ($C_8H_{16}O$) in its defensive glands. 4-Methyl-3-heptanone is also synthesized by *L. ventricosum* and *Hadrobonus maculosus* (Jones *et al.*, 1976b). The dimethyl alkenones (*E*)-4,6-dimethyl-6-octen-3-one ($C_{10}H_{18}O$) and (*E*)-4,6-dimethyl-6-nonen-3-one ($C_{11}H_{22}O$) are present in the defensive glands of *L. calcar* (Jones *et al.*, 1976b, 1977) and *L. longipes* (Jones *et al.*, 1977), respectively.

(*E*)-4-Methyl-4-hexen-3-one (*E*)-4-Methyl-4-hepten-3-one

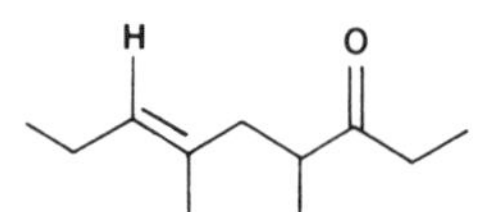

(*E*)-4,6-Dimethyl-6-nonen-3-one

Males and females of the velvet ant *Dasymutilla occidentalis* (Hymenoptera:Mutillidae) generate 4,6-dimethyl-3-octanone ($C_{10}H_{20}O$) and 4,6-dimethyl- 3-nonanone ($C_{11}H_{22}O$) in their capacious mandibular glands (Fales *et al.*, 1980).

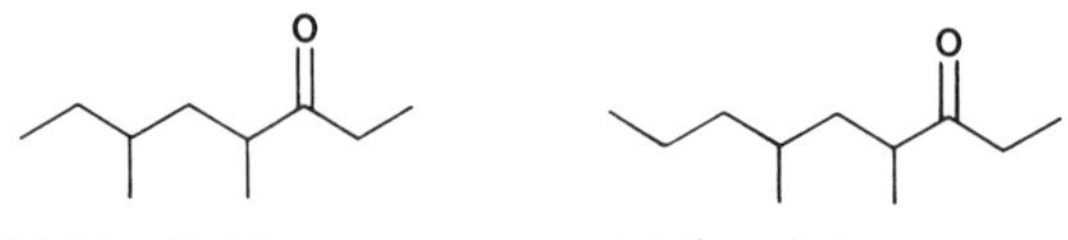

4,6-Dimethyl-3-octanone

4,6-Dimethyl-3-nonanone

Disturbed males, females, and workers of the bumblebee *Bombus lapidarius* emit a mandibular gland secretion fortified with a large series of methyl ketones. 2-Pentanone ($C_5H_{10}O$), 2-heptanone ($C_7H_{14}O$), 2-nonanone ($C_9H_{18}O$), 2-undecanone ($C_{11}H_{22}O$), and 2-tridecanone ($C_{13}H_{26}O$) form part of the defensive exudate of this species (Cederberg, 1977).

The frontal gland secretions of both major and minor soldiers of the termite *Schedorhinotermes lamanianus* (Rhinotermitidae) contain 1-dodecen-3-one ($C_{12}H_{22}O$), 3-dodecanone ($C_{12}H_{24}O$), and 1,13-tetradecadien-3-one ($C_{14}H_{24}O$) (Prestwich, 1975; Prestwich *et al.*, 1975).

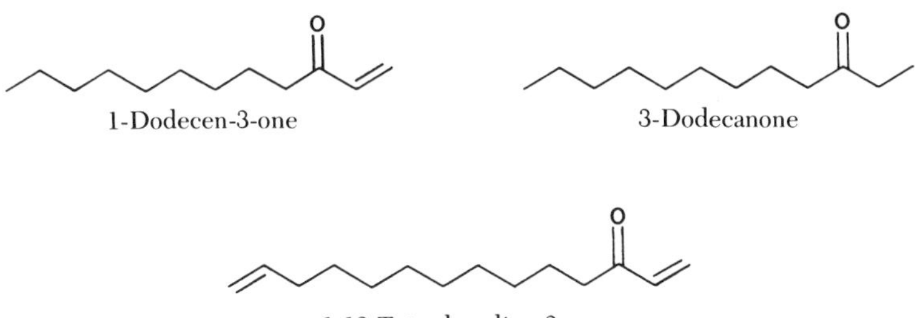

Females of the bees (Andrenidae) *Andrena praecox*, *A. denticulata*, and males of *A. fuscipes* produce 6-methyl-5-hepten-2-one ($C_8H_{14}O$) in their mandibular glands (Tengö and Bergström, 1976). 3-Octanone ($C_8H_{16}O$) is present in the exudates of males and females of *A. praecox* and *A. helvola*, as well as females of *A. clarkella*. 2-Nonanone ($C_9H_{18}O$) was detected as a glandular constituent of males of *A. helvola*. Both males and females of *A. fucata* and *A. clarkella* produce 7-methyl-4-octanone ($C_9H_{18}O$) in their mandibular glands. 3-Decanone ($C_{10}H_{20}O$) is a characteristic glandular product of males and females of *A. praecox*, *A. helvola*, and females of *A. clarkella*. Males of *A. helvola* synthesize 2-undecanone ($C_{11}H_{22}O$) in their mandibular glands, whereas another C_{11} ketone, 2-methyl-5-decanone, is produced by males and females of *A. fucata* as well as females of *A. clarkella*. Females of *A. clarkella* also produce 3-tetradecanone ($C_{14}H_{28}O$) and 4-pentadecanone ($C_{15}H_{30}O$) (Tengö and Bergström, 1976).

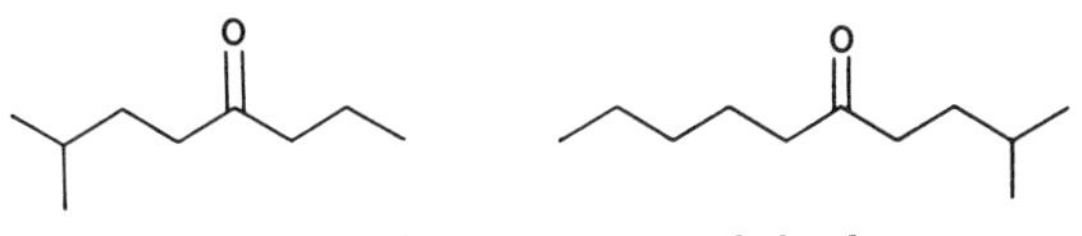

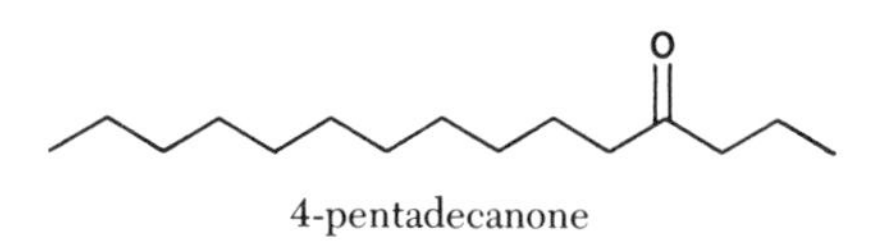

4-pentadecanone

2-Pentanone ($C_5H_{10}O$) and 2-heptanone ($C_7H_{14}O$) constitute pygidial gland products of the carabid beetle *Dyschirius wilsoni* (Moore and Brown, 1979).

Two diterpenes, 3β-hydroxy-7β-kemp-8(9)-en-6-one($C_{20}H_{30}O_2$) and 2β-acetoxy-3β-hydroxy-7β-kemp-8(9)-en-6-one($C_{22}H_{32}O_4$), have been characterized as frontal gland constituents of the termite *Nasutitermes octopilis* (Prestwich *et al.*, 1979).

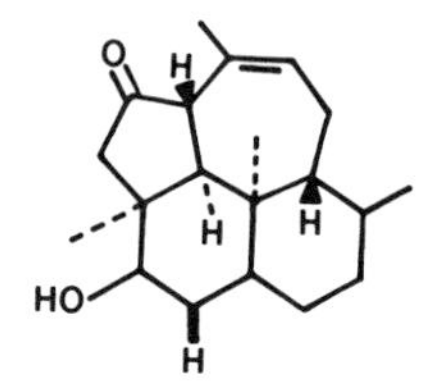

3β-Hydroxy-7β-kemp-8(9)-en-6-one

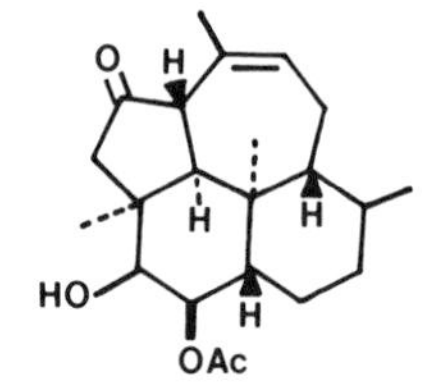

2β-Acetoxy-3β-hydroxy-7β-kemp-8(9)-en-6-one

Soldiers of the termite *Amitermes unidentatus* discharge a frontal gland secretion that is dominated by 2-undecanone ($C_{11}H_{22}O$), 2-tridecanone ($C_{13}H_{26}O$), 2-pentadecanone ($C_{15}H_{30}O$), and 2-heptadecanone ($C_{17}H_{34}O$) (Meinwald *et al.*, 1978).

Camphor ($C_{10}H_{16}O$) has been identified in worker ant extracts of *Formica rufa* and *F. polyctena* (Bühring *et al.*, 1976). The glandular origin of this ketonic terpene is unknown.

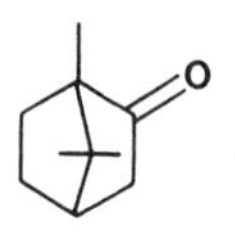

Camphor

Piperitone ($C_{10}H_{16}O$) has been characterized in the terpene-rich metathoracic gland secretion of the true bug *Niesthrea louisianica* (Rhopalidae) (Aldrich *et al.*, 1979).

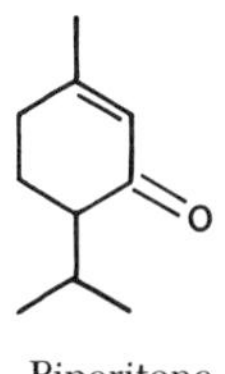

Piperitone

2-Tridecanone ($C_{13}H_{26}O$) has been identified as a minor component in the anterior portion of the prothoracic defensive gland of larvae of *Schizura concinna* (Notodontidae) (Weatherston *et al.*, 1979).

V. ALCOHOLS

The metasternal scent gland secretions of the adult bugs (Coreidae) *Holopterna allata*, *Anoplocnemis dallasiana*, and *A. montandorii* contain 1-butanol ($C_4H_{10}O$) and 1-hexanol ($C_6H_{14}O$) as minor constituents (Prestwich, 1976). 1-Hexanol was the only alcohol detected in the secretion of *Acanthocoris obscuricornis*.

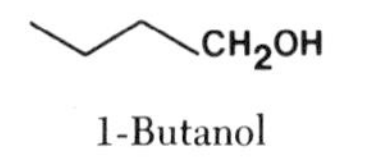

1-Butanol

Major workers from one population of the ant *Oecophylla longinoda* contain, in addition to the previously identified 1-hexanol, both 1-octanol ($C_8H_{18}O$) and 1-nonanol ($C_9H_{20}O$) in their mandibular gland secretions (Bradshaw *et al.*, 1975). 2-Undecanol ($C_{11}H_{24}O$) is the dominant constituent in the mandibular gland secretion of the ant *Tetramorium termitobium* (Longhurst *et al.*, 1979b).

6-Methyl-5-hepten-2-ol ($C_8H_{16}O$) has been identified as a minor constituent in the mandibular gland secretions of workers of the ants *Formica perpilosa* and *F. nitidiventris* (Duffield *et al.*, 1977).

The paired defensive glands of the opilionid (Phalangiidae) *Leiobunum nigripalpi* produce 4-methyl-3-hexanol ($C_7H_{16}O$), whereas (*E*),(*E*)-2,4-dimethylhexa- 2,4-dien-1-ol ($C_8H_{14}O$) is present in the secretions of *L. leiopenis* and *L. calcar* (Jones *et al.*, 1977). (*E*),(*E*)-2,4- Dimethylhepta-2,4-dien-1-ol ($C_9H_{16}O$) has been identified in the exudate of *L. leiopenis*.

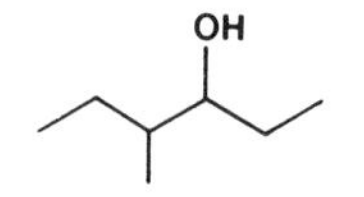

4-Methyl-3-hexanol

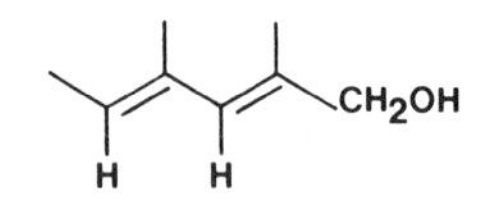

(*E*),(*E*)-2,4-Dimethylhexa-2,4-dien-1-ol

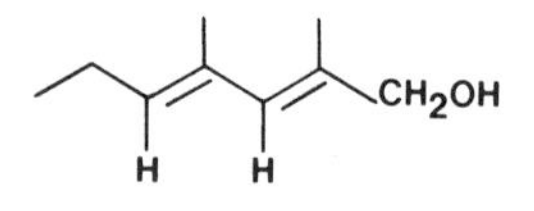

(*E*),(*E*)-2,4-Dimethylhepta-2,4-dien-1-ol

2-Decen-1-ol ($C_{10}H_{16}O$) has been characterized in the mandibular gland secretion of the sphecid wasp *Sceliphron caementarium* (Hefetz and Batra, 1979).

2-Decen-1-ol

2-Phenylethanol ($C_8H_{10}O$), 1-tridecanol ($C_{13}H_{28}O$), and farnesol ($C_{15}H_{26}O$) constitute minor products in the mandibular gland exudates of workers of the ant *Atta sexdens* (Schildknecht, 1976). The mandibular gland secretion of the ant *Calomyrmex* sp. contains nerol ($C_{10}H_{18}O$) and geraniol ($C_{10}H_{18}O$) (Brown and Moore, 1979).

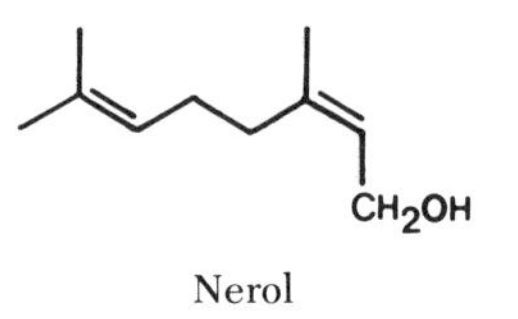

Nerol

The terpene-pure metathoracic gland secretion of the rhopalid bug *Niesthrea louisianica* contains terpinen-4-ol ($C_{10}H_{18}O$) and thymol ($C_{10}H_{14}O$); perilla alcohol ($C_{10}H_{18}O$) dominates the dorsal abdominal gland exudate (Aldrich *et al.*, 1979).

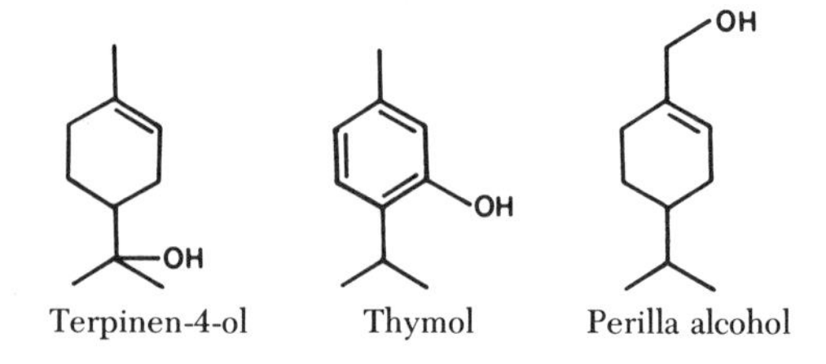

Terpinen-4-ol Thymol Perilla alcohol

The mandibular gland secretion of females of the bee *Andrena helvola* (Andrenidae) contains nerol, whereas geraniol is present in the exudates of female *A. helvola* and males and females of *A. labiata* (Tengö and Bergström, 1976). Citronellol ($C_{10}H_{20}O$) occurs in the secretions of females of *A. helvola* and *A. fucata;* both males and females of *A. labiata* produce this terpene. 1-Dodecanol ($C_{12}H_{26}O$) is detectable as a glandular product of males and females of *A. labiata*, whereas 1-tetradecanol ($C_{14}H_{30}O$) is produced by *A. labiata* males. *trans*-Farnesol ($C_{15}H_{26}O$) is a major glandular constituent in the secretions of *A. fuscipes* and *A. nigriceps* (Tengö and Bergström, 1976).

The mandibular gland secretion of males, females, and workers of the bumblebee *Bombus lapidarius* contains 2-nonanol ($C_9H_{20}O$), 2-undecanol ($C_{11}H_{24}O$), geraniol ($C_{10}H_{18}O$), and citronellol ($C_{10}H_{20}O$) (Cederberg, 1977). These alcohols form part of the defensive exudate utilized by *B. lapidarius*.

Three monoterpene alcohols have been identified in worker ant extracts of *Formica rufa* and *F. polyctena* (Bühring *et al.*, 1976). Borneol ($C_{10}H_{18}O$) and isoborneol ($C_{10}H_{18}O$) were detected in extracts of the combined head–thorax, and may constitute mandibular gland products. Isopulegol ($C_{10}H_{18}O$), on the other hand, was detected in abdominal extracts as well.

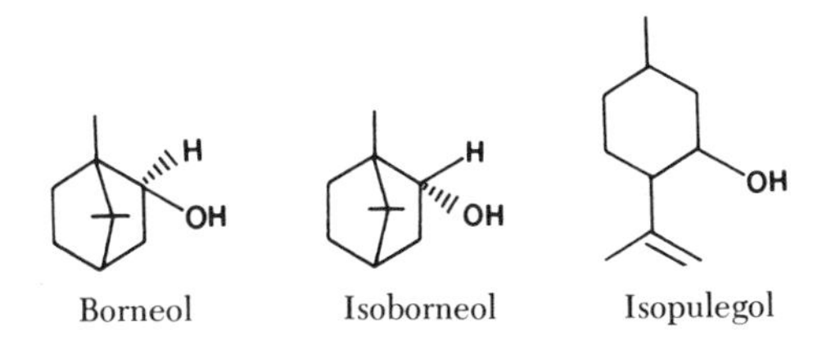

Borneol Isoborneol Isopulegol

Platambin, $7\alpha H,5\beta,10\alpha$-eudesm-4(14)-ene-$1\alpha,6\alpha$-diol ($C_{15}H_{26}O_2$), was identified as a trace constituent of the prothoracic gland secretion of the water beetle (Dytiscidae) *Platambus maculatus* (Schildknecht *et al.*, 1975a).

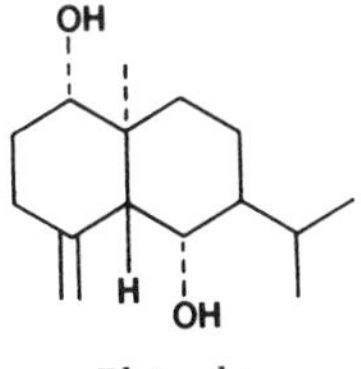

Platambin

A series of novel diterpene alcohols have been identified as frontal gland products of the termites *Trinervitermes gratiosus* and *T. bettonianus* (Prestwich *et al.*, 1976a). The compounds, the "trinervitenes," include trinervi-9β-ol ($C_{20}H_{32}O$), trinervi-2β,3α-diol ($C_{20}H_{32}O_2$), and isotrinervi-2β,3α-diol ($C_{20}H_{32}O_2$). Trinervi-9β-ol and trinervi-2β,3α-diol have also been identified in the frontal gland exudates of soldiers of *Nasutitermes rippertii* (Vrkoč *et al.*, 1977). Another monohydroxytrinervitene, isotrinervi-2β-ol ($C_{20}H_{32}O$), was identified in soldiers of *T. gratiosus;* both mono- and diterpene concentrations were variable in different populations of these species (Prestwich, 1978). Recently, a monohydroxykempene ($C_{20}H_{32}O$) was characterized in the secretion of *N. ephratae* (Prestwich, 1979).

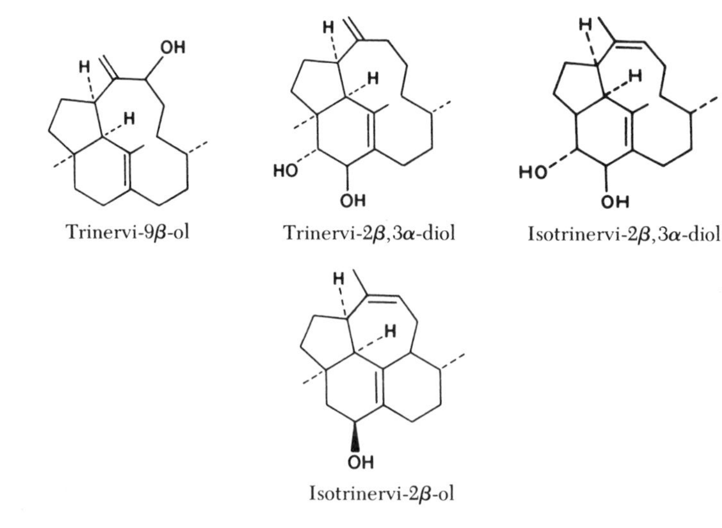

VI. ACIDS

Formic acid (CH_2O_2) has been identified in the cervical gland secretion of larvae of the notodontids *Catochria catocaloides* (Geertsema *et al.*, 1976), *Notodonta anceps*, *Cerura* (=*Dicranura*) *vinula* (Hintze, 1969), and *Schizura concinna* (Weatherston *et al.*, 1979).

The pygidial gland secretion of the beetle (Staphylinidae) *Zyras humeralis* contains 3-methylbutyric acid ($C_5H_{10}O_2$) (Kolbe and Proske, 1973). This acid is also produced in a sternal abdominal gland of the staphylinids *Eusphalerum longipenne*, *E. anale*, *E. minutum*, and *E. abdominale* (Klinger and Maschwitz, 1977).

Acetic acid ($C_2H_4O_2$) is present in the metasternal scent gland secretions of the adult bugs (Coreidae) *Holopterna allata, Anoplocnemis dallasiana, A. montandorii,* and *Acanthocoris obscuricornis* (Prestwich, 1976). Hexanoic acid ($C_6H_{12}O_2$) has been detected in the defensive exudates of *H. allata* and *A. montandorii.*

Butyric acid ($C_4H_8O_2$) is the major compound present in the mandibular gland exudates of females, workers, and males of *Bombus lapidarius* (Cederberg, 1977). This compound is also present in the secretions of all *Bombus* and *Psithyrus* species examined.

Benzoic acid ($C_7H_6O_2$) is in the defensive discharge of the centipede *Geophilus vittatus* (Geophilomorpha:Geophilidae) (Jones *et al.*, 1976a).

Neric ($C_{10}H_{16}O_2$), geranic ($C_{10}H_{16}O_2$), and decanoic ($C_{10}H_{20}O_2$) acids are synthesized in the mandibular glands of workers of the ant *Atta sexdens* (Schildknecht, 1976).

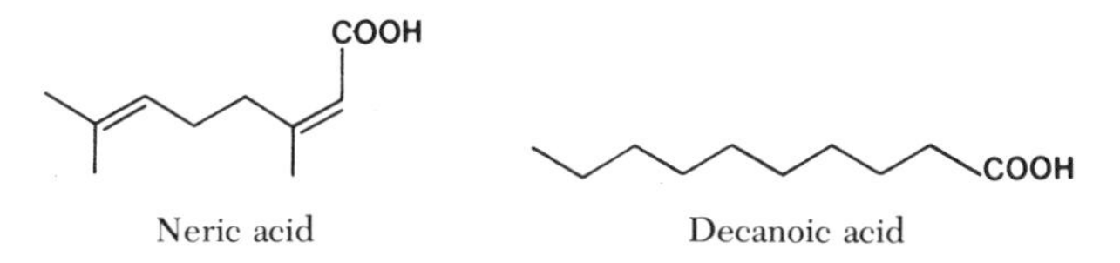

Neric acid Decanoic acid

Dettner and Schwinger (1977) identified phenylacetic acid ($C_8H_8O_2$) and 3-indoleacetic acid ($C_{10}H_9NO_2$) as pygidial gland products of dytiscid beetles in six genera. Recently, Dettner (1979) analyzed the glandular secretions of 45 species of water beetles and reported that these two acids were produced by several additional dytiscid species mainly in the subfamily Hydroporinae. In addition to these acids, hydroporine and noterine species also commonly produced phenylpyruvic acid ($C_9H_8O_3$) and *p*-hydroxyphenylacetic acid ($C_8H_8O_3$). Species in the subfamily Colymbetinae characteristically synthesized benzoic acid ($C_7H_6O_2$) and *p*-hydroxybenzoic acid ($C_7H_6O_3$). *Ilybius* species were distinctive in producing tiglic acid ($C_5H_8O_2$). Phenylacetic acid was detected in a haliplid species whereas the secretions of hygrobiid species were unusual in containing α-hydroxyhexanoic acid ($C_6H_{12}O_2$), S-methyl-2-hydroxy-4-mercaptobutanoic acid ($C_5H_{10}O_2S$), and related products (Dettner 1979; Dettner and Schwinger, 1980).

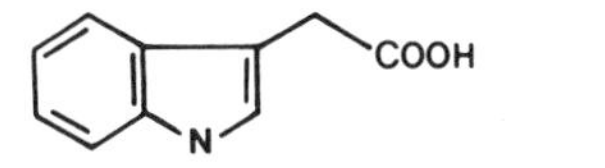

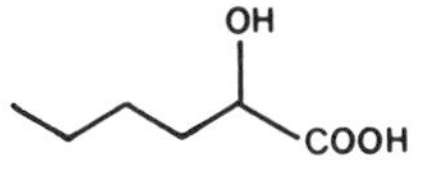

3-Indoleacetic acid α-Hydroxyhexanoic acid

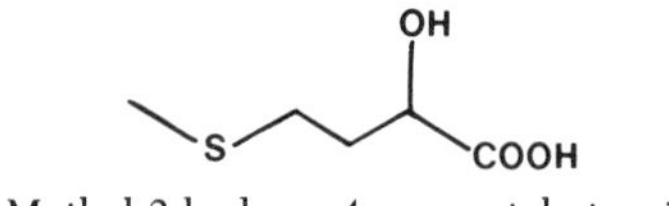

S-Methyl-2-hydroxy-4-mercaptobutanoic acid

In addition to benzoic ($C_7H_6O_2$) and 3-methylbutyric ($C_5H_{10}O_2$) acids, the defensive secretion from the lateral paired glands of the millipede *Pseudopolydesmus serratus* (Polydesmida:Polydesmidae) contains myristic ($C_{14}H_{28}O_2$) and stearic ($C_{16}H_{32}O_2$) acids (Conner *et al.*, 1977).

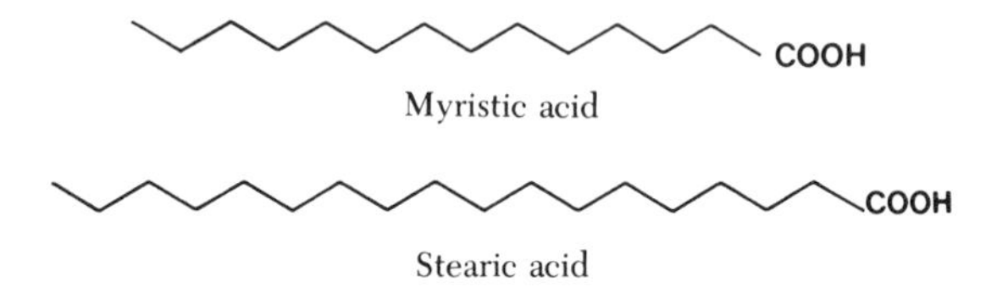

Myristic acid

Stearic acid

The osmeterial secretion of fifth-instar larvae of *Papilio protenor* (Papilionidae) contains isobutyric acid ($C_4H_8O_2$) and 2-methylbutyric acid ($C_5H_{10}O_2$) (Honda, 1980).

VII. ALDEHYDES

The metasternal scent gland secretions of the stink bugs (Pentatomidae) *Aspongopus* sp., *Delegorguella lautus, Caura rufiventris,* and *Veterna patula* are characterized by the presence of (*E*)-2-hexenal ($C_6H_{10}O$) and 4-oxo-(*E*)-2-hexenal ($C_6H_8O_2$) (Prestwich, 1976). The exudates from the dorsal abdominal glands of larvae of the leaf-footed bugs (Coreidae) *Holopterna allata* and *Anoplocnemis dallasiana* also contain these two enals. Adults of *H. allata, A. dallasiana,* and *Acanthocoris obscuricornis* produce hexanal ($C_6H_{12}O$) in their metasternal scent glands (Prestwich, 1976).

Hexa-2,4-dienal (C_6H_8O) and octa-2,4-dienal ($C_8H_{12}O$) accompany the corresponding α,β-unsaturated aldehydes in the metasternal scent gland secretions of adults of *Oncopeltus fasciatus* (Hemiptera:Lygaeidae) (Games and Staddon, 1973a).

CHO

Hexa-2,4-dienal

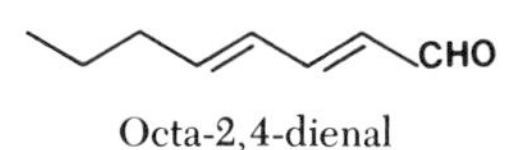

Octa-2,4-dienal

Benzaldehyde (C_7H_6O) is one of the aromatic constituents in the defensive exudate of the centipede *Geophilus vittatus* (Geophilomorpha: Geophilidae) (Jones *et al.*, 1976a).

(*E*)-2-Hexenal ($C_6H_{10}O$) is produced in a sternal abdominal gland of the staphylinids *Eusphalerum longipenne, E. anale, E. minutum*, and *E. abdominale* (Klinger and Maschwitz, 1977).

The secretions of the millipedes *Pseudopolydesmus erasus, Euryurus maculatus, Cherokia georgiana ducilla, C. g. georgiana, C. g. latassa, Motyxia tularea*, and *Sigmoria nantahalae* contain a high concentration of benzaldehyde (C_7H_6O) (Duffey *et al.*, 1977).

p-Hydroxybenzaldehyde ($C_7H_6O_2$) has been identified as a pygidial gland product of dytiscid beetles in the subfamilies Colymbetinae and Dytiscinae (Dettner, 1979; Newhart and Mumma, 1979b).

One of the minor constituents in the metasternal gland secretion of *Phoracantha synonyma* (Cerambycidae) is 2-hydroxy-6-methylbenzaldehyde ($C_8H_8O_2$) (Moore and Brown, 1976).

2-Hexenal is a metathoracic gland product of the bugs (Cydnidae) *Adrisa magna* and *Macrocystus japonensis* (Hayashi *et al.*, 1976). These two cydnids, along with *Aethus nigritus*, also produce 2-octenal ($C_8H_{14}O$) in this gland. Both *M. japonensis* and *A. nigritus* synthesize the C_{10} enal—2-decenal ($C_{10}H_{18}O$)—as well. Another cydnid, *Adrisa numeensis*, produced 4-oxo-(*E*)-2-hexenal in its scent gland (Smith, 1978).

Adults of the stink bug *Vitellus insularis* (Pentatomidae) discharge metasternal scent gland products consisting mostly of 4-oxo-(*E*)-2-hexenal ($C_6H_8O_2$) and (*E*)-2-decenal (Smith, 1974); the latter has also been identified in the secretion of another pentatomid, *Lamprophara bifasciata* (Smith, 1978).

Benn *et al.* (1977) identified 2-hexenal and 2-octenal in the abdominal secretion of the cockroach *Platyzosteria novaeseelandiae*.

The opilionid *Leiobunum nigripalpi* (Phalangida:Phalangiidae) emits an exudate fortified with (*E*),(*E*)-2,4-dimethylhexa-2,4-dienal ($C_8H_{12}O$) (Jones *et al.*, 1977).

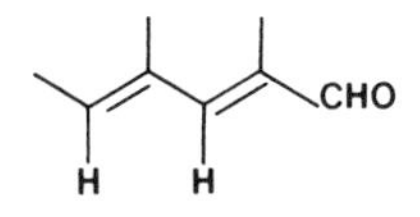

(*E*),(*E*)-2,4-Dimethylhexa-2,4-dienal

Nonanal ($C_9H_{18}O$), decanal ($C_{10}H_{20}O$), and tetradecanal ($C_{14}H_{28}O$) constitute the aldehydic compounds in the mandibular gland secretion of the ant (workers) *Atta sexdens* (Schildknecht, 1976).

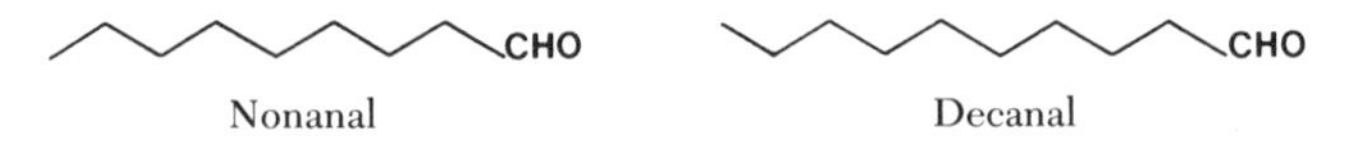

Nonanal Decanal

(*E*)-2-Hexenal ($C_6H_{10}O$) is present in the dorsal abdominal gland secretions of the rhopalid bugs *Niesthrea louisianica* and *Jadera haematoloma;* (*E*)-2-octenal ($C_8H_{14}O$) also occurs as a product of the latter species (Aldrich *et al.*, 1979).

The metathoracic scent gland secretion of the bug *Leptoglossus phyllopus* (Coreidae) contains several aldehydes including 2-*n*-butyloct-2-enal ($C_{12}H_{22}O$) and both the cis and trans isomers of hexanal trimer ($C_{18}H_{36}O_3$) (Aldrich *et al.*, 1978).

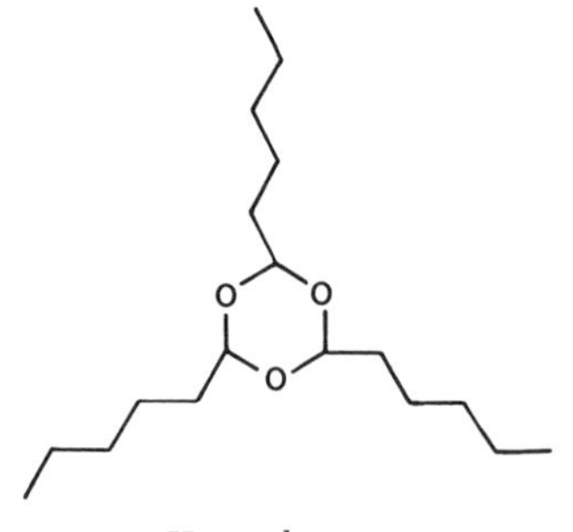

Hexanal trimer

Citronellal ($C_{10}H_{18}O$), has been identified as a mandibular gland product of the female bee (Andrenidae) *Andrena helvola* (Tengö and Bergström, 1976). Females of *A. nigriceps*, males of *A. helvola*, and both males and females of *A. praecox*, *A. labiata*, *A. fuscipes*, and *A. denticulata* produce both isomers of citral ($C_{10}H_{16}O$)—neral and geranial—in their mandibular glands. Dodecanal ($C_{12}H_{24}O$) has been detected in the glandular exudate of males and females of *A. labiata* (Tengö and Bergström, 1976).

The mandibular gland secretion of the stingless bee *Trigona denoiti* essentially contains only neral and geranial, the two stereoisomers of citral (Crewe and Fletcher, 1976).

Nonanal ($C_9H_{18}O$) has been identified in worker ant extracts of *Formica rufa* (Bühring *et al.*, 1976). Two unusual aldehydes, 2-(1-methylethyl)-4-methylhex-2-enal($C_{10}H_{18}O$) and 2-(1-methylethyl)-5-methylhex-

2-enal($C_{10}H_{18}O$), have been characterized in the mandibular gland secretion of the ant *Calomyrmex* species (Brown and Moore, 1979).

2-(1-Methylethyl)-4-methylhex-2-enal 2-(1-Methylethyl)-5-methylhex-2-enal

Iridodial ($C_{10}H_{16}O_2$) is a major product in the pygidial gland exudate of the carabid beetle *Dyschirius wilsoni* (Moore and Brown, 1979).

Chrysomelidial ($C_{10}H_{14}O_2$), an isomer of dolichodial, has been identified in the secretions form the eversible thoracic and abdominal glands of the beetle larvae (Chrysomelidae) *Gastrophysa cyanea* (Blum *et al.*, 1978), *G. atrocyanea* (Sugawara *et al.*, 1979a), *Plagiodera versicolora, P. v. distincta* (Meinwald *et al.*, 1977; Sugawara *et al.*, 1979b), *Phaedon brassicae* (Sugawara *et al.*, 1979a), and *Linaeidea aenea* (Sugawara *et al.*, 1979b). A probable absolute configuration of chrysomelidial has been recently assigned (Meinwald and Jones, 1978). The secretion of *P. v. distincta* also contains two other cyclopentanoid monoterpenes—plagiodial ($C_{10}H_{14}O_2$) and epichrysomelidial ($C_{10}H_{14}O_2$). Recently, Matsuda and Sugawara (1980) identified benzaldehyde (C_7H_6O) and salicylaldehyde ($C_7H_6O_2$) in the secretion of *Chrysomela viginpunctata costella*, whereas that of *C. populi* only contained salicylaldehyde.

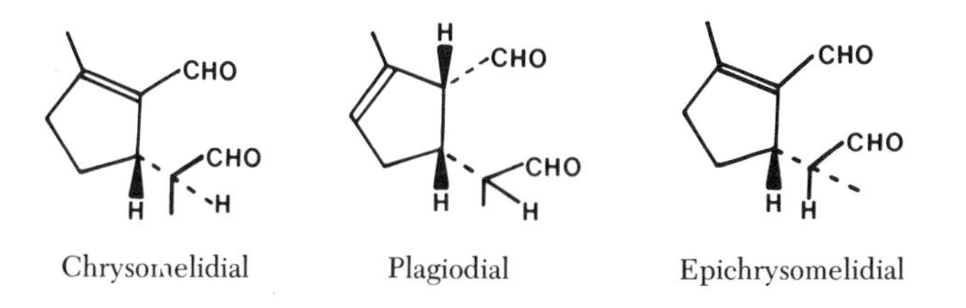

Chrysomelidial Plagiodial Epichrysomelidial

The major constituent in the secretion of minor soldiers of *Ancistrotermes cavithorax* has been identified as ancistrodial ($C_{15}H_{22}O_2$),

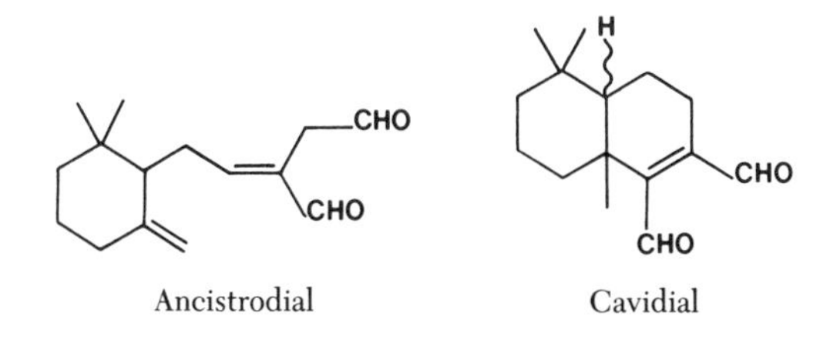

Ancistrodial Cavidial

whereas cavidial ($C_{15}H_{22}O_2$) is one of several unusual compounds produced by major soldiers (Evans *et al.*, 1979).

Isomeric iridodials ($C_{10}H_{16}O_2$) are the major constituents in the defensive secretion derived from the prothoracic glands of the coconut stick insect, *Graeffa crouani* (Smith *et al.*, 1979).

Gyrinidal ($C_{14}H_{18}O_3$) and isogyrinidal ($C_{14}H_{18}O_3$) are pygidial gland products of the gyrinid *Gyrinus frosti* (Newhart and Mumma, 1978).

VIII. ESTERS

Hexa-2,4-dienyl acetate ($C_8H_{12}O_2$) and octa-2,4-dienyl acetate ($C_{10}H_{16}O_2$) are major constituents in the metasternal scent gland secretions of males of the milkweed bug (Lygaeidae) *Oncopeltus fasciatus* (Games and Staddon, 1973b; Everton and Staddon, 1979). The esters are quantitatively less important in the secretions of female bugs.

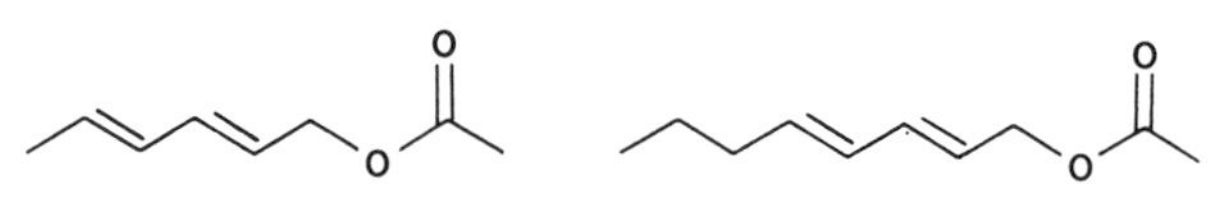

Hexa-2,4-dienyl acetate Octa-2,4-dienyl acetate

Isopentyl acetate ($C_7H_{14}O_2$) an atypical defensive product of hemipterans, has been identified in extracts of *Adrisa numeensis* (Cydnidae) (Smith, 1978).

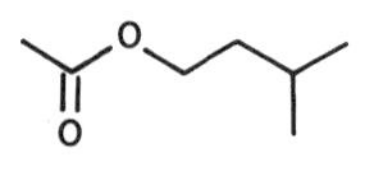

Isopentyl acetate

The stink bugs (Pentatomidae) *Aspongopus* sp., *Delegorguella lautus*, *Caura rufiventris*, and *Veterna patula* discharge metasternal scent gland secretions containing minor amounts of 2-octenyl acetate ($C_{10}H_{18}O_2$) (Prestwich, 1976). 2-Decenyl acetate ($C_{12}H_{22}O_2$) is also present in the excudates of *D. lautus*, *C. rufiventris*, and *V. patula*. The saturated esters *n*-butyl butyrate ($C_8H_{16}O_2$), *n*-hexyl acetate ($C_8H_{16}O_2$), *n*-hexyl butyrate ($C_{10}H_{20}O_2$), and *n*-octyl acetate ($C_{10}H_{20}O_2$) are present in the secretions of the following leaf-footed bugs (Coreidae): *Holopterna allata*, *Anoplocnemis dallasiana*, *A. montandorii*, and *Acanthocoris obscuricornis* (Prestwich, 1976).

The metasternal scent gland secretions of the guava bug (Pentatomidae)

Vitellus insularis and the single-spot bug *Lamprophara bifasciata* contain (*E*)-2-decenyl acetate ($C_{12}H_{22}O_2$) (Smith, 1974, 1978).

Methyl 2-methylbutyrate ($C_6H_{12}O_2$) and ethyl 2-methylbutyrate ($C_7H_{14}O_2$) have been identified as metasternal gland products of *Phoracantha synonyma* (Cerambycidae) (Moore and Brown, 1976). These two esters are also present in the osmeterial secretion of larvae of *Papilio protenor*, along with methyl isobutyrate ($C_5H_{10}O_2$) and ethyl isobutyrate ($C_6H_{12}O_2$) (Honda, 1980).

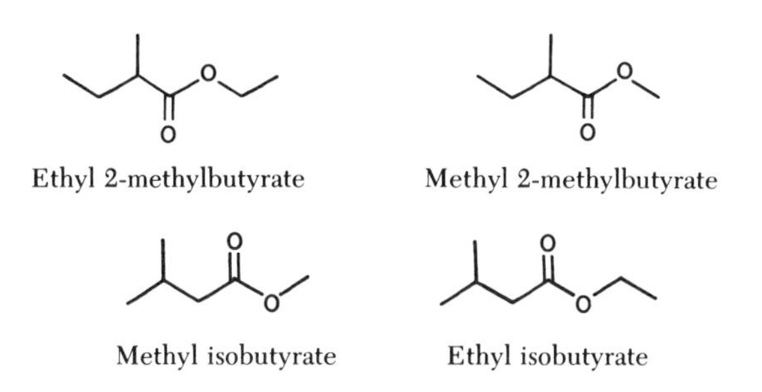

Ethyl 2-methylbutyrate Methyl 2-methylbutyrate

Methyl isobutyrate Ethyl isobutyrate

Hexadecyl acetate ($C_{18}H_{38}O_2$), octadecyl acetate ($C_{20}H_{40}O_2$), and (Z)-11-eicosenyl acetate ($C_{22}H_{42}O_2$) have been identified in the exudates from the eversible glands of larvae of the chrysomelid beetles *Gastrophysa atrocyanea* (Sugawara *et al.*, 1978), *Plagiodera versicolora*, and *Linaeidea aenea* (Sugawara *et al.*, 1979b).

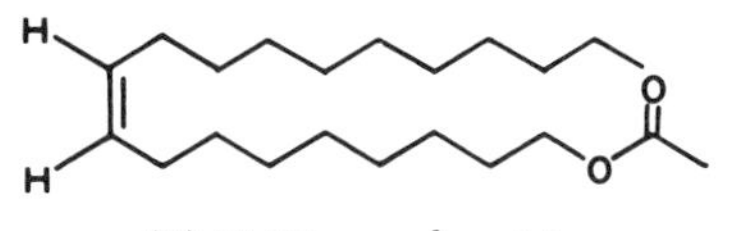

(Z)-11-Eicosenyl acetate

The anterior portion of the prothoracic defensive gland of the notodontid larva *Schizura concinna* produces decyl acetate ($C_{12}H_{24}O_2$) and dodecyl acetate ($C_{14}H_{28}O_2$) as concomitants of the major compound in the secretion, formic acid (Weatherston *et al.*, 1979).

2-Decenyl acetate ($C_{12}H_{22}O_2$) has been identified as a sting-derived alarm pheromone of *Apis dorsata* and *A. florea* (Veith *et al.*, 1978; Koeniger *et al.*, 1979). Octyl acetate ($C_{10}H_{20}O_2$) is also produced by these species.

Mandelonitrile benzoate ($C_{15}H_{11}NO_2$) has been identified in the glandular exudate of the millipede (Polydesmida:Polydesmidae) *Pseudopolydesmus serratus* (Jones *et al.*, 1977).

The pygidial gland secretions of dytiscid beetles in the subfamilies

Colymbetinae and Dytiscinae are a rich source of aromatic esters. Methyl 3,4-dihydroxybenzoate ($C_8H_8O_4$) is widely distributed in the secretions of these species and is infrequently accompanied by methyl 2,5-dihydroxyphenylacetate ($C_9H_{10}O_4$) (Dettner, 1979). The secretions of *Acilius semisulcatus*, *A. sylvanus*, and *A. mediatus* contain methyl *p*-hydroxybenzoate ($C_8H_8O_3$) (Newhart and Mumma, 1979b).

Methyl 3-isopropyl pentanoate ($C_9H_{18}O_2$) is a probable mandibular gland product that has been identified in head–thorax extracts of workers of the ants *Formica rufa* and *F. polyctena* (Bühring *et al.*, 1976).

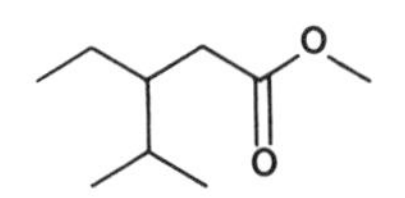

Methyl 3-isopropyl pentanoate

Citronellyl acetate ($C_{12}H_{22}O_2$) and geranyl acetate ($C_{12}H_{22}O_2$) constitute part of the defensive secretion produced in the mandibular glands of the anthophorid bee *Pithitis smaragdula* (Hefetz *et al.*, 1979). The same secretion of the sphecid wasp *Scelliphron caementarium* also contains geranyl acetate (Hefetz and Batra, 1979).

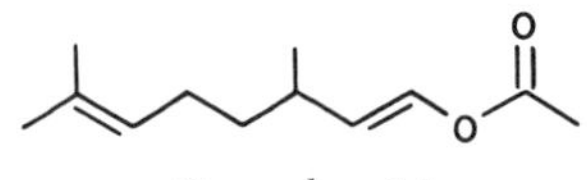

Geranyl acetate

Recently, methyl 2-hydroxy-6-methylbenzoate ($C_9H_{10}O_3$) has been identified in the unusual pygidial gland secretion of the carabid *Dyschirius wilsoni* (Moore and Brown, 1979).

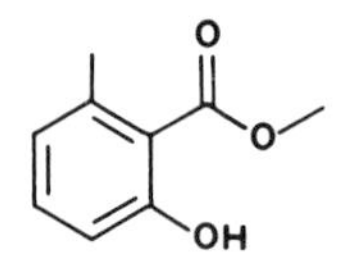

Methyl 2-hydroxy-6-methylbenzoate

Three novel diterpene esters have been identified as part of the frontal gland exudates of soldiers of two termite species. Both *Trinervitermes gratiosus* and *T. bettonianus* produce trinervi-2β,3α,9α-triol 9-*O*-acetate ($C_{22}H_{34}O_4$), whereas trinervi-2β,3α,9α-triol 2,3-*O*-diacetate ($C_{24}H_{36}O_5$) is synthesized by minor soldiers of *T. gratiosus* and trinervi-2β,3α,17-triol 17-*O*-acetate ($C_{22}H_{34}O_4$) by *T. bettonianus* soldiers (Prestwich *et al.*,

1976a,b). Vrkoč *et al.* (1977) identified four additional trinervitene esters in the frontal gland secretions of soldiers of *Nasutitermes rippertii*. The major ester present is trinervi-2β,3α,13β-triol 2,3,13-*O*-triacetate ($C_{26}H_{38}O_6$); trinervi-2β,3α-diol 3-*O*-acetate ($C_{22}H_{34}O_3$), trinervi-2β,3α-diol 2,3-*O*-diacetate ($C_{24}H_{36}O_4$), and trinervi-13-oxo-2β,3α-diol 2,3-*O*-diacetate ($C_{24}H_{34}O_5$) are minor concomitants.

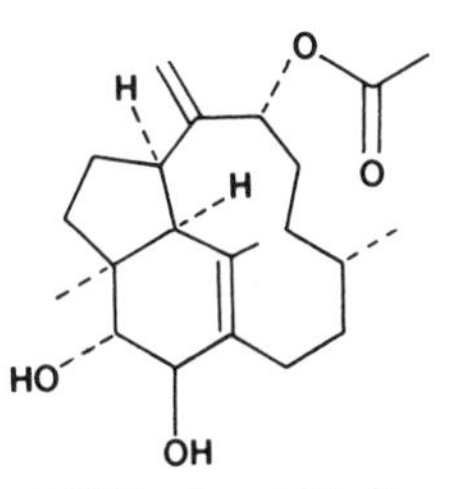

Trinervi-2β,3α,9α-triol 9-*O*-acetate

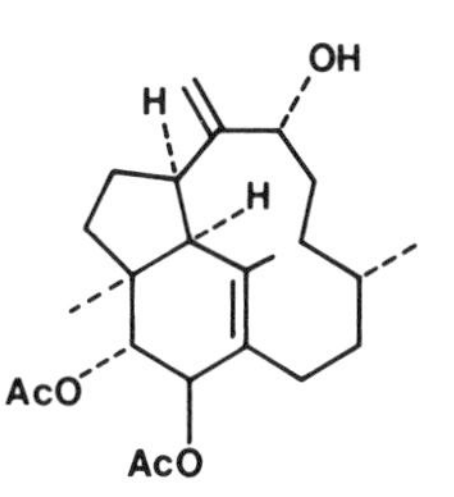

Trinervi-2β,3α,9α-triol 2,3-*O*-diacetate

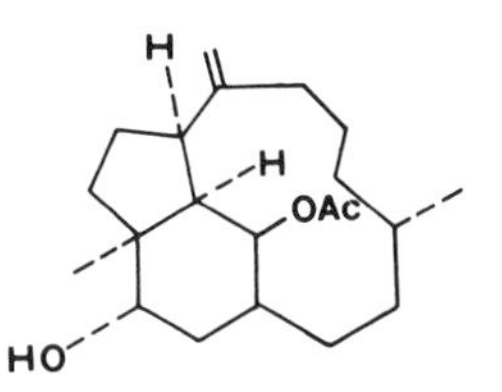

Trinervi-2β,3α-17-triol 17-*O*-acetate

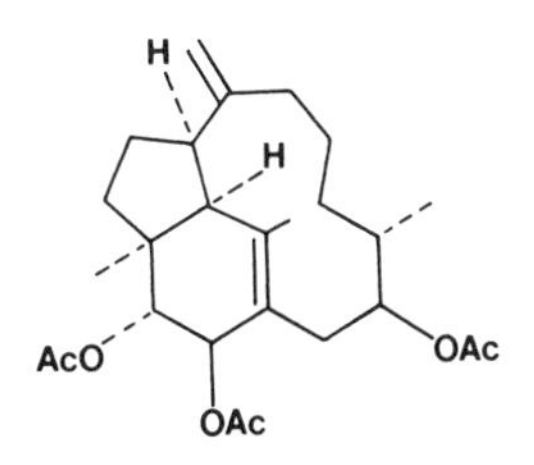

Trinervi-2β,3α,13β-triol 2,3,13-*O*-triacetate

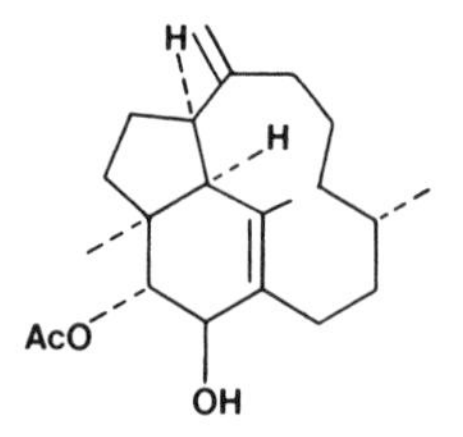

Trinervi-2β,3α-diol 3-*O*-acetate

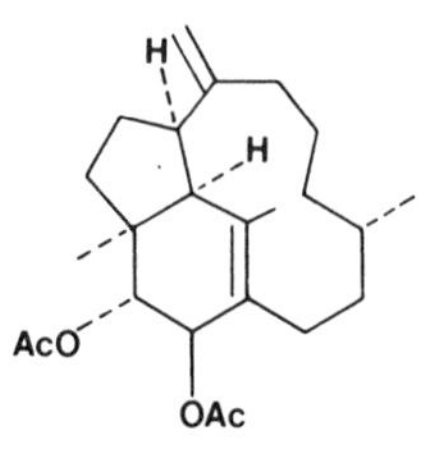

Trinervi-2β,3α-diol 2,3-*O*-diacetate

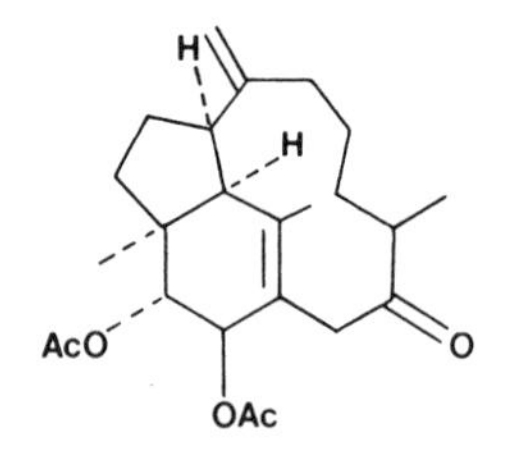

Trinervi-13-oxo-2β,3α-diol 2,3-*O*-diacetate

Two novel diterpene esters have been identified in the frontal gland secretions of soldiers of the termitid *Nasutitermes kempae* (Prestwich *et al.*, 1977b). These compounds, possessing novel tetracyclic cembrene-derived carbon skeletons, have been designated as kempene-1 ($C_{24}H_{34}O_4$) and kempene-2 ($C_{22}H_{30}O_3$).

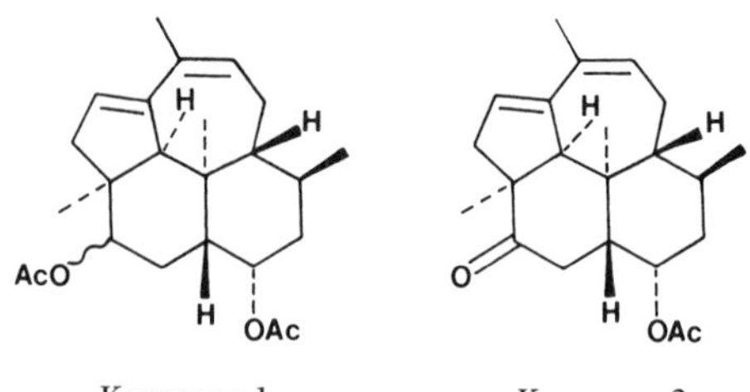

Kempene-1 Kempene-2

IX. HYDROCARBONS

The frontal gland secretion of soldiers of the termite (Termitidae) *Trinervitermes gratiosus* is enriched with the monoterpene hydrocarbons ($C_{10}H_{16}$) α-pinene, β-pinene, camphene, myrcene, and limonene (Prestwich, 1975, 1977). Whereas the exudate from major soldiers lacks β-pinene, that from minor soldiers lacks myrcene; similar qualitative differences are also reported for the frontal gland secretions of major and minor soldiers of *T. bettonianus* (Prestwich, 1975). The glandular exudate of major soldiers contains α-pinene and myrcene; minor soldiers also produce limonene in their frontal glands. Lloyd *et al.* (1978b) analyzed a different population of *T. bettonianus* soldiers and identified, in addition to limonene and myrcene, two additional monoterpene hydrocarbons, β-ocimene X ($C_{10}H_{16}$) and β-ocimene Y ($C_{10}H_{16}$).

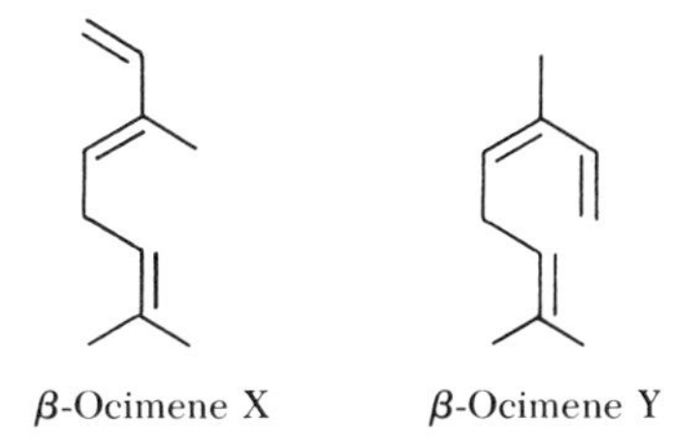

β-Ocimene X β-Ocimene Y

Vrkoč *et al.* (1973) identified a host of hydrocarbons in the frontal gland secretions of soldiers of *Nasutitermes rippertii* and *N. costalis*. Both species eject α-pinene, β-pinene, camphene, α-phellandrene, α-terpinene, γ-terpinene, limonene, myrcene, terpinolene, and Δ^3-carene (all $C_{10}H_{16}$) from their capacious cephalic glands.

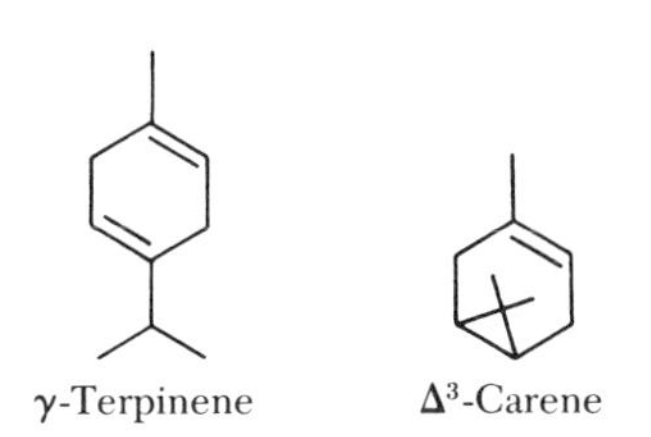

γ-Terpinene Δ^3-Carene

α-Pinene, β-pinene, myrcene, terpinolene, and limonene (all $C_{10}H_{16}$) are present in the metathoracic gland secretion of the rhopalid bug *Niesthrea louisianica;* most of these terpenes are also produced in the dorsal abdominal gland of the rhopalid *Jadera haematoloma* (Aldrich *et al.*, 1979).

β-Pinene ($C_{10}H_{16}$) is a minor constituent in the mandibular gland exudate of workers of the ant *Atta sexdens* (Schildknecht, 1976).

The frontal gland secretions of soldiers of the termitid *Macrotermes subhyalinus* is enriched with an extensive series of aliphatic hydrocarbons (Prestwich *et al.*, 1977a). The major olefins identified were (Z)-9-heptacosene ($C_{17}H_{34}$) and (Z)-9-nonacosene ($C_{19}H_{38}$). These alkenes were accompanied by *n*-tricosane ($C_{23}H_{48}$), *n*-tetracosane ($C_{24}H_{50}$), *n*-pentacosane ($C_{25}H_{52}$), pentacosene ($C_{25}H_{50}$), 3-methylpentacosane ($C_{26}H_{54}$), 5-methylpentacosane ($C_{26}H_{54}$), *n*-heptacosane ($C_{27}H_{56}$), 3-methylheptacosane ($C_{28}H_{58}$), 5-methylheptacosane ($C_{28}H_{58}$), 7-methylheptacosane ($C_{28}H_{58}$), (Z)-9-octacosene ($C_{28}H_{56}$), and (Z)-9-hentriacontene ($C_{31}H_{62}$).

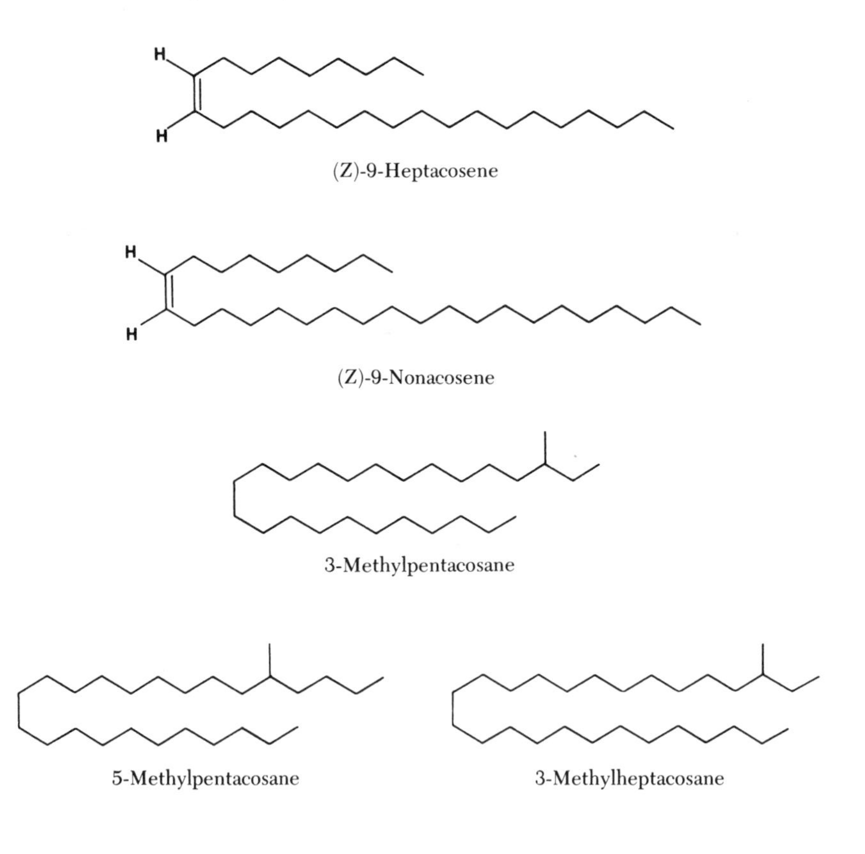

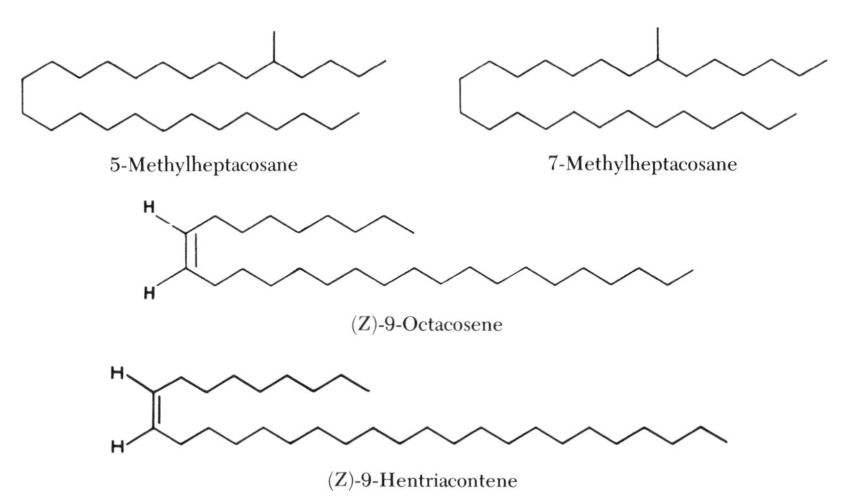

5-Methylheptacosane 7-Methylheptacosane

(Z)-9-Octacosene

(Z)-9-Hentriacontene

Cubitene ($C_{20}H_{32}$), the first example of a diterpene based on a twelve-membered carbocyclic ring, has been identified as the major frontal gland product of *Cubitermes umbratus* (Prestwich *et al.*, 1978). In addition to cubitene, several other novel diterpene hydrocarbons including cembrene A ($C_{20}H_{32}$), (3Z)-cembrene A ($C_{20}H_{32}$) (Wiemer *et al.*, 1979), and biflora-4,10(19),15-triene ($C_{20}H_{32}$) (Wiemer *et al.*, 1980) have been characterized as glandular constituents of the defensive exudate of *C. umbratus*.

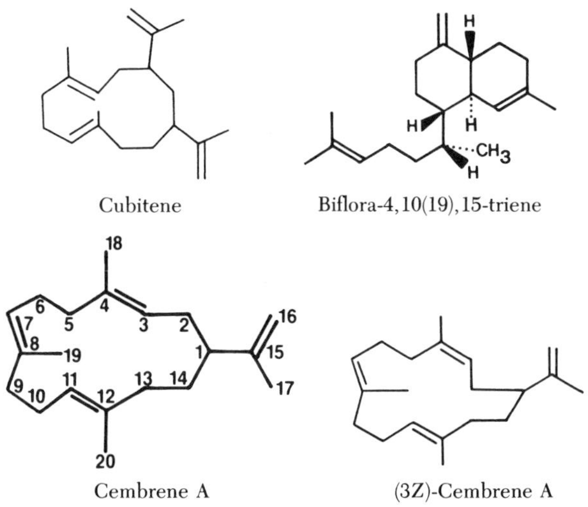

Cubitene Biflora-4,10(19),15-triene

Cembrene A (3Z)-Cembrene A

Δ^3-Carene, along with terpinolene, limonene, myrcene, α-pinene, and β-pinene (all $C_{10}H_{16}$), has also been identified as a frontal gland product of *Hospitalitermes monoceros* (Kistner *et al.*, 1977). In contrast, soldiers of a *Termes* sp. produce a secretion that contains only three monoterpenes—limonene, α-terpinene, and terpinolene (Kistner *et al.*, 1977). *Longipedetermes longipes* soldiers produce a distinctive frontal gland exudate that contains, in addition to α-pinene, β-pinene, terpinolene, α-terpinene, β-phellandrene, camphene, and limonene, *p*-cymene as well (Blum *et al.*, 1977a). Soldiers of a *Bulbitermes* sp. produce these same compounds along with Δ^3-carene and sabinene ($C_{10}H_{14}$) (Blum *et al.*, 1977a). Soldiers of *Amitermes messinae* produce limonene as a minor constituent of their frontal gland secretion (Meinwald *et al.*, 1978).

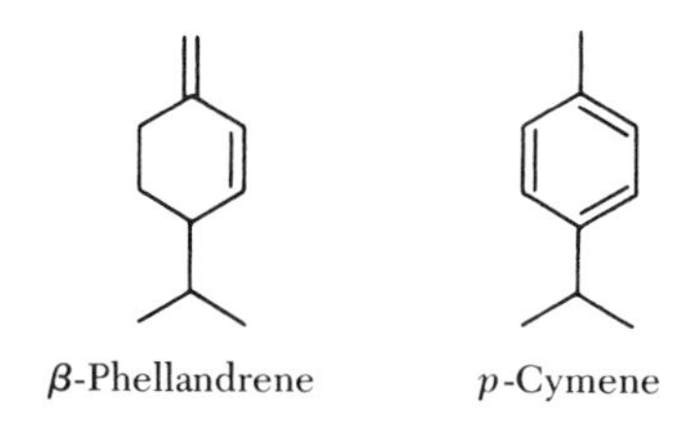

β-Phellandrene p-Cymene

Minor amounts of α-selinene ($C_{15}H_{24}$) and *cis*-β-ocimene (β-ocimene X) ($C_{10}H_{16}$) fortify the frontal gland secretion of soldiers of *Amitermes evuncifer* (Evans *et al.*, 1979). β- and α-Cyclogeraniolenes (C_9H_{16}), along with toluene (C_7H_8), represent minor constituents in the frontal gland secretion of *Ancistrotermes cavithorax* (Evans *et al.*, 1979).

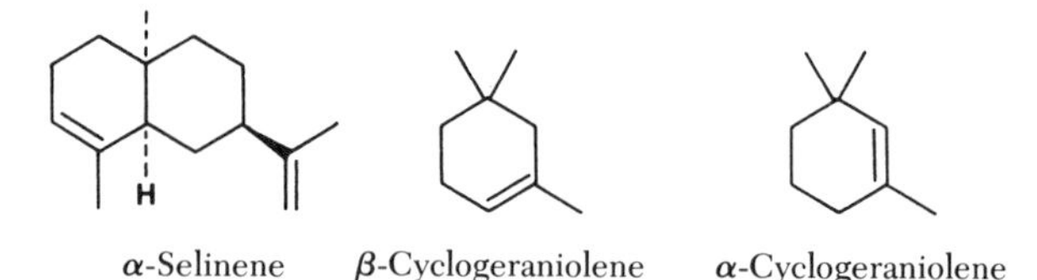

α-Selinene β-Cyclogeraniolene α-Cyclogeraniolene

n-Undecane ($C_{11}H_{24}$) is a concomitant of the aldehyde-rich secretions of the stink bugs (Pentatomidae) *Musgraveia sulciventris* and *Biprorulus bibax* (MacLeod *et al.*, 1975).

The metasternal scent gland secretions of the stink bugs *Aspongopus* sp., *Delegorguella lautus*, *Caura rufiventris*, and *Veterna patula* contain *n*-undecane, *n*-dodecane ($C_{12}H_{26}$), *n*-tridecane ($C_{13}H_{28}$), 1-tridecene ($C_{13}H_{26}$), and 1-tetradecene ($C_{14}H_{28}$) (Prestwich, 1976). All species except *Aspongopus* sp. also produce *n*-tetradecane ($C_{14}H_{30}$).

n-Undecane ($C_{11}H_{24}$), *n*-dodecane ($C_{12}H_{26}$), and *n*-tridecane ($C_{13}H_{28}$) have been identified in the defensive exudate of adult guava bugs (Pentatomidae) *Vitellus insularis* (Smith, 1974). *n*-Dodecane and *n*-tridecane are produced by the pentatomid *Lamprophara bifasciata* and the cydnid *Adrisa numeensis* (Smith, 1978).

Hayashi *et al.* (1976) reported that the metasternal scent gland secretions of the bugs (Cydnidae) *Macrocystus japonensis* and *Aethus nigritus* contained *n*-tridecane. The exudate of *A. nigritus* was dominated by *n*-pentadecane ($C_{15}H_{32}$).

Germacrene A ($C_{15}H_{24}$), the proposed precursor of several bicyclic sesquiterpene hydrocarbons, has been identified as the alarm pheromone in the cornicle secretion of the aphid (Aphididae) *Therioaphis maculata* (Bowers *et al.*, 1977; Nishino *et al.*, 1977). Its known distribution is limited to species in the subfamily Drepanosiphinae. Germacrene A and B have been detected in the osmeterial secretion of third and fourth instar larvae of *Papilio protenor* (Papilionidae) (Honda, 1980).

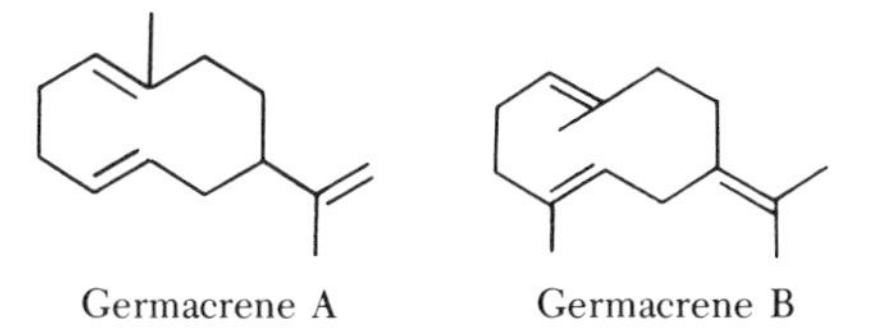

Germacrene A Germacrene B

Molested larvae of the papilionid *Parides arcas* emit an osmeterial secretion containing the sesquiterpene hydrocarbon β-elemene ($C_{15}H_{24}$) (Blum *et al.*, 1977b). Third and fourth instar larvae of another papilionid species, *Papilio protenor*, also produce this compound along with β-caryophyllene ($C_{15}H_{24}$), β-ocimene X and Y, β-phellandrene, myrcene, α-pinene, and limonene (all $C_{10}H_{16}$) (Honda, 1980).

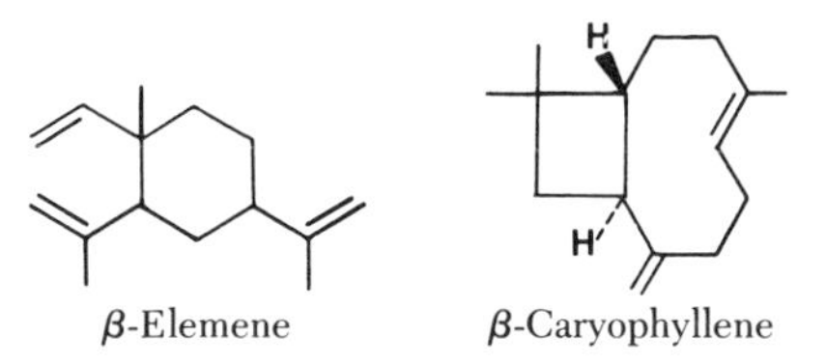

β-Elemene β-Caryophyllene

The abdominal gland secretions of *Tribolium confusum* and *T. castaneum* (Coleoptera:Tenebrionidae) are characterized by the presence of 1-tetradecene ($C_{14}H_{28}$), 1-pentadecene ($C_{15}H_{30}$), 1,6-pentadecadiene ($C_{15}H_{28}$), 1-hexadecene ($C_{16}H_{32}$), 1-heptadecene ($C_{17}H_{34}$), 1,8-

heptadecadiene ($C_{17}H_{32}$), and heptadecatriene ($C_{17}H_{30}$) (Suzuki *et al.*, 1975b).

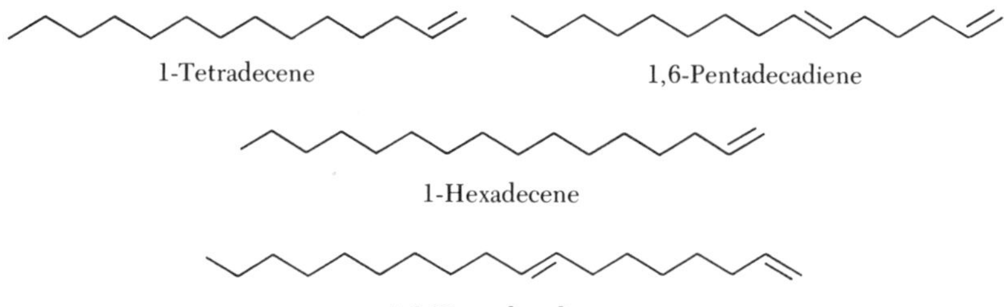

The hydrocarbons *n*-heptadecane ($C_{17}H_{36}$), heptadecene ($C_{17}H_{34}$), *n*-nonadecane ($C_{19}H_{40}$), nonadecene ($C_{19}H_{38}$), *n*-tricosane ($C_{23}H_{48}$), and tricosene ($C_{23}H_{46}$) have been identified in the mandibular gland secretion of the bee (Andrenidae) *Andrena fuscipes* (Tengö and Bergström, 1976). The secretion of *A. fucata* only contains one hydrocarbon—tricosene.

Camphene ($C_{10}H_{16}$) has been identified in worker ant extracts of *Formica rufa* and *F. polyctena* (Bühring *et al.*, 1976). This monoterpene hydrocarbon may constitute a mandibular gland product.

Eisner *et al.* (1977) and Roach *et al.* (1979) have detected hydrocarbons in the pygidial gland secretions of bombardier beetles in several subfamilies. *n*-Tridecane ($C_{13}H_{28}$) is produced by *Metrius contractus* and *Goniotropis nicaraguensis*. The secretions of these two species, along with a *Pachyteles* species, also contain *n*-tetradecane ($C_{14}H_{30}$). Several species of these carabids—*Brachinus quadripennis, B. sublaevis, M. contractus, Platycerozaena panamensis, G. nicaraguensis, G.* sp., *Pachyteles* spp., *P. striola, P. longicornis, Physea hirta*, and *Ozaena magna*—discharge exudates containing *n*-pentadecane ($C_{15}H_{32}$) (Eisner *et al.*, 1977; Roach *et al.*, 1979).

X. MISCELLANEOUS COMPOUNDS IN ARTHROPOD SECRETIONS

The web of a garden spider, *Aranea diadema*, contains KH_2PO_4, KNO_3, and pyrrolidone (C_4H_7NO) (Schildknecht *et al.*, 1972a; Schildknecht, 1976). KH_2PO_4 is believed to form protons by dissociation and thus to contribute to the inhibition of bacterial growth. KNO_3 is presumed to function by keeping the glue in solution by salting. Pyrrolidone is regarded as an agent for hindering the glue on the web from drying out.

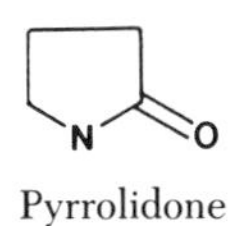

Pyrrolidone

The osmeterial secretion of third- and fourth-instar larvae of *Papilio protenor* contains caryophyllene oxide ($C_{15}H_{24}O$) as a major constituent (Honda, 1980).

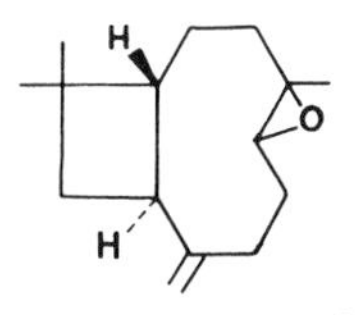

Caryophyllene oxide

Hydrogen cyanide (CHN) has been detected in the glandular secretions of the millipedes *Euryurus maculatus, Cherokia georgiana ducilla, C. g. georgiana, C. g. latassa, Motyxia tularea, Sigmoria nantahalae,* and *Pseudopolydesmus erasus* (Duffey *et al.*, 1977).

Benzoyl cyanide (C_8H_5NO) and mandelonitrile (C_8H_7NO) have been identified in the glandular exudates of the millipedes (Polydesmida:Polydesmidae) *Pseudopolydesmus serratus, Apheloria corrugata,* and *A. trimaculata* (Connor *et al.*, 1977).

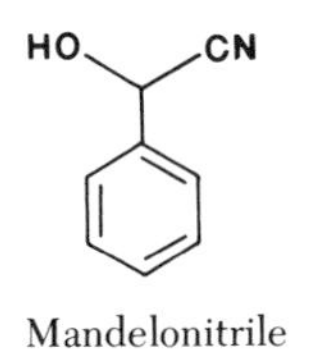

Mandelonitrile

The mandibular gland secretion of the ponerine ant *Megaponera foetens* is dominated by dimethyldisulfide ($C_2H_6S_2$), dimethyltrisulfide ($C_2H_6S_3$), and benzylmethyl sulfide ($C_8H_{10}S$) (Longhurst *et al.*, 1979a).

Benzylmethyl sulfide

The alkaloid actinidine ($C_{10}H_{13}N$) has been demonstrated to be an anal gland product of two ant species in the genus *Conomyrma* (Wheeler *et al.*, 1977b).

4,11-Epoxy-*cis*-eudesmane ($C_{15}H_{26}O$) is the major constituent present in the frontal gland secretion of soldiers of the termite *Amitermes messinae* (Meinwald *et al.*, 1978).

N-Ethyl-3-(2-methylbutyl)piperidine ($C_{12}H_{25}N$) (stenusine) has been isolated from the larger pygidial glands of the beetle (Staphylinidae) *Stenus comma* (Schildknecht *et al.*, 1976). This compound is believed to be primarily responsible for propelling the beetles over the surface of the water (*Entspannungsschwimmen*).

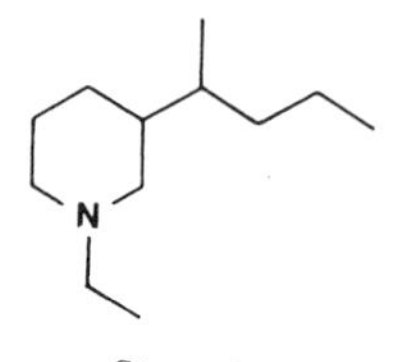

Stenusine

A series of 2,6- and 2,5-dimethyl-3-alkylpyrazines were identified as mandibular gland products of the ponerine ants *Odontomachus troglodytes*, *Anochetus sedilloti*, and *Brachyponera sennaarensis* (Longhurst *et al.*, 1978). Workers and males of *O. troglodytes* produce 2,6-dimethyl-3-ethylpyrazine ($C_8H_{12}N_2$), 2,6-dimethyl-3-butylpyrazine ($C_{10}H_{16}N_2$), 2,6-dimethyl-3-pentylpyrazine ($C_{11}H_{18}N_2$), and 2,6-dimethyl-3-hexylpyrazine ($C_{12}H_{20}N_2$), whereas workers of *B. sennaarensis* produce only the butyl- and pentylpyrazines. On the other hand, both 2,6- and 2,5-dimethyl-3-alkylpyrazines were identified in the secretions of *A. sedilloti*. In addition to 2,6-dimethyl-3-(1-methylpropyl)pyrazine ($C_{10}H_{16}N_2$), 2,6-dimethyl-3-(2-methylpropyl)pyrazine ($C_{10}H_{16}N_2$), and 2,6-dimethyl-3-butylpyrazine, the exudate of these workers contained 2,5-dimethyl-3-(1-methylpropyl)pyrazine ($C_{10}H_{16}N_2$), 2,5-dimethyl-3-(2-methylpropyl)pyrazine, ($C_{10}H_{16}N_2$), and 2,5-dimethyl-3-pentylpyrazine ($C_{11}H_{18}N_2$). Brown and Moore (1979) identified 3-(2-methylpropyl)-2,5-dimethylpyrazine ($C_{10}H_{16}N_2$) as one of three pyrazines in the unusual mandibular gland secretion of a *Calomyrmex* species.

2,6-Dimethyl-3-hexylpyrazine

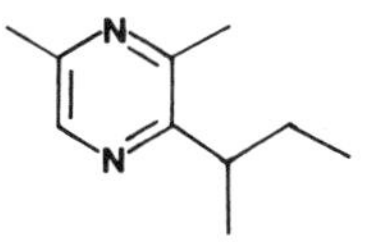

2,6-Dimethyl-3-(1-methylpropyl)pyrazine

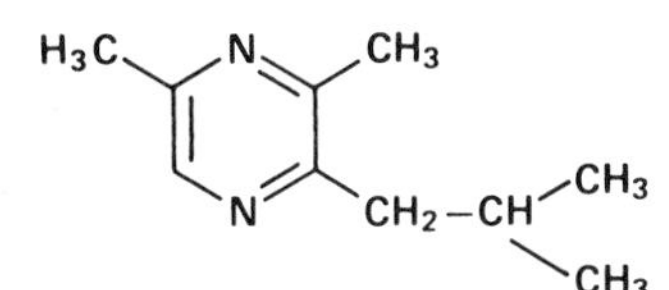

2,6-Dimethyl-3-(2-methylpropyl)pyrazine

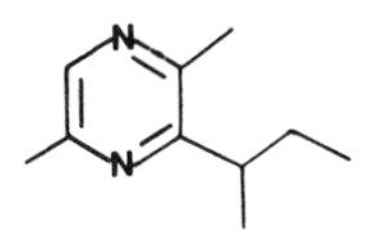

2,5-Dimethyl-3-(1-methylpropyl)pyrazine

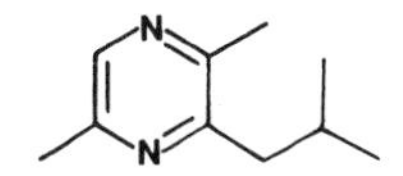

2,5-Dimethyl-3-(2-methylpropyl)pyrazine

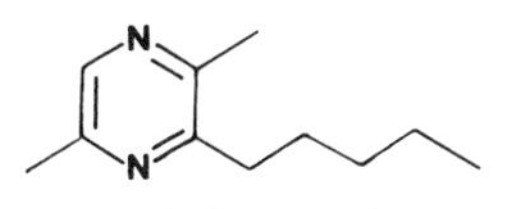

2,5-Dimethyl-3-pentylpyrazine

2-(3-Furyl)-4,4,7α-trimethyloctahydrobenzofuran (ancistrofuran) (C_{15}-$H_{22}O_2$) has been identified as a frontal gland product of soldiers of the termite *Ancistrotermes cavithorax* (Evans *et al.*, 1977, 1979). In addition, this species produces both *epi*-caparrapi oxide ($C_{15}H_{26}O$) and caparrapi oxide ($C_{15}H_{26}O$) in this unusual secretion (Evans *et al.*, 1979).

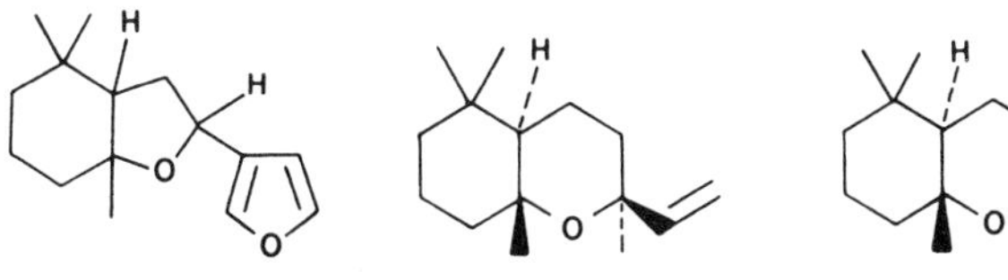

Ancistrofuran *epi*-Caparrapi oxide Caparrapi oxide

XI. STEROIDS

Cardiac glycosides, not sequestered from the host plant, have been identified in the defensive exudates of adult chrysomelid beetles (Pasteels and Daloze, 1977). These steroids, which are produced by glands located along the external edges of the elytra and anterior corners of the pronotum, constitute the first examples of this class of compounds synthesized *de novo* by animals. Beetles in the genera *Chrysolina, Chrysochloa,* and *Dlochrysa* synthesize cardenolides consisting of monohydroxylated digitoxigenins combined with pentoses such as arabinose and xylose.

Daloze and Pasteels (1979) identified the cardenolides produced by adults of *Chrysolina coerulans* and *C. dydimata.* The secretion of *C. coerulans* contains three cardenolide aglycones—periplogenin ($C_{23}H_{34}O_5$), sarmentogenin ($C_{23}H_{34}O_5$), and bipindogenin ($C_{23}H_{34}O_6$)—whereas only one cardenolide, sarmentogenin, is detectable in the secretion of *C. dydimata.* In addition to these aglycones, the exudate of *C. coerulans* contained their corresponding xylosides, sarmentogenin-3β-xylopyranoside ($C_{28}H_{42}O_9$), periplogenin- 3β-xylopyranoside ($C_{28}H_{42}O_9$), and bipindogenin-3β-xylopyranoside ($C_{28}H_{42}O_{10}$). *Chrysolina herbacea* is reported to produce sarmentogenin, bipindogenin, and their corresponding xylosides (Pasteels *et al.*, 1979). Larvae of these species, which lack defensive glands, also produce cardenolides, and these compounds are present in the eggs as well. Cardiac glycosides are not secreted by several other species of *Chrysolina.*

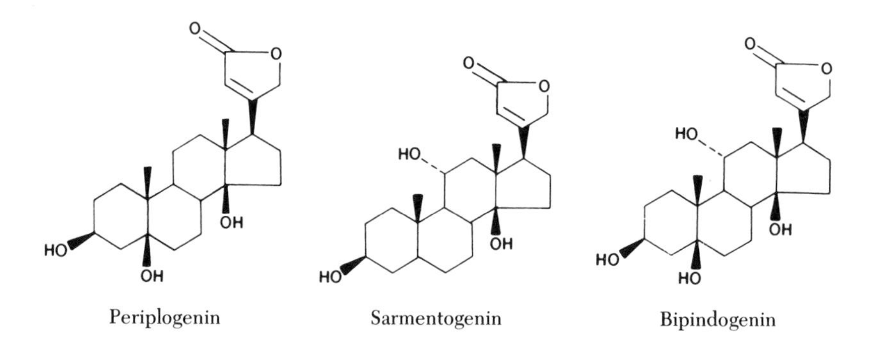

Periplogenin Sarmentogenin Bipindogenin

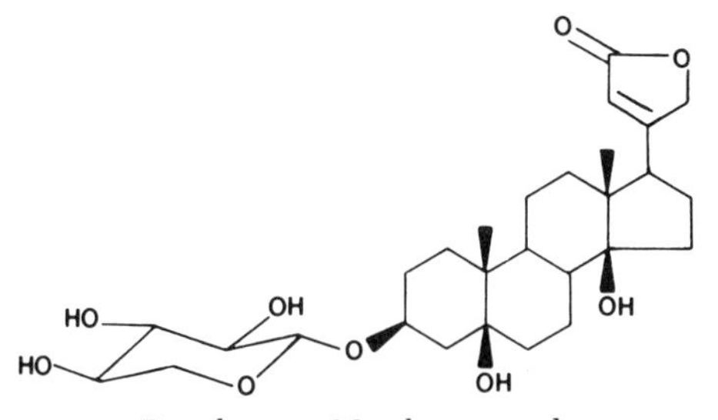

Periplogenin-3β-xylopyranoside

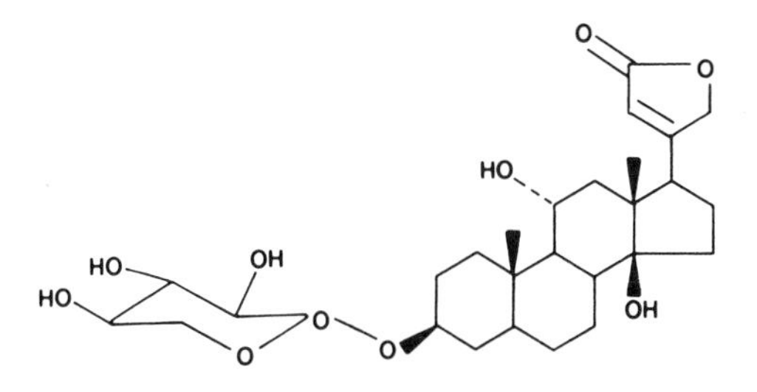

Sarmentogenin-3β-xylopyranoside

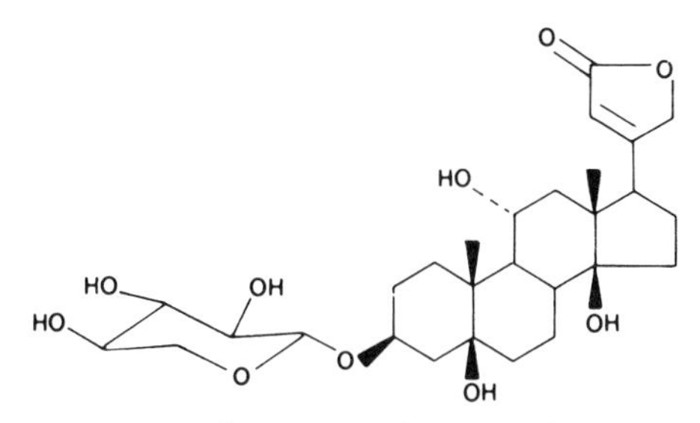

Bipindogenin-3β-xylopyranoside

Eisner *et al.* (1978) reported adult fireflies (*Photinus* spp.) synthesized steroidal pyrones (lucibufagins) that are related to the cardiotonic bufadienolides, characteristic of some toads and plants. The major defensive steroids in *P. pyralis* were identified as 12-oxo-2β,3β-di-*O*-acetyl-5β,11α-dihydroxybufalin ($C_{28}H_{36}O_{10}$), 12-oxo-2β-*O*-acetyl-5β,11α-dihydroxybufalin ($C_{26}H_{34}O_9$), 12-oxo-3β-*O*-acetyl-2β,5β,11α-trihydroxybufalin ($C_{26}H_{34}O_9$), 12-oxo-2β-*O*-acetyl-3β-*O*-isobutyryl-5β,11α-dihydroxybufalin ($C_{30}H_{40}O_{10}$), and 12-oxo-2β-*O*-acetyl-3β-*O*-propionyl-5β,11α-dihydroxybufalin ($C_{29}H_{38}O_{10}$) (Meinwald *et al.*, 1979). Additional bufadienolides have been characterized from extracts of *P. ignitus* and *P. marginellus* (Goetz *et al.*, 1979). These include 12-oxo-3β-*O*-acetyl-5β,11α-dihydroxybufalin ($C_{26}H_{34}O_8$), 12-oxo-5β,11α-dihydroxybufalin ($C_{24}H_{32}O_7$), 11-oxo-3β-*O*-acetyl-5β,12β-dihydroxybufalin ($C_{26}H_{34}O_8$), and 11-oxo-5β,12β-dihydroxybufalin ($C_{24}H_{32}O_7$).

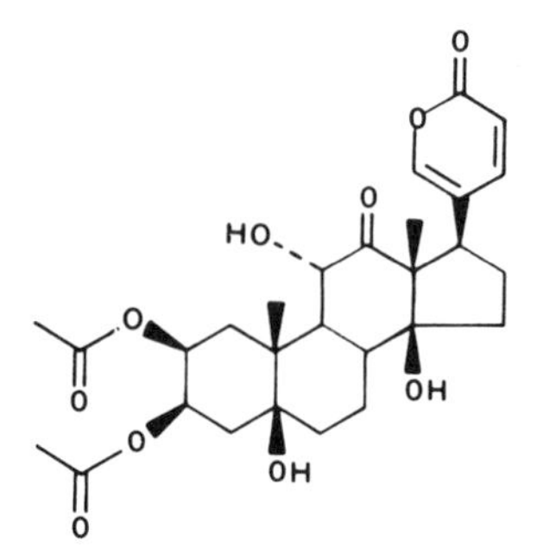

12-Oxo-2β,3β-di-*O*-acetyl-5β,11α-dihydroxybufalin

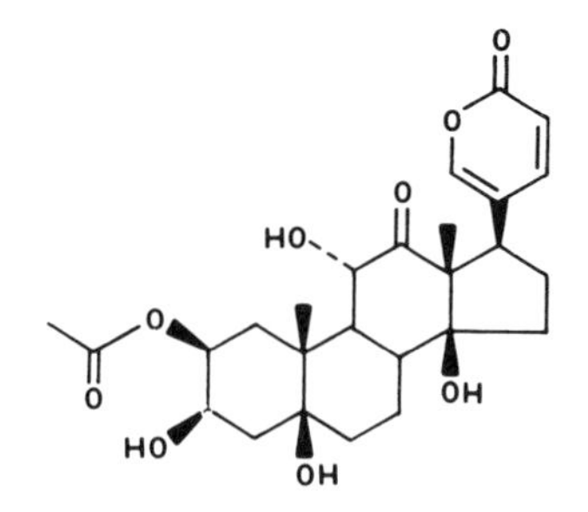

12-Oxo-2β-*O*-acetyl-5β,11α-dihydroxybufalin

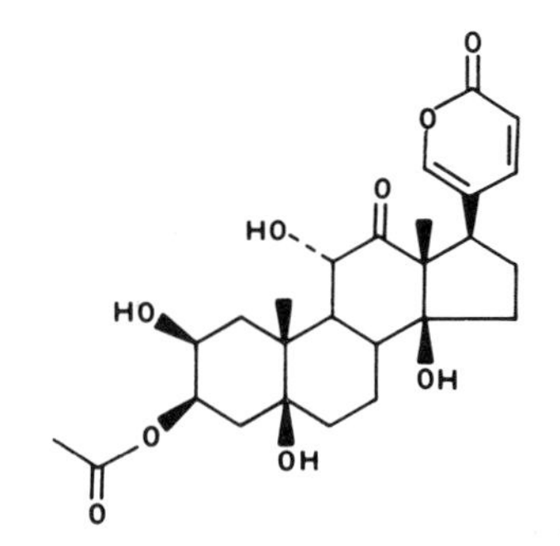

12-Oxo-3β-*O*-acetyl-2β,5β,11α-trihydroxybufalin

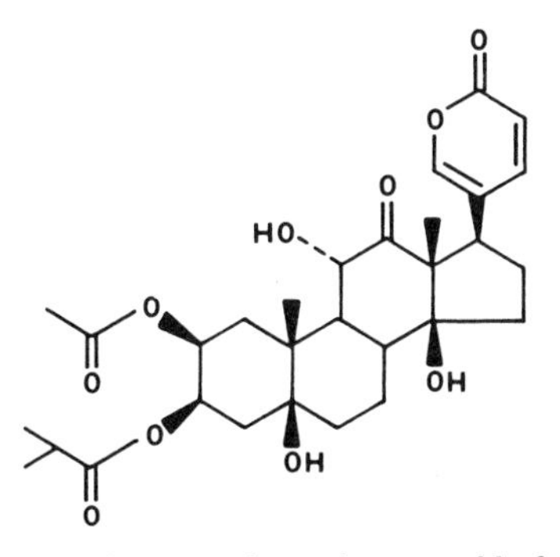

12-Oxo-2β-O-acetyl-3β-O-isobutyryl-5β,11α-dihydroxybufalin

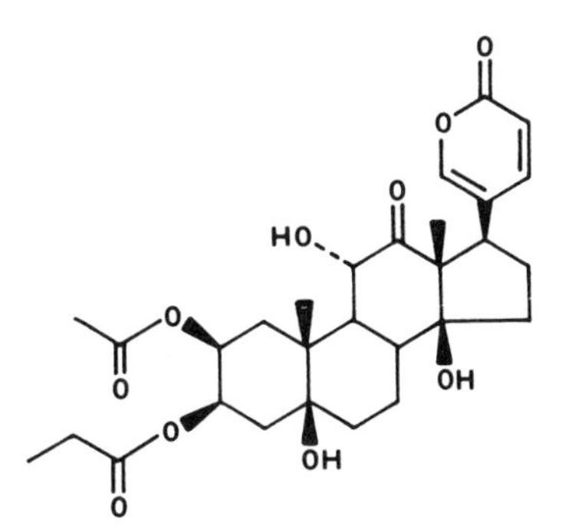

12-Oxo-2β-O-acetyl-3β-O-propionyl-5β,11α-dihydroxybufalin

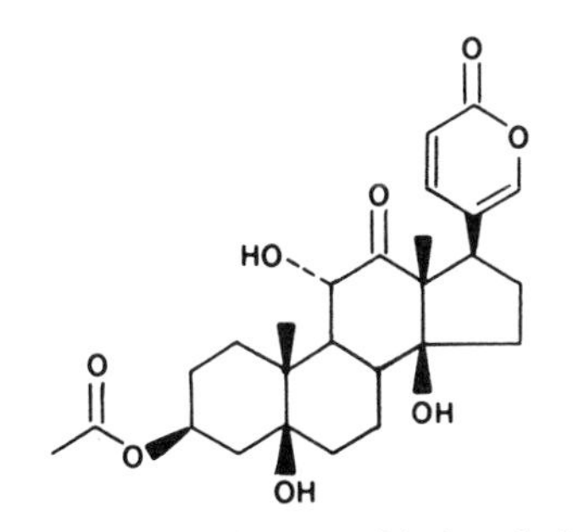

12-Oxo-3β-O-acetyl-5β,11α-dihydroxybufalin

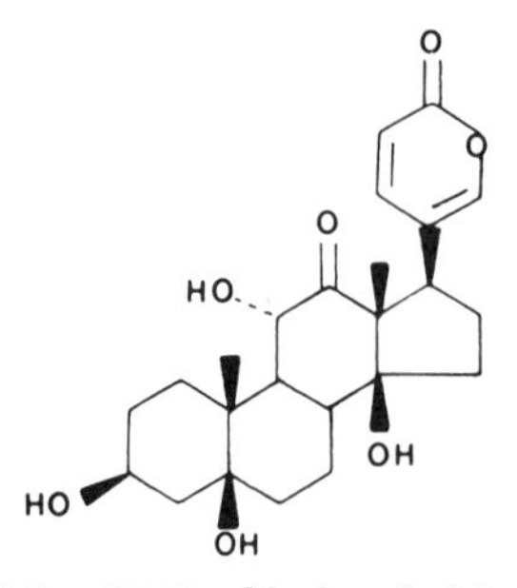

12-Oxo-5β,11α-dihydroxybufalin

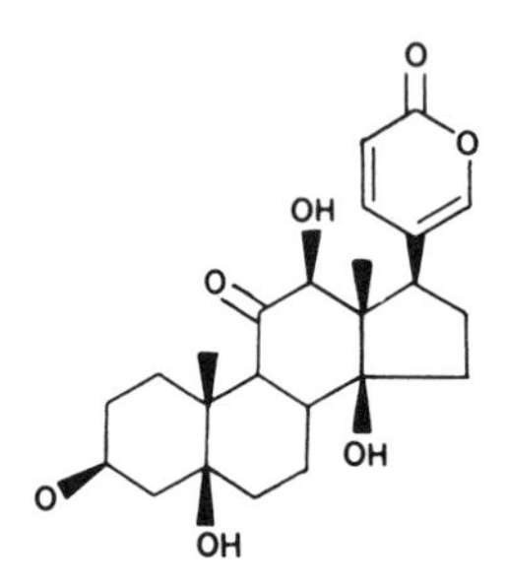

11-Oxo-3β-*O*-acetyl-5β,12β-dihydroxybufalin

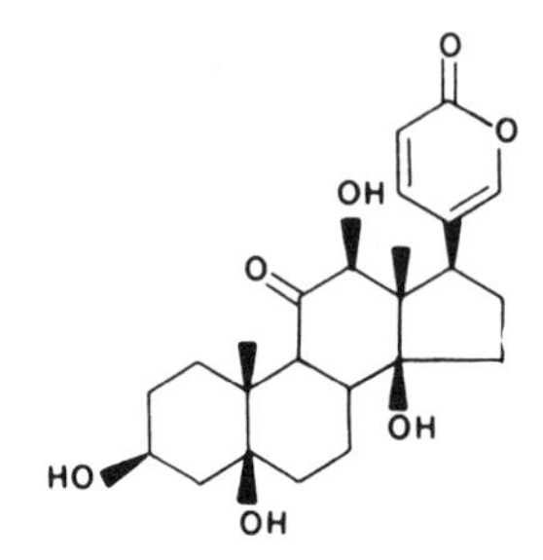

11-Oxo-5β,12β-dihydroxybufalin

XII. NONPROTEINACEOUS CONSTITUENTS IN ARTHROPOD VENOMS

Histamine ($C_5H_9N_3$) has been identified in the venom of the scorpion *Palamneus gravimanus* (Scorpionidae) (Ismail *et al.*, 1975).

The low molecular-weight constituents in the venom of the Sydney funnel web spider, *Atrax robustus*, were recently determined by Duffield *et al.* (1979). The major pharmacologically active compounds detected were γ-aminobutyric acid ($C_4H_9NO_2$), spermidine ($C_7H_{19}N_3$), spermine ($C_{10}H_{26}N_4$), tyramine ($C_8H_{11}NO$), octopamine ($C_8H_{11}NO_2$), and in female venom only, 5-methoxytryptamine ($C_8H_{11}NO$).

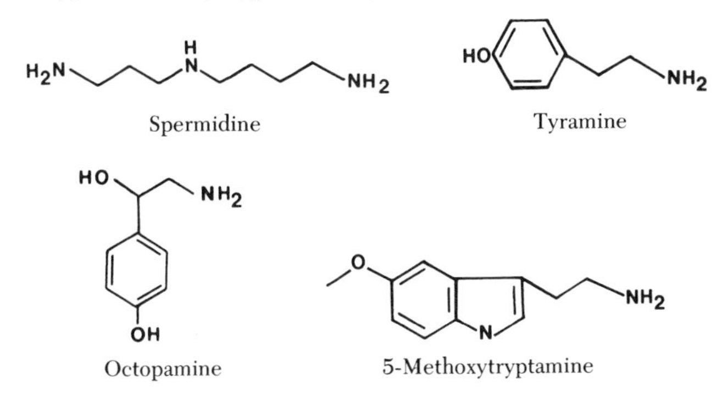

Spermidine Tyramine Octopamine 5-Methoxytryptamine

Several additional alkaloids have been identified as poison gland constituents of the ant *Monomorium pharaonis*. 2-(5′-Hexenyl)-5-pentylpyrrolidine ($C_{15}H_{29}N$), 2-(5′-hexenyl)-5-heptylpyrrolidine ($C_{17}H_{33}N$), 5-methyl-3-(3′-hexenyl)octahydroindolizine ($C_{19}H_{33}N$), and 2-(5′-hexenyl)-5-nonylpyrrolidine ($C_{19}H_{37}N$) are present in the venom of this myrmicine species (Ritter *et al.*, 1975).

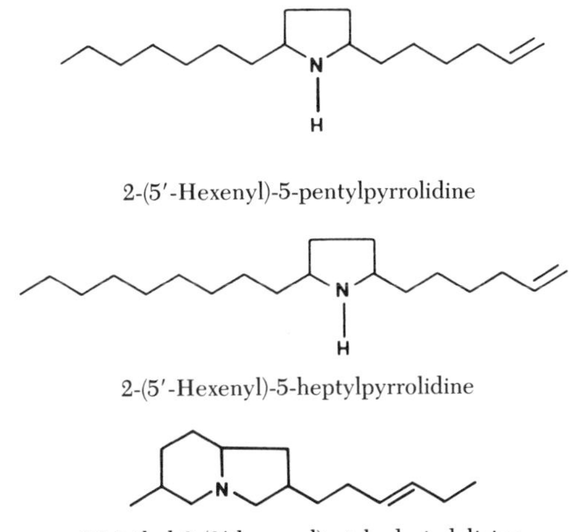

2-(5′-Hexenyl)-5-pentylpyrrolidine

2-(5′-Hexenyl)-5-heptylpyrrolidine

5-Methyl-3-(3′-hexenyl)octahydroindolizine

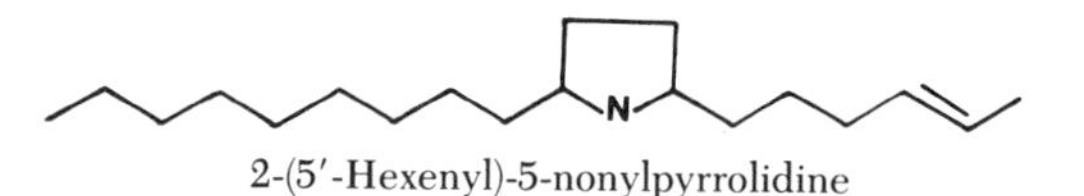

2-(5′-Hexenyl)-5-nonylpyrrolidine

A novel alkaloid, 3-heptyl-5-methylpyrrolizidine ($C_{15}H_{29}N$), has been characterized in the venom of the ant *Solenopsis tennesseensis* (Jones *et al.*, 1980). Recently, Wheeler *et al.* (1981) identified anabaseine ($C_{10}H_{12}N_2$) as a poison gland product of two *Aphaenogaster* species.

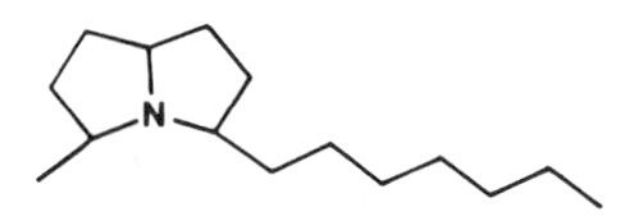

3-Heptyl-5-methylpyrrolizidine

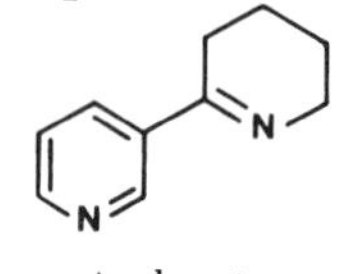

Anabaseine

XIII. NONEXOCRINE DEFENSIVE COMPOUNDS OF ARTHROPOD ORIGIN*

The alkaloids precoccinelline ($C_{13}H_{23}N$) and coccinelline ($C_{13}H_{23}NO$) have been identified in the beetle *Coccinella transversoguttata* (Ayer *et al.*, 1976). Hippodamine ($C_{13}H_{23}N$) and convergine ($C_{13}H_{23}NO$) are present in the blood of *Hippodamia caseyi*. Moore and Brown (1978) identified precoccinelline, hippodamine, and propyleine ($C_{13}H_{21}N$) in the secretion from the prothoracic and abdominal glands of the soldier beetle (Cantharidae) *Chauliognathus pulchellus*. Two new alkaloids, hippocasine ($C_{13}H_{21}N$) and hippocasine *N*-oxide ($C_{13}H_{21}NO$), were also identified in extracts of *H. caseyi* (Ayer *et al.*, 1976).

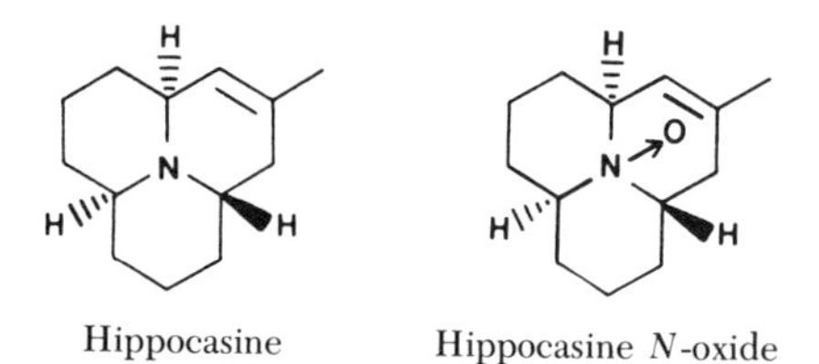

Hippocasine Hippocasine *N*-oxide

Tursch *et al.* (1976) reports that the identification of propyleine in the boll weevil (*Anthonomus grandis*) (Hedin *et al.*, 1974) is, in all probability, incorrect. This compound is probably myrrhine ($C_{13}H_{23}N$). Similarly,

*See Section XI for discussion of the steroidal pyrones (lucibufagins) of the Lampyridae.

the identification of precoccinelline in the ladybird *Coleomegilla maculata* (Henson *et al.*, 1975) may be in error, and the alkaloid may also be myrrhine (Tursch *et al.*, 1976).

Mueller and Thompson (1980) synthesized propyleine and demonstrated that the natural product actually consisted of this compound and 75% of an unknown isomer, isopropyleine ($C_{13}H_{21}N$).

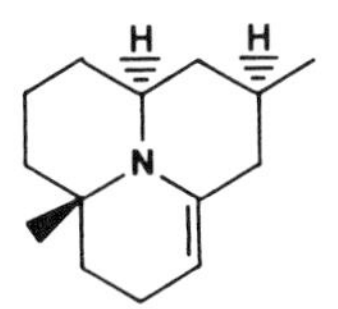

Isopropyleine

Chapter 19

Arthropod Defensive Compounds

I. AN OVERVIEW

Although the chemistry of a wide range of arthropod defensive substances has been established, the selective advantages of producing particular compounds or blends of compounds are virtually unknown. In the absence of information on the specific agents that may have been responsible for the distributional peculiarities that characterize the defensive allomones of arthropod species, it is impossible to comprehend how these chemical defenses evolved. Furthermore, the significance of the quantitative and frequent qualitative variations in defensive secretions between individuals or populations of species can only be understood with some knowledge of the evolution of these compounds. In essence, although the compositional idiosyncrasies that often characterize defensive secretions demonstrate that these animals are versatile if not unpredictable chemists, they tell us nothing about the factors responsible for the generation of such characteristic defensive allomones.

It would seem that for many species of arthropods, natural products chemistry "makes life possible." An examination of the properties and characteristic distributions of a variety of these compounds can provide some possible insight into their adaptiveness as defensive agents. In addition, such an analysis serves to focus on major research problems that will have to be resolved before it will be possible to appreciate why particular allomones have been selected by so many arthropod species as key defensive agents.

II. DEFENSIVE ALLOMONES AS PHEROMONES OR CRYPTIC "PHEROMONES"

The structural identicalness of many arthropod defensive compounds and pheromones has led to the suggestion that in many cases the former may have been antecedents of the latter (Blum, 1974). For example, *n*-undecane and benzaldehyde, two compounds that have a wide distribution in the defensive secretions of nonsocial arthropods, are utilized as pheromones by bees and ants. Indeed, it would seem highly adaptive for social insects to employ defensive compounds as pheromones, since these exocrine compounds would now possess dual functions. Obviously such a development would require the evolution of a preprogrammed responsiveness to the pheromone, mediated by the integration of signals in the central nervous system. On the other hand, no new metabolic pathways would have to be evolved for pheromonal biosynthesis. Furthermore, proteins for recognizing the pheromone would already be present as part of the biosynthetic apparatus, and conceivably, units of these macromolecules, or conformationally related forms, could be incorporated directly into the antennal receptor sites of these social insects. The ability of these invertebrates to exploit defensive allomones to function simultaneously as volatile information-bearing agents would constitute a major development in the evolution of sociality.

Selection will favor the production of effective doses of these "allomonal" pheromones because of their proved value as predator deterrents. This appears to be the case for a variety of pheromonal stimuli that include formic acid and citral. Both of these compounds, which are utilized defensively by a variety of beetles (Moore and Wallbank, 1968; Wheeler *et al.*, 1972a; Bellas *et al.*, 1974), are also produced in large quantities by ants and bees, serving as pheromones for these hymenopterans. This energetically parsimonious utilization of defensive compounds as pheromones by ants and bees appears to be typical of many hymenopterous species (Blum and Brand, 1972).

The structural congruence of exocrine products in nonsocial and social arthropods appears to also be highly adaptive for many of the former species as well. Many normally solitary arthropods produce defensive compounds that are either identical to or very similar in structure to the alarm pheromones of ants. Perception by ants of their alarm pheromones usually results in their initially exhibiting characteristic erratic and rapid dispersive movements. These defensive secretions momentarily disrupt attacking ants while enabling their possessors to flee the scene of the encounter. In a sense, the defensive allomones of these solitary arthropods are functioning as chemical mimics of the ants' own alarm pheromones by releasing alarm behavior in the formicids. In reality, when directed at ants by nonsocial arthropods, these mimetic compounds serve as cryptic "pheromones" that enable the latter to figuratively "hide" from the predatory formicids when confrontations occur. Significantly, most of the species that produce these cryptic "pheromones" are rapidly moving arthropods that inhabit environments containing large populations of predatory ants. Survival in these formicid-dominated environments selected for defensive strategies capable of blunting the attacks of these formidable predators. The combination of cryptic alarm pheromones and rapid escape behavior would appear to be two major evolutionary adaptations for successfully occupying ant-rich habitats (Blum, 1980). Widespread examples of this duality in defensive behaviors demonstrates that it occurs frequently among nonsocial arthropods.

Many species of opilionids generate defensive secretions that bear a remarkable chemical similarity to the pheromonal blends widely utilized by ants to signal alarm. When tactually stimulated, these harvestmen can discharge secretions fortified with ketones such as 4-methyl-3-hexanone and 4-methyl-3-heptanone (Jones *et al.*, 1976b, 1977; Blum and Edgar, 1971). Significantly, 4-methyl-3-heptanone is one of the most widespread alarm pheromones produced by ants (Blum and Brand, 1972), and in *Manica* species this alkanone is accompanied by 4-methyl-3-hexanone (Fales *et al.*, 1972). The utilization of 4,6-dimethyl-4-octen-3-one as another powerful releaser of alarm behavior by *Manica* species (Fales *et al.*, 1972) demonstrates that compounds very similar to another opilionid defensive allomone, 4,6-dimethyl-6-octen-3-one (Meinwald *et al.*, 1971), are also utilized by ants for signaling. In effect, these harvestmen are unleashing a host of ant alarm pheromones at their formicid adversaries.

Mutillids constitute another group of ground-dwelling arthropods that probably have frequent encounters with ants. These parasitic wasps spend their lives moving across the ground searching for hosts, a lifestyle that almost guarantees that they intrude on the foraging territories of ants. Aposematic mutillids in the genus *Dasymutilla* and related genera, in common with opilionids, secrete copious amounts of 4-methyl-3-heptanone when disturbed (Schmidt and Blum, 1977; Fales *et al.*, 1980). Since *Dasymutilla* species occur in areas particularly abundant in 4-methyl-3-heptanone-producing species, e.g., *Pogonomyrmex badius* and *Camponotus floridanus*, it would appear to be highly adaptive for these mutillids, which possess effective nonchemical defenses as well, to be able to challenge ants with this powerful releaser of formicid alarm behavior. In addition, some mutillids also produce compounds such as 4,6-dimethyl-3-octanone (Fales *et al.*, 1980), a ketone that bears an obvious structural relationship to another ant alarm pheromone, 4,6-dimethyl-4-octen-3-one (Fales *et al.*, 1972). Since alarm pheromones appear to be the least specific of the chemical releasers of behavior (Blum, 1969), it is obvious that an alarm signal can be effectively generated with compounds that bear a structural similarity to ant alarm pheromones.

Cockroaches constitute another group of insects that commonly share their living space with ants. Many species of cockroaches eject defensive secretions when disturbed, and not surprisingly, some of these exudates are dominated by compounds identical to well-known ant alarm pheromones. 2-Heptanone, a common alarm releaser of dolichoderine ants (Blum *et al.*, 1963), is the major constituent in the defensive spray produced by the cockroach *Platyzosteria armata* (Wallbank and Waterhouse, 1970). Another common methyl ketone utilized by dolichoderines to generate an alarm signal, 6-methyl-5-hepten-2-one (Trave and Pavan, 1956), is the primary product in the defensive secretion of the cockroach *Neostylopyga rhombifolia* (Alsop *et al.*, 1980). Since the spray of *N. rhombifolia* releases exaggerated alarm behavior in dolichoderines (M. S. Blum, unpublished, 1980), it appears that this defensive secretion is capable of functioning as a cryptic alarm pheromone.

Ground beetles also utilize a well-known formicid alarm pheromone as the mainstay of their chemical defenses. Many species of carabids produce formic acid in their pygidial glands, a ubiquitous poison gland product of formicine ants. This carboxylic acid, produced in large quantities by many formicines, releases sustained alarm behavior in the aggressive workers of species in several genera (Maschwitz, 1964). In common with opilionids, mutillids and cockroaches, ground beetles are rapid-moving arthropods that can easily abandon the scene of an encounter after inducing alarm behavior in ants with their defensive secretion. It thus appears that the evolution of cryptic "pheromones" has occurred widely in the Arthropoda as an antipredator device directed primarily against predatory ants.

III. MODES OF ACTION OF DEFENSIVE ALLOMONES

While the term defensive compound is useful in describing the context in which an exocrine product is utilized, it tells us nothing about its mode of action as a predator deterrent. One possible approach to determining the role of exocrine products is to examine some of the chemical and physiological properties of the compounds that are frequently utilized as antagonistic allomones. The widespread use of products such as hydrocarbons, aldehydes, and 1,4-benzoquinones as defensive compounds indicates that they possess a distinct evolutionary or ecological advantage based on their molecular properties. If, for example, these chemical deterrents function by disrupting olfactory processes, then their defensive efficacy must reflect site-specific action on the chemoreceptors of predatory species. An examination of some of the properties of these major classes of defensive compounds can prove useful in probing their *raison d'être* as defensive allomones *par excellence.*

Hydrocarbons have been evolved as defensive allomones by a wide variety of arthropods despite the fact that these compounds are not generally considered to be very stimulatory olfactants. It has been suggested that hydrocarbons could function as a lipid solvent and spreading agent for the more polar constituents in a defensive exudate (Waterhouse and Gilby, 1964). However, these compounds may also possess a more subtle defensive function that may be at least as important as their putative role as additives capable of disrupting cuticular lipids. Blum and Brand (1972) have cited evidence that indicates that hydrocarbons can temporarily alter the generator characteristics of antennal chemoreceptors, an effect that could result in a partial and temporary anosmia for the predator. Under these circumstances the hydrocarbon emitter would be able to "hide" from its aggressor, whose predatory behavior would be effectively neutralized because of the disruption of its normal chemosensory input. In essence, hydrocarbons may represent some of the cryptic odors of Haldane (1955) which, notwithstanding their weak organoleptic properties, are eminently capable of altering the olfactory signals relayed to the central nervous system.

In an adumbration of the roles of antagonistic allomones in marine systems, Kittredge *et al.* (1974) visualized the possible existence of cryptic odors as a consequence of at least two neurophysiological phenomena. Cryptic odors may represent either the chemical equivalent of a "white noise" by generating an "uncoded" array of spikes in the chemosensory neurons, or a "negative odor" that blocks the generator potential of the dendritic membrane after altering its characteristics. Many of the multicomponent defensive secretions of arthropods may constitute devices for overstimulating the chemosensory neurons of predators and thus triggering bursts of uncoded information into the central nervous system. Such a cryptic odor would be particularly effective if it were composed of a complex mixture of compounds belonging to several chemical classes and thus able to stimulate a variety of chemoreceptors. The resultant sensory input would constitute a highly disruptive chemical overload and the "scrambled" message would in general be maladaptive for predators that use olfactory cues or overide visual cues when hunting prey. The Dufour's gland secretion of the ant *Camponotus herculeanus* would seem to constitute such a cryptic odor, since it is composed of at least 44 fairly volatile compounds including alcohols, esters, ketones, alkanes, and alkenes (Bergström and Löfqvist, 1973). The chemical diversity which characterizes so many arthropod defensive secretions may be identified with the discharge of a chemical "smokescreen" that "conceals" its emitter by producing an undecipherable sensory input as a result of stimulating a variety of generalist receptors.

Along with hydrocarbons, both 1,4-quinones and conjugated aldehydes may also function as cryptic odors but, unlike hydrocarbons, these two classes of compounds are highly stimulatory olfactants. Furthermore, such quinones and aldehydes also differ from hydrocarbons because they are highly reactive, a fact that may be of considerable importance in analyzing their roles as defensive compounds. Since both types of carbonyl compounds are conjugated, they are capable of rapidly reacting with nucleophilic compounds by Michael addition, and forming new compounds which are substituted β to the carbonyl function. Conceivably, exposed chemoreceptor proteins in the olfactory pores and pore filaments are amenable to nucleophilic attack on the electrophilic quinones and α,β-unsaturated aldehydes. The attack by the nucleophilic groups of the antennal proteins on the electrophiles may temporarily inactivate the affected chemoreceptors and reduce olfactory acuity drastically, enabling the quinonoid emitter to "hide" from its aggressor.

Provided that the reaction environment is suitable, 1,4-quinones are capable of being attacked by nucleophiles with great celerity. For example, Moore (1968) has demonstrated that *p*-benzoquinone, the major defensive compound in the frontal gland of the termite *Mastotermes darwiniensis*, reacts almost instantly with discharged proteins to form a rubberlike polymer that can entangle small assailants. The attack on electrophiles such as 1,4-benzoquinone by nucleophilic compounds occurs very rapidly at extreme pH's. The defensive exudates of many tenebrionids in the tribe Eleodini are distinguished by the presence of octanoic acid (Tschinkel, 1975b), a compound that could play a role in accelerating the reaction of the quinones with nucleophilic antennal proteins. The minor acidic constituents in defensive secretions may serve as important devices for producing cryptic odors by promoting the speed of reaction of electrophilic compounds with a predator's olfactory proteins.

Although *p*-quinones are one of the major classes of arthropod defensive compounds, it is curious that these invertebrates do not utilize *o*-quinones, which may have potential as cryptic odors, in the same way. *o*-Quinones are much more reactive than *p*-quinones, and significantly, pathways for the synthesis of the former are present in invertebrates for the formation of cuticular products. Normally, these quinones are produced near the time of molting and their generation is under the control of molting hormones. The apparent absence of *o*-quinones from the defensive secretions of arthropods is in contrast to that of another group of invertebrates, the cephalopods. The ink of the octopus is made up of *o*-quinones which rapidly polymerize when discharged into sea water after the cephalopod has been molested. MacGintie and MacGintie (1968) have demonstrated that the "smokescreen" generated by the ink actually serves to produce an olfactory "blindness" in the octopus's main predator, the moray eel. Therefore, these quinones appear to be eminently well suited to function as deterrents to aggressors that utilize olfactory cues when hunting prey.

The apparent absence of this system from arthropod defensive glands is intriguing and serves to emphasize our lack of comprehension of the bases for utilizing selected compounds as antagonistic allomones. The elucidation of such questions should provide us with considerable insight into the biochemical *raison d'être* of many defensive substances.

IV. SELECTED ARTHROPODS AS NATURAL PRODUCTS CHEMISTS *PAR EXCELLENCE*

Although defensive compounds have a fairly widespread distribution among arthropod species, selected groups of these invertebrates are demonstrably versatile producers of these allomones. An analysis of the natural products emphases of these arthropods serves to illuminate the magnitude of this versatility while at the same time focusing on arthropod groups that constitute good candidates for further natural products investigations.

A. Millipedes

While polydesmoid and spiroboloid millipedes produce benzaldehyde-HCN and 1,4-benzoquinones, respectively, studies on species in other orders indicate that a variety of other interesting and novel natural products are produced by diplopods. For example, the defensive exudate of a glomerid species contains two quinazolinones that constitute uncommon animal natural products (Y. Meinwald *et al.*, 1966; Schildknecht *et al.*, 1966c). More recently, the allomones of a polyzoniid millipede—a nitrogen-containing monoterpene (polyzonimine) (Smolanoff *et al.*, 1975b) and a related tricyclic compound containing a nitro group (nitropolyzonamine) (Meinwald *et al.*, 1975)—have been identified as new animal natural products. Analyses of the defensive exudates of millipedes in other orders should produce further chemical surprises.

B. Termites

Termites are another group of arthropods that produce a diversity of natural products. Although the chemistry of the defensive exudates of soldiers in relatively few termite genera has been examined, there are already good grounds for regarding these isopterans as versatile allomonal chemists. Among insects, termitid soldiers clearly excel in the biosynthesis of monoterpene hydrocarbons (Moore, 1968: Vrkoč *et al.*, 1973) but in addition, these soldiers produce sesquiterpene hydrocarbons such as α-selinene (Evans *et al.*, 1979) and diterpene hydrocarbons such as cubitene (Prestwich *et al.*, 1978).

However, it is in the production of oxygenated sesqui- and diterpenes that the biosynthetic versatility of termite soldiers becomes apparent. Highly distinctive eudesmane sesquiterpenes are produced by *Amitermes* species (Wadhams *et al.*, 1974), whereas the characteristic sesquiterpenes ancistrofuran, caparrapi oxide, ancistrodial, and cavidial have been identified from the frontal gland of *Ancistrotermes* species (Evans *et al.*, 1979) (see Chapter 18). Other termitid species produce a wide variety of diterpene alcohols and esters (Prestwich *et al.*, 1976a; Vrkoč *et al.*, 1977)—the trinervitenes (see Chapter 18). In addition, other novel diterpene esters, the kempenes, have also been identified (Prestwich *et al.*, 1977b).

Termites species in other families have also proved to be a rich source of distinctive defensive allomones. Analyses of the defensive products of a few rhinotermitid species have resulted in the identification of several novel insect exocrine compounds. The only nitro compound identified as an insect exocrine product, 1-nitro-1-pentadecene, has been demonstrated to be a frontal gland constituent of soldiers of *Prorhinotermes simplex* (Vrkoč and Ubik, 1974). Two *Schedorhinotermes* species produce an unusual series of 3-alkanones and 3-alkenones (Quennedey *et al.*, 1973; Prestwich *et al.*, 1975) which are characteristic natural products of members of this genus. Considering the fact that the cephalic secretions of soldiers in most termite genera have not been analyzed, it seems likely that a host of interesting defensive compounds await structural elucidation.

C. Hymenoptera

Defensive allomones synthesized in the exocrine glands of ants and bees show great structural diversity and many are produced only in this order. In particular, Hymenoptera are versatile terpene specialists producing a large variety of oxygenated mono-, sesqui-, and diterpenes. Citronellal, geraniol, and a variety of geranyl esters are typical of the monoterpene allomones that fortify the defensive secretions of many bees and ants. The furanoterpene perillene is a particularly distinctive compound whose known arthropod distribution is limited to a few genera of ants (Bernardi *et al.*, 1967; Longhurst *et al.*, 1979a). However, it is in the production of iridoids by dolichoderine ants that the monoterpenoid versatility of hymenopterans is most evident.

Several iridoids have a scattered distribution in the defensive exudates of beetles (Bellas *et al.*, 1974; Meinwald *et al.*, 1977; Blum *et al.*, 1978) and phasmids (Smith *et al.*, 1979), but the majority of these compounds have been identified as anal gland products of dolichoderines. Excluding iridodial isomers (McGurk *et al.*, 1968), five of the cyclopentanoid monoterpenes—iridomyrmecin, isoiridomyrmecin, isodihydronepetelactone, iridodial, and dolichodial—are known to be produced by dolichoderine ants in a variety of genera (Blum and Hermann, 1978b). Considering the virtuosity of these ants as iridoid chemists, it would seem highly worthwhile to examine species in a variety of other dolichoderine genera for the presence of these distinctive monoterpenes.

In addition, Hymenoptera produce characteristic sesquiterpenes that have not yet been detected in the exudates of any nonhymenopterous species. For example, farnesal (Bernardi *et al.*, 1967), dihydrofarnesal (Bergström and Löfqvist, 1970; Bergström and Svensson, 1973), and their corresponding acetates (Kullenberg *et al.*, 1970; Bergström and Löfqvist,

1972) are widely distributed natural products of formicine ants and bees. The unusual furanosesquiterpene dendrolasin (Bernardi *et al.*, 1967) further typifies the ability of ants to biosynthesize novel C_{15} compounds. However, these hymenopterans are no less versatile when it comes to the synthesis of acyclic diterpenes.

Five diterpenes identified as exocrine products of ants and bees are not known to be produced by any nonhymenopterous species. Geranylgeranial, geranylcitronellol, their corresponding aldehydes, and geranylgeranyl acetate, constitute the only acyclic diterpenes yet identified as arthropod natural products (Blum and Hermann, 1978a). The distribution of these compounds is limited to a few species of formicine ants and apid bees, suggesting that investigations of the exocrine products of other species in these large taxa would be very worthwhile. Indeed, studies of the defensive compounds of nonformicine ants demonstrate that these insects produce a remarkable variety of natural products.

In terms of exocrine diversity, species in the ant subfamily Ponerinae constitute ideal candidates for allomonal studies. Although the defensive compounds produced by species in relatively few ponerine genera have been characterized, it is quite evident this primitive ant taxon is a virtual exocrine gold mine. Ponerine species are the only known arthropod source of alkyl and aryl sulfides (Casnati *et al.*, 1967; Longhurst *et al.*, 1979b), but in addition, these formicids synthesize distinctive compounds belonging to several other chemical classes. A variety of 2,5- and 2,6-dimethyl-3-alkylpyrazines are generated in the mandibular glands of some ponerines (Wheeler and Blum, 1973; Longhurst *et al.*, 1978) whereas others produce methyl 6-methylsalicylate (Duffield and Blum, 1975a); in contrast, one species in another ponerine genus produces 4-methyl-3-heptanone in the same glands (Duffield and Blum, 1973). These results indicate that future investigations of ponerine natural products may yield further chemical surprises.

Like the mandibular glands, the poison glands of ants are a good source of unusual defensive allomones; products of these glands vary from monoterpenes (Quilico *et al.*, 1962; Brand *et al.*, 1974) to an impressive array of alkaloids. Alkaloidal-rich venoms are characteristic of fire ants in the genus *Solenopsis*, the members of which produce a large variety of 2,6-dialkylpiperidines (MacConnell *et al.*, 1971). Other species of *Solenopsis* synthesize venoms that are dominated by 2,5-dialkylpyrrolidines and -pyrrolines (Pedder *et al.*, 1976; Jones *et al.*, 1979). In addition to the 2,5-dialkylpyrrolidines, a *Monomorium* species synthesizes two indolizines in its poison gland (Ritter *et al.*, 1973, 1975). These species, which are all members of the subfamily Myrmicinae, demonstrate that ants produce a wide range of alkaloids in their venoms which often contain only traces of proteins. Ant venoms would appear to be outstanding candidates for investigations of new arthropod alkaloids.

The demonstrated diversity of natural products already identified as hymenopterous defensive allomones should act to stimulate additional studies on the behavioral and ecological correlates of the secretions of these insects. Although virtually all investigations have been undertaken with ants and bees, it will not prove surprising if wasps are also determined to be a rich source of distinctive exocrine defensive compounds.

D. Coleoptera

When it comes to allomonal multifariousness in the Arthropoda, beetles are preeminent (Weatherston and Percy, 1978b). Furthermore, the most complex nonproteinaceous constituents identified as arthropod natural products have been isolated from beetles as abundant testimony to the biosynthetic versatility possessed by these insects. However, it is the great structural diversity in both exocrine and nonexocrine defensive compounds that identifies beetles as the arthropod chemists *par excellence.* A brief examination of the allomonal virtuosity of these coleopterans should provide strong grounds for considering them as excellent candidates for future studies on the chemistry of defensive products.

Beetles synthesize a much greater variety of aromatic exocrine compounds than arthropods in any other order. For example, aromatic esters, aldehydes, and acids are produced in the pygidial glands of a wide variety of dytiscid species (Schildknecht, 1970). Additional aldehydes such as salicylaldehyde and benzaldehyde have been identified in the secretions of larval and adult beetles in the families Chrysomelidae and Carabidae (Wain, 1943; Moore and Wallbank, 1968; Moore and Brown, 1971a). A variety of phenols are present in the defensive exudates of tenebrionids (Tschinkel, 1969), carabids (Schildknecht *et al.*, 1968a), and, in the case of cerambycid species, the phenols accompany an aromatic hydrocarbon, toluene (Moore and Brown, 1971b). Aromatic allomones clearly constitute an important group of coleopterous defensive compounds.

Among the arthropods, only beetles have been demonstrated to synthesize steroidal defensive allomones. A wide variety of dytiscids produce steroids which are sometimes identical to vertebrate hormones such as testosterone and estrone (Schildknecht, 1970). In other cases, these prothoracic glandular products constitute unique steroids which may be esterified with isobutyric or pentenoic acids (Schildknecht, 1968, 1971). In contrast to these steroids, chrysomelid adults synthesize cardenolides, and lampyrids produce steroidal pyrones related to the bufadienolides of toads. The chrysomelid compounds, which consist of xylosides and their corresponding aglycones, are secreted from pronotal and elytral glands (Daloze and Pasteels, 1979). The lampyrid steroids, termed lucibufagins (Meinwald *et al.*, 1979), are not secreted from glands but can be liberated by reflex bleeding. No other animals are known to synthesize either the chrysomelid cardenolides or the lampyrid lucibufagins.

Although the triterpenoid steroids of these beetles are almost certainly derived from ingested sterolic precursors (Schildknecht, 1970), mostly novel mono- and sesquiterpenes are probably biosynthesized from simple precursors by a variety of beetle species. Iridodial, a monoterpene known as a defensive compound of ants, has been identified as a glandular product of both staphylinid and cerambycid beetles (Bellas *et al.*, 1974; Vidari *et al.*, 1973), and rose oxide sometimes accompanies it. Chrysomelidial, an isomer of dolichodial, is produced in the larval exudates a variety of chrysomelid larvae (Meinwald *et al.*, 1977; Blum *et al.*, 1978; Sugawara *et al.*, 1979a,b), often being accompanied by related lactones such as plagiolactone (Meinwald and Jones, 1978). In contrast to the monoterpenoid emphasis of these beetles, that expressed by gyrinid species is novelly sesquiterpenoid. Four norsesquiterpenes have been identified in the pygidial gland exudates of whirligig beetles (Meinwald *et al.*, 1972; Schildknecht *et al.*, 1972b; Wheeler *et al.*, 1972b; Miller *et al.*, 1975) and these compounds constitute some of the most distinctive terpenes produced by arthropods.

When it comes to generating short-chain fatty acids, beetles have no arthropod peers. Fifteen of these compounds have been identified in their defensive exudates (see Table 6.2), mostly in the pygidial gland secretions of carabids. Conjugated acids such as isocrotonic, crotonic, and angelic acid are limited in their arthropod distribution to carabid exudates (Moore and Wallbank, 1968), and these secretions are frequently characterized by great acidic diversity. The defensive secretions of cantharid species are especially distinctive because they are dominated by dihydromatricaria acid, the only acetylenic compound identified as an arthropod natural product. The identification of S-methyl-2-hydroxy-4-mercaptobutanoic acid and related compounds as pygidial gland products of hygrobiid species (H. Schildknecht and G. Krebs, in Dettner, 1977) demonstrates that beetles are also capable of producing sulfur-containing compounds.

Beetles, particularly in the family Tenebrionidae, produce more 1,4-quinones than any other group of arthropods. At least ten 1,4-quinones have been detected in the exudates from the abdominal sternal glands of tenebrionids, and in some cases novel arthropod products such as 2-propyl-1,4-benzoquinone are produced (Tschinkel, 1975b). Tenebrionids are also distinctive in synthesizing the only 1,4-naphthoquinones identified as insect exocrine products (Tschinkel, 1969). However, the quinone-fortified exudates of the beetles frequently also contain other classes of novel compounds as well.

The exudate from the sternal glands of *Eleodes beameri* contains a dazzling variety of carbonyl compounds that include typical hemipterous products such as hexanal and (*E*)-2-hexenal as well as unique constituents such as 1-nonen-3-one (Tschinkel, 1975a). The composition of the abdominal secretion of another tenebrionid, *Apsena pubescens*, is equally surprising since it contains two isocoumarins in addition to the usual tenebrionid 1,4-benzoquinones (Lloyd *et al.*, 1978a).

Other uncommon exocrine products have been identified as defensive allomones of a variety of beetles. The cerambycid *Phoracantha semipunctata* produces a highly characteristic secretion in its metasternal scent gland that contains alicyclic aldehydes such as 5-ethylcyclopent-1-enecarbaldehyde and 2-ethylcyclopentanecarbaldehyde (Moore and Brown, 1972). Another species in this genus generates an unusual series of macrocyclic lactones in the same gland (Moore and Brown, 1976). The identification of methyl 8-hydroxyquinoline-2-carboxylate in the prothoracic glandular secretion of a dytiscid species (Schildknecht *et al.*, 1969a) further emphasizes the natural products diversity that may characterize coleopterous allomones.

Among the most distinctive defensive compounds synthesized by coleopterans are those that are not liberated from exocrine glands but rather are present in the blood and tissues. The terpenoid anhydride cantharidin appears to be an ubiquitous natural product of meloid species (Dixon *et al.*, 1963) which has not been detected in members of any other arthropod group. Nine alkaloids have been identified as products of a variety of coccinellid beetles (Pasteels, 1977; Ayer *et al.*, 1976) and these compounds constitute some of the most characteristic compounds produced by arthropodans. The most complex compound identified as an arthropod defensive allomone is pederin, a secondary amide (Table 13.2) that enriches the blood of a variety of *Paederus* species (Cardani *et al.*, 1965a,b). Two related amides are also present, providing these staphylinids with a potent vesicatory system that has probably been evolved as a defense against vertebrates. The pederins are representative of the idiosyncratic allomones that characterize the defensive compounds produced by coleopterans in a wide variety of taxa.

V. THE LARGELY UNEXPLORED DEFENSIVE ALLOMONE CHEMISTRY OF THE ARTHROPODA

The last 30 years have witnessed an explosive increase in our knowledge of animal natural products, chiefly as a result of investigations of arthropodous species. Although a relatively large number of compounds have been identified as allomones of these invertebrates, it is safe to say that the defensive chemistry of this group of invertebrates is largely *terra incognita*. The defensive secretions of species in many arthropod taxa have not been studied and it will not prove surprising if a host of new natural products are identified when these exudates are analyzed. Therefore, it would seem appropriate to briefly discuss some groups of arthropods that would appear to be good candidates for natural product investigations.

While the exocrine products of a relatively small number opilionid species in few families have been examined, these invertebrates have already proved to be a rich source of interesting natural products. A variety of 1,4-benzoquinones (Fieser and Ardao, 1956) 1,4-naphthoquinones (Wiemer *et al.*, 1978), aliphatic aldehydes and ketones (Jones *et al.*, 1976b, 1977), and phenols have been identified in the defensive secretions of opilionids. Considering the fact that allomones are widely produced by harvestmen, these arthropods should continue to constitute outstanding candidates for natural products investigations.

Chemical investigations on the defensive chemistry of millipedes have been essentially limited to species in spiroboloid and polydesmoid orders. In the few studies of the exocrine products of species in other orders, novel defensive compounds have been identified. Millipedes will probably yield a host of new natural products when the chemical defenses of species in other orders are studied.

Although a plethora of distinctive allomones have been identified in hymenopterous secretions, relatively few eusocial species have been studied. Among ants, ponerine species should continue to be a treasure trove of natural products, along with the dolichoderines and their novel anal gland products. Myrmicine venoms may yield a diversity of new alkaloids as a variety of genera in this large subfamily are analyzed. Eusocial bees, especially apid species, synthesize a variety of ketones, esters, and aromatic aldehydes (Blum, *et al.*, 1973b; Luby *et al.*, 1973) and investigations of additional species seem especially worthwhile. Beyond these eusocial species, there is essentially nothing known about the defensive chemistry of nonsocial bees and wasps.

The wealth of novel compounds already characterized from the cephalic secretions of termite soldiers (Section IV,B) should act as a spur to examine additional species, especially those in families that have not been studied or have received relatively little attention (e.g., Rhinotermitidae). The same can be said for the exocrine products of coleopterous species. Species in this, the largest of the insectan orders, have already been demonstrated to possess outstanding prowess as natural products chemists. This is particularly true of allomonal steroids, a dazzling variety of which are produced by dytiscid, chrysomelid, and lampyrid species. The demonstrated steroid biosynthetic versatility of these coleopterans, which has not been detected in arthropod species in any other order, makes it seem highly likely that this class of defensive compounds will be detected as products of beetles in additional families. However, for large families such as the Cerambycidae and Staphylinidae, the defensive exudates of relatively few species have been studied. The abilities of both larvae and adults of a wide variety of coleopterous families to synthesize defensive allomones mark the beetles as ideal subjects for further natural products investigations. The same can be said for hemipterans. Although saturated and conjugated aldehydes and esters generally dominate their defensive exudates (Weatherston and Percy, 1978a), the secretions of species in some families may be characterized by the presence of novel terpenes (Aldrich *et al.*, 1979).

Beyond the arthropod taxa that have already been cited, others, scarcely investigated, hold great promise as candidates for studies on the chemistry of allomones. Lepidopterous larvae have been demonstrated to produce defensive exudates containing distinctive alcohols (Trave *et al.*, 1966) and eudesmane sesquiterpenes (Eisner *et al.*, 1971b). Since many lepidipterous species produce deterrent secretions, these insects may provide many natural products surprises when their allomonal chemistry is known. Trichopterous adults would also appear to be good subjects for study, especially in view of the fact that the defensive exudate of only one species has been examined (Duffield *et al.*, 1977). The same can be said for isopods, for which defensive compounds of only two species have yet been identified (Cavill *et al.*, 1966).

In short, the multifarious arthropods should continue to be an outstanding source of natural products for many years. For the chemist and chemical ecologist, the best is yet to come. Hopefully, the identification of these invertebrate allomones will act as a spur for investigating the evolutionary and ecological bases of arthropod chemical defenses. For biologists, the wondrous world of arthropods will surely become a little more comprehensible once they appreciate the evolutionary bases for the remarkable diversity of their natural products.

References

Abou-Donia, S. A., Fish, L. J., and Pattenden, G. (1971). *Tetrahedron Lett.*, 4038.

Abushama, F. T., and Ahmed, A. A. (1976). *Z. Angew. Entomol.* **80**, 206.

Adam, K. R., and Weiss, C. (1956). *Nature (London)* **178**, 421.

Adam, K. R., and Weiss, C. (1959). *Nature (London)* **183**, 1398.

Adrouny, G. A., Derbes, V. J., and Jung, R. C. (1959). *Science* **130**, 449.

Agrén, L., Cederberg, B., and Svensson, B. G. (1979). *Zoon* **7**, 1.

Aldrich, J. R., and Yonke, T. R. (1975). *Ann. Entomol. Soc. Am.* **68**, 955.

Aldrich, J. R., Blum, M. S., Hefetz, A., Fales, H. M., Lloyd, H. A., and Roller, P. (1978). *Science* **201**, 452.

Aldrich, J. R., Blum, M. S., Lloyd, H. A., Evans, P. H., and Burkhard, D. R. (1979). *Entomol. Exp. Appl.* **26**, 323.

Alexander, P., and Barton, D. H. P. (1943). *Biochem. J.* **37**, 463.

Allalouf, D., Ber, A., and Ishay, J. (1972). *Comp. Biochem. Physiol. B* **43B**, 119.

Alsop, D. W., Blum, M. S., and Fales, H. M. (1980). Unpublished data.

Alston, R. E., and Turner, B. L. (1963). "Biochemical Systematics." Prentice-Hall, Englewood Cliffs, New Jersey.

Andersson, C.-O., Bergström, G., Kullenberg, B., and Ställberg-Stenhagen, S. (1967). *Ark. Chem.* **26**, 191.

Aplin, R. T., and Rothschild, M. (1972). *In* "Toxins of Animal and Plant Origin" (A. de Vries and K. Kochva, eds.), pp. 579–595. Gordon & Breach, New York.

Aplin, R. T., Benn, M. H., and Rothschild, M. (1968). *Nature (London)* **219**, 747.

Aplin, R. T., Ward, R. d'Arcy, and Rothschild, M. (1975). *J. Entomol., Ser. A: Physiol. & Behav.* **50**, 73.

Ardao, M. I., Perdomo, C. S., and Pellaton, M. G. (1966). *Nature (London)* **209**, 1139.

Atsatt, P. R., and O'Dowd, D. J. (1976). *Science* **193**, 24.

Attias, J., Schlesinger, M. J., and Schlesinger, S. (1969). *J. Biol. Chem.* **244**, 3810.

Ayer, W. A., Bennett, M. J., Browne, L. M., and Purdham, J. T. (1976). *Can. J. Chem.* **54**, 1807.

Ayre, G. L., and Blum, M. S. (1971). *Physiol Zool.* **44**, 77.

Babin, D. R., Watt, D. D., Goos, S. M., and Mlejnek, R. V. (1974). *Arch. Biochem. Biophys.* **164,** 694.
Babin, D. R., Watt, D. D., Goos, S. M., and Mlejnek, R. V. (1975). *Arch. Biochem. Biophys.* **166,** 125.
Bachmayer, H., Kreil, G., and Suchanek, G. (1972). *J. Insect Physiol.* **18,** 1515.
Baer, H., Liu, D. T., Hooton, M., Blum, M. S., James, F., and Schmid, W. H. (1977). *Ann. Allergy* **38,** 378.
Baer, H., Liu, T.-Y., Anderson, M. C., Blum, M. S., Schmid, W. H., and James, F. J. (1979). *Toxicon* **17,** 397.
Baggini, A., Bernardi, R., Casnati, G., Pavan, M., and Ricca, A. (1966). *Eos* **42,** 7.
Baker, J. T., and Jones, P. A. (1969). *Aust. J. Chem.* **22,** 1793.
Baker, J. T., and Kemball, P. A. (1967). *Aust. J. Chem.* **20,** 395.
Baker, J. T., Blake, J. D., MacLeod, J. K., and Ironside, D. A. (1972). *Aust. J. Chem.* **25,** 393.
Baker, R., Briner, P. H., and Evans, D. A. (1978). *J. Chem. Soc., Chem. Commun.* p. 410.
Balozet, L. (1962). *Arch. Inst. Pasteur Alger.* **40,** 149.
Banks, B. E. C., Hanson, J. M., and Sinclair, N. M. (1976). *Toxicon* **14,** 117.
Baptist, B. A. (1941). *J. Microsc. Sci.* **83,** 91.
Barbetta, M., Casnati, G., and Pavan, M. (1966). *Mem. Entomol. Ital.* **45,** 169.
Barbier, M. (1959). *J. Chromatogr.* **2,** 649.
Barbier, M., and Lederer, E. (1957). *Biokhimiya (Moscow)* **22,** 236.
Barker, S. A., Bayyuk, S. I., Brimacombe, J. S., and Palmer, J. D. (1963). *Nature (London)* **199,** 693.
Beard, R. L. (1962). *Proc. Int. Congr. Entomol., 11th, 1960* Vol. 3, p. 44.
Béhal, A., and Phisalix, M. C. (1901). *Bull. Soc. Chim. Fr.* **25,** 88.
Bell, E. A. (1971). *In* "Chemotaxonomy of the Leguminosae" (J. B. Harborne, D. Boulter, and B. L. Turner, eds.), pp. 179–206. Academic Press, New York.
Bell, E. A., Lackey, J. A., and Polhill, R. M. (1978). *Biochem. Syst. Ecol.* **6,** 201.
Bellas, T. E., Brown, W. V., and Moore, B. P. (1974). *J. Insect Physiol.* **20,** 277.
Benfied, E. F. (1974). *Ann. Entomol. Soc. Am.* **67,** 739.
Benn, M. H., Lencucha, A., Maxie, S., and Telang, S. A. (1973). *J. Insect Physiol.* **19,** 2173.
Benn, M. H., Hutchins, R. F. N., Folwell, R., and Cox, J. (1977). *J. Insect Physiol.* **23,** 1281.
Benton, A. W. (1967). *J. Apic. Res.* **6,** 91.
Bergström, G. (1974). *Chem. Scr.* **5,** 39.
Bergström, G., and Löfqvist, J. (1968). *J. Insect Physiol.* **14,** 995.
Bergström, G., and Löfqvist, J. (1970). *J. Insect Physiol.* **16,** 2353.
Bergström, G., and Löfqvist, J. (1971). *In* "Chemical Releasers in Insects" (A. S. Tahori, ed.), pp. 195–223. Gordon & Breach, New York.
Bergström, G., and Löfqvist, J. (1972). *Entomol. Scand.* **3,** 225.
Bergström, G., and Löfqvist, J. (1973). *J. Insect Physiol.* **19,** 887.
Bergström, G., and Svensson, B. G. (1973). *Zoon, Suppl.* **1,** 61.
Bergström, G., and Tengö, J. (1973). *Zoon, Suppl.* **1,** 55.
Bergström, G., and Tengö, J. (1974). *Chem. Scr.* **5,** 28.
Bergström, G., Kullenberg, B., and Ställberg-Stenhagen, S. (1973). *Chem. Scr.* **4,** 174.
Bernardi, R., Cardani, C., Ghiringhelli, D., Selva, A., Baggini, A., and Pavan, M. (1967). *Tetrahedron Lett.* p. 3893.
Bernays, E., Edgar, J. A., and Rothschild, M. (1977). *J. Zool.* **182,** 85.

Bernheimer, A. W., Avigad, L. S., and Schmidt, J. O. (1980). *Toxicon* **18,** 271.
Bettini, S., ed. (1978). "Arthropod Venoms," Handb. Exp. Pharmakol. [N. S], Vol. 48. Springer-Verlag, Berlin and New York.
Bettini, S., and Toschi-Frontali, N. (1962). *Proc. Int. Congr. Entomol., 11th, 1960* Vol. 3, p. 115.
Bevan, C. W. L., Birch, A. J., and Caswell, H. (1961). *J. Chem. Soc.* 488.
Bhoola, K. D., Calle, J. D., and Schachter, M. (1961). *J. Physiol. (London)* **159,** 167.
Billingham, M. E. J., Morley, J., Hanson, J. M., Shipolini, R. A., and Vernon, C. A. (1973). *Nature (London)* **245,** 163.
Bisset, G. W., Frazer, J. F. D., Rothschild, M., and Schachter, M. (1960). *Proc. R. Soc. London, Ser. B* **152,** 255.
Blagoveshchenskii, A. V. (1955). "Die Biochemischen Grundlagen des Evolutionsprozesses der Pflanzen." Akademie-Verlag, Berlin.
Blakley, R. L. (1969). *In* "The Biochemistry of Folic Acid and Related Pteridines" (A. Neuberger and E. L. Tatum, eds.), pp. 188–218. North-Holland Publ., Amsterdam.
Blau, P. A., Feeny, P., Contardo, L., and Robson, D. S. (1978). *Science* **200,** 1296.
Blest, A. D., Collett, T. S., and Pye, J. D. (1963). *Proc. R. Soc. London, Ser. B* **158,** 196.
Blum, M. S. (1961). *Ann. Entomol. Soc. Am.* **54,** 410.
Blum, M. S. (1964). *Ann. Entomol. Soc. Am.* **57,** 600.
Blum, M. S. (1966). *Proc. R. Entomol. Soc. London, Ser. A* **41,** 155.
Blum, M. S. (1969). *Annu. Rev. Entomol.* **14,** 57.
Blum, M. S. (1970). *In* "Chemicals Controlling Insect Behavior" (M. Beroza, ed.), pp. 61–94. Academic Press, New York.
Blum, M. S. (1974). *J. N.Y. Entomol. Soc.* **82,** 141.
Blum, M. S. (1980). *In* "Animals and Environmental Fitness" (R. Gilles, ed.), pp. 207–222. Pergamon, Oxford.
Blum, M. S., and Bohart, G. E. (1972). *Ann. Entomol. Soc. Am.* **65,** 274.
Blum, M. S., and Brand, J. M. (1972). *Am. Zool.* **12,** 553.
Blum, M. S., and Crain, R. D. (1961). *Ann. Entomol. Soc. Am.* **54,** 474.
Blum, M. S., and Edgar, A. L. (1971). *Insect Biochem.* **1,** 181.
Blum, M. S., and Hermann, H. R. (1978a). *Handb. Exp. Pharmakol.* [N.S.] **48,** 801–869.
Blum, M. S., and Hermann, H. R. (1978b). *Handb. Exp. Pharmakol.* [N.S.] **48,** 871–894.
Blum, M. S., and Sannasi, A. (1974). *J. Insect Physiol.* **20,** 451.
Blum, M. S., and Traynham, J. G. (1962). *Proc. Int. Congr. Entomol., 11th, 1960* Vol. 3, p. 48.
Blum, M. S., and Warter, S. L. (1966). *Ann. Entomol. Soc. Am.* **59,** 774.
Blum, M. S. and Wheeler, J. W. (1974). Unpublished data.
Blum, M. S., and Wheeler, J. W. (1978). *In* "Insect Chemoreception" (A. Skirkevicius, ed.), Vol. 3, pp. 17–21. Acad. Sci. Lith. SSR, Vilnius.
Blum, M. S., and Woodring, J. P. (1962). *Science* **138,** 512.
Blum, M. S., Walker, J. R., Callaham, P. S., and Novak, A. F. (1958). *Science* **128,** 306.
Blum, M. S., Novak, A. F., and Taber, S., III (1959). *Science* **130,** 452.
Blum, M. S., Traynham, J. G., Chidester, J. B., and Boggus, J. D. (1960). *Science* **132,** 1480.
Blum, M. S., Crain, R. D., and Chidester, J. B. (1961). *Nature (London)* **189,** 245.
Blum, M. S., Warter, S. L., Monroe, R. S., and Chidester, J. C. (1963). *J. Insect Physiol.* **9,** 881.
Blum, M. S., Padovani, F., and Amante, E. (1968a). *Comp. Biochem. Physiol.* **26,** 291.

Blum, M. S., Padovani, F., Hermann, H. R., Jr., and Kannowski, P. B. (1968b). *Ann. Entomol. Soc. Am.* **61**, 1354.
Blum, M. S., Crewe, R. M., Sudd, J. H., and Garrison, A. W. (1969a). *J. Ga. Entomol. Soc.* **4**, 145.
Blum, M. S., Padovani, F., Curley, A., and Hawk, R. E. (1969b). *Comp. Biochem. Physiol.* **29**, 461.
Blum, M. S., Crewe, R. M., Kerr, W. E., Keith, L. H., Garrison, A. W., and Walker, M. M. (1970). *J. Insect Physiol.* **16**, 1637.
Blum, M. S., Crewe, R. M., and Pasteels, J. M. (1971). *Ann. Entomol. Soc. Am.* **64**, 975.
Blum, M. S., Brand, J. M., Wallace, J. B., and Fales, H. M. (1972). *Life Sci.* **11**, 525.
Blum, M. S., MacConnell, J. G., Brand, J. M., Duffield, R. M., and Fales, H. M. (1973a). *Ann. Entomol. Soc. Am.* **66**, 235.
Blum, M. S., Fales, H. M., and Kerr, W. E. (1973b). Unpublished data.
Blum, M. S., Brand, J. M., Duffield, R. M., and Snelling, R. R. (1973c). *Ann. Entomol. Soc. Am.* **66**, 702.
Blum, M. S., Wallace, J. B., and Fales, H. M. (1973d). *Insect Biochem.* **3**, 353.
Blum, M. S., Wheeler, J. W., Fales, H. M., and Holmberg, R. G. (1973e). Unpublished data.
Blum, M. S., Brand, J. M., and Amante, E. (1974a). Unpublished data.
Blum, M. S., Wheeler, J. W., and Kerr, W. E. (1974b). Unpublished data.
Blum, M. S., Fales, H. M., and Peck, S. (1975). Unpublished data.
Blum, M. S., Kistner, D. H., and Lloyd, H. A. (1977a). Unpublished data.
Blum, M. S., Young, A. M., and Lloyd, H. A. (1977b). Unpublished data.
Blum, M. S., Wallace, J. B., Duffield, R. M., Fales, H. M., and Sokoloski, E. A. (1978). *J. Chem. Ecol.* **4**, 47.
Blum, M. S., Jones, T. H., Hölldobler, B., Fales, H. M., and Jaouni, T. (1980). *Naturwissenschaften* **67**, 144.
Blunck, H. (1917). *Z. Wiss. Zool.* **117**, 205.
Blyth, A. W., and Blyth, M. W. (1920). "Poisons: Their Effects and Detection," 5th ed. Griffin, London.
Boch, R., and Shearer, D. A. (1967). *Z. Vergl. Physiol.* **54**, 1.
Boit, H.-G. (1961). "Engebnisse der Alkaloid-Chemie bis 1960." Akademie-Verlag, Berlin.
Bowers, W. S., Nault, L. R., Webb, R. E., and Dutky, S. R. (1972). *Science* **177**, 1121.
Bowers, W. S., Nishino, C., Montgomery, M. E., and Nault, L. R. (1977). *J. Insect Physiol.* **23**, 697.
Bradshaw, J. W. S., Howse, P. E., and Baker, R. (1973). *Proc., Int. Congr.—Int. Union Study Soc. Insects, 7th, 1973* p. 45.
Bradshaw, J. W. S., Baker, R., and Howse, P. E. (1975). *Nature (London)* **258**, 230.
Bradshaw, J. W. S., Baker, R., and Howse, P. E. (1979). *Physiol. Entomol.* **4**, 15.
Brand, J. M. (1978). *Biochem. Syst. Ecol.* **6**, 337.
Brand, J. M., Blum, M. S., Fales, H. M., and MacConnell, J. G. (1972). *Toxicon* **10**, 259.
Brand, J. M., Blum, M. S., and Barlin, M. R. (1973a). *Toxicon* **11**, 325.
Brand, J. M., Blum, M. S., Fales, H. M., and Pasteels, J. M. (1973b). *J. Insect Physiol.* **19**, 369.
Brand, J. M., Blum, M. S., and Ross, H. H. (1973c). *Insect Biochem.* **3**, 45.
Brand, J. M., Duffield, R. M., MacConnell, J. G., Blum, M. S., and Fales, H. M. (1973d). *Science* **179**, 388.
Brand, J. M., Fales, H. M., Sokoloski, E. A., MacConnell, J. G., Blum, M. S., and Duffield, R. M. (1973e). *Life Sci.* **13**, 201.

Brand, J. M., Blum, M. S., Lloyd, H. A., and Fletcher, D. J. C. (1974). *Ann. Entomol. Soc. Am.* **67**, 525.

Brand, J. M., Young, J. C., and Silverstein, R. M. (1979). *Prog. Chem. Org. Nat. Prod.* **37**, 1.

Brattsten, L. B. (1979). *Drug Metab. Rev.* **10**, 35.

Brattsten, L. B., Wilkinson, C. F., and Eisner, T. (1977). *Science* **196**, 1349.

Brockmann, H., Pohl, F., Maier, K., and Haschad, M. N. (1942). *Justus Liebig's Ann. Chem.* **553**, 1.

Brodie, B. B., Gillette, J. R., and LaDu, B. N. (1958). *Annu. Rev. Biochem.* **27**, 427.

Brophy, J. J., Cavill, G. W. K., and Shannon, J. S. (1973). *J. Insect Physiol.* **19**, 791.

Brossut, R., and Roth, L. M. (1977). *J. Morphol.* **151**, 259.

Brower, L. P. (1969). *Sci. Am.* **220**, 22.

Brower, L. P., and Brower, J. V. Z. (1964). *Zoologica (N.Y.)* **49**, 137.

Brower, L. P., and Glazier, S. C. (1975). *Science* **188**, 19.

Brower, L. P., and Moffitt, C. M. (1974). *Nature (London)* **249**, 280.

Brower, L. P., Brower, J. V. Z., and Corvino, J. M. (1967). *Proc. Natl. Acad. Sci. U.S.A.* **57**, 893.

Brower, L. P., McEvoy, P. B., Williamson, K. L., and Flannery, M. A. (1972). *Science* **177**, 426.

Brower, L. P., Edmunds, M., and Moffitt, C. M. (1975). *J. Entomol., Ser. A: Gen. Entomol.* **49**, 183.

Brown, K. S. (1967). *Syst. Zool.* **16**, 213.

Brown, W. V., and Moore, B. P. (1979). *Insect Biochem.* **9**, 451.

Brüschweiler, F., Stöcklin, W., Stöckel, K., and Reichstein, T. (1969). *Helv. Chim. Acta* **52**, 2086.

Bücherl, W. (1946). *Mem. Inst. Butantan, Sao Paulo* **19**, 181.

Bücherl, W., and Buckley, E. E., eds. (1971). "Venomous Animals and Their Venoms," Vol. 3. Academic Press, New York.

Buckley, E. E., and Porges, N., eds. (1956). "Venoms," Publ. No. 44. Am. Assoc. Adv. Sci., Washington, D.C.

Buffkin, D. C., and Russell, F. E. (1971). *Proc. West. Pharmacol. Soc.* **14**, 166.

Buffkin, D. C., and Russell, F. E. (1972). *Toxicon* **10**, 526.

Bühring, M., Francke, W., and Heemann, V. (1976). *Z. Naturforsch., C: Biosci.* **31C**, 748.

Buren, W. F. (1958). *J. N.Y. Entomol. Soc.* **66**, 119.

Buren, W. F., Allen, G. E., Whitcomb, W. H., Lennartz, F. E., and Williams, R. N. (1974). *J. N.Y. Entomol. Soc.* **81**, 113.

Burger, B. V., Roth, M., Roux, M. L., Spies, H. S. C., Truter, V., and Geertsema, H. (1978). *J. Insect Physiol.* **24**, 803.

Busgen, M. (1891). *Jena. Z. Naturwiss.* **25**, 339.

Butenandt, A., and Rembold, H. (1957). *Hoppe-Seyler's Z. Physiol. Chem.* **308**, 284.

Butenandt, A., and Tam, N. (1957). *Hoppe-Seyler's Z. Physiol. Chem.* **308**, 277.

Butenandt, A., Linzen, B., and Lindauer, M. (1959). *Arch. Anat. Microsc. Morphol. Exp.* **48**, 13.

Caius, J. F., and Mhaskar, K. S. (1932). *Indian Med. Res. Mem.* **24**, 1.

Calam, D. H. (1969). *Nature (London)* **221**, 856.

Calam, D. H., and Scott, G. C. (1969). *J. Insect Physiol.* **15**, 1695.

Calam, D. H., and Youdeowei, A. (1968). *J. Insect Physiol.* **14**, 1147.

Cantore, G. P., and Bettini, S. (1958). *Rend. Ist. Super. Sanita (Ital. Ed.)* **21**, 794.

Cardani, C., Ghiringhelli, D., Mondelli, R., Pavan, M., and Quilico, A. (1965a). *Ann. Soc. Entomol. Fr.* **1**, 813.
Cardani, C., Ghiringhelli, D., Mondelli, R., and Quilico, A. (1965b). *Tetrahedron Lett.* p. 2537.
Cardani, C., Ghiringhelli, D., Mondelli, R., and Quilico, A. (1966). *Gazz. Chim. Ital.* **96**, 3.
Cardani, C., Ghiringhelli, D., Quilico, A., and Selva, A. (1967). *Tetrahedron Lett.* p. 4023.
Cardani, C., Fuganti, C., Ghiringhelli, D., Grasselli, P., Pavan, M., and Valcurone, M. D. (1973). *Tetrahedron Lett.* p. 2815.
Carpenter, G. D. H. (1921). *Trans. Ent. Soc. London* **1921**, 1.
Carpenter, G. D. H. (1938). *Proc. Zool. Soc. London, Ser. A* **108**, 17.
Carrel, J. E., and Eisner, T. (1974). *Science* **183**, 755.
Casnati, G., Nencini, G., Quilico, A., Pavan, M., Ricca, A., and Salvatori, T. (1963). *Experientia* **19**, 409.
Casnati, G., Pavan, M., and Ricca, A. (1964). *Boll. Soc. Entomol. Ital.* **44**, 147.
Casnati, G., Pavan, M., and Ricca, A. (1965). *Ann. Soc. Entomol. Fr.* **1**, 705.
Casnati, G., Ricca, A., and Pavan, M. (1967). *Chim. Ind. (Milan)* **49**, 57.
Castellani, A. A., and Pavan, M. (1966). *Boll. Soc. Ital. Biol. Sper.* **42**, 221.
Castellani, A. A., Gabba, A., Leterza, L., and Pavan, M. (1969). *Boll. Soc. Entomol. Ital.* **48**, 147.
Cavill, G. W. K. (1970). *J. Proc. Soc. N. S. W.* **103**, 109.
Cavill, G. W. K., and Clark, D. V. (1967). *J. Insect Physiol.* **13**, 131.
Cavill, G. W. K., and Ford, D. L. (1960). *Aust. J. Chem.* **13**, 296.
Cavill, G. W. K., and Hinterberger, H. (1960). *Aust. J. Chem.* **13**, 514.
Cavill, G. W. K., and Hinterberger, H. (1961). *Aust. J. Chem.* **14**, 143.
Cavill, G. W. K. and Hinterberger, H. (1962). *Proc. Int. Congr. Entomol., 11th, 1960* Vol. 3, p. 53.
Cavill, G. W. K., and Houghton, E. (1973). *Aust. J. Chem.* **26**, 1131.
Cavill, G. W. K., and Houghton, E. (1974a). *Aust. J. Chem.* **27**, 879.
Cavill, G. W. K., and Houghton, E. (1974b). *J. Insect Physiol.* **20**, 2049.
Cavill, G. W. K., and Williams, P. J. (1967). *J. Insect Physiol.* **13**, 1097.
Cavill, G. W. K., Ford, D. L., and Locksley, H. D. (1956a). *Chem. Ind. (London)* p. 465.
Cavill, G. W. K., Ford, D. L., and Locksley, H. D. (1956b). *Aust. J. Chem.* **9**, 288.
Cavill, G. W. K., Ford, D. L., Hinterberger, H., and Solomon, D. H. (1961). *Aust. J. Chem.* **14**, 276.
Cavill, G. W. K., Robertson, P. L., and Whitfield, F. B. (1964). *Science* **146**, 79.
Cavill, G. W. K., Clark, O. V., and Hinterberger, H. (1966). *Aust. J. Chem.* **19**, 1495.
Cavill, G. W. K., Williams, P. J., and Whitfield, F. B. (1967). *Tetrahedron Lett.*, 2201.
Cavill, G. W. K., Clark, D. V., and Whitfield, F. B. (1968). *Aust. J. Chem.* **21**, 2819.
Cavill, G. W. K., Clark, D. V., Howden, M. E. H., and Wyllie, S. G. (1970). *J. Insect Physiol.* **16**, 1721.
Cederberg, B. (1977). *Proc. Int. Congr. Int. Union Study Soc. Insects, 8th,* 1977 p. 77.
Chadha, M. S., Eisner, T., and Meinwald, J. (1961a). *J. Insect Physiol.* **7**, 46.
Chadha, M. S., Eisner, T., and Meinwald, J. (1961b). *Ann. Entomol. Soc. Am.* **54**, 642.
Chadha, M. S., Eisner, T., Monro, A., and Meinwald, J. (1962). *J. Insect Physiol.* **8**, 175.
Chadha, M. S., Joshi, N. K., Mamdapur, V. R., and Sipahimalani, A. T. (1970). *Tetrahedron Lett.*, 2061.
Chan, T. K., Geren, C. R., Howell, D. E., and Odell, G. V. (1975). *Toxicon* **13**, 61.
Cheng, E. Y., Cutkomp, L. K., and Koch, R. B. (1977). *Biochem. Pharmacol.* **26**, 1179.

Chhatwal, G. S., and Habermann, E. (1980). *IRCS Med. Sci.* **8,** 517.
Choudhuri, D. K., and Das, K. K. (1968). *Indian J. Entomol.* **30,** 203.
Clark, K. J., Fray, G. I., Jaeger, R. H., and Robinson, Sir R. (1959). *Tetrahedron Lett.*, 217.
Cmelik, S. (1969). *Hoppe-Seyler's Z. Physiol. Chem.* **350,** 1076.
Colledge, W. C. (1910). *Pharm. J.* **84,** 674.
Collins, R. P. (1968). *Ann. Entomol. Soc. Am.* **61,** 1338.
Collins, R. P., and Drake, T. H. (1965). *Ann. Entomol. Soc. Am.* **58,** 764.
Conner, W. E., Jones, T. H., Eisner, T., and Meinwald, J. (1977). *Experientia* **33,** 206.
Contardi, A., and Latzer, P. (1928). *Biochem. Z.* **197,** 222.
Creighton, W. S. (1950). *Bull. Mus. Comp. Zool.* **104,** 1.
Crewe, R. M., and Blum, M. S. (1970a). *Z. Vergl. Physiol.* **70,** 363.
Crewe, R. M., and Blum, M. S. (1970b). *J. Insect Physiol.* **16,** 141.
Crewe, R. M., and Blum, M. S. (1971). *Ann. Entomol. Soc. Am.* **64,** 1007.
Crewe, R. M., and Blum, M. S. (1972). *J. Insect Physiol.* **18,** 31.
Crewe, R. M., and Fletcher, D. J. C. (1974). *J. Entomol. Soc. S. Afr.* **37,** 291.
Crewe, R. M., and Fletcher, D. J. C. (1976). *S. Afr. J. Sci.* **72,** 119.
Crewe, R. M., and Ross, F. P. (1975a). *Nature (London)* **254,** 448.
Crewe, R. M., and Ross, F. P. (1975b). *Insect Biochem.* **5,** 839.
Crewe, R. M., Brand, J. M., and Fletcher, D. J. C. (1969). *Ann. Entomol. Soc. Am.* **62,** 1212.
Crewe, R. M., Brand, J. M., Fletcher, D. J. C., and Eggers, S. H. (1970). *J. Ga. Entomol. Soc.* **5,** 42.
Crewe, R. M., Blum, M. S., and Collingwood, C. A. (1972). *Comp. Biochem. Physiol. B* **43B,** 703.
Crossley, A. C., and Waterhouse, D. F. (1969). *Tissue Cell* **1,** 525.
Crout, D. H. G., Curtis, R. F., Hassall, C. H., and Jones, T. L. (1963). *Tetrahedron Lett.* p. 63.
Crout, D. H. G., Hassall, C. H., and Jones, T. L. (1964). *J. Chem. Soc.* p. 2187.
Cuénot, L. (1890). *Bull. Soc. Zool. Fr.* **15,** 126.
Cuénot, L. (1894). *C. R. Seances Soc. Biol. Ses Fil.* **118,** 875.
Cuénot, L. (1896a). *C. R. Seances Soc. Biol. Ses Fil.* **122,** 328.
Cuénot, L. (1896b). *Arch. Zool. Exp. Gen.* **4,** 655.
Curasson, G. (1934). *Bull. Acad. Vet. Fr.* **7,** 377.
Dahlman, D. L. (1977). *Entomol. Exp. Appl.* **22,** 123.
Dahlman, D. L., and Rosenthal, G. A. (1975). *Comp. Biochem. Physiol. A* **51A,** 33.
D'Ajello V., Zlotkin, E., Miranda, F., Lissitzky S., and Bettini, S. (1972). *Toxicon* **10,** 399.
Daloze, D., and Pasteels, J. M. (1979). *J. Chem. Ecol.* **5,** 63.
Darlington, P. J., Jr. (1938). *Trans. R. Entomol. Soc. London* **87,** 681.
Dateo, G. P., and Roth, L. M. (1967a). *Science* **155,** 88.
Dateo, G. P., and Roth, L. M. (1967b). *Ann. Entomol. Soc. Am.* **60,** 1025.
Davenport, D., Wootton, D. M., and Cushing, J. F. (1952). *Biol. Bull. (Woods Hole, Mass.)* **102,** 100.
Davis, R., and Nahrstedt, A. (1979). *Comp. Biochem. Physiol. B* **64B,** 395.
de Jong, M. C. J. M., and Bleumink, E. (1977a). *Arch. Dermatol. Res.* **259,** 247.
de Jong, M. C. J. M., and Bleumink, E. (1977b). *Arch. Dermatol. Res.* **259,** 263.
de la Lande, I. S., Thomas, D. W., and Tyler, M. J. (1963). *Biochem. Pharmacol.* **12,** Suppl., 187.

del Moral, R. (1972). *Oecologia* **9**, 289.
Dethier, V. G. (1939). *J. N.Y. Entomol. Soc.* **47**, 131.
Dettner, K. (1977). *Biochem. Syst. Ecol.* **7**, 129.
Dettner, K. (1979). *Biochem. Syst. Ecol.* **7**, 129.
Dettner, K., and Schwinger, G. (1977). *Naturwissenschaften* **64**, 42.
Dettner, K., and Schwinger, G. (1980). *Biochem. Syst. Ecol.* **8**, 89.
Devakul, V., and Maarse, H. (1964). *Anal. Biochem.* **7**, 269.
Diniz, C. R. (1963). *An. Acad. Bras. Cienc.* **35**, 283.
Diniz, C. R., and Gomez, M. V. (1968). *Mem. Inst. Butantan, Sao Paulo* **33**, 899.
Diniz, C. R., and Gonçalves, J. M. (1960). *Biochim. Biophys. Acta* **41**, 470.
Dixon, A. F. G. (1958). *Trans. R. Entomol. Soc. London* **10**, 319.
Dixon, A. F. G., Martin-Smith, M., and Smith, S. J. (1963). *Can. Pharm. J.* **96**, 501.
Dixon, C. A., Erickson, J. M., Kellett, D. N., and Rothschild, M. (1978). *J. Zool.* **185**, 437.
Dolinger, P. M., Ehrlich, P. R., Fitch, W. L., and Breedlove, D. E. (1973). *Oecologia* **13**, 191.
Doyen, J. T. (1972). *Quaest. Entomol.* **8**, 357.
Drainas, D., Moores, G. R., and Lawrence, A. J. (1978). *FEBS Lett.* **86**, 49.
Duffey, S. S. (1970). *Science* **169**, 78.
Duffey, S. S. (1977). *Proc. Int. Congr. Entomol., 15th,* pp. 323–394.
Duffey, S. S. (1980). *Annu. Rev. Entomol.* **25**, 447.
Duffey, S. S., and Blum, M. S. (1976). Unpublished data.
Duffey, S. S., and Blum, M. S. (1977). *Insect Biochem.* **7**, 57.
Duffey, S. S., and Scudder, G. G. E. (1972). *J. Insect Physiol.* **18**, 63.
Duffey, S. S., and Scudder, G. G. E. (1974). *Can. J. Zool.* **52**, 283.
Duffey, S. S., Underhill, E. W., and Towers, G. H. N. (1974). *Comp. Biochem. Physiol. B* **47B**, 753.
Duffey, S. S., Blum, M. S., Fales, H. M., Evans, S. L., Roncadori, R. W., Tieman, D. L., and Nakagawa, Y. (1977). *J. Chem. Ecol.* **3**, 101.
Duffey, S. S., Blum, M. S., Isman, M. B., and Scudder, G. G. E. (1978). *J. Insect Physiol.* **24**, 639.
Duffield, P. H., Duffield, A. M., Carroll, P. R., and Morgans, D. (1979). *Biomed. Mass Spectrom.* **6**, 105.
Duffield, R. M., and Blum, M. S. (1973). *Ann. Entomol. Soc. Am.* **66**, 1357.
Duffield, R. M., and Blum, M. S. (1975a). *Experientia* **31**, 466.
Duffield, R. M., and Blum, M. S. (1975b). *Comp. Biochem. Physiol. B* **51B**, 281.
Duffield, R. M., and Blum, M. S. (1975c). Unpublished data.
Duffield, R. M., Blum, M. S., and Brand, J. M. (1974). *Ann. Entomol. Soc. Am.* **67**, 821.
Duffield, R. M., Blum, M. S., Wallace, J. B., Lloyd, H. A., and Regnier, F. E. (1977). *J. Chem. Ecol.* **3**, 649.
Dyar, H. G. (1891). *Entomol. News* **2**, 50.
Dyar, H. G. (1915). *Proc. U.S. Natl. Mus.* **47**, 174.
Edery, H., Ishay, J., Lass, I., and Gitter, S. (1972). *Toxicon* **10**, 13.
Edwards, J. S. (1961). *J. Exp. Biol.* **38**, 61.
Edwards, J. S. (1962). *Proc. Int. Congr. Entomol., 11th, 1960* Vol. 3, p. 259.
Edwards, J. S. (1966). *Nature (London)* **211**, 73.
Edwards, L. J., Siddall, J. B., Dunham, L. L., Uden, P., and Kislow, C. J. (1973). *Nature (London)* **241**, 126.
Ehrlich, P. R., and Raven, P. H. (1965). *Evolution* **18**, 586.

Eisner, H. E., Eisner, T., and Hurst, J. J. (1963). *Chem. Ind. (London)* p. 124.
Eisner, H. E., Alsop, D. W., and Eisner, T. (1967). *Psyche* **74,** 107.
Eisner, H. E., Wood, W. F., and Eisner, T. (1975). *Psyche* **82,** 20.
Eisner, T. (1970). *In* "Chemical Ecology" (E. Sondheimer and J. B. Simeone, eds.), pp. 157–217. Academic Press, New York.
Eisner, T. (1972). *Verh. Dtsch. Zool. Gesl.* **65,** 123.
Eisner, T., and Meinwald, Y. C. (1965). *Science* **150,** 1733.
Eisner, T., McKittrick, F., and Payne, R. (1959). *Pest Control* **27,** 11.
Eisner, T., Meinwald, J., Monro, A., and Ghent, R. (1961). *J. Insect Physiol.* **6,** 272.
Eisner, T., Eisner, H. E., Hurst, J. J., Kafatos, F. C., and Meinwald, J. (1963a). *Science* **129,** 1218.
Eisner, T., Hurst, J. J., and Meinwald, J. (1963b). *Psyche* **70,** 94.
Eisner, T., Swithenbank, C., and Meinwald, J. (1963c). *Ann. Entomol. Soc. Am.* **56,** 37.
Eisner, T., McHenry, F., and Salpeter, M. M. (1964). *J. Morphol.* **115,** 355.
Eisner, T., Hurst, J. J., Keeton, W. T., and Meinwald, Y. (1965). *Ann. Entomol. Soc. Am.* **58,** 247.
Eisner, T., Van Tassell, E., and Carrel, J. E. (1967). *Science* **158,** 1471.
Eisner, T., Meinwald, Y. C., Alsop, D. W., and Carrel, J. E. (1968). *Ann. Entomol. Soc. Am.* **61,** 610.
Eisner, T., Silberglied, R. E., Aneshansley, D., Carrel, J. E., and Howland, H. C. (1969). *Science* **166,** 1172.
Eisner, T., Pliske, T. E., Ikeda, M., Owen, D. F., Vázquez, L., Pérez, H., Franclemont, J. G., and Meinwald, J. (1970). *Ann. Entomol. Soc. Am.* **63,** 914.
Eisner, T., Hendry, L. B., Peakall, D. B., and Meinwald, J. (1971a). *Science* **172,** 277.
Eisner, T., Kluge, A. F., Carrel, J. E., and Meinwald, J. (1971b). *Science* **173,** 650.
Eisner, T., Kluge, A. F., Ikeda, M. I., Meinwald, Y. C., and Meinwald, J. (1971c). *J. Insect Physiol.* **17,** 245.
Eisner, T., Kluge, A. F., Carrel, J. E., and Meinwald, J. (1972). *Ann. Entomol. Soc. Am.* **65,** 765.
Eisner, T., Aneshansley, D., Eisner, M., Rutowski, R., Chong, B., and Meinwald, J. (1974a). *Psyche* **81,** 189.
Eisner, T., Johnessee, J. S., Carrel, J., Hendry, L. B., and Meinwald, J. (1974b). *Science* **184,** 996.
Eisner, T., Jones, T. H., Aneshansley, D. J., Tschinkel, W. R., Silberglied, R. E., and Meinwald, J. (1977). *J. Insect Physiol.* **23,** 1383.
Eisner, T., Wiemer, D. F., Haynes, L. W., and Meinwald, J. (1978). *Proc. Natl. Acad. Sci. U.S.A.* **75,** 905.
Eisner, T., Nowicki, S., Goetz, M., and Meinwald, J. (1980). *Science* **208,** 1039.
El-Asmar, M. F., Ibrahim, S. A., and Rabie, F. (1972). *Toxicon* **10,** 73.
Emmelin, N., and Feldberg, W. (1947). *J. Physiol. (London)* **106,** 440.
Engelhardt, M., Rapoport, H., and Sokoloff, A. (1965). *Science* **150,** 632.
Ercoli, A. (1940). *In* "Handbuch der Enzymologie" (F. F. Nord and R. Weidenhagen, eds.), Vol. 1, pp. 480–494. Akad. Verlagsges., Leipzig.
Erickson, J. M. (1973). *Psyche* **80,** 230.
Erickson, J. M., and Feeny, P. (1974). *Ecology* **55,** 103.
Ernst, E. (1959). *Rev. Suisse Zool.* **66,** 289.
Erspamer, V. (1954). *Pharmacol. Rev.* **6,** 425.

Evans, D. A., Baker, R., Briner, P. H., and McDowell, P. G. (1977). *Proc. Int. Congr. Int. Union Study Soc. Insects, 8th, 1977* pp. 46–47.
Evans, D. A., Baker, R., and Howse, P. E. (1979). *In* "Chemical Ecology: Odour Communication in Animals" (F. Ritter, ed.), pp. 213–224. Elsevier, Amsterdam.
Everton, I. J., and Staddon, B. W. (1979). *J. Insect Physiol.* **25**, 133.
Everton, I. J., Games, D. E., and Staddon, B. W. (1974). *Ann. Entomol. Soc. Am.* **67**, 815.
Everton, I. J., Knight, D. W., and Staddon, B. W. (1979). *Comp. Biochem. Physiol. B* **63B**, 157.
Fales, H. M., Blum, M. S., Crewe, R. M., and Brand, J. M. (1972). *J. Insect Physiol.* **18**, 1077.
Fales, H. M., Jaouni, T. M., Schmidt, J. O., and Blum, M. S. (1980). *J. Chem. Ecol.* **6,** 895.
Feir, D., and Suen, Jin-Shwu (1971). *Ann. Entomol. Soc. Am.* **64,** 1173.
Fieser, L. F., and Ardao, M. I. (1956). *J. Am. Chem. Soc.* **78,** 774.
Fischer, F. G., and Bohn, H. (1957a). *Justus Liebigs Ann. Chem.* **603,** 232.
Fischer, F. G., and Bohn, H. (1957b). *Hoppe-Seyler's Z. Physiol. Chem.* **306,** 265.
Fish, L. J., and Pattenden, G. (1975). *J. Insect Physiol.* **21,** 741.
Fishelson, L. (1960). *Eos* **36,** 41.
Fraenkel, G. (1959). *Science* **129,** 1466.
Frankenburg, W. G., and Gottscho, A. M. (1955). *J. Am. Chem. Soc.* **77,** 5728.
Frazer, J. F. D., and Rothschild, M. (1962). *Proc. Int. Congr. Entomol., 11th, 1960* Vol. 3, p. 249.
Fredholm, B., and Haegermark, O. (1967). *Acta Physiol. Scand.* **69,** 304.
Freeman, M. A. (1968). *Comp. Biochem. Physiol.* **26,** 1041.
Frontali, N., and Grasso, A. (1964). *Arch. Biochem. Biophys.* **106,** 213.
Furusaki, A., Watanabe, T., Matsumoto, T., and Yanagiya, M. (1968). *Tetrahedron Lett.* p. 6301.
Fusco, R., Trave, R., and Vercellone, A. (1955). *Chim. Ind. (Milan)* **37,** 251.
Gadamer, J. (1914). *Arch. Pharm. (Weinheim, Ger.)* **252,** 636.
Games, D. E., and Staddon, B. W. (1973a). *Experientia* **29,** 532.
Games, D. E., and Staddon, B. W. (1973b). *J. Insect Physiol.* **19,** 1527.
Games, D. E., Schofield, C. J., and Staddon, B. W. (1974). *Ann. Entomol. Soc. Am.* **67,** 820.
Garb, G. (1915). *Entomol. Zool.* **8,** 88.
Gauldie, J., Hanson, J. M., Rumjanik, F. D., Shipolini, R. A., and Vernon, C. A. (1976). *Eur. J. Biochem.* **61,** 369.
Gauldie, J., Hanson, J. M., Shipolini, R. A., and Vernon, C. A. (1978). *Eur. J. Biochem.* **83,** 405.
Geertsema, H., Burger, B. V., le Roux, M., and Spies, H. S. C., (1976). *J. Insect Physiol.* **22,** 1369.
Geller, R. G., Yoshida, H., Beaven, M. A., Horakova, Z., Atkins, F. L., Yamabe, H., and Pisano, J. J. (1976). *Toxicon* **14,** 27.
Ghent, R. L. (1961). Ph.D. Thesis, Cornell University, Ithaca, New York.
Gilbo, C. M., and Coles, N. W. (1964). *Aust. J. Biol. Sci.* **17,** 758.
Gilby, A. R., and Waterhouse, D. F. (1964). *Aust. J. Chem.* **17,** 1311.
Gilby, A. R., and Waterhouse, D. F. (1965). *Proc. R. Soc.* London, *Ser. B* **162,** 105.
Gilby, A. R., and Waterhouse, D. F. (1967). *Nature* (*London*) **216,** 90.
Gilchrist, T. L., Stansfield, F., and Cloudsley-Thompson, J. L. (1966). *Proc. R. Entomol. Soc.* London, *Ser. A* **41,** 55.

Gillette, J. R., Conney, A. H., Cosmides, G. J., Estabrook, R. W., Fouts, J. R., and Mannering, G. J., eds. (1969). "Microsomes and Drug Oxidations." Academic Press, New York.
Goetz, M., Wiemer, D., Haynes, L. R. W., Meinwald, J., and Eisner, T. (1979). *Helv. Chim. Acta* **62,** 1396.
Goff, A. M., and Nault, L. R. (1974). *Environ. Entomol.* **3,** 565.
Goldman, L., Sayer, F., Levine, A., Goldman, J., Goldman, S., and Spinanger, J. (1960). *J. Invest. Dermatol.* **34,** 67.
Goodfellow, R. D., Huang, Y. S., and Radtke, H. E. (1972). *Insect Biochem.* **2,** 467.
Goodfellow, R. D., Radtke, H. E., Huang, Y. S., and Liv, G. C. K. (1973). *Insect Biochem.* **3,** 61.
Gordon, H. T., Waterhouse, D. F., and Gilby, A. R. (1963). *Nature (London)* **197,** 818.
Graham, J. D. P., and Staddon, B. W. (1974). *J. Entomol., Ser. A: Gen. Entomol.* **48,** 177.
Graham, R. A., Brand, J. M., and Markovetz, A. J. (1979). *Insect Biochem.* **9,** 331.
Granata, F., Paggi, P., and Frontali, N. (1972). *Toxicon* **10,** 551.
Grasso, A. (1976). *Biochim. Biophys. Acta* **439,** 406.
Green, C. R., Chan, T. K., Howell, D. E., and Odell, G. V. (1975). *Toxicon* **13,** 233.
Green, C. R., Chan, T. K., Howell, D. E., and Odell, G. V. (1976). *Arch. Biochem. Biophys.* **174,** 90.
Guldensteeden-Egeling, C. (1882). *Arch. Gesamte Physiol. Menschen Tiere* **28,** 576.
Guthrie, F. E., Ringler, R. L., and Bowery, T. G. (1957). *J. Econ. Entomol.* **50,** 821.
Haase, E. (1896). "Researches on Mimicry on the Basis of a Natural Classification of the Papilionidae, pt. 2, Researches on Mimicry." Nägele, Stuttgart.
Habermann, E. (1958). *Z. Exp. Med.* **129,** 436.
Habermann, E. (1971). *In* "Venomous Animals and their Venoms" (W. Bücherl and E. E. Buckley, eds.), Vol. 3, pp. 61–93. Academic Press, New York.
Habermann, E. (1972). *Science* **177,** 314.
Habermann, E., and Jentsch, J. (1967). *Hoppe-Seyler's Z. Physiol. Chem.* **348,** 37.
Hackman, R. H., Pryor, M. G. M., and Todd, A. R. (1948). *Biochem. J.* **43,** 474.
Haldane, J. B. S. (1955). *Sci. Prog. (Oxford)* **43,** 385.
Hall, F. R., Hollingworth, R. M., and Shankland, D. L. (1971). *Comp. Biochem. Physiol. B* **38B,** 723.
Hansel, R. (1956). *Arch. Pharm. (Weinheim, Ger.)* **259,** 619.
Happ, G. M. (1967). *Ann. Entomol. Soc. Am.* **60,** 279.
Happ, G. M. (1968). *J. Insect Physiol.* **14,** 1821.
Happ, G. M., and Eisner, T. (1961). *Science* **134,** 329.
Happ, G. M., and Meinwald, J. (1965). *J. Am. Chem. Soc.* **87,** 2507.
Harborne, J. B. (1972). *Recent Adv. Phytochem.* **4,** 107.
Hartter, P., and Weber, U. (1975). *Hoppe-Seyler's Z. Physiol. Chem.* **356,** 693.
Haux, P. (1969). *Hoppe-Seyler's Z. Physiol. Chem.* **350,** 536.
Hayashi, N., Komae, H., and Hiyama, H. (1973a). *Z. Naturforsch., C: Biosci.* **28C,** 226.
Hayashi, N., Komae, H., and Hiyama, H. (1973b). *Z. Naturforsch., C: Biosci.* **28C,** 626.
Hayashi, N., Yamamura, Y., Ôhama, S., Yokochô, K., Komae, H., and Kuwahara, Y. (1976). *Experientia* **32,** 418.
Hedin, P. A., Gueldner, R. C., Henson, R. D., and Thompson, A. C. (1974). *J. Insect Physiol.* **20,** 2135.
Hefetz, A., and Batra, S. W. T. (1979). *Experientia* **35,** 1138.
Hefetz, A., and Blum, M. S. (1978a). *Science* **201,** 454.

Hefetz, A., and Blum, M. S. (1978b). *Biochim. Biophys. Acta* **543,** 484.
Hefetz, A., Blum, M. S., Eickwort, G. C., and Wheeler, J. W. (1978). *Comp. Biochem. Physiol. B* **61B,** 129.
Hefetz, A., Batra, S. W. T., and Blum, M. S. (1979). *J. Chem. Ecol.* **5,** 753.
Heitz, J. R., and Norment, B. R. (1974). *Toxicon* **12,** 181.
Henson, R. D., Thompson, A. C., Hedin, P. A., Nichols, P. R., and Neel, W. W. (1975). *Experientia* **31,** 145.
Hepburn, H. R., Berman, N. J., Jacobson, H. J., and Fatti, L. P. (1973). *Oecologia* **12,** 373.
Hermann, H. R., and Blum, M. S. (1968). *Psyche* **75,** 216.
Hikino, H., Ohizumi, Y., and Takemoto, T. (1971). *J. Chem. Soc., Chem. Commun.* **17,** 1036.
Hikino, H., Ohizumi, Y., and Takemoto, T. (1975a). *J. Insect Physiol.* **21,** 1953.
Hikino, H., Ohizumi, Y., and Takemoto, T. (1975b). *Hoppe-Seyler's Z. Physiol. Chem.* **356,** 309.
Hintze, C. (1969). *Z. Morphol. Tiere* **64,** 9.
Hirai, Y., Yasuhara, T., Yoshida, H., Terumi, N., Fujino, M., and Kitada, C. (1979a). *Chem. Pharm. Bull.* **27,** 1942.
Hirai, Y., Kuwada, M., Yasuhara, T., Yoshida, H., and Nakajima, T. (1979b). *Chem. Pharm. Bull.* **27,** 1945.
Hoch, J. H. (1961). "A Survey of Cardiac Glycosides and Genins." Univ. of South Carolina Press, Columbia.
Hoffman, D. R. (1978). *Ann. Allergy* **40,** 171.
Holdstock, D. J., Mathias, A. P., and Schachter, M. (1957). *Br. J. Pharmacol. Chemother.* **12,** 149.
Hollande, A.-C. (1909). *Ann. Univ. Grenoble, Sect. Sci.-Med.* **21,** 459.
Hollande, A.-C. (1911). *Arch. Anat. Microsc.* **13,** 171.
Hölldobler, B., and Maschwitz, U. (1965). *Z. Vergl. Physiol.* **50,** 551.
Honda, K. (1980). *J. Insect Physiol.* **26,** 39.
Hrdý, I., Křeček, J., and Vrkoč, J. (1977). *Proc. Int. Congr. Int. Union Study Soc. Insects, 8th, 1977* pp. 303–304.
Hsiao, T. H., and Fraenkel, G. (1969). *Toxicon* **7,** 119.
Hurst, J. J., Meinwald, J., and Eisner, T. (1964). *Ann. Entomol. Soc. Am.* **57,** 44.
Ikan, R., Cohen, E., and Shulov, A. (1970). *J. Insect Physiol.* **16,** 2201.
Ishay, J., Abraham, Z., Grunfeld, Y., and Gitter, S. (1974). *Comp. Biochem. Physiol. A* **48A,** 369.
Ismail, M., El-Asmar, M. F., and Osman, O. H. (1975). *Toxicon* **13,** 49.
Isman, M. B. (1977). *J. Insect Physiol.* **23,** 1183.
Isman, M. B., Duffey, S. S., and Scudder, G. G. E. (1977a). *Can. J. Zool.* **55,** 1024.
Isman, M. B., Duffey, S. S., and Scudder, G. G. E. (1977b). *J. Chem. Ecol.* **3,** 613.
Isoe, S., Katsumura, S., Hyeon, S. B., and Sakan, T. (1971). *Tetrahedron Lett.* p. 1089.
Janzen, D. H. (1975). *In* "Evolutionary Strategies of Parasitic Insects and Mites" (P. W. Price, ed.), p. 154. Plenum Press, New York.
Jaques, R. (1955). *Helv. Physiol. Pharmacol. Acta* **13,** 113.
Jaques, R. (1956). *In* "Venoms" (E. Buckley and N. Porges, eds.), Publ. No. 44, pp. 291–293. Am. Assoc. Adv. Sci., Washington, D.C.
Jaques, R., and Schachter, M. (1954). *Br. J. Pharmacol. Chemother.* **9,** 53.
Jentsch, J. (1969). *Proc. Int. Congr. Int. Union Study Soc. Insects, 6th, Bern 1969* p. 69.
Jentsch, J., and Dielenberg, D. (1972). *Justis Liebigs Ann. Chem.* **757,** 187.
Jentsch, J., and Mücke, H. (1977). *Int. J. Pept. Protein Res.* **9,** 78.

Jones, C. G. and Blum, M. S. (1979). Unpublished data.
Jones, D. A., Parsons, J., and Rothschild, M. (1962). *Nature (London)* **193,** 52.
Jones, F. M. (1937). *Proc. R. Entomol. Soc. London, Ser. A* **12,** 74.
Jones, T. H., Conner, W. E., Meinwald, J., Eisner, H. E., and Eisner, T. (1976a). *J. Chem. Ecol.* **2,** 421.
Jones, T. H., Conner, W. E., Kluge, A. F., Eisner, T., and Meinwald, J. (1976b). *Experientia* **32,** 1234.
Jones, T. H., Meinwald, J., Hicks, K., and Eisner, T. (1977). *Proc. Natl. Acad. Sci. U.S.A.* **74,** 419.
Jones, T. H., Blum, M. S., and Fales, H. M. (1979). *Tetrahedron Lett.* p. 1031.
Jones, T. H., Blum, M. S., Fales, H. M., and Thompson, C. R. (1980). *J. Org. Chem.* **45,** 4778.
Jong, Y. S., Norment, B. R., and Heitz, J. R. (1979). *Toxicon* **17,** 307.
Jouvenaz, D. P., Blum, M. S., and MacConnell, J. G. (1972). *Antimicrob. Agents Chemother.* **2,** 291.
Jungreis, A. M., and Vaughan, G. L. (1977). *J. Insect Physiol.* **23,** 503.
Kaiser, E. (1953). *Monatsh. Chem.* **84,** 482.
Kaiser, E. (1956). *In* "Venoms" (E. E. Buckley and N. Porges, eds.), Publ. No. 44, pp. 91–93. Am. Assoc. Adv. Sci., Washington, D.C.
Kaplan, N. O., Ciohi, M. M., Hamolsky, M., and Bieber, R. E. (1960). *Science* **131,** 392.
Karlsson, R., and Losman, D. (1972). *J. Chem. Soc., Chem. Commun.* p. 626.
Karpas, A. B., Baer, H., Hooton, M. L., and Evans, R. (1977). *J. Allergy Clin. Immunol.* **60,** 155.
Kaschnitz, R., and Kreil, G. (1978). *Biochem. Biophys. Res. Commun.* **83,** 901.
Kawamoto, F., and Kumada, N. (1979). *In* "Kinins-II: Biochemistry, Pathophysiology, and Clinical Aspects" (S. Fujii, H. Moriya, and T. Suzuki, eds.), pp. 51–55. Plenum, New York.
Kendall, D. A. (1968). *Trans. R. Entomol. Soc. London* **120,** 139.
Kendall, D. A. (1971). *Entomologist* p. 233.
Kerr, W. E., Blum, M. S., and Fales, H. M. (1973). Unpublished data.
Keville, R., and Kannowski, P. B. (1975). *J. Insect Physiol.* **21,** 81.
Kim, Y. H., Brown, G. B., Mosher, H. S., and Fuhrman, F. A. (1975). *Science* **189,** 151.
Kindås-Mügge, I., Lane, C. D., and Kreil, G. (1974). *J. Mol. Biol.* **74,** 451.
Kindås-Mügge, I., Frasel, L., and Diggelmann, H. (1976). *J. Mol. Biol.* **105,** 177.
Kindl, H., and Underhill, E. W. (1968). *Phytochemistry* **1,** 145.
Kishimura, H., Yasuhara, T., Yoshida, H., and Nakajima, T. (1976). *Chem. Pharm. Bull.* **24,** 2896.
Kistner, D. H., and Blum, M. S. (1971). *Ann. Entomol. Soc. Am.* **64,** 589.
Kistner, D. H., Lloyd, H. A., and Blum, M. S. (1977). Unpublished data.
Kittredge, J. S., Takahashi, F. T., Lindsey, J., and Lasker, R. (1974). *Fish. Bull.* **72,** 1.
Klinger, R., and Maschwitz, U. (1977). *J. Chem. Ecol.* **3,** 401.
Kluge, A. F., and Eisner, T. (1971). *Ann. Entomol. Soc. Am.* **64,** 314.
Kobert, R. (1906). "Lehrbuch der Intoxikationen." Enke, Stuttgart.
Koch, R. B., and Desaiah, D. (1975). *Life Sci.* **17,** 1315.
Koch, R. B., Desaiah, D., Foster, D., and Ahmed, K. (1977). *Biochem. Pharmacol.* **26,** 983.
Koeniger, N., Weiss, J., and Maschwitz, U. (1979). *J. Insect Physiol.* **25,** 467.
Kolbe, W., and Proske, M. G. (1973). *Entomolog bl. Biol. syst. Kaefer,* **69,** 57.
Kreil, G. (1973a). *FEBS Lett.* **33,** 241.
Kreil, G. (1973b). *Eur. J. Biochem.* **33,** 558.

Kreil, G. (1975). *FEBS Lett.* **54,** 100.
Kreil, G., and Bachmayer, H. (1971). *Eur. J. Biochem.* **20,** 344.
Kreil, G., and Kreil-Kiss, G. (1967). *Biochem. Biophys. Res. Commun.* **27,** 275.
Kreil, G., Suchanek, G., and Kindås-Mügge, I. (1977). *Fed. Proc., Fed. Am. Soc. Exp. Biol.* **36,** 2081.
Krieger, R. I., Feeny, P. P., and Wilkinson, C. F. (1971). *Science* **172,** 579.
Kugler, C. (1978). *Insectes Soc.* **25,** 267.
Kugler, C. (1979). *Ann. Entomol. Soc. Am.* **72,** 532.
Kullenberg, B., Bergström, G., and Ställberg-Stenhagen, S. (1970). *Acta Chem. Scand.* **24,** 1481.
Kullenberg, B., Bergström, G., Bringer, B., Carlberg, B., and Cederberg, B. (1973). *Zoon, Suppl.* **1,** 23.
Lane, C. (1959). *Entomol. Mon. Mag.* **95,** 93.
Law, J. H., Wilson, E. O., and McCloskey, J. A. (1965). *Science* **149,** 544.
Lawrence, A. J., and Moores, G. R. (1975). *FEBS Lett.* **49,** 287.
Lazarovici, P., Lester, D., Menashè, M., Hochman, Y., and Zlotkin, E. (1979). *Toxicon* **17,** Suppl. 1, 97.
Lee, C. K., Chan, T. K., Ward, B. C., Howell, D. E., and Odell, G. V. (1974). *Arch. Biochem. Biophys.* **164,** 341.
Levin, D. A. (1976). *Annu. Rev. Ecol. Syst.* **7,** 121.
Lewis, J. C. and de la Lande, I. S. (1967). *Toxicon* **4,** 225.
Lewis, J. C., Day, A. J., and de la Lande, I. S. (1968). *Toxicon* **6,** 109.
Linsenmair, K. E., and Jander, R. (1963). *Naturwissenschaften* **50,** 231.
Lloyd, H. A., Blum, M. S., and Duffield, R. M. (1975). *Insect Biochem.* **5,** 489.
Lloyd, H. A., Evans, S. L., Khan, A. H., Tschinkel, W. R., and Blum, M. S. (1978a). *Insect Biochem.* **8,** 333.
Lloyd, H. A., Kistner, D. H., and Blum, M. S. (1978b). Unpublished data.
Loconti, J. D., and Roth, L. M. (1953). *Ann. Entomol. Soc. Am.* **46,** 281.
Longhurst, C., Baker, R., Howse, P. E., and Speed, W. (1978). *J. Insect Physiol.* **24,** 833.
Longhurst, C., Baker, R., and Howse, P. E. (1979a). *J. Chem. Ecol.* **5,** 703.
Longhurst, C., Baker, R., and Howse, P. E. (1979b). *Experientia* **35,** 870.
López, A., and Quesnel, V. C. (1970). *Caribb. J. Sci.* **10,** 5.
Lowy, P. H., Sarmiento, L., and Mitchell, H. K. (1971). *Arch. Biochem. Biophys.* **145,** 338.
Lowy, P. H., Mitchell, H. K., and Tracy, U. W. (1976). *Toxicon* **14,** 203.
Lübke, K., Matthes, S., and Kloss, G. (1971). *Experientia* **27,** 765.
Luby, J. M., Regnier, F. E., Clarke, E. T., Weaver, E. C., and Weaver, N. (1973). *J. Insect Physiol.* **19,** 1111.
McAlister, W. H. (1960). *Tex. J. Sci.* **12,** 17.
MacConnell, J. G., Blum, M. S., and Fales, H. M. (1970). *Science* **168,** 840.
MacConnell, J. G., Blum, M. S., and Fales, H. M. (1971). *Tetrahedron* **26,** 1129.
MacConnell, J. G., Williams, R. N., Brand, J. M., and Blum, M. S. (1974). *Ann. Entomol. Soc. Am.* **67,** 134.
MacConnell, J. G., Blum, M. S., Buren, W. F., Williams, R. N., and Fales, H. M. (1976). *Toxicon* **14,** 69.
McCrone, J. D. (1969). *Am. Zool.* **9,** 153.
McCrone, J. D., and Hatala, R. J. (1967). *In* "Animal Toxins" (F. E. Russell and P. R. Saunders, eds.), pp. 29–34. Pergamon, Oxford.
McCullough, T. (1966a). *Ann. Entomol. Soc. Am.* **59,** 410.
McCullough, T. (1966b). *Ann. Entomol. Soc. Am.* **59,** 1018.

McCullough, T. (1966c). *Ann. Entomol. Soc. Am.* **59**, 1020.
McCullough, T. (1967a). *Ann. Entomol. Soc. Am.* **60**, 861.
McCullough, T. (1967b). *Ann. Entomol. Soc. Am.* **60**, 862.
McCullough, T. (1968). *Ann. Entomol. Soc. Am.* **61**, 1044.
McCullough, T. (1969). *Ann. Entomol. Soc. Am.* **62**, 673.
McCullough, T. (1971). *Ann. Entomol. Soc. Am.* **64**, 1191.
McCullough, T. (1972a). *Ann. Entomol. Soc. Am.* **65**, 275.
McCullough, T. (1972b). *Ann. Entomol. Soc. Am.* **65**, 772.
McCullough, T. (1973). *Ann. Entomol. Soc. Am.* **66**, 231.
McCullough, T. (1974a). *Ann. Entomol. Soc. Am.* **67**, 298.
McCullough, T. (1974b). *Ann. Entomol. Soc. Am.* **67**, 300.
McCullough, T., and Weinheimer, A. J. (1966). *Ann. Entomol. Soc. Am.* **59**, 410.
MacGintie, G. E., and MacGintie, N. (1968). "Natural History of Marine Animals." McGraw-Hill, New York.
McGurk, D. J., Frost, J., Eisenbraun, E. J., Vick, K., Drew, W. A., and Young, J. (1966). *J. Insect Physiol.* **12**, 1435.
McGurk, D. J., Frost, J., Waller, G. R., Eisenbraun, E. J., Vick, K., Drew, W. A., and Young, J. (1968). *J. Insect Physiol.* **14**, 841.
MacLeod, J. K., Howe, I., Cable, J., Blake, J. D., Baker, J. T., and Smith, D. (1975). *J. Insect Physiol.* **21**, 1219.
Marchesini, A., Garanti, L., and Pavan, M. (1969). *Ric. Sci.* **39**, 874.
Marie, Z. A., and Ibrahim, S. A. (1976). *Toxicon* **14**, 93.
Marion, L. (1939). *Can. J. Res.* **17**, 21.
Marsh, N. A., and Rothschild, M. (1974). *J. Zool.* **174**, 89.
Marsh, N. A., Clarke, C. A., Rothschild, M., and Kellett, D. N. (1977). *Nature (London)* **268**, 726.
Marshall, G. A. K. (1902). *Trans. R. Entomol. soc. London* p. 287.
Maschwitz, U. (1964). *Z. Vergl. Physiol.* **47**, 596.
Maschwitz, U. (1971). *Naturwissenschaften* **57**, 1.
Maschwitz, U. (1974). *Oecologia* **16**, 303.
Maschwitz, U. (1975). *In* "Pheromones and Defensive Secretions in Social Insects" (C. Noirot, P. E. Howse, and G. Le Masne, eds.), pp. 41–45. Univ. Dijon Press.
Maschwitz, U., and Maschwitz, E. (1974). *Oecologia* **14**, 289.
Maschwitz, U., and Tho, Y. P. (1974). *Insectes Soc.* **21**, 15.
Maschwitz, U., and Tho, Y. P. (1978). *J. Chem. Ecol.* **4**, 375.
Maschwitz, U., Koob, K., and Schildknecht, H. (1970). *J. Insect Physiol.* **16**, 387.
Maschwitz, U., Jander, R., and Burkhardt, D. (1972). *J. Insect Physiol.* **18**, 1715.
Master, R. W. P., Rao, S. S., and Soman, P. D. (1963). *Biochim. Biophys. Acta* **71**, 422.
Mathias, A. P., and Schachter, M. (1958). *Br. J. Pharmacol. Chemother.* **13**, 326.
Matsuda, K., and Sugawara, F. (1980). *Appl. Entomol. Zool.* **15**, 316.
Matsumoto, T., Yanagiya, M., Maeno, S., and Yasuda, S. (1968). *Tetrahedron Lett.* p. 6297.
Meinwald, J., and Jones, T. H. (1978). *J. Am. Chem. Soc.* **100**, 1883.
Meinwald, J., Chadha, M. S., Hurst, J. J., and Eisner, T. (1962). *Tetrahedron Lett.* p. 29.
Meinwald, J., Happ, G. M., Labows, J., and Eisner, T. (1966a). *Science* **151**, 79.
Meinwald, J., Koch, K. F., Rogers, J. E., Jr., and Eisner, T. (1966b). *J. Am. Chem. Soc.* **88**, 1590.
Meinwald, J., Erickson, K., Hartshorn, M., Meinwald, Y. C., and Eisner, T. (1968a). *Tetrahedron Lett.* p. 2959.
Meinwald, J., Meinwald, Y. C., Chalmers, A. N., and Eisner, T. (1968b). *Science* **160**, 890.

Meinwald, J., Kluge, A. F., Carrel, J. E., and Eisner, T. (1971). *Proc. Natl. Acad. Sci. U.S.A.* **68**, 1467.
Meinwald, J., Opheim, K., and Eisner, T. (1972). *Proc. Natl. Acad. Sci. U.S.A.* **69**, 1208.
Meinwald, J., Smolanoff, J., McPhail, A. T., Miller, R. W., Eisner, T., and Hicks, K. (1975). *Tetrahedron Lett.* p. 2367.
Meinwald, J., Jones, T. H., Eisner, T., and Hicks, K. (1977). *Proc. Natl. Acad. Sci. U.S.A.* **74**, 2189.
Meinwald, J., Prestwich, G. D., Nakanishi, K., and Kubo, I. (1978). *Science* **199**, 1167.
Meinwald, J., Wiemer, D. F., and Eisner, T. (1979). *J. Am. Chem. Soc.* **101**, 3055.
Meinwald, Y. C., and Eisner, T. (1964). *Ann. Entomol. Soc. Am.* **57**, 513.
Meinwald, Y. C., Meinwald, J., and Eisner, T. (1966). *Science* **154**, 390.
Melander, A. L., and Brues, C. T. (1906). *Wis. Nat. Hist. Soc. Bull.* **4**, 22.
Meyer, D., Schlatter, C., Schlatter-Lang, I., Schmid, H., and Bovey, P. (1968). *Experientia* **24**, 995.
Miller, J., and Mumma, R. (1973). *J. Insect Physiol.* **19**, 917.
Miller, J. R., Hendry, L. B., and Mumma, R. O. (1975). *J. Chem. Ecol.* **1**, 59.
Miranda, F., Rochat, H., and Lissitzky, S. (1964). *Toxicon* **2**, 51.
Miranda, F., Rochat, H., Rochat, C., and Lissitzky, S. (1966). *Toxicon* **4**, 145.
Miranda, F., Kupeyan, C., Rochat, H., Rochat, C., and Lissitzky, S. (1970). *Eur. J. Biochem.* **16**, 514.
Mollay, C., and Kreil, G. (1974). *FEBS Lett.* **46**, 141.
Monro, A., Chadha, M., Meinwald, J., and Eisner, T. (1962). *Ann. Entomol. Soc. Am.* **55**, 261.
Monteiro, H. (1961). *An. Assoc. Bras. Quim.* **20**, 29.
Monterosso, B. (1928). *Arch. Zool. Ital.* **12**, 63.
Montgomery, M. E., and Nault, L. R. (1977a). *Entomol. Exp. Appl.* **22**, 236.
Montgomery, M. E., and Nault, L. R. (1977b). *Ann. Entomol. Soc. Am.* **70**, 669.
Moore, B. P. (1964). *J. Insect Physiol.* **10**, 371.
Moore, B. P. (1967). *J. Aust. Entomol. Soc.* **6**, 36.
Moore, B. P. (1968). *J. Insect Physiol.* **14**, 33.
Moore, B. P., and Brown, W. V. (1971a). *J. Aust. Entomol. Soc.* **10**, 142.
Moore, B. P., and Brown, W. V. (1971b). *J. Aust. Entomol. Soc.* **10**, 230.
Moore, B. P., and Brown, W. V. (1972). *Aust. J. Chem.* **25**, 591.
Moore, B. P., and Brown, W. V. (1976). *Aust. J. Chem.* **29**, 1365.
Moore, B. P., and Brown, W. V. (1978). *Insect Biochem.* **8**, 393.
Moore, B. P., and Brown, W. V. (1979). *J. Aust. Entomol. Soc.* **18**, 123.
Moore, B. P., and Wallbank, B. E. (1968). *Proc. R. Entomol. Soc. London, Ser. B* **37**, 62.
Moore, R. M., Williams, J. D., and Chia, J. (1967). *Aust. J. Biol. Sci.* **20**, 1131.
Morgan, E. D., and Wadhams, L. J. (1972). *J. Insect Physiol.* **18**, 1125.
Mori, H. (1919). *Jpn. Med. Lit.* **4**, 10.
Mori, K. (1973). *Tetrahedron Lett.* p. 723.
Morley, J., and Schachter, M. (1963). *J. Physiol. (London)* **168**, 706.
Morrow, P. A., Bellas, T. E., and Eisner, T. (1976). *Oecologia* **24**, 193.
Morse, R. A., Shearer, D. A., Boch, R., and Benton, A. W. (1967). *J. Apic. Res.* **6**, 113.
Moser, J. C., Brownlee, R. G., and Silverstein, R. M. (1968). *J. Insect Physiol.* **14**, 529.
Mosher, H. S., Fuhrman, F. A., Buchwald, H. G., and Fischer, H. G. (1964). *Science* **144**, 1100.
Moussatché, H., Cuadra, J. L., Ramos, P. R., Perissé, A. C. M., Salles, C. A., and Loureiros, E. G. (1969). *Rev. Brasil. Biol.* **29**, 25.

Mueller, R. H., and Thompson, M. E. (1980). *Tetrahedron Lett.* **21**, 1097.
Mukerji, S. K., and Sharma, H. L. (1966). *Tetrahedron Lett.* p. 2479.
Munjal, D., and Elliot, W. B. (1971). *Toxicon* **9**, 403.
Nair, R. B., and Kurup, P. A. (1973a). *Ind. J. Biochem. Biophys.* **10**, 133.
Nair, R. B., and Kurup, P. A. (1973b). *Ind. J. Biochem. Biophys.* **10**, 230.
Nair, R. B., and Kurup, P. A. (1975). *Biochim. Biophys. Acta* **381**, 165.
Nault, L. R., and Bowers, W. S. (1974). *Entomol. Exp. Appl.* **17**, 455.
Nault, L. R., Edwards, L. J., and Styer, W. E. (1973). *Environ. Entomol.* **2**, 101.
Nault, L. R., Montgomery, M. E., and Bowers, W. S. (1976). *Science* **192**, 1349.
Nayler, L. S. (1964). *J. Entomol. Soc. South. Afr.* **27**, 62.
Nelson, D. A., and O'Connor, R. (1968). *Can. J. Biochem.* **46**, 1221.
Newhart, A. T., and Mumma, R. O. (1978). *J. Chem. Ecol.* **4**, 503.
Newhart, A. T., and Mumma, R. O. (1979a). *Ann. Entomol. Soc. Am.* **72**, 427.
Newhart, A. T., and Mumma, R. O. (1979b). *J. Chem. Ecol.* **5**, 643.
Nishino, C., Bowers, W. S., Montgomery, M. E., Nault, L. R., and Nielson, M. W. (1977). *J. Chem. Ecol.* **3**, 349.
Nishio, S., and Blum, M. S. (1980). Unpublished data.
Noguchi, T., and Hashimoto, Y. (1973). *Toxicon* **11**, 305.
Noirot, C. (1969). *In* "Biology of Termites" (K. Krishna and F. M. Weesner, eds.), Vol. 1, pp. 89–123. Academic Press, New York.
Norment, B. R., Jong, Y. S., and Heitz, J. R. (1979). *Toxicon* **17**, 539.
Nutting, W. L., Blum, M. S., and Fales, H. M. (1974). *Psyche* **81**, 167.
O'Connor, R., Henderson, G., Nelson, D., Parker, R., and Peck, M. L. (1967). *In* "Animal Toxins" (F. E. Russell and P. R. Saunders, eds.), pp. 17–22. Pergamon, Oxford.
Okita, G. T. (1967). *Fed. Proc. Fed. Am. Soc. Exp. Biol.* **26**, 1125.
Ornberg, R. L., Smyth, T., Jr., and Benton, A. W. (1976). *Toxicon* **14**, 329.
O'Rourke, F. (1950). *Ann. Entomol. Soc. Am.* **43**, 437.
Osman, M. F. H., and Brander, J. (1961). *Z. Naturforsch., B: Anorg. Chem., Org. Chem., Biochem., Biophys., Biol.* **16B**, 749.
Owen, M. D. (1971). *Experientia* **27**, 544.
Owen, M. D. (1978). *J. Insect Physiol.* **24**, 433.
Owen, M. D. (1979a). *Toxicon* **17**, 94.
Owen, M. D. (1979b). *Toxicon* **17**, 519.
Owen, M. D., and Braidwood, J. L. (1974). *Can. J. Zool.* **52**, 387.
Owen, M. D., and Bridges, A. R. (1976). *Toxicon* **14**, 1.
Owen, M. D., Braidwood, J. L., and Bridges, A. R. (1977). *J. Insect Physiol.* **23**, 1031.
Pallares, E. S. (1946). *Arch. Biochem.* **9**, 105.
Pansa, M. C., Natalizi, G. M., and Bettini, S. (1973). *Toxicon* **11**, 283.
Park, R. J., and Sutherland, M. D. (1962). *Aust. J. Chem.* **15**, 172.
Parker, R. (1972). *Toxicon* **10**, 79.
Parsons, J. A. (1965). *J. Physiol.* **178**, 290.
Pasteels, J. M. (1977). *Proc. Int. Congr. Entomol. 15th, 1976* pp. 281–293.
Pasteels, J. M., and Daloze, D. (1977). *Science* **197**, 70.
Pasteels, J. M., Deroe, C., Tursch, B., Braekman, J. C., Daloze, D., and Hootele, C. (1973). *J. Insect Physiol.* **19**, 1771.
Pasteels, J. M., Daloze, D., van Dorsser, W., and Roba, J. (1979). *Comp. Biochem. Physiol. C: Biosci.* **63C**, 117.
Pattenden, G., and Staddon, B. W. (1968). *Experientia* **24**, 1092.
Pattenden, G., and Staddon, B. W. (1970). *Ann. Entomol. Soc. Am.* **63**, 900.

Pattenden, G., and Staddon, B. W. (1972). *Ann. Entomol. Soc. Am.* **65**, 1240.
Pavan, M. (1952). *Arch. Int. Pharmacodyn. Ther.* **84**, 223.
Pavan, M. (1953). *Arch. Zool. Ital.* **38**, 157.
Pavan, M. (1959). "*Proc. Int. Congr. Biochem., 4th, 1958* Vol. 4, pp. 15–36.
Pavan, M., and Dazzini, M. V. (1971). *In* "Chemical Zoology" (M. Florkin and B. T. Scheer, eds.), Vol. 6, pp. 365–409. Academic Press, Inc. New York.
Peck, M. L., and O'Connor, R. (1974). *J. Agric. Food Chem.* **22**, 51.
Pedder, D. J., Fales, H. M., Jaouni, T., Blum, M. S., MacConnell, J., and Crewe, R. M. (1976). *Tetrahedron* **32**, 2275.
Pereira Lima, F. A., and Schenberg, S. (1964). *Cienc. Cult. (Sao Paulo)* **16**, 187.
Perissé, A. C. M., and Salles, C. A. (1970). *Atas Soc. Biol. Rio de Janeiro* **13**, 95.
Perret, B. A. (1977). *Toxicon* **15**, 505.
Pese, H., and Delgado, A. (1971). *In* "Venomous Animals and their Venoms" (W. Bücherl and E. E. Buckley, eds.), Vol. 3, pp. 119–156. Academic Press, New York.
Peter, M. G., Woggon, W. D., Dürsteler-Meier, A., and Schmid, H. (1972). *Adv. Study Inst. Biochem. Insects Sci. Affairs Div. NATO, 1972* p. 14.
Petersen, B. (1964). *J. Lepidopt. Soc.* **18**, 165.
Picarelli, Z. P., and Valle, J. R. (1971). *In* "Venomous Animals and their Venoms" (W. Bücherl and E. E. Buckley, eds.), Vol. 3, pp. 103–118. Academic Press, New York.
Piek, T., Spanjer, W., Nijo, K. D., Veenendaal, R. L., and Mantel, P. (1974). *J. Insect Physiol.* **20**, 2307.
Pinder, A. R., and Staddon, B. W. (1965a). *Nature (London)* **205**, 106.
Pinder, A. R., and Staddon, B. W. (1965b). *J. Chem. Soc.* **530**, 2955.
Pisano, J. J. (1970). *Handb. Exp. Pharmakol.* [N. S.] **25**, 589–595.
Plattner, H., Salpeter, M., Carrel, J. E., and Eisner, T. (1972). *Z. Zellforsch. Mikrosk. Anat.* **125**, 45.
Possani, L. D., Alagón, A. C., Fletcher, P. L., Jr., and Erickson, B. W. (1977). *Arch. Biochem. Biophys.* **180**, 394.
Poulton, E. B. (1888). *Br. Assoc. Adv. Sci., Rep.* **5**, 765.
Poulton, E. B. (1916). *Proc. Entomol. Soc. London* p. 65.
Prado, J. L., Tamura, Z., Furano, E., Pisano, J. J., and Udenfriend, S. (1966). *In* "Hypotensive Peptides" (E. G. Erdös, N. Baek, and F. Sicuteri, eds.), pp. 93–104. Springer-Verlag, Berlin and New York.
Prestwich, G. D. (1975). *In* "Pheromones and Defensive Secretions in Social Insects" (C. Noirot, P. E. Howse, and G. Le Masne, eds.), pp. 149–152. Univ. Dijon Press.
Prestwich, G. D. (1976). *Ann. Entomol. Soc. Am.* **69**, 812.
Prestwich, G. D. (1977). *Insect Biochem.* **7**, 91.
Prestwich, G. D. (1978). *Experientia* **34**, 682.
Prestwich, G. D. (1979). *Biochem. Syst. Ecol.* **7**, 211.
Prestwich, G. D., Kaib, M., Wood, W. F., and Meinwald, J. (1975). *Tetrahedron Lett.*, 4701.
Prestwich, G. D., Tanis, S. P., Springer, J. P., and Clardy, J. (1976a). *J. Am. Chem. Soc.* **98**, 6061.
Prestwich, G. D., Tanis, S. P., Pilkiewicz, F. G., Miura, I., and Nakanishi, K. (1976b). *J. Am. Chem. Soc.* **98**, 6062.
Prestwich, G. D., Bierl, B. A., Devilbiss, E. D., and Chaudhury, M. F. B. (1977a). *J. Chem. Ecol.* **3**, 579.

Prestwich, G. D., Solheim, B. A., Clardy, J., Pilkiewicz, F. G., Miura, I., Tanis, S. P., and Nakanishi, K. (1977b). *J. Am. Chem. Soc.* **99**, 8082.

Prestwich, G. D., Wiemer, D. F., Meinwald, J., and Clardy, J. (1978). *J. Am. Chem. Soc.* **100**, 2560.

Prestwich, G. D., Lauher, J. W., and Collins, M. S. (1979). *Tetrahedron* p. 3827.

Quennedey, A., Brulè, G., Rigaud, J., Dubois, P., and Brossut, R. (1973). *Insect Biochem.* **3**, 67.

Quilico, A., Piozzi, F., and Pavan, M. (1957a). *Rend., Ist. Lomb. Accad. Sci. Lett. B* **91**, 271.

Quilico, A., Piozzi, F., and Pavan, M. (1957b). *Tetrahedron Lett.* **1**, 177.

Quilico, A., Grünanger, P., and Pavan, M. (1962). *Proc. Int. Congr. Entomol., 11th, 1960* Vol. 3, pp. 66–68.

Rafaeli-Berstein, A., and Mordue, W. (1978). *Physiol. Entomol.* **3**, 59.

Ralph, C. P. (1976). *Oecologia* **26**, 157.

Read, G. W., Lind, N. K., and Oda, C. S. (1978). *Toxicon* **16**, 361.

Redfield, A. C. (1936). *Am. Nat.* **70**, 110.

Rees, C. J. C. (1969). *Entomol. Exp. Appl.* **12**, 565.

Regnier, F. E., and Wilson, E. O. (1968). *J. Insect Physiol.* **14**, 955.

Regnier, F. E., and Wilson, E. O. (1969). *J. Insect Physiol.* **15**, 893.

Regnier, F. E., and Wilson, E. O. (1971). *Science* **172**, 267.

Regnier, F. E., Nieh, M., and Hölldobler, B. (1973). *J. Insect Physiol.* **19**, 981.

Rehr, S. S., Bell, E. A., Janzen, D. H., and Feeney, P. P. (1973). *Biochem. Syst. Ecol.* **1**, 63.

Reichstein, T., von Euw, J., Parsons, J. A., and Rothschild, M. (1968). *Science* **161**, 861.

Riddiford, L. M. (1976). *In* "The Juvenile Hormones" (L. I. Gilbert, ed.), p. 199. Plenum, New York.

Riley, R. G., Silverstein, R. M., and Moser, J. C. (1974). *J. Insect Physiol.* **20**, 1629.

Ring, B., Abramov, I., Ishay, J., and Slor, H. (1978a). *Toxicon* **16**, 77.

Ring, B., Slor, H., Perna, B., and Ishay, J. S. (1978b). *Toxicon* **16**, 473.

Ritter, F. J. and Wientjens, W. H. J. M. (1967). *TNO Nieuws* **22**, 381.

Ritter, F. J., Rotgans, I. E. M., Talman, E., Verwiel, P. E. J., and Stein, F. (1973). *Experientia* **29**, 530.

Ritter, F. J., Brüggemann-Rotgan, I. E. M., Verkuil, E., and Persoons, C. J. (1975). *In* "Pheromones and Defensive Secretions in Social Insects" (C. Noirot, P. E. Howse, and G. Le Masne, eds.), pp. 99–103. Univ. Dijon Press.

Roach, B., Dodge, K. R., Aneshansley, D. J., Wiemer, D., Meinwald, J., and Eisner, T. (1979). *Coleop. Bull.* **33**, 17.

Roach, B., Eisner, T., and Meinwald, J. (1980). *J. Chem. Ecol.* **6**, 511.

Robinson, W. G., Bachhaus, B. K., and Coon, M. J. (1956). *J. Biol. Chem.* **218**, 391.

Rochat, C., Rochat, H., Miranda, F., and Lissitzky, S. (1967). *Biochemistry* **6**, 578.

Rochat, H., Rochat, C., Kupeyan, C., Miranda, F., Lissitzky, S., and Edman, P. (1970). *FEBS Lett.* **10**, 349.

Roeske, C. N., Seiber, J. N., Brower, L. P., and Moffitt, C. M. (1976). *In* "Biochemical Interactions between Plants and Insects" (J. M. Wallace and R. L. Mansell, eds.), pp. 93–167. Plenum, New York.

Rosenberg, P., Ishay, J., and Gitter, S. (1977). *Toxicon* **15**, 141.

Rosenbrook, W., Jr., and O'Connor, R. (1964). *Can. J. Biochem.* **42**, 1005.

Rosenthal, G. A., and Dahlman, D. L. (1975). *Comp. Biochem. Physiol. A* **52A**, 105.

Rosenthal, G. A., Dahlman, D. L., and Janzen, D. H. (1976). *Science* **192,** 256.
Rosenthal, G. A., Dahlman, D. L., and Robinson, G. W. (1977a). *J. Biol. Chem.* **252,** 3679.
Rosenthal, G. A., Janzen, D. H., and Dahlman, D. L. (1977b). *Science* **196,** 658.
Rosenthal, G. A., Dahlman, D. L., and Janzen, D. H. (1978). *Science* **202,** 528.
Rosin, R. (1969). *Toxicon* **6,** 225.
Roth, L. M. (1961). *Ann. Entomol. Soc. Am.* **54,** 900.
Roth, L. M., and Stahl, W. H. (1956). *Science* **13,** 798.
Roth, L. M., and Stay, B. (1958). *J. Insect Physiol.* **1,** 305.
Roth, L. M., and Willis, E. R. (1960). *Smithson. Misc. Collect.* **141,** 1.
Roth, L. M., Niegisch, W. D., and Stahl, W. H. (1956). *Science* **123,** 670.
Rothschild, M. (1970). *Animals* **1970,** 402.
Rothschild, M. (1972). *Symp. R. Entomol. Soc. London* **6,** 59–83.
Rothschild, M., and Aplin, R. T. (1971). *In* "Chemical Releasers in Insects" (A. S. Tahori, ed.), Vol. 3, pp. 177–182. Gordon & Breach, New York.
Rothschild, M., and Haskell, P. T. (1966). *Proc. R. Entomol. Soc. London, Ser. A* **41,** 167.
Rothschild, M., and Reichstein, T. (1976). *Nova Acta Leopold. Suppl.* **7,** 507.
Rothschild, M., von Euw, J., and Reichstein, T. (1970a). *J. Insect Physiol.* **16,** 1141.
Rothschild, M., Reichstein, T., von Euw, J., Aplin, R., and Harman, R. R. M. (1970b). *Toxicon* **8,** 293.
Rothschild, M., von Euw, J., and Reichstein, T. (1973a). *Proc. R. Soc. London, Ser. B* **183,** 227.
Rothschild, M., von Euw, J., and Reichstein, T. (1973b). *J. Entomol., Ser. A: Gen. Entomol.* **48,** 89.
Rothschild, M., von Euw, J., Reichstein, T., Smith, D. A. S., and Pierre, J. (1975). *Proc. R. Soc. London, Ser. B* **190,** 1.
Rothschild, M., Rowan, M. G., and Fairburn, J. W. (1977). *Nature (London)* **266,** 650.
Russell, F. E. (1965). *In* "Marine Toxins and Venomous and Poisonous Marine Animals" (F. E. Russell, ed.), Vol. 3, pp. 255–384. Academic Press, New York.
Russell, F. E. (1966). *Toxicon* **4,** 153.
Russell, F. E. (1967). *Clin. Pharmacol. Ther.* **8,** 849.
Russell, F. E., and Saunders, P. R., eds. (1967). "Animal Toxins." Pergamon, Oxford.
Said, E. E. (1960). *Bull. Soc. Entomol. Egypte* **44,** 167.
Sakan, T., Isoe, S., and Hyeon, S. B. (1970). *In* "Control of Insect Behavior by Natural Products" (D. L. Wood, R. M. Silverstein, and M. Nakajima, eds.), pp. 237–247. Academic Press, New York.
Sampieri, F., and Habersetzer-Rochat, C. (1975). *Toxicon* **13,** 120.
Schachtele, C. F., and Rogers, P. (1965). *J. Mol. Biol.* **14,** 474.
Schachter, M., and Thain, E. M. (1954). *Br. J. Pharmacol. Chemother.* **9,** 352.
Schall, C. (1892). *Ber. Dtsch. Chem. Ges.* **25,** 1489.
Schanbacher, F. L., Lee, C. K., Hall, J. E., Wilson, I. B., Howell, D. E., and Odell, G. V. (1973). *Toxicon* **11,** 21.
Schenberg, S., and Periera Lima, F. A. (1971). *In* "Venomous Animals and their Venoms" (W. Bücherl and E. E. Buckley, eds.), Vol. 3, pp. 279–297. Academic Press, London.
Schildknecht, H. (1957). *Angew. Chem.* **69,** 62.
Schildknecht, H. (1968). *Nachr. Chem. Tech.* **18,** 311.
Schildknecht, H. (1970). *Angew. Chem. Int. Ed. Engl.* **9,** 1.
Schildknecht, H. (1971). *Endeavour* **3,** 136.

Schildknecht, H. (1976). *Angew. Chem. Int. Ed. Engl.* **15,** 214.
Schildknecht, H., and Birringer, H. (1969a). *Chem. Ber.* **102,** 1859.
Schildknecht, H., and Birringer, H. (1969b). *Z. Naturforsch., B: Anorg. Chem., Org. Chem., Biochem., Biophys., Biol.* **24B,** 1529.
Schildknecht, H., and Holoubek, K. (1961). *Angew. Chem.* **73,** 1.
Schildknecht, H., and Hotz, D. (1967). *Angew. Chem. Int. Ed. Engl.* **6,** 881.
Schildknecht, H., and Koob, K. (1970). *Angew. Chem.* **82,** 181.
Schildknecht, H., and Koob, K. (1971). *Angew. Chem.* **83,** 110.
Schildknecht, H., and Körnig, W. (1968). *Angew. Chem. Int. Ed. Engl.* **7,** 62.
Schildknecht, H., and Krämer, H. (1962). *Z. Naturforsch., B: Anorg. Chem., Org. Chem., Biochem., Biophys., Biol.* **17B,** 701.
Schildknecht, H., and Schmidt, H. (1963). *Z. Naturforsch., B: Anorg. Chem., Org. Chem., Biochem., Biophys., Biol.* **18B,** 585.
Schildknecht, H., and Tacheci, H. (1970). *Chem.-Ztg.* **94,** 101.
Schildknecht, H., and Weis, K. H. (1960a). *Z. Naturforsch., B* **15b,** 200.
Schildknecht, H., and Weis, K. H. (1960b). *Z. Naturforsch., B* **15b,** 755.
Schildknecht, H., and Weis, K. H. (1961). *Z. Naturforsch., B* **16b,** 810.
Schildknecht, H., and Weis, K. H. (1962a). *Z. Naturforsch., B* **17b,** 448.
Schildknecht, H., and Weis, K. H. (1962b). *Z. Naturforsch., B* **17b,** 452.
Schildknecht, H., and Wenneis, W. F. (1967). *Tetrahedron Lett.* p. 1815.
Schildknecht, H., Weis, K. H., and Vetter, H. (1962). *Z. Naturforsch., B: Anorg. Chem., Org. Chem., Biochem., Biophys., Biol.* **17B,** 350.
Schildknecht, H., Holoubek, K., Weis, K. H., and Krämer, H. (1964). *Angew. Chem. Int. Ed. Engl.* **3,** 73.
Schildknecht, H., Siewerdt, R., and Maschwitz, U. (1966a). *Angew. Chem. Int. Ed. Engl.* **5,** 421.
Schildknecht, H., Siewerdt, R., and Maschwitz, U. (1966b). *Angew. Chem.* **78,** 392.
Schildknecht, H., Wenneis, W. F., Weis, K. H., and Maschwitz, U. (1966c). *Z. Naturforsch., B: Anorg. Chem., Org. Chem., Biochem., Biophys., Biol.* **21B,** 121.
Schildknecht, H., Birringer, H., and Maschwitz, U. (1967a). *Angew. Chem. Int. Ed. Engl.* **6,** 558.
Schildknecht, H., Hotz, D., and Maschwitz, U. (1967b). *Z. Naturforsch., B: Anorg. Chem., Org. Chem., Biochem., Biophys., Biol.* **22B,** 938.
Schildknecht, H., Maschwitz, U., and Wenneis, W. F. (1967c). *Naturwissenschaften* **54,** 196.
Schildknecht, H., Siewerdt, R., and Maschwitz, U. (1967d). *Justus Liebigs Ann. Chem.* **703,** 182.
Schildknecht, H., Maschwitz, U., and Winkler, H. (1968a). *Naturwissenschaften* **55,** 112.
Schildknecht, H., Maschwitz, U., and Krauss, D. (1968b). *Naturwissenschaften* **55,** 230.
Schildknecht, H., Winkler, H., Krauss, D., and Maschwitz, U. (1968c). *Z. Naturforsch., B: Anorg. Chem., Org. Chem., Biochem., Biophys., Biol.* **23B,** 46.
Schildknecht, H., Winkler, H., and Maschwitz, U. (1968d). *Z. Naturforsch., B: Anorg. Chem., Org. Chem., Biochem., Biophys., Biol.* **23B,** 637.
Schildknecht, H., Birringer, H., and Krauss, D. (1969a). *Z. Naturforsch., B: Anorg. Chem., Org. Chem., Biochem., Biophys., Biol.* **24B,** 38.
Schildknecht, H., Tacheci, H., and Maschwitz, U. (1969b). *Naturwissenschaften* **56,** 37.

Schildknecht, H., Körnig, W., Siewerdt, R., and Krauss, D. (1970). *Justus Liebigs Ann. Chem.* **734**, 116.
Schildknecht, H., Kunzelmann, P., Kraub, D., and Kuhn, C. (1972a). *Naturwissenschaften* **59**, 98.
Schildknecht, H., Neumaier, H., and Tauscher, B. (1972b). *Justus Liebigs Ann. Chem.* **756**, 155.
Schildknecht, H., Tauscher, B., and Kraub, D. (1972c). *Chem. Ztg.* **96**, 33.
Schildknecht, H., Holtkotte, H., Kraub, D., and Tacheci, H. (1975a). *Justus Liebigs Ann. Chem.* **1975**, 1850.
Schildknecht, H., Krauss, D., Connert, J., Essenbreis, H., and Orfanides, N. (1975b). *Angew. Chem.* **87**, 421.
Schildknecht, H., Berger, D., Krauss, D., Connert, J., Gehlhaus, J., and Essenbreis, H. (1976). *J. Chem. Ecol.* **2**, 1.
Schlatter, C., Waldner, E. E., and Schmid, H. (1968). *Experientia* **24**, 994.
Schlunnegger, P., and Leuthold, R. H. (1972). *Insect Biochem.* **2**, 150.
Schmidt, J. O., and Blum, M. S. (1977). *Entomol. Exp. Appl.* **21**, 99.
Schmidt, J. O., and Blum, M. S. (1978a). *Science* **200**, 1064.
Schmidt, J. O., and Blum, M. S. (1978b). *Comp. Biochem. Physiol. C: Biosci.* **61C**, 239.
Schmidt, J. O., and Blum, M. S. (1978c). *Toxicon* **16**, 645.
Schreuder, G. D., and Brand, J. M. (1972). *J. Ga. Entomol. Soc.* **7**, 188.
Schröder, E., Lübke, K., Lehmann, M., and Beetz, I. (1971). *Experientia* **27**, 764.
Scott, P. D., Hepburn, H. R., and Crewe, R. M. (1975). *Insect Biochem.* **5**, 805.
Scudder, G. G. E., and Duffey, S. S. (1972). *Can. J. Zool.* **50**, 35.
Seiber, J. N., Tuskes, P. M., Brower, L. P., and Nelson, C. J. (1980). *J. Chem. Ecol.* **6**, 321.
Seitz, A. (1925). "Macrolepidoptera of the American Region," Vol. 6, p. 424. A. Kernen, Stuttgart.
Selander, R. B. (1960). *Univ. Ill. Biol. Monogr.* **28**, 1.
Selander, R. B., and Mathieu, J. M. (1969). *Univ. Ill. Biol. Monogr.* **41**, 1.
Self, L. S., Guthrie, F. E., and Hodgson, E. (1964a). *Nature (London)* **204**, 300.
Self, L. S., Guthrie, F. E., and Hodgson, E. (1964b). *J. Insect Physiol.* **10**, 907.
Seligman, I. M. (1972). *In* Seligman and Doy (1973).
Seligman, I. M., and Doy, F. A. (1972). *Comp. Biochem. Physiol. B* **42B**, 341.
Seligman, I. M., and Doy, F. A. (1973). *Insect Biochem.* **3**, 205.
Shearer, D. A., and Boch, R. (1965). *Nature (London)* **206**, 530.
Shipolini, R. A., Bradbury, A. F., Callewaert, G. L., and Vernon, C. A. (1967). *J. Chem. Soc., Chem. Commun.* p. 679.
Shipolini, R. A., Callewaert, G. L., Cotrell, R. C., and Vernon, C. A. (1971a). *FEBS Lett.* **17**, 39.
Shipolini, R. A., Callewaert, G. L., Cotrell, R. C., Doonan, S., and Banks, B. E. C. (1971b). *Eur. J. Biochem.* **20**, 459.
Shkenderov, S. (1973). *FEBS Lett.* **33**, 343.
Shkenderov, S. (1975). *Toxicon* **13**, 124.
Shkenderov, S. (1976). *In* "Animal, Plant, and Microbial Toxins" (A. Ohsaka, K. Hayashi, and Y. Sawai, eds.), Vol. 1, p. 263. Plenum, New York.
Sicuteri, F., Franchi, P. L., Del Bianco, P. L., and Fanciullaci, M. (1966). *In* "Hypotensive Peptides" (E. G. Erdös, N. Back, and F. Sicuteri, eds.), pp. 522–533. Springer Publ., New York.
Sierra, J. R., Woggon, W.-D., and Schmid, H. (1976). *Experientia* **32**, 142.

Sipahimalani, A. T., Mamdapur, V. R., Joshi, N. K., and Chadha, M. S. (1970). *Naturwissenschaften* **57**, 40.
Slater, J. W. (1877). *Trans. R. Entomol. Soc. London* p. 205.
Slor, H., Ring, B., and Ishay, J. (1976). *Toxicon* **14**, 427.
Smith, D. A. S. (1977). *Experientia* **34**, 845.
Smith, R. M. (1974). *N. Z. J. Zool.* **1**, 375.
Smith, R. M. (1978). *N. Z. J. Zool.* **5**, 821.
Smith, R. M., Brophy, J. J., Cavill, G. W. K., and Davies, N. W. (1979). *J. Chem. Ecol.* **5**, 727.
Smolanoff, J., Demange, J. M., Meinwald, J., and Eisner, T. (1975a). *Psyche* **82**, 78.
Smolanoff, J., Kluge, A. F., Meinwald, J., McPhail, A., Miller, R. W., Hicks, K., and Eisner, T. (1975b). *Science* **188**, 734.
Sobotka, A. K., Franklin, R. M., Adkinson, N. F., Valentine, M., Baer, H., and Lichtenstein, L. M. (1976). *Allergy Clin. Immunol.* **57**, 29.
Sondheimer, E., and Simeone, J. B., eds. (1970). "Chemical Ecology." Academic Press, New York.
Staddon, B. W. (1973). *Entomologist* **126**, 253.
Staddon, B. W., and Weatherston, J. (1967). *Tetrahedron Lett.* p. 4567.
Stahnke, H. L., and Johnson, B. D. (1967). *In* "Animal Toxins" (F. E. Russell and P. R. Saunders, eds.), pp. 35–39. Pergamon, Oxford.
Ställberg-Stenhagen, S. (1970). *Acta Chem. Scand.* **24**, 358.
Stewart, J. M. (1968). *Fed. Proc., Fed. Am. Soc. Exp. Biol.* **27**, 534.
Strong, F. E. (1967). *Ann. Entomol. Soc. Am.* **60**, 668.
Stumper, R. (1922). *C. R. Hebd. Seances Acad. Sci.* **174**, 413.
Stumper, R. (1951). *C. R. Hebd. Seances Acad. Sci.* **233**, 1144.
Stumper, R. (1952). *C. R. Hebd. Seances Acad. Sci.* **234**, 149.
Suchanek, G., and Kreil, G. (1977). *Proc. Natl. Acad. Sci. U.S.A.* **74**, 975.
Suchanek, G., Kindås-Mügge, I., Kreil, G., and Schreier, M. H. (1975). *Eur. J. Biochem.* **60**, 309.
Sugawara, F., Matsuda, K., Kobayashi, A., and Yamashita, K. (1978). *Agric. Biol. Chem.* **42**, 678.
Sugawara, F., Matsuda, K., Kobayashi, A., and Yamashita, K. (1979a). *J. Chem. Ecol.* **5**, 635.
Sugawara, F., Matsuda, K., Kobayashi, A., and Yamashita, K. (1979b). *J. Chem. Ecol.* **5**, 929.
Suzuki, T., Suzuki, T., Huynh, V. M., and Muto, T. (1975b). *Agric. Biol. Chem.* **39**, 2207.
Svensson, B. G., and Bergström, G. (1977). *Insectes Soc.* **24**, 212.
Swynnerton, C. F. M. (1919). *J. Linn. Soc. London, Zool.* **33**, 203.
Tabor, H., Mehler, A. H., Hayaishi, O., and White, J. (1951). *J. Biol. Chem.* **296**, 121.
Takahashi, S., and Kitamura, C. (1972). *Appl. Entomol. Zool.* **7**, 199.
Talman, E., Ritter, F. J., and Verwiel, P. E. J. (1974). *Proc. Int. Symp. Mass Spectrom. Biochem. Med., 1973* pp. 197–217.
Tapper, B. A., Zilg, H., and Conn, E. E. (1972). *Phytochemistry* **11**, 1047.
Tarabini-Castellani, G. (1938). *Arch. Ital. Med. Sper.* **2**, 969.
Teas, H. J. (1967). *Biochem. Biophys. Res. Commun.* **26**, 686.
Teas, H. J., Dyson, J. G., and Whisenant, B. R. (1966). *J. Ga. Entomol. Soc.* **1**, 21.
Tengö, J., and Bergström, G. (1976). *Comp. Biochem. Physiol. B* **55B**, 179.
Tetsch, C., and Wolff, K. (1936). *Biochem. Z.* **288**, 126.

Thomson, R. H. (1957). "Naturally Occurring Quinones." Butterworth, London.
Toledo, D., and Neves, A. G. A. (1976). *Comp. Biochem. Physiol. B* **55B**, 249.
Towers, G. H. N., Duffey, S. S., and Siegel, S. M. (1972). *Can. J. Zool.* **50**, 1047.
Trave, R., and Pavan, M. (1956). *Chim. Ind. (Milan)* **38**, 1015.
Trave, R., Garanti, L., and Pavan, M. (1959). *Chim. Ind. (Milan)* **41**, 19.
Trave, R., Garanti, L., Marchesini, A., and Pavan, M. (1966). *Chim. Ind. (Milan)* **48**, 1167.
Tricot, M.-C., Pasteels, J. M., and Tursch, B. (1972). *J. Insect Physiol.* **18**, 499.
Tschinkel, W. R. (1969). *J. Insect Physiol.* **15**, 191.
Tschinkel, W. R. (1972). *J. Insect Physiol.* **18**, 711.
Tschinkel, W. R. (1975a). *J. Insect Physiol.* **21**, 659.
Tschinkel, W. R. (1975b). *J. Insect Physiol.* **21**, 753.
Tseng, Yueh-chu L., Davidson, J. A., and Menzer, R. E. (1971). *Ann. Entomol. Soc. Am.* **64**, 425.
Tsuyuki, T., Ogata, Y., Yamamoto, I., and Shimi, K. (1965). *Agric. Biol. Chem.* **29**, 419.
Tumlinson, J. H., Silverstein, R. M., Moser, J. C., Brownlee, R. G., and Ruth, J. M. (1971). *Nature (London)* **234**, 348.
Tursch, B., Daloze, D., Dupont, M., Hootele, C., Kaisin, M., Pasteels, J. M., and Zimmermann, D. (1971a). *Chimia* **25**, 307.
Tursch, B., Daloze, D., Dupont, M., Pasteels, J. M., and Tricot, M.-C. (1971b). *Experientia* **27**, 1380.
Tursch, B., Daloze, D., and Hootele, C. (1972a). *Chimia* **26**, 74.
Tursch, B., Daloze, D., Pasteels, J. M., Cravador, A., Braekman, J. C., Hootele, C., and Zimmermann, D. (1972b). *Bull. Soc. Chim. Belg.* **81**, 649.
Tursch, B., Braekman, J. C., Daloze, D., Hootele, C., Losman, D., Karlsson, R., and Pasteels, J. M. (1973). *Tetrahedron Lett.* p. 201.
Tursch, B., Daloze, D., Braekman, J. C., Hootele, C., and Pasteels, J. M. (1975). *Tetrahedron Lett.*, 1541.
Tursch, B., Braekman, J. C., and Daloze, D. (1976). *Experientia* **32**, 401.
Valcurone, M. L., and Baggini, A. (1957). Boll. Inst. Sieroter **36**, 283.
Valle, J. R., Picarelli, Z. P., and Prado, J. L. (1954). *Arch. Int. Pharmacodyn. Ther.* **98**, 324.
Vanderplank, F. L. (1958). *J. Entomol. Soc. South. Afr.* **21**, 308.
Vanderzant, E. S., and Chremos, J. H. (1971). *Ann. Entomol. Soc. Am.* **64**, 480.
van Rietschoten, J., Granier, C., Rochat, H., Lissitzky, S., and Miranda, F. (1975). *Eur. J. Biochem.* **56**, 35.
van Zijp, C. (1917). *Pharm. Weekbl.* **54**, 295.
van Zijp, C. (1922). *Pharm. Weekbl.* **59**, 285.
Vaughan, G. L., and Jungreis, A. M. (1977). *J. Insect Physiol.* **23**, 585.
Veith, H. J., Weiss, J., and Koeniger, N. (1978). *Experientia* **34**, 423.
Verhoeff, K. W. (1925). *In* "Klassen und Ordnungen des Tierreiches" (H. G. Bronn, ed.), Vol. 5, pp. 351–365. Akad. Verlagsges., Leipzig.
Vick, K., Drew, W. A., McGurk, D. J., Eisenbraun, E. J., and Waller, G. R. (1969). *Ann. Entomol. Soc. Am.* **62**, 723.
Vidari, G., de Bernardi, M., Pavan, M., and Raggozino, L. (1973). *Tetrahedron Lett.* p. 4065.
Viehoever, A., and Capen, R. G. (1923). *J. Assoc. Off. Agric. Chem.* **6**, 489.
Visser, B. J., Spanjer, W., de Klonia, H., Piek, T., van der Meer, C., van der Drift, A. C. M. (1976). *Toxicon* **14**, 357.
Vitzthum, H. G. (1929). *Z. Parasitenkd.* **2**, 223.
Vogt, W. (1970). *Toxicon* **8**, 251.

von Endt, D. W. and Wheeler, J. W. (1971). *Science* **172,** 60.
von Euw, J., Fishelson, L., Parsons, J. A., Reichstein, T., and Rothschild, M. (1967). *Nature (London)* **214,** 35.
von Euw, J., Reichstein, T., and Rothschild, M. (1968). *Isr. J. Chem.* **6,** 659.
von Euw, J., Reichstein, T., and Rothschild, M. (1971). *Insect Biochem.* **1,** 373.
Vosseler, J. (1893). *Jh. Ver. Vaterl. Naturk. Württ.* **49,** 87.
Vrkoč, J., and Ubik, K. (1974). *Tetrahedron Lett.* p. 1463.
Vrkoč, J., Ubik, K., Dolejš, L., and Hrdý, I. (1973). *Acta Entomol. Bohemoslov.* **70,** 74.
Vrkoč, J., Ubik, K., Zdarek, J., and Kontev, C. (1977). *Acta Entomol. Bohemoslov.* **74,** 205.
Vrkoč, J., Křeček, J., and Hrdý, I. (1978). *Acta Entomol. Bohemoslov.* **75,** 1.
Wadhams, L. J., Baker, R., and Howse, P. E. (1974). *Tetrahedron Lett.* p. 1697.
Wain, R. L. (1943). *Rep.—Long Ashton Res. Stn.* p. 108.
Waldner, E. E., Schlatter, C., and Schmid, H. (1969). *Helv. Chim. Acta* **52,** 15.
Waley, S. G. (1969). *Comp. Biochem. Physiol.* **30,** 1.
Wallace, J. B., and Blum, M. S. (1969). *Ann. Entomol. Soc. Am.* **62,** 503.
Wallace, J. B., and Blum, M. S. (1971). *Ann. Entomol. Soc. Am.* **64,** 1021.
Wallace, J. W., and Mansell, R. L., eds. (1976). "Biochemical Interaction between Plants and Insects," Recent Adv. Phytochem., Vol. 10. Plenum, New York.
Wallbank, B. E. and Waterhouse, D. F. (1970). *J. Insect Physiol.* **16,** 2081.
Wanstall, J. C., and de la Lande, I. S. (1974). *Toxicon* **12,** 649.
Watanabe, M., Yasuhara, T., and Nakajima, T. (1975). *Toxicon* **13,** 130.
Watanabe, M., Yasuhara, T., and Nakajima, T. (1976). *In* "Animal, Plant, and Microbial Toxins" (A. Ohsaka, K. Hayaslu, and Y. Sawai, eds.), Vol. 2, p. 105. Plenum, New York.
Waterhouse, D. F., and Gilby, A. R. (1964). *J. Insect Physiol.* **10,** 977.
Waterhouse, D. F., and Wallbank, B. E. (1967). *J. Insect Physiol.* **13,** 1657.
Waterhouse, D. F., Forss, D. A., and Hackman, R. H. (1961). *J. Insect Physiol.* **6,** 113.
Weatherston, J. (1971). *In* "Chemical Releasers in Insects" (A. Tahori, ed.), pp. 183–193. Gordon & Breach, New York.
Weatherston, J. and Cheesman, J. (1975). Unpublished data.
Weatherston, J., and Gardiner, E. J. (1973). *Can. Entomol.* **105,** 1375.
Weatherston, J. and Gardiner, E. J. (1975). Unpublished data.
Weatherston, J., and Percy, J. E. (1969). *Can. J. Zool.* **47,** 1389.
Weatherston, J., and Percy, J. E. (1978a). *Handb. Exp. Pharmakol.* [N.S.] **48,** 489–509.
Weatherston, J., and Percy, J. E. (1978b). *Handb. Exp. Pharmakol.* [N.S.] **48,** 511–554.
Weatherston, J., Tyrrell, D., and Percy, J. E. (1971). *Chem. Phys. Lipids* **7,** 98.
Weatherston, J., Percy, J. E., MacDonald, L. M., and MacDonald, J. A. (1979). *J. Chem. Ecol.* **5,** 165.
Weber, N. A. (1958). *Proc. Int. Cong. Entomol., 10th, 1956* Vol. 2, pp. 459–473.
Wiemer, D. F., Hicks, K., Meinwald, J., and Eisner, T. (1978). *Experientia* **34,** 969.
Wiemer, D. F., Meinwald, J., Prestwich, G. D., and Miura, I. (1979). *J. Org. Chem.* **44,** 3950.
Wiemer, D. F., Meinwald, J., Prestwich, G. D., Solheim, B. A., and Clardy, J. (1980). *J. Org. Chem.* **45,** 191.
Wellhöner, H.-H. (1969). *Naunyn-Schmiedebergs Arch. Pharmakol. Exp. Pathol.* **262,** 29.
Welsh, J. H. (1964). *Annu. Rev. Pharmacol.* **4,** 293.
Welsh, J. H., and Batty, C. S. (1963). *Toxicon* **1,** 165.
Welsh, J. H., and Moorhead, M. (1960). *J. Neurochem.* **6,** 146.
Wender, S. H. (1970). *Recent Adv. Phytochem.* **3,** 1.

Wheeler, J. W. (1975). Unpublished data.

Wheeler, J. W., and Blum, M. S. (1973). *Science* **182**, 501.

Wheeler, J. W., Meinwald, J., Hurst, J. J., and Eisner, T. (1964). *Science* **144**, 540.

Wheeler, J. W., Chung, R. H., Oh, S. K., Benfield, E. F., and Neff, S. E. (1970). *Ann. Entomol. Soc. Am.* **63**, 469.

Wheeler, J. W., Happ, G. M., Araujo, J., and Pasteels, J. M. (1972a). *Tetrahedron Lett.*, 4635.

Wheeler, J. W., Oh, S. K., Benfield, E. F., and Neff, S. E. (1972b). *J. Am. Chem. Soc.* **94**, 7589.

Wheeler, J. W., Evans, S. L., Blum, M. S., and Torgerson, R. L. (1975). *Science* **187**, 254.

Wheeler, J. W., Blum, M. S., Daly, H. V., Kislow, C. J., and Brand, J. M. (1977a). *Ann. Entomol. Soc. Am.* **70**, 635.

Wheeler, J. W., Olagbemiro, T., Nash, A., and Blum, M. S. (1977b). *J. Chem. Ecol.* **3**, 241.

Wheeler, J. W., Olubajo, O., Storm, C. B., and Duffield, R. M. (1981). *Science* **211**, 1051.

Wheeler, W. M. (1890). *Psyche* **5**, 442.

Wheeling, C. H., and Keegan, H. L. (1972). *Toxicon* **10**, 305.

Whitfield, F. G. S. (1925). *Proc. Zool. Soc. London* p. 599.

Wientjens, W. H. J. M., Lakwijk, A. C., and van der Marel, T. (1973). *Experientia* **29**, 658.

Wilbrandt, W. (1962). *Proc. Int. Pharmacol. Meet., 1st, 1961* Vol. 3.

Wilson, E. O., and Regnier, F. E. (1971). *Am. Nat.* **105**, 279.

Wood, W. F. (1974). *Ann. Entomol. Soc. Am.* **67**, 988.

Wood, W. F., Truckenbrodt, W., and Meinwald, J. (1975). *Ann. Entomol. Soc. Am.* **68**, 359.

Woodhead, S., and Cooper-Driver, G. (1979). *Biochem. Syst. Ecol.* **7**, 309.

Woodring, J. P., and Blum, M. S. (1965). *J. Morphol.* **116**, 99.

Wray, J. (1670). *Philos. Trans. R. Soc. London* **5**, 2063.

Wright, K. P., Elgert, K. D., Campbell, B. J., and Barrett, J. T. (1973). *Arch. Biochem. Biophys.* **159**, 415.

Yasuhara, T., Yoshida, H., and Nakajima, T. (1977). *Chem. Pharm. Bull.* **25**, 936.

Yeh, J. Z., Narahashi, T., and Almon, R. R. (1975). *J. Pharmacol. Exp. Ther.* **194**, 373.

Yoshida, H., Geller, R. G., and Pisano, J. J. (1976). *Biochemistry* **15**, 61.

Youdeowei, A., and Calam, D. H. (1969). *Proc. R. Entomol. Soc. London, Ser. A* **44**, 38.

Zerachia, T., Shulov, A., and Bergmann, F. (1972a). *Toxicon* **10**, 536.

Zerachia, T., Bergmann, F., and Shulov, A. (1972b). *Toxicon* **10**, 537.

Zlotkin, E., Rochat, H., Kopeyan, C., Miranda, F., and Lissitzky, S. (1971a). *Biochimie* **53**, 1073.

Zlotkin, E., Miranda, F., Kupeyan, C., and Lissitzky, S. (1971b). *Toxicon* **9**, 9.

Zlotkin, E., Miranda, F., and Lissitzky, S. (1972a). *Toxicon* **10**, 207.

Zlotkin, E., Miranda, F., and Lissitzky, S. (1972b). *Toxicon* **10**, 211.

Zlotkin, E., Martinez, G., Rochat, H., and Miranda, F. (1975). *Insect Biochem.* **5**, 243.

Zlotkin, E., Miranda, F., and Rochat, H. (1978). *Handb. Exp. Pharmakol.* [N.S.] **48**, 317–369.

Zlotkin, E., Teitelbaum, Z., Lester, D., and Lazarovici, P. (1979a). *Toxicon* **17**, 208.

Zlotkin, E., Teitelbaum, Z., Rochat, H., and Miranda, F. (1979b). *Insect Biochem.* **9**, 347.

Empirical Formula Index*

*For compounds in Tables and in Chapter 18.

Animal and Plant Index

C

D

E

F

G

M

N

T

U

V

Subject Index

D

E

F